PHOSPHATE AND MINERAL HOMEOSTASIS

ADVANCES IN EXPERIMENTAL MEDICINE AND BIOLOGY

Recent Volumes in this Series

Volume 201
LIPOPROTEIN DEFICIENCY SYNDROMES
Edited by Aubie Angel and Jiri Frohlich

Volume 202
INFECTIONS IN THE IMMUNOCOMPROMISED HOST
Laboratory Diagnosis and Treatment
Edited by Paul Actor, Alan Evangelista, James Poupard, and Eileen Hinks

Volume 203
EXCITATORY AMINO ACIDS AND EPILEPSY
Edited by Robert Schwarcz and Yehezkel Ben-Ari

Volume 204
NEUROBIOLOGY OF CENTRAL D_1-DOPMAINE RECEPTORS
Edited by George R. Breese and Ian Creese

Volume 205
MOLECULAR AND CELLULAR ASPECTS OF REPRODUCTION
Edited by Dharam S. Dhindsa and Om P. Bahl

Volume 206
ESSENTIAL NUTRIENTS IN CARCINOGENESIS
Edited by Lionel A. Poirier, Paul Newberne, and Michael W. Pariza

Volume 207
THE MOLECULAR AND CELLULAR BIOLOGY OF FERTILIZATION
Edited by Jerry L. Hedrick

Volume 208
PHOSPHATE AND MINERAL HOMEOSTASIS
Edited by Shaul G. Massry, Michel Olmer, and Eberhard Ritz

PHOSPHATE AND MINERAL HOMEOSTASIS

Edited by

Shaul G. Massry

University of Southern California
Los Angeles, California

Michel Olmer

University of Aix-Marseille
Marseille, France

and

Eberhard Ritz

University of Heidelberg
Heidelberg, Federal Republic of Germany

PLENUM PRESS • NEW YORK AND LONDON

Library of Congress Cataloging in Publication Data

International Workshop on Phosphate and Other Minerals (7th: 1985: Marseille, France)
Phosphate and mineral homeostasis.

(Advances in experimental medicine and biology; v. 208)
"Proceedings of the Seventh International Workshop on Phosphate Metabolism, held September 1–4, 1985, in Marseille, France"—T.p. verso.
Bibliography: p.
Includes index.
1. Phosphates—Metabolism—Congresses. 2. Calcium—Metabolism—Congresses. 3. Magnesium—Metabolism—Congresses. 4. Homeostasis—Congresses. I. Massry, Shaul G. II. Olmer, Michael. III. Ritz, Eberhard. IV. Title. V. Series.
QP535.P1I57 1985 599′.0133 86-30431
ISBN 978-1-4684-5208-2 ISBN 978-1-4684-5206-8 (eBook)
DOI 10.1007/978-1-4684-5206-8

Proceedings of the Seventh International Workshop on Phosphate Metabolism, held September 1–4, 1985, in Marseille, France

Softcover reprint of the hardcover 1st edition 1986
A Division of Plenum Publishing Corporation
233 Spring Street, New York, N.Y. 10013

TO OUR WIVES

Meira Massry
Helene Olmer
Christina Ritz

AND TO OUR CHILDREN

PREFACE

We are pleased to present to our readers the proceedings of the Seventh International Workshop on Phosphate and Other Minerals which was held in Marsielle, France during September 1-4, 1985. It was hosted by Professor Michel Olmer, the Chief of the Divison of Nephrology in Hopital de la Conception of the University of D'Aix-Marsielle II.

The workshop was attended by 250 scientists from 17 countries including Algeria, Austria, Belgium, Canada, Czechoslovakia, Democratic Republic of Germany, Egypt, England, France, Israel, Italy, Japan, Spain, Switzerland and the United States of America. There were 28 invited presentations by leading scientists and 40 oral and 75 poster presentations selected from over 200 abstracts submitted to the Organizing Committee. This meeting provided an excellent opportunity for interested scientists from interrelated disciplines including nephrology, endcrinology, physiology, biochemistry and nutrition to get together and discuss recent advances in the field of phosphate and mineral metabolism.

We would like to express our thanks and gratitude to all those who stimulated, encouraged and supported us to hold the Seventh International Workshop in Marseille. This endeavor could not have been possible without the generous financial support of many companies and agencies in France including Bayer Pharma, Boehringer Ingelheim, Ciba, Cobe, Conseil general des Bouches du Rhone, Conseil Regional Provence-Alpes-Cote d'Azur, Eaux Minerales D'Evian, Fournier-Dijon, Gambro, Hemofrance, Hoechst, Hospal, I.C.I. Pharma, LEO, Merck Sharp & Dohme-Chibret, Municipalite de Marseille, Organon Teknika, Robert Et Carriere, Roussel, Sandoz, Servier Medical, Theraplix, Travenol, Wyeth-Byla and of others in the United States of America including Hoffman LaRoche, Inc., Marion Laboratories, Inc., Plenum Publishing Corp., Proctor and Gamble Company, Salt Institute.

We would like to express our deepest appreciation to Ms. Lilian Neel for her tremendous and valuable contribution to the organization of the Workshop and to Ms. Haidi Samouha for her important organizational assistance during the workshop.

Shaul G. Massry
Michel Olmer
Eberhard Ritz

CONTENTS

I. SPECIAL PRESENTATIONS

II. RENAL HANDLING OF PHOSPHATE

III. RENAL HANDLING OF CALCIUM AND MAGNESIUM

IV. INTESTINAL ABSORPTION OF PHOSPHATE AND CALCIUM

V. BONE METABOLISM

VI. PARATHYROID HORMONE

VII. MINERALS AND BLOOD PRESSURE

VIII. NEPHROLITHIASIS

IX. MISCELLANEOUS

SPECIAL PRESENTATIONS

HIGH ENERGY PHOSPHATES, PHOSPHOLIPIDS, AND CALCIUM IN ISCHEMIC RENAL TUBULAR CELL INJURY

H. David Humes, Vo D. Nguyen, and Deborah A. Hunt

Departments of Internal Medicine,
Veterans Administration Medical Center and
University of Michigan
Ann Arbor, Michigan

It is becoming increasingly clear that the final common pathogenetic pathway for the development of ischemic acute renal failure is renal tubular cell injury.[1,2] Acute renal failure developing from an ischemic insult occurs from tubular cell injury that produces segmental necrosis in renal tubules, so that a patchy distribution of frankly necrotic lesions appears to be the rule rather than the exception in the pathology of acute renal failure.[3] The segmental, patchy renal tubular cell necrosis initiates a variety of factors responsible at the nephronal level for excretory failure of the kidney. These factors include intratubular obstruction, backleak of glomerular filtrate, and glomerular hemodynamic alterations. Ultimately, the understanding of the pathogenesis of acute tubular necrosis resides in the understanding of the biochemical alterations responsible for renal tubular cell injury. In this regard, declines in levels of high energy phosphates within the cell, alterations in cellular calcium metabolism, and degradation of membrane phospholipids appear to be metabolic derangements induced by ischemia critical in the evolution of cell injury.

HIGH ENERGY PHOSPHATES IN ISCHEMIC ACUTE RENAL FAILURE

Because ischemia is associated with limited oxygen availability, declines in oxidative phosphorylation and high energy phosphate stores are invariable consequences during ischemia. Multiple studies have demonstrated that renal cell ATP levels fall rapidly with the onset of ischemia and then reach a low sustained level.[4,5] Along with this decline in ATP content, a persistent loss of total adenine nucleotide content also occurs during intermediate to long durations of ischemia. The decline in total adenine nucleotide pool size is due to degradation of ATP to its breakdown products of adenosine, inosine and hypoxanthine which may be lost through the plasma membrane.[6,7] This degradation and loss of adenine nucleotide pool must be considered to be of major importance in the ability of the renal tubular cell to recover from an ischemic insult because of the high energy cost of replenishing the pool. The importance of this high energy phosphate pool in the development of ischemia-induced renal tubular cell injury is highlighted by the finding that infusions of exogenous adenine nucleotides are protective against ischemic acute renal failure.[8,9] Furthermore, these infusions also lessen the recovery time for the kidney damaged by an ischemic insult to return to normal renal function.[10] A recent in vitro study has demonstrated that exogenous administration of ATP-$MgCl_2$ directly influences in a major way intracellular adenine nucleotide metabolism in renal tubular cells, so that the size of the tubule adenine nucleotide pool is preserved after hypoxic stress.[11,12] Thus, preservation of high energy phosphate stores during an ischemic event protects renal tubular cells from lethal cell injury, supporting a critical role for these compounds in developing cell injury.

CALCIUM IN ISCHEMIC ACUTE RENAL FAILURE

Tissue calcium levels invariable increase when lethal cell injury develops in a tissue which is perfused or reperfused with extracellular fluid containing calcium.[13,14] This increase in tissue calcium levels is due predominantly to mitochondrial accumulation and sequestration which occurs even in the environment of severe cell disruption. That calcium plays a critical role in developing cell injury has been supported by a number of observations. Multiple studies in various tissues have demonstrated that cellular calcium overload produces lethal cell injury.[15,16] Intramitochondrial accumulation of moderate amounts of calcium alters the permeability characteristics of the inner mitochondrial membrane which permit it to engage in efficient oxidative phosphorylaltion,[17-19] and leads to marked structural and functional derangements. Furthermore, calcium channel blockers have been demonstrated to protect against the norepinephrine model of ischemic acute renal failure.[20] It is not yet resolved, however, whether the mechanism of this protective effect of calcium channel blockers relates to an alteration of the vasoconstrictive effects of norepinephrine or to a direct cellular effect to diminish the progression from sublethal to lethal cell injury.[11,21]

Calcium thus appears to be a critical pathogenetic mediator of ischemic cell injury[13,14] but the precise mechanism by which alterations in cellular calcium most critically influence cell viability remains to be agreed upon. Several potentially important intracellular sites of action and mechanisms of action at these sites have been identified for a number of tissues. Major sites of deleterious calcium action in the cell include the plasma membrane, the mitochondrion, the endoplasmic reticulum, and the cytoskeleton. Mechanisms by which calcium promotes injury include activation of membrane-bound and free phospholipases, alterations of membrane permeability properties as a result of both direct effects on permeability pathways and activation of membrane phospholipases, and effects on intracellular contractile and cytoskeletal structures. Of these sites and mechanisms, calcium induced mitochondrial dysfunction and calcium activation of phospholipases presently appear to be most important.[13,14]

PHOSPHOLIPIDS IN ISCHEMIC ACUTE RENAL FAILURE

Phospholipids provide the major structural framework for cell membranes[22] and also participate in the regulation of membrane enzyme activity,[23,24] permeability[25,26] and hormone activation.[27] Ischemic injury to the kidney, as well as to other tissues, is accompanied by depletion of the major structural phospholipids, phosphatidylcholine and phosphatidylethanolamine, because of activation of phospholipases.[28] It has been proposed that the loss of these phospholipids accounts for the membrane permeability changes leading to cellular calcium overload and cell death in ischemia.[29,30]

Support for an important role for phospholipid degradation and depletion in ischemic cell injury emanates from two key observations. Treatment of animals with chlorpromazine as a phospholipase inhibitor prevents the decrease in tissue phospholipid content following liver and myocardial ischemia and protects against the occurrence of tissue calcium overload in these tissues and the development of lethal cell injury.[28,31] In addition, the in vitro treatment of suspensions of isolated proximal tubule segments with exogenous phospholipase leads to substantial tubule cell injury. Cell injury develops, however, only if phospholipid degradation resulted in significant declines in phospholipid content.[32] If synthetic rates were able to maintain phospholipid content during phospholipase-induced phospholipid degradation, demonstrable cell injury did not occur.

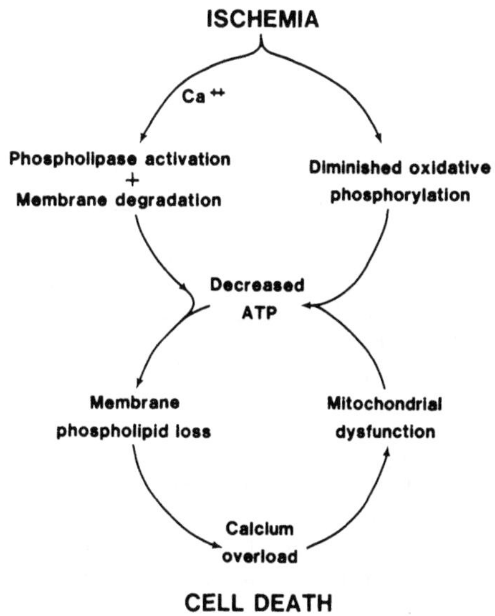

Figure 1. Proposed scheme delineating the role for high energy phosphates, calcium, and phospholipids in the pathogenesis of ischemic cell injury.

SUMMARY

Substantial data is available supporting critical roles for high energy phosphates, calcium, and phospholipids in ischemic cell injury. Although it has been proposed that one of these compounds may be most critical and defines the "final common pathway" for lethal cell injury, it is more accurate to define the pathogenesis of ischemic cell injury as the simultaneous derangement of several critical metabolic process which act in concert to produce a cascade of events that finally lead to plasma and subcellular membrane dysfunction incompatible with the maintenance of cell viability and integrity. As detailed in Figure 1, a tentative scheme is proposed interrelating high energy phosphate depletion, cellular calcium derangements, and membrane phospholipid degradation and loss. Ischemia directly leads to declines in the rate of oxidative phosphorylation due to a lack of oxygen availability. A fall in cellular ATP levels develops. Ischemia also promotes a redistribution of intracellular calcium pools and results in phospholipase activation and phospholipid degradation. When phospholipid degradation occurs concomitantly with a decline in high energy phosphate levels, phospholipid synthesis cannot keep pace with phospholipid degradation and net membrane phospholipid loss occurs. An increase in plasma membrane calcium permeability develops and the influx of calcium down its electrochemical gradient from extracellular to intracellular spaces occurs. This calcium is taken up and sequestered in mitochondria and causes further alterations in mitochondrial structure and function, leading to further declines in cellular ATP content. What, therefore, eventuates is an unremitting vicious cycle of progresssive cellular membrane deterioration, cellular energy store depletion, and deranged cellular bioenergetics until reparative and synthetic processes cannot keep pace with degradative events. Irreversible and lethal cell injury is the ultimate outcome. This proposed scheme is only tentative but establishes a thesis which can be directly tested and, hopefully, lead to further insights into this important disease process.

REFERENCES

1. H. D. Humes and J. M. Weinberg, Alterations in renal tubular cell metabolism during acute renal failure, Min. Elect. Metab. 9: 290 (1983).
2. H. D. Humes and J. M. Weinberg, Cellular energetics in acute renal failure, in: "Acute Renal Failure," B. M. Brenner and J. M. Lazarus, eds., W. B. Saunders, Philadelphia (1983).
3. J. I. Kriesberg, E. Matthys, and M. A. Venkatachalam, Morphologic factors in acute renal failure, in: "Acute Renal Failure," B. M. Brenner and J. M. Lazarus, eds., W. B. Saunders, Philadelphia (1983).
4. D. Hems and J. Brosnan, Effects of ischaemia on content of metabolites in rat liver and kidney in vivo, Biochem. J. 120: 105 (1970).
5. M. Kahng, I. Berezesky, and B. Trump, Metabolic and ultrastructural response of rat kidney cortex to in vitro ischemia, Exp. Molec. Path. 29: 183 (1978).
6. H. Osswald, H.J. Schmitz, and R. Kemper, Tissue content of adenosine, inosine and hypoxanthine in the rat kidney after ischemia and postischemic recirculation, Pflugers Arch. 317: 45 (1977).
7. W. Miller, R. Thomas, R. Berne, and R. Rubio, Adenosine production in the ischemic kidney, Circulation Res. 43: 390 (1978).
8. N. J. Siegel, W. B. Glazier, I. H. Chaudry, K. M. Gaudio, B. Lytton, A. E. Baue, and M. Kashgarian, Enhanced recovery from acute renal failure by the postischemic infusion of adenine nucleotides and magnesium chloride in rats, Kidney Int. 17: 338 (1980).
9. K. M. Gaudio, M. R. Taylor, I. H. Chaudry, M. Kashgarian, and N. J. Siegel, Accelerated recovery of single nephron function by the postischemic infusion of ATP-$MgCl_2$, Kidney Int. 22: 13 (1982).
10. K. M. Gaudio, T. A. Ardito, H. F. Reilly, M. Kashgarian, and N. J. Siegel, Accelerated cellular recovery after an ischemic renal injury, Am. J. Pathol. 112: 338 (1983).
11. H. D. Humes, D. A. Hunt, M. J. Clark, M. D. White, and J. M. Weinberg, Cellular mechanisms of protective maneuvers in nephrotoxic and ischemic acute renal failure, in: "Nephrology," R. R. Robinson, ed., Springer-Verlag, New York (1984).
12. J. M. Weinberg, M. Clark, and H. D. Humes, Effects of exogenous ATP on tubule cell ATP levels, Clin. Res. 32: 459A (1984).

13. H. D. Humes, Role of calcium in the pathogenesis of acute renal failure, Am. J. Physiol. In press.
14. J. M. Weinberg, Calcium as a mediator of renal tubule cell injury, Semin. Nephrol. 4: 178 (1984).
15. F. Schanne, A. Kane, E. Young, and J. Farber, Calcium dependence of toxic cell death: a final common pathway, Science 206: 700 (1979).
16. H. D. Humes and J. M. Weinberg, Ionophore A23187 induced reductions in toad urinary bladder epithelial cell oxidative phosphorylation and viability, Pflugers Arch. 388: 217 (1980).
17. D. Hunter, R. Haworth, and J. Southard, Relationship between configuration, function, and permeability in calcium-treated mitochondria, J. Biol. Chem. 251: 506 (1976).
18. D. Pfeiffer, R. Kauffman, and H. Lardy, Effects of N-ethylmaleimide on the limited uptake of Ca^{++}, Mn^{++}, and Sr^{++} by rat liver mitochondria, J. Biol. Chem. 253: 4165 (1978).
19. D. Pfeiffer, P. Schmid, M. Beatrice, and H. Schmid, Intramitochondrial phospholipase activity and the effects of Ca^{++} plus N-ethylmaleimide on mitochondrial function, J. Biol. Chem. 254: 11485 (1979).
20. T. J. Burke, P. E. Arnold, J. A. Gordon, R. E. Bulger, D. C. Dobyan, and R. W. Schrier, Protective effect of intrarenal calcium membrane blockers before or after renal ischemia: functional, morphological and mitochondrial studies, J. Clin. Invest. 74: 1830 (1984).
21. C. D. Malis, J. Y. Cheung, L. Alexander, and J. Bonventre, Effects of verapamil in models of ischemic acute renal failure in the rat, Am. J. Physiol. 245: F735 (1983).
22. S. Singer and G. Nicolson, The fluid mosaic model of the structure of cell membranes, Science 175: 720 (1972).
23. J. Dahl and L. Hokin, The sodium-potassium adenosine triphosphatase, Ann Rev. Biochem. 43: 327 (1974).
24. B. Rubalcava and M. Rodbell, The role of acidic phospholipids in glucagon action on rat liver adenylate cyclase, J. Biol. Chem. 248: 3831 (1973).
25. D. Green, M. Fry, and G. Blondin, Phospholipids as the molecular instruments of ion and solute transport in biological membranes, Proc. Nat. Acad. Sci. USA 77: 257 (1980).
26. P. Cullis, B. deKruijff, M. Hope, R. Nayar, and S. Schmid, Phospholipids and membrane transport, Can. J. Biochem. 58: 1091 (1980).
27. L. Jones, S. Cockcroft, and R. Mitchell, Stimulation of phosphatidylinositol turnover in various tissues by cholinergic and adrenergic agonists, by histamine and by coerulein, Biochem. J. 182: 669 (1979).
28. K. Chien, J. Abrams, A. Serroni, J. Martin, and J. Farber, Accelerated phospholipid degradation and associated membrane dysfunction in irreversible, ischemic liver cell injury, J. Biol. Chem. 253: 4809 (1978).
29. J. Farber, K. Chien, and S. Mittnacht, The pathogenesis of irreversible cell injury in ischemia, Am. J. Pathol. 102: 271 (1981).
30. J. Farber, The role of calcium in cell death, Life Sci. 29: 1289 (1981).
31. K. R. Chien, R. G. Pfau, and J. L. Farber, Ischemic myocardial cell injury: prevention by chlorpromazine of an accelerated phospholipid degradation and associated membrane dysfunction, Am. J. Pathol. 97: 505 (1979).
32. V. D. Nguyen, D. A. Hunt, P. A. Weinhold, and H. D. Humes, Injurious effects of exogenous phospholipase on renal proximal tubule segments, Kidney Int. In press.

CALCIUM, CELL FUNCTION AND CELL DEATH

Louis V. Avioli

Division of Bone and Mineral Diseases
Washington University School of Medicine
216 S. Kingshighway, St. Louis, Missouri 63110

The nature of the biologic mechanism(s) which initiate and perpetuate the aging process is still under intensive investigation. Although those with vested interests in "aging research" would agree that the programming of the aging phenomenon is related to specific acquired and/or genetic molecular dysfunctions which condition DNA and protein synthesis, and post-translational events in cells, the fundamental "trigger" mechanism(s) which activates mammalian senescence has yet to be discovered. A number of theories have been promulgated to account for the molecular cellular dysfunctions which attend the senescent process. These include: (a) alterations in host-defense systems and cell-mediated immunity; (b) progressive accumulations of circulating "toxic" substances which damage DNA; (c) an acquired inability to destroy "free-radicals", the latter resulting in the cellular accumulation of superoxide and hydrogen peroxide type reactants and a gradual decline in cellular integrity; (d) cellular "commitment" theories which stress programmed aging and finite replicative life-span of cells; and (e) gradual alterations in those mechanisms which regulate cytosolic calcium concentration resulting in molecular disarray, decrease in, cellular viability and cell death.

The calcium ion plays a unique role as both activator and regulator of a myriad of biological processes (1). This control depends on the continued maintenance of a low cystosolic Ca^{2+} concentration in the range of 0.05-0.5 µM by a complex array of membrane and cytosolic transport processes in the presence of a extracellular fluid concentration of 1-10 mM. It, therefore, appears appropriate to review some of those facts which argue in favor of the theories that changes in the exquisite temporal control of cellular Ca^{2+} occur with increasing age, and that Ca^{2+} accumulation within the cell is detrimental to continued viability.

Intracellular calcium (Ca^{2+}) increases significantly as human fibroblasts age in culture (2). Observations that diploid human fibroblasts have a limited life-span in culture have been made by many laboratories and associated with acquired alterations in enzyme function (3). Although the potential role of the increment in Ca^{2+} in the *in vitro* human fibroblast aging model is still uncertain, it is appropriate to note that skin calcium, but not magnesium and phosphate increases with age (4). Ca^{2+} content of fibroblasts from subjects with cystic fibrosis is also greater than those obtained from control subjects (2). Recognizing that intracellular Ca^{2+} and premature aging has been incriminated in the pathogenesis of cystic fibrosis (5,6), one might infer from these observations that the change in cellular Ca^{2+} is one of the fundamental events which characterizes cell aging. Of additional note in this regard is that the effects of endotoxin, a lipopolysaccharide of gram-negative bacterial cell walls, on decreasing the proliferation of human fibroblasts is attended by increments in Ca^{2+} influx and total cellular Ca^{2+} content (7).

The cardiac myocyte is dependent on Ca^{2+} in the extracellular fluid for excitation - contraction coupling and for the structural integrity of its surface membrane. Calcium influx and intracellular calcium release determine the force of contraction of myocardial cells. Age related "physiological" alterations in myocardial function and reserve have been associated with increments in the calcium content of cardiac musculature. Recent studies, designed to analyze the phenomenon of calcium-induced cell death in cardiac myocytes demonstrate an age-related pattern of sensitivity of the cells to the toxic effects of calcium with decreased sensitivity to the "calcium toxin" noted in cardiac myocytes obtained from the hearts of younger animals (8). Other observations reveal that calcium causes increased excitability and death in cardiac cells (9). Agents which reduce the transmembrane influx of calcium ions such as the "calcium channel blocking" drugs verapamil, nifedipine and diltiazem, have been approved for certain types of angina pectoris and supraventricular arrhythmias in the United States (10). It seems appropriate to consider the possibility that the continued application of agents with the "calcium channel blocking" potential to individuals with the aforementoned clinical disorders might ultimately prevent the irreversible accumulation of calcium in myocardial musculative, the physiological aging of the myocardium, and the undesirable sequellae.

We are still confronted with the fact that cardiovascular disease remains the leading cause of death in industrial nations. The principal underlying cause is atherosclerosis, the latter classically equated with elevations in circulating cholesterol and/or a decrease in high density lipoproteins. Recent studies in macaque monkeys revealed that, despite atherogenic diets and hypercholesterolemia, the concomitant administration of the inorganic calcium channel "blocker", lanthanum, prevented the formation of the aortic fibrotic atherosclerotic lesion (11). The results of the study support the concept that calcium is also an endothelial cell toxin and suggests that calcium channel blocking agents, capable of modifying endothelial as well as myocardial calcium content, may prevent atherosclerosis despite abnormalities in circulating cholesterol and lipoprotein concentrations.

The slow Ca^{2+} channel blocking agent diltiazem, has therapeutic benefit for patients with ischemic heart disease, primarily because its ability to inhibit Ca^{2+} flux across coronary and systemic arterial smooth muscle cells (12-14). This agent may not only prove promising in the management of myocardial ischemic because of its specific effect on myocardial or endothelial cells, but also because of its ability to inhibit Ca^{2+} flux in platelets, thus decreasing platelet activation and thromboxane generation (15). Since blood platelets have many features in common with smooth muscle cells, one might presume that a predisposition to spontaneous platelet activation as a result of age-related increments in platelet calcium concentration might predispose aging individuals to the atherosclerotic process. This hypothesis is presently untenable since studies in platelets obtained from patients ranging in age from 17-69 years revealed no correlation between age and platelet Ca^{2+} levels (16). However, a close correlation does exist between platelet Ca^{2+} levels and both systolic and diastolic blood pressure (16). The administration of calcium channel blockers verapamil, nifedipine or nitrendipine result in a reduction in platelet Ca^{2+} levels, the latter correlating well with the fall in blood pressure (16). These observations were subsequently confirmed in studies with platelets obtained from other patients with essential hypertension, and rats with spontaneous hypertension (17). Although the accumulated results are provocative, we must still entertain the possibility that the increase in platelet Ca^{2+} represents a consequence rather than a cause of the elevated blood pressure.

The precise relationships between calcium transport, the metabolism of vascular smooth muscle cells, smooth muscle contraction and vasoconstruction is still not completely understood. It has been established, however, that angiotensin II and vasopressin, two physiologically important vasocontrictor hormones, rapidly raise cytosolic Ca^{2+} in vascular smooth muscle cells (18,19). Conversely, the vasodilator, nitroglycerine, decreases cytosolic Ca^{2+} in vascular smooth muscle cells (20). If a gradual accumulation of calcium in myocardial vascular smooth muscle and endothelial results in age-related cardiomyopathy, loss of aortic elasticity, atherosclerosis and hypertension, we must invoke a common pathological perturbation which not only attends the senescent process, but it also functions as a common calcium ionophore.

A most likely candidate for consideration in this regard is parathyroid hormone (PTH). Although there are occasional reports to the contrary, there appears to be overwhelming evidence that circulating parathyroid hormone <u>and its biological effects</u> increase with age (21-24). PTH stimulation resuls in an increase in cytosolic calcium in a variety of tissues (9,25-30). It has also been implicated as a prominent uremic toxin in patients with chronic renal failure, specifically as a contributing cause of the cardiomyopathy, peripheral neuropathy, dementia, soft tissue calcification and anemia (31). The "toxic" effects of PTH in the chronic uremic failure syndrome have been largely ascribed to a calcium ionophore effect of the hormone (9,31-33) although one should also consider the effect of mobilization of intracellular Ca^{2+} stores by a PTH induced increment in inositol triphosphate generation (34), as recently described for angiotensin II (35). In studies designed to evaluate the relationship between alterations in vascular smooth muscle tone, coronary spasm, and cardiac sclerosis, plasma PTH levels were highest in those patients with coronary heart disease affecting three vessels and in patients with evidence of myocardial infarction (36). Finally, it also is becoming increasingly apparent that the incidence of coincidental hypertension in patients with primary hyperparathyroidism exceeds that anticipated in normal subjects of similar age and sex distributions (37).

Recent reports equating low dietary calcium with a tendency to hypertension must also be considered (38-41). Untreated patients with hypertension on low calcium intakes are characterized by elevations in PTH, hypercalciuria, decreased circulating ionized calcium and hypophosphatemia (41). Calcium supplements when administered to hypertensive patients result in progressive decreases in both diastolic and systolic blood pressure (39,42). Chronic ingestion of low calcium diets could only function as a secretagogue for PTH resulting in chronic, albeit, intermittent elevations in circulating levels. One might easily also ascribe the increments in the Ca^{2+} content of platelets to subtle, but progressive elevations in circulating PTH in the calcium deficient population.

However, many pieces of the puzzle are still unaccounted for. Not only is the PTH molecule hypotensive and vasodilatory in hypertensive models (43,44), but the hypotensive and the hypercalcemic effects of PTH appear to represent two separate properties of the same molecule (44). Moreover, the <u>in vitro</u> response of vascular smooth muscular cells to PTH is an acute decrease in cAMP content (45). This latter in vitro observation warrants careful scrutiny and corroboration since there are presently no reports which correlate a decrease in cAMP with vasodilation. Since an increment in cell calcium results in an increased contraction of the arterial smooth muscle, a reduction in cell calcium as can be obtained with calcium channel blockers should lower blood pressure. Although it has been suggested that a reduction in circulating ionized calcium per se facilitates Ca^{2+} entry via the Ca^{2+} voltage dependent channels (19), the explanation for the decreased circulating Ca^{2+} in hypertensive subjects and the hypotensive effect of calcium supplements is still conjectural at best.

No theory which incriminates calcium as an age related cellular toxin would be complete without inclusion of the senile dementia syndromes. Although recent years have witnessed major advances in our understanding of senile dementia and Alzheimer's disease, we are still confronted with concepts of "non-specific cellular atrophy" to account for simple senile dementia; a wealth of theories which include autoimmunity, aluminum toxicity and slow virus infection have been advanced to explain the pathogenesis and etiology of Alzheimer's disease. Basal ganglia calcification, which develops gradually in patients after the fifth decade, is noted much earlier in individuals with either Down's syndrome (trisomy 21) or progeria and considered a marker of premature aging. Studies designed to evaluate the possibility that calcium is elevated in brain neurons during aging in rats are also consistent with the hypothesis that altered Ca^{2+} homeostasis may play a significant role in normal brain aging (46). Recent investigations have been designed to analyze the distribution of calcium and aluminum in the neurofibrillary tangle of the hippocampus in patients with parkinsonism-dementia who were also subjected to chronic calcium deficiency (47). The authors concluded: "the findings support the hypothesis that secondary hyperparathyroidism resulting from the low environmental calcium and magnesium in the high-incidence focus of amyotrophic lateral sclerosis and parkinsonism-dementia on Guam had led to abnormal deposition of calcium and aluminum in the central nervous system" (47). Although provocative but far from conclusive, this latter statement demands consideration. Recently, in a simple report of "Reversible Long-Standing Dementia with Normocalcemic Hyperparathyroidism (in a 77 year old woman), the authors' noted: "this patient's history of gradual mental deterioration, with episodic exacerbation of confusion associated with acute illnesses, was assumed to be due to Alzeimer's disease, but in fact was due to hyperparathyroidism" (48). Thus, in spite of peristently normal calcium levels, elevated levels of PTH and the biological sequallae are apparently sufficient to cause a senile dementia syndrome which mimics Alzheimer's disease. Are these data not reminescent of those obtained in other studies cited earlier which incriminate PTH as a central nervous system toxin and as a causative agent for peripheral neuropathy (49,50)?

Research which provides insight into processes which regulate cellular proliferation, viability and senescence must assume a top priority. Evidence, albeit still primarily circumstantial and woefully inadequate, is accumulating which appears to incriminate Ca^{2+} as a toxin for cardiovascular and neural functions. Approximately 2000 years ago Virgil noted: "age carries all things, even the mind away" (51). Some have commented that "remarkably little has changed since Virgil wrote these words" (52). This opinion may have been premature since new horizons and developments of molecular biochemistry are presently being kindled at an accelerated pace by the urgency to insure the viability and health of our ever increasing aged population.

References

1. Racker, E.: Fluxes of Ca^{2+} and concepts. Federation Proc. 39: 2422-2426, 1980.
2. Shapiro, B.L., Lam, L.F-H.: Calcium and Age in Fibroblasts from Control Subjects and Patients with Cystic Fibrosis. Science 216: 417-419, 1982.
3. Holliday, R., Tarrant, G.M.: Altered Enzymes in Aging Human Fibroblasts. Nature 238: 26-30, 1972.
4. Subryan, V.L., Solomons, C.C., Gordon, S.G., Neldaer, K.H., Reeve, E.B.: A micromethod for the quantitative analysis of calcium magnesium and phosphorus in human skin. J. Lab. Clin. Med. 86: 1056-1060, 1975.
5. Shapiro, B.L., Lam, L.F-H., Fast, L.H.: Premature Senescence in Cultured Skin Fibroblasts from Subjects with Cystic Fibrosis. Science 203: 1251-1253, 1979.
6. Lam, L.F-H., Shapiro, B.L.: Differential Incorporation of ^{3}H-Thymidine into DNA in Cultured Skin Fibroblasts Derived from Patients with Cystic Fibrosis and Control. Life Sci. 24: 2483-2489, 1979.
7. Subramoniam, A., Padh, H., Aleo, J.J.: Factors Influencing Calcium Influx in Endotoxin-Challenged Fibroblasts. Proc. of the Soc. for Exp. Biol. and Med. 178: 50-59, 1985.
8. Chizzonite, R.A., Zak, R.: Calcium-Induced Cell Death: Susceptibility of Cardiac Myocytes Is Age-Dependent. Science 213: 1508-1510, 1981.
9. Bogin, E., Massry, S.G., Harary, I.: Effect of Parathyroid Hormone on Rat Heart Cells. J. Clin. Invest. 67: 1215-1227, 1981.
10. Council Report. Calcium Channel Blocking Agents. JAMA 250: 2522-2524, 1983.
11. Kramsch, D.M., Aspen, A.J., Apstein, C.S.: Suppression of Experimental Atherosclerosis by Ca^{++}-Antagonist Lanthanum. Possible Role of Calcium in Atherogenesis. J. Clin. Invest. 65: 967-981, 1980.
12. Nakajima, H., Hoshiyama, M., Yamashita, K., Kiyomoto, A.: Effect of diltiazem on electrical and mechanical activity of isolated cardiac ventricular muscle of guinea pig. Jpn. J. Pharmacol. 25: 383-387, 1975.
13. Nago, T., Ikeo, T., Sato, M.: Influence of calcium ions in response to diltiazem in coronary arteries. Jpn. J. Pharmacol. 27: 330-334, 1977.
14. Magaribuchi, T., Nakajima, H., Kiyomoto, A.: Effects of diltiazem in electrical and mechanical activity of isolated guinea pig atrial cell. Jpn. J. Pharmacol. 27: 361-364, 1977.
15. Mehta, P., Mehta, J., Ostrowski, N., Brigmon, L.: Inhibitory effects of diltiazem on platelet activation caused by ionophore A23187 plus ADP or epinephrine in subthreshold concentrations. J. Lab. Clin. Med. 102: 332-339, 1983.
16. Erne, P., Bolli, P., Bürgisser, E., Bühler, F.R.: Correlation of Platelet Calcium with Blood Pressure. N. Engl. J. Med. 310: 1084-1088, 1984.
17. Bruschi, G., Bruschi, M.E., Caroppo, M., Orlandini, G., Spaggiari, M., Cavatorta, A.: Cytoplasmic free [Ca^{2+}] is increased in the platelets of spontaneously hypertensive rats and essential hypertensive patients. Clin. Sci. 68: 179-184, 1985.
18. Capponi, A.M., Lew, P.D., Valloton, M.B.: Cytosolic Free Calcium Levels in Monolayers of Cultured Rat Aortic Smooth Muscle Cells. Effects of Angiotensin II and Vasopressin. J. Biol. Chem. 260: 7836-7842, 1985.

19. Nabika, T., Velletri, P.A., Lovenberg, W., Beaven, M.A.: Increase in Cytosolic Calcium and Phosphoinositide Metabolism Induced By Angiotensin II and [Avg] Vasopressin in Vascular Smooth Muscle Cells. J. Biol. Chem. 260: 4461-4670, 1985.
20. Kobayashi, S., Kanaide, H., Nakamura, M.: Cytosolic-Free Calcium Transients In Cultured Vascular Smooth Muscle Cells. Science 229: 553-556, 1985.
21. Insogna, K.I., Lewis, A.M., Lipinski, B.A., Bryant, C., Baran, D.: Effect of Age on Serum Immunoreactive Parathyroid Hormone and Its Biological Effects. J. Clin. Endocrinol. Metab. 53: 1072-1080, 1981.
22. Gallagher, J.C., Riggs, B.L., Jerpbak, C.M., Arnaud, C.: Effect of Age on Serum Immunoreactive Parathyroid Hormone in Normal and Osteoporotic Women. J. Lab. Clin. Med. 95: 373-385, 1980.
23. Wiske, P.S., Epstein, S., Bell, N., Queener, S.F., Edmondson, J., Johnston, C.C. Jr.: Increases in Immunoreactive Parathyroid Hormone With Age. N. Engl. J. Med. 300: 1419-1421, 1979.
24. Roof, B.S., Piel, C.F., Hansen, J., Fadenberg, H.H.: Serum Parathyroid Hormone Levels and Serum Calcium Levels From Birth to Senescence. Mech. Ageing Dev. 5: 289-294, 1976.
25. Brautban, N., Chakroborty, J., Coats, J., Massry, S.G.: Calcium, Parathyroid Hormone and Phospholipid Turnover of Human Red Blood Cells. Min. Elect. Metab. 11: 111-116, 1985.
26. Parsons, J.A., Neer, R.M., Potts, J.T.: Initial fall of plasma calcium after intravenous injection of parathyroid hormone. Endocrinology 89: 735-740, 1971.
27. Rasmussen, H.: Calcium and cAMP stimuli-response coupling. Ann. NY. Acad. Sci. 356: 346-353, 1980.
28. Farese, R.V., Bidot-Lopez, P., Sabir, M.A., Larson, R.E.: The phosphatidate-polyphosphoinositide cycle: activation by parathyroid hormone and dibutyryl-cAMP in rabbit kidney cortex. Ann. NY Acad. Sci. 372: 539-550, 1981.
29. Chausmer, A.B., Sherman, B.S., Wallach, S.: Effect of Parathyroid Hormone on Hepatic Cell Transport of Calcium. Endocrinol. 90: 663-672, 1972.
30. Borle, A.B.: Calcium Metabolism at the Cellular Level. Fed. Proc. 32: 1944-1947, 1973.
31. Massry, S.G.: Role of Parathyroid Hormone in the Pathogenesis of the Uremic Manifestation. Klin. Wochenschr. 57: 1085-1088, 1979.
32. Bogin, E., Massry, S.G., Levi, J., Djaldelti, M., Bristol, G., Smith, J.: Effect of Parathyroid Hormone on Osmotic Fragility of Human Erythrocytes. J. Clin. Invest. 69: 1017-1025, 1982.
33. Bogin, E., Leir, J., Harary, I., Massry, S.G.: Effects of Parathyroid Hormone on Oxidative Phosphorylation of Heart Mitochondria. Min. Elec. Metab. 7: 151-156, 1982.
34. Lopez-Bidot, P., Farese, R.V., Subiv, M.A.: Parathyroid Hormone and Kenosine-3'5'-Monophosphate Acutely Increase Phospholipids of the Phosphatidate-Polyphosphoinesitide Pathway in Rabbit Kidney Cortex Tubules in vitro by a Cedoheximide Sensitive Process. Endocrinol. 108: 2078-2081, 1981.
35. Smith Bingham, J., Smith, L., Renee Brown, E., Barnes, D., Sabiv, M.A., Davis, J.S., Farese, R.V.: Angiotensin II Rapidly Increases Phosphatidate-Phosphoinositide Synthesis and Phosphoinositide Hydrolysis and Mobilizes Intracellular Calcium in Cultured Arterial Muscle Cells. Proc. Nat. Acad. Sci. 81: 7812-7816, 1984.
36. Herrmann, G., Hehrmann, R., Scholz, H., Atkinson, M., Lichten, P., Hesch, R-D.: Parathyroid hormone in coronary artery disease (submitted for publication).

37. Avioli, L.V.: Stones, bones, abdominal groans, psychic moans, and hypertones. Cardiovascular Medicine 3: 835-842, 1978.
38. Ackley, S., Barrett-Connor, E., Suarez, L.: Dairy products, calcium, and blood pressure. Am. J. Clin. Nutr. 38: 457-461, 1983.
39. Belizan, J.M., Villar, J., Zalazar, A. et al: Preliminary evidence of the effect of calcium supplementation on blood pressure in normal pregnant women. Am. J. Obstet. Gynecol. 146: 175-180, 1983.
40. Garcia-Palmieri, M.R., Costas, Jr., R., Cruz-Vidal, M., Sorlie, P.D., Tillotson, J., Havlik, R.J.: Milk Consumption, Calcium Intake, and Decreased Hypertension in Puerto Rico. Puerto Rico Heart Health Program Study. Hypertension 6: 322-328, 1984.
41. Kolata, G.: Does a Lack of Calcium Cause Hypertension? Several recent studies link calcium and high blood pressure but the medical community is divided on how to interpret them. Science 225: 705-706. 1984.
42. Belizan, J.M., Villar, J., Pineda, O., Gonzalez, A.E., Sainz, E., Garrera, G., Sibrian, R.: Reduction of Blood Pressure with Calcium Supplementation in Young Adults. JAMA 249: 1061-1164, 1983.
43. McCarron, D.A., Ellison, D.H., Anderson, S.: Vasodilation mediated by human PTH 1-34 in the spontaneously hypertensive rat. Am. J. Physiol. 246: (Renal Fluid Electrolyte Physiol. 15): F96-F100, 1984.
44. Pang, P.K.T., Yang, M.C.M., Keutmann, H.T., Kenny, A.D.: Structure activity relationship of parathyroid hormone: Separation of the hypotensive and the hypercalcemic properties. Endocrinology 112: 284-289, 1983.
45. Stanton, R.C., Plant, S.B., McCarron, D.A.: cAMP response of vascular smooth muscle cells to bovine parathyroid hormone. Am. J. Physiol. 247: (Endocrinol. Metab. 10): E822-E826, 1985.
46. Landfield, P.W., Pitler, T.A.: Prolonged Ca^{2+}-Dependent Afterhyperpolarizations in Hippocampal Neurons of Aged Rats. Science 226: 1089-1091, 1984.
47. Garruto, R.M., Fukatsu, R. Yanagihara, R., Gajdusek, D.C., Hook, G., Fiori, C.E.: Imaging of calcium and aluminum in neurofibrillary tangle-bearing neurons in parkinsonism-dementia of Guam. Proc. Natl. Acad. Sci. 81: 1875-1879, 1984.
48. Luxenberg, J., Feigenbaum, L.Z., Aron, J.M.: Reversible long-standing dementia with normocalcemic hyperparathyroidism. Case Report. J. Amer. Geriatr. Soc., p. 546-547, 1984.
49. Goldstein, D.A., Chui, L.A., Massry, S.G.: Effect of parathyroid hormone and uremia on peripheral nerve calcium and motor nerve conduction velocity. J. Clin. Invest. 62: 88-93, 1978.
50. Akmal, M., Goldstein, D.A., Multani, S., Massry, S.G.: Role of uremia, brain calcium, and parathyroid hormone on changes in electroencephalogram in chronic renal failure. Am. J. Physiol. 246: (Renal Fluid Electrolyte Physiol. 15): F575-F579, 1984.
51. Virgil: Eclogues IX, line 51.
52. Marx, J.L.: Aging Research (I): Cellular Theories of Senescence. Science 186: 1105-1107, 1974.

PRIMARY HYPERPARATHYROIDISM (PHP) REVISITED

Charles R. Kleeman[1], Saleh-Salehmoghaddam[2], and Gabriel Hertz[3]

Division of Nephrology, Department of Medicine
UCLA School of Medicine, Los Angeles, California

"In the final analysis very little is known about anything, and much that seems true today turns out to be only partly true tomorrow, but as things go in medicine our knowledge of the interrelation one to the other of all the sequelae which result from the action of the parathyroid hormone is probably clearer than that for any other hormone." Fuller Albright, Chapter 1, Page 1, THE PARATHYROID GLANDS AND METABOLIC BONE DISEASE, 1948

In 1904, Askanazy (8)* first noted the association between a parathyroid adenoma and Von Recklinghausen's Disease - Generalized Osteitis Fibrosa Cystica. The association was noted at autopsy many times subsequently but not understood. It was generally felt that the parathyroid lesion was somehow secondary to the bone disease, a compensation

---

* The numbers for each reference correspond to their number in the chronology of the Parathyroid glands, etc. which follows the text as part of the manuscript.

1) Charles R. Kleeman, M.D. - Professor of Medicine and Nephrology
Max Factor Family Foundation Professor of Nephrology and Medicine
2) Saleh-Salehmoghaddam, M.D. - Assistant Professor of Medicine
3) Gabriel Hertz, M.D. - Senior Research Fellow

or response to the bone disease. Thus, in 1925, Mandel(20) tried to help a 38 year old man with severe Osteitis Fibrosa by transplanting four or five human parathyroids into him. This did not help and he decided to explore his neck. He removed a large adenoma and the patient recovered remarkably. In the United States, about the same time (1923-25), four separate investigators were preparing extracts of the parathyroid glands.(16-19) The most intensive studies of the extract was carried out by Collip,(17) who produced a stable hydrochloric acid extract of bovine (PTE) glands and assayed it in dogs. He produced the first example of experimental hyperparathyroidism with all the symptoms of marked hypercalcemia. From his work he concluded: "while it may later develop that these glands have other functions, there would seem to be no need at present to ascribe any function to them other than that of regulators of calcium metabolism."

Albright's group quickly saw the potential of this new hormone and Hunter and Aub (21) reported their results in 1926 of the use of PTE extract in the mobilization of lead from the skeleton of patients with chronic lead poisoning. During these studies they inadvertantly caused, in one of these patients, the first reported case of severe parathyroid intoxication.

At this time Dr. Du Bois in New York was investigating a young man in his thirties with hypercalcemia and progressive severe bone disease and concluded that he had what Collip and Hunter and Aub had described. He sent the patient, Captain Martell, to Albright at the Massachusetts General Hospital in 1926 for the first of six explorations for a parathyroid tumor.(26) Finally, at Captain Martell's request, in 1932 they carried out a mediastinal exploration and found the large solitary adenoma.

In 1929, Barr, Bulger and Dixon (24) in St. Louis investigated a woman with hypercalcemia, severe bone disease and severe hypotonia. They successfully removed a large adenoma and stated: "The finding of the parathyroid tumor and the effects that followed its removal seemed to complete the chain of reasoning necessary to implicate the parathyroids in the causation. In the literature, instances of similar difficulties have been reported under a great variety of titles. With the assembling of all the evidence it seems justifiable to introduce the name of hyperparathyroidism and to consider under this heading the various cases as constituting a clinical entity as definite and distinct as parathyroid tetany or exopthalmic goiter."

In 1934 Albright, Aub and Bauer (29) described their experiences with their first seventeen operated cases of PHP and emphasized the polymorphic clinical picture. They had patients with bone disease and kidney stones, bone disease without kidney stones, kidney stones without bone disease and without bone diseases or kidney stones; the total spectrum that we see today. On the basis of these cases and the size of their parathyroid tumors he made this remarkable statement about mass to secretory relationship: "It is now quite clear, however, that the tumor tissue is no more potent per gram of tissue than normal tissue."

We would define primary hyperparathyroidism as a disorder of the parathyroid glands characterized always by three fundamental features: 1) an increased mass of parathyroid tissue, 2) an increased secretion of PTH and as a consequence an increased level of biological active hormone in the circulation, and 3) an increase in the free calcium ion (Ca^{++}) concentration in the blood unless there is some associated specific disorder such as severe vitamin D deficiency or a decline in renal function which may prevent hypercalcemic primary hyperparathyroidism. All physiologic and pathologic characteristics of the disorder are directly or indirectly the consequence of these three fundamental features.

At present we do not know what are the basic genetic and/or environmental factors responsible for the abnormal cell multiplication that results in an adenoma or diffuse hyperplasia of all the glands. What is clear is that the cells are secreting too much parathyroid hormone despite concommitant hypercalcemia. Yet we now know these glands are not simply autonomous, and from many in vitro and in vivo experiments it is clear that they can be suppressed. Murray, Peácock, Powell et al in 1972 (75) clearly showed the striking suppressibility of the great majority of these lesions regardless of their pathology. In fact, they observed that the rate of fall of I-PTH during a hypercalcemic infusion did not differ from the rate of fall of the I-PTH in the same patient in the 4 hours after parathyroidectomy. This lack of autonomy has been demonstrated now in many laboratories.

Thus why the increased secretion? The studies by Gittes and Radde (55) demonstrated clearly in rats that a simple increase in mass of normal parathyroid cells transplanted to an intramuscular site could result in sustained chronic hyperparathyroidism and hypercalcemia. They made this prophetic comment in the beginning of the discussion of their findings: "The most striking finding in these experiments is that normal tissue which normally regulates its secretory activity to maintain calcium homeostasis has been demonstrated to be incapable of cutting back beyond a certain point and the aggregate secretion of many such normal glands transplanted into an isologous host has led to an obligate state of systemic hyperparathyroidism. It seem that the complex machinery involved in the production of the specialized parathyroid hormone may be turned down to a slow idling speed but not off completely." Subsequent studies have proved this statement to be true.

Then in 1969, (63) Parfitt formulated a theoretical model to explain the sustained hypersecretion and hypercalcemia caused by a progressive increase in the number or mass of normal cells secreting a small residual of the hormone. In 1971, Targovnik, Rodman and Sherwood (70) showed it was not possible to completely inhibit PTH output from normal bovine glands maintained in organ culture, and in 1976 Mayer, Habener and Potts (87) demonstrated a sustained nonsuppressible component of secretion in vivo in calves during continuous hypercalcemia.

The other mechanism that has been proposed to explain the defect of control is that higher levels of Ca^{++} are necessary to suppress hormone secretion, A "set point error." This concept must be considered undecided because of the conflicting results from the experimental studies.

The basic frequency and pathologic characteristics of the various lesions causing PHP have not changed since their earliest description. This is clear from the review every ten years of the Massachusetts General Hospital cases since 1930. (Habener and Potts in Chapter 53, Endocrinolgoy Vol. @, Ed. Degroot, 1979.) Further of great interest is the fact that since Mandel's first report, the disease has been detected with increasing frequency with each passing decade, and it seems almost with each passing year. PHP is being diagnosed more frequently than any other endocrinologic disorder other than diabetes, and thyroid disease.

While a progressive increase was observed in the 1940s, 50s and 60s, the consecutive clinic admission studies of Boonstra and Jackson in 1971, (68) and Heath, Hodson and Kennedy in 1981, (98) emphasized that the almost exponential increase of the past 5-20 years has accompanied the universal introduction of routine automated chemistry which including calcium determinations. The present rates of PHP is now about 1/1000 cases for men and 2/1000 cases for women.

From it earliest diagnosis it was a disorder predominantly of the fifth, sixth and seventh decades; women predominating over men, 2-3/1; the 3:1 ratio was seen after the menopause with the loss of ovarian function, and its effect on the clinical expression of the disease.

In 1947 Norris (35) reviewed all the reported cases of parathyroid adenomas in the world literature discovered at autopsy or at surgery between 1903 and 1947, a total of 322 cases. 60% of them had clinically evident generalized osteitis fibrosa. 31% had bone disease and nephrolithiasis, 5% has stones alone and only 1.5% were free of both bone disease and kidney stones. Also the mean weight of the adenoma in Norris' review was 8.0 grams, strikingly heavier than has been observed in subsequent years to the present. In recent years we are seeing progressively more cases that have neither clinically evident bone or stone disease and supposedly fewer symptoms. However despite the automated routine calciums and supposedly earlier detection of the PHP these more recent patients are not young and it is not likely that they have had the disorder for a shorter period of time. Also why in the cases reviewed by Norris was the mean weight of their tumors so much larger, osteomalacia frequently described as a component of the bone disease, and the serum inorganic phosphorus of the patients lower? I would like to speculate or suggest that the striking difference in the clinical syndrome prior to 1945 or 1950 was not the difficulty or lateness of detection but something different in the lives of these earlier patients, in their environment or in their nutrition. I believe it is quite possible that they had a far greater incidence of subclinical vitamin D deficiency or marginal intake of the vitamin whose ultimate metabolite 1,25 $(OH)_2D_3$ is an important modulator of parathyroid secretion and its peripheral action. There probably still were many mild cases of PHP in the community, but they were going through life and dying without being recognized or treated.

In 1968 Martin Lloyd (60) wrote a remarkable paper on PHP after analyzing retrospectively 138 cases from University College Hospital, London seen personally by the late Professor Charles Dent between the years 1950-1965. Lloyd concluded that the cases could be divided into 2 main types. Type 1 had larger tumors (mean weight 5.9 grams) higher serum calciums, lower rate of calcium excretion (333 vs 408 mg.) shorter histories (3.6 vs 6.9 years), overt bone disease, and much less stone disease (in short much higher growth and secretory rate of the tumors). In contrast, Type II cases had lower serum calciums, higher urinary excretion of calcium/24 hours, smaller tumors, longer histories, much more stone disease but very little bone disease (lesser growth and secretion rate of their tumor). Could Lloyd's type I cases also have been cases of marginal Vit. D deficiency intake relative to that of his Type II cases? Because both Type I and Type II cases in Lloyd's series were probably detected and treated earlier than those before 1945 described by Norris, the latter series would be expected to have more severe manifestations of the disease. I suggest that for research or even clinical purposes it would be of value to measure 1,25 $(OH)_2D_3$ on as many consecutive cases of PHP as possible.* Slatopolsky and his associates have demonstrated clearly that intravenous infusions of 1,25 $(OH)_2D_3$ to patients with severe secondary hyperparathyoidism definitely inhibits PTH secretion. Would this happen in primary hyperparathyroidism as well?

With regard to other manifestations of the disease. Albright and his associates, in 1929, (29) emphasized the striking effect of primary hyperparathyroidism on the renal excretion of phosphorus and actually concluded that this was the primary and fundamental action of the hormone. He also showed, but never referred to, the obvious fall in calcium clearance which occurred in this study following the administration of parathyroid tract. His belief in the primacy of the phosphate changes was further strengthened when he showed that giving phosphorus could reverse all the biochemical abnormalities of PHP.

Crawford and Talbott in 1950 (37) were the first to suggest the clinical usefulness of the fractional tubular reabsorption of phosphorus to assess the level of parathyroid function and there followed many papers using phosphate clearance, T.R.P., P.E.I. with and without phosphate and calcium challenges, to assist in diagnosis. However, it was not until the late 60's that Bijvoet (64) carried out the experiments and did the mathematical analysis of the tubular transport of phosphate in humans which showed Tm PO4/GFR, to be equivalent to a threshold plasma concentration. This parameter probably measures a fundamental property of reabsorptive activity and by its innate characteristics is a main

---

*In a personal communication from Dr. Michel Paillard and his associates, who participated in the Phosphate Workshop in Marseille, they have informed me (C.R.K.) that they have made measurements of 1,25 $(OH)_2D_3$ on 102 randomly sampled patients with PHP from a total group of 306 seen between 1979 and 1984. They found that the patients with severe bone disease, the highest I - PTH and the lowest serum inorganic phosphorus had the lowest 1,25 $(OH)_2D_3$, and low 25$(OH)D_3$. These observations lend added weight to the postulation described above.

determinant of serum PO4 concentration. Subsequent studies in his and others laboratories has shown it to be the most sensitive and accurate measure of the effect of PTH on the tubular reabsorption of inorganic phosphorus, and I believe equal to that of urinary cyclic AMP measurements.

In 1954-56, Talmage (41-43) and his group conclusively showed in the rat that hypoparathyroidism increased the renal clearance of calcium and hyperparathyroidism or PTH administration decreased Ca clearance. In 1961 our laboratory confirmed these observations in two other species, the human and the dog.(49) It is clear that this action of PTH assists in the support of the serum calcium and it can be said that PHP is the only hypercalcemic disorder, except familial hypocalciuric hypercalcemia, in which, in the presence of normal kidney function, calcium excretion is often normal. Then we may ask why is urinary Ca excretion so elevated in most cases of so called "normocalcemic" PHP often with recurrent nephrolithiasis? The term normocalcemic PHP is, for any given patient, a misnomer, except in the absence of a specific cause such as Vit. D deficiency or renal failure that is lowering the serum calcium. To my knowledge there is no reported case of "normocalcemic" PHP in which the serum calcium remains the same or does not fall significantly after successful parathyroid surgery.

After Rose in 1957 (44) described a practical chemical method for free calcium ion (Ca^{++}) determination he and Lloyd in 57 and 58 (46) applied it to cases of PHP and other hypercalcemic disorders. They amazingly found that calcium was bound less to plasma proteins in PHP than in other hypercalcemic states and that for any given level of total calcium the Ca^{++} was higher in PHP. Transbol in 1970 (67) clearly confirmed Lloyd and Rose's observations on a large number of borderline and obvious hypercalcemic patients with PHP. In 1973 and 76 Muldowney (81) showed that a number of cases of so called idiopathic hypercalciuric stone formers, with normal total serum calcium, actually had both elevated Ca^{++} and elevated PTH by RIA. Surgery on these patients confirmed the primary hyperparathyroidism and post operatively the total and the Ca^{++} fell to normal (for that patient) and there was a marked decrease in urinary calcium.

Broadus and his associates in the late 70s studied calcium absorption from the gastrointestinal tract and measured 1,25 $(OH)_2D_3$ levels in the blood of PHP patients with borderline or only slightly elevated serum calcium. (97) They found that the incidence of recurrent stone formation was much greater in those with elevated 1,25 $(OH)_2D_3$. Also the stone forming group had greater gastrointestinal absorption of calcium and much higher urinary excretion of calcium, while all other biochemical parameters of PHP were similar in the 2 groups. Now we can return to our earlier question. Why in the PHP patients with borderline elevations of serum calcium or, in the so called "normocalcemic" ones, do we see such marked hypercalciuria when, on the basis of the action of PTH on the renal clearance of calcium, we would expect lower urinary excretion of calcium? The answer, I believe is found in the elevated 1,25 $(OH)_2D_3$ in the plasma of the patients, and its metabolic consequences as described by Broadus and his associates.

The 1,25 $(OH)_2D_3$, in turn, was increased because the rise in circulatory PTH stimulated the renal 1-hydroxylase enzyme system. It is conceivable that the higher circulating level of 1,25 $(OH)_2D_3$ in these cases could "dampen" through feedback inhibition a part of the excess of PTH secretion from the abnormal gland or glands. A new steady state would develop with a milder form of PHP with lower serum calcium but higher urinary calcium.

We have come to take for granted the measurement of plasma immunoreactive PTH but it was Berson and Yalow (51) who revolutionized the whole field of parathyroid physiology when they reported for the first time in 1963 the radioimmunoassay of circulating parathyroid hormone, and in 1966 (54) applied it to cases of primary and secondary hyperparathyroidism and normocalcemic bronchogenic carcinoma. Then in 1968 (59) they let us know that it was probable that their RIA curves indicated heterogencity of the circulating hormone and possibly the presence of fragments of the intact hormone. From then until now these fragments have plagued the field of PTH radioimmunoassay yet stimulated an enormous amount of outstanding immunochemistry on the intact hormone and its metabolic fragments, and the development of accurate and extremely sensitive bioassays (93, 94, 103) and from simultaneous use of the latter it has been possible to develop RIAs that have measured the intact hormone in the circulation at concentration closely approximating that of the true biologically active molecule present in the circulation both in normal and abnormal states of parathyroid gland activity. (89, 103)

During the same period as Berson and Yalow's work, Chase and Aurbach reported for the first time in 1967 (58) that physiologic amounts of purified PTH injected into rats caused a striking rise in cyclic AMP excretion that preceded the rise in phosphate clearance. Inhibition of PTH secretion with a calcium infusion caused a sharp drop in urinary cyclic AMP. They proved that the primary source of urinary cAMP was the kidney itself. Between 1970 and 1980 Broadus and his associates, (91, 99)Murad and Pak, (78) as well as many others, clearly delineated the clinical physiology of urinary cyclic AMP and its use as a most practical, accurate and sensitive, in vivo bioassay of physiologically active circulating PTH when carried out and expressed properly.

During the past fifteen years we have witnessed a progressive evolution (with respect) to the etiology of the syndrome of Humoral Hypercalcemia of Malignancy (HHM). We have evolved from the near certainty that the hypercalcemia in most cases was due to ectopic production of PTH to the near certainty in recent years that the syndrome is not caused by biologically active PTH but by a substance, possibly a peptide, that has receptor binding capability and bioactivity in part resembling that of intact PTH. The clarification of this difficult differential diagnostic problem between HHM and PHP has been made possible by the use of cyclic AMP (90, 99) and 1,25 $(OH)_2D_3$ (88) assays, urinary calcium excretion and the use of remarkably sensitive and accurate plasma bioassays of PTH (93, 94) ; the latter utilizing cytochemical techniques, or the in vitro stimulation of the adenylcyclase -cAMP system. Also by combining the bioassay of PTH with new RIA of intact PTH it has been possible to develop sensitive immunoassays for PTH that measure the intact molecule of PTH at the very low physiologic concentration that are detected by the bioassays.

It is reasonable that tests would develop to assist the clinician and the surgeon to localize the site of the parathyroid adenoma, especially as imaging techniques have become more and more sophisticated and reasonably accurate (62, 85). The use with ever increasing frequency prior to the first operation of CAT scans, ultrasound, technetium -thallium radioisotope subtraction scan, digital subtraction angiography, and venography with collection of selected samples of blood from the thyroparathyroid area. Their value to an excellent surgeon who knows thoroughly his anatomy of the glands, the usual and unusual sites for an adenoma, and who carries out systematic dissection of all gland areas has not been proven. The best localizing procedure is still the dissection by an experienced and knowledgeable surgeon. We have to prove that the technology has not gone beyond what is necessary to do the best and most cost effective job.

The advance of technology has created for us another problem. The patient with "asymptomatic" and uncomplicated P.H.P. who was detected through routine multichannel biochemical screening. It has been our policy that once a definitive diagnosis of PHP has been made, to carry out the curative surgery on the patient. In almost every instance the so called asymptomatic patient realizes post operatively that they had subtle changes in their behavior or well being preoperatively that they were unaware of. Furthermore cure of the disease removed the possibility of later complication secondary to continued hypercalcemia and excess parathyroid hormone.

However it remained for Scholz and Purnell (100) of the Mayo Clinic to carry out a 10 year prospective study on 133 patients with PHP. The asymptomatic cases has serum calciums between 10.0 and 11.0 All those with calciums above 11.0 were immediately taken to surgery. In the 10 year period 23 percent required surgery because of progression and 32 died, 14 from atherosclerotic heart disease and/or hypertension, and 6 died of cancer. They concluded "Our recommendation for patients whose clinical and laboratory studies support the diagnosis of PHP has been and continues to be surgical exploration by an experienced parathyroid surgeon . . ." More recently Van't Hogg and associates (100) took issue with the conclusions of Scholz and Purnell and tried to make the case for a nonsurgical approach. They studied only 32 patients for 4.2 years. Despite their conclusion it is of great interest that 11, or 16% with hypertension (mean 150/100) at the start of observations increased to 16, or 50% by the end of their study.

It is exciting and most encouraging to realize that with such tests as 1. free Ca ion, 2. Highly sensative and accurate bioassays and RIAs of PTH, 3. cyclic AMP, 4. 1,25 $(OH)_2D_3$ assay, 5. Tmp_{O4}/gFR, that we are now in a period when these measurements can be readily brought together for clinical purposes to make accurate pathophysiologic diagnoses and to delineate and clarify the various presentations of PHP. What Albright said on page one of his book in 1948, we can say is certainly true today.

CHRONOLOGY OF THE PARATHYROID GLANDS, THEIR ANATOMY, PHYSIOLOGY, AND THE SYNDROME PRIMARY HYPERPARATHYROIDISM

A Selected List 1862 - 1983

1862 1) Richard Owen - first recognized the parathyroid glands during dissection of an Indian rhinoceros. Trans. Zool. Soc. Lond. 4:31, 1862.

1863 2) Virchow, R. Die Krankenhaften Geschwulste. Berlin: A. Hirschwald, 1963. The first to describe the parathyroid glands in man.

3) Sandstrom, I., Upsala Lakaref. Forh. 15:441, 1880. An excellent description of the anatomy of the parathyroid glands in man and other mammals. Translated by Hammar. Bull. Hist. Med. 6:179, 1938. First relatively complete comparative description of the gland which he gave its present name.

1891 4) Gley E. Compt. rend. Soc. de Biol. Paris. 3:250, 567, 583, 1891. He demonstrated that they were anatomically distinct from the thyroid gland, but he felt they were functionally connected.

1896, 1897 & 1898 5) Vassale and Generali. In a series of articles in Arch. Ital. de Biol. 1897, 1898, Vol. 25, 26, 30,33 and Reforma Med. Napoli, 2:235, 631, 1897 and 3:77, 1897. Clearly showed that the removal of the parathyroid glands (thyroid intact) caused fetal tetany in various mammals, but that if one gland was left intact, no tetany occurred. They also noted that i.v. injection of fresh extract of bovine parathyroid afforded temporary relief from tetany. Oral administration was without effect.

6) Welsh, J. Anat. & Physiol. 32:38, 1898. Carried out first extensive anatomic gross and microscopic studies of human parathyroid glands. 40 cases in man, infancy to advanced age. Concluded the glands were totally separate from the thyroid with no embryonic relationship. Also, cat experiments: Total 4 gland PTX caused rapid fatal tetany, 3 gland PTX caused transient tetany, 2 gland PTX caused no symptoms. No benefit feeding cat fresh parathyroid glands.

1903 7) Erdeim, J. Brit Path. Anat. 33:158, 1903. Probably the earliest record of a case that can be accepted as a parathyroid adenoma.

1904 8) Askanazy, M. Arb. Gen. Pathol. Anat. Inst. Tubingen, Leipz. 4:398, 1904. First to report a parathyroid adenoma (found at autopsy) in a patient with Von Recklinghausen's Disease (osteitis fibrosa cystica generalisata).

9) Erdheim, J., Wien. Klin. Wehnschr 19:716, 1906 and Grenzgeb d Med u Chir. Jena 16:632, 1906. Extensive studies in various animals of PTX and tetany. Found acute symptoms worse in 24 hours. Carried out serial sections of 3 patients who died of postop. tetany. In none could he find any trace of parathyroid tissue.

1907 10) Halsted, and Evans. Ann. Surg. 46:489, 1907. The first complete description of the surgical anatomy of the parathyroid glands and their blood supply in relation to their preservation at thyroid surgery.

1908 & 1909 11) Mac Callum and Voegtlin Johns Hopkins Hosp. Bull. 19:91, 1908 and

12) Mac Callum and Voegtlin, J. Exp. Med. 11:118, 1909. Gave the most complete and vivid description of tetany and all of its manifestations in the dog. Found for the first time that blood calcium decreased to 1/2 normal and that various calcium salts were most effective. They were the first to note the probable acute increase in calcium excretion in PTXed fasting dogs. They formulated the hypothesis that in some way the parathyroid glands are critical for the maintenance of blood calcium and that tetany is secondary to the hypocalcemia.

1916 13) Biedl, Innere Sekretion, Berlin & Vienna 3rd Ed, part 1, 136. Transplanted parathyroid glands of a dog into the spleen of the same animal, then removed the thyroid without effect, but splenectomy caused fetal tetany in 24 hours.

1918 14) Howland and Marriott, Q.J. Med 11:289, 1918 found normal blood calcium in humans 9.2-11.3 mg/dl and that patients in tetany had a mean blood calcium of 5.6 mg/dl.

1919 15) Borchees, Zentr. Chir. 46:34, 907, 1919. Demonstrated the temporary relief of tetany by parathyroid grafts in humans.

1923 16) Berman, Proc. Soc. Exp Biol and Med. 21:456, 1923-24. Made parathyroid extract from ox parathyroid glands, with acidified alcohol which removed most protein and lipid and left a crystalline material, "which increases definitely the calcium content of the blood."

1925 17) Collip, J. Biolchem. 63:395, 1925. Produced a potent, very stable extract of bovine parathyroid glands with hot dilute hydrochloric acid. This extract prevented hypocalcemia in PTX dogs and when given in excess amounts produced the first example of experimental hyperparathyroidism with marked hypercalcemia. Collip stated "while it may later develop that these glands have other functions there would seem to be no need at present to ascribe any function to them other than that of regulators of calcium metabolism."

18) Hanson, Military Surgeon 52:280, 1923 and ibid 54:218, 1924 prepared a similar hydrochloric acid extract to that of Collip's without knowledge of the latter's work. Although he described the technique of preparation in detail, there is no evidence that he carried out any experiments with his material.

19) Hjort, Robison and Tendick J. Biochem 65:117, 1925 were able to prepare extracts comparable to Collip. Berman, and Hansen. They confirmed all of Collip's experimental observations in the dog. They made these interesting observations "most of the protein bulk can be removed without appreciably affecting potency of the extract." "The parathyroid hormone is very stable when preserved in 0.1 N HCL for a period of over a year."

20) Mandl, Wein Klin Wschr. 38:1343, 1925. Carried out the first surgical removal of a parathyroid adenoma from a 38 year old male with severe osteitis fibrosa cystica with remarkable improvement in his clinical picture.

1926 21) Hunter and Aub. Quart J of Med 20:123, 1926/27. Produced the first case of experimental "parathyroid toxicosis" in a human subject when they gave parathyroid extract (100/units/day) to humans to treat chronic lead poisoning. This patient developed marked hypercalcemia, reversed in stopping the parathyroid extract.

22) Greenwald and Gross, J. Biolchem. 68:325, 1926. Demonstrated the marked phosphaturic effect of Collip parathyroid extract without a rise in serum phosphate. They also concluded that the hypercalcemia and marked hypercalciuria could only come from the release of calcium from bone.

23) The first case of primary hyperparathyroidism operated on in the USA was carried out at the Mass. Gen. Hospital in May 1926. It was unsuccessful. Further cervical explorations were also unsuccessful until at the demand of the patient, a mediastinal exploration revealed a 3 x 3 cm. typical parathyroid adenoma in the anterior mediastinum in November 1932. Bauer and Federman. Hyperparathyroidism Epitomized. The case of Capt. Charles E. Martell. Metab. 71:21, 1962.

<u>1929</u> 24) Barr, Bulger and Dixon J Am A 92:951, 1929. Were the first to recommend the name "Hyperparathyroidism" for the syndrome due to the chronic excess of parathyroid hormone in humans.

25) Albright and Ellsworth J. Clin. Invest. 7:183, 1929. Gave Collip parathyroid extract to patient with idiopathic hyperparathyroidism. They observed striking phosphaturia, a progressive fall in serum phosphorus, a rise in serum calcium and a <u>decrease</u> in urinary calcium. The authors did not remark about the latter, although it clearly demonstrated the hormone-induced <u>decrease</u> in calcium clearance.

<u>1930</u> 26) Bauer, Albright and Aub, J. Clin Invest. 8:229, 1930 reported their first biochemical and metabolic observations on Captain Martell prior to this first surgery. They concluded "Because the pathological physiology in this case is similar to that in an individual receiving large doses of parathormone. We feel justified in concurring with the diagnosis of hyperparathyroidism, suggesting by Dr. Eugene F. Dubois."

1932 27) Albright, Bauer, Claffin et al stated conclusively that all the biochemical manifestations of PHP could be reversed by phosphate ingestion for a number of days and suggested that hypercalcemia was secondary to hypophosphatemia.

28) Thomson and Pugsley, Amer. J. Physiol., 102:350, 1932, concluded that parathyroid hormone directly stimulates cellular elements in the bone to increase osteoclastic activity when they showed that in the dog the rise in serum calcium after PTH injection is NOT secondary to any fall in serum phosphate.

1934 29) Albright, Aub, and Bauer JAMA 102:1276, 1934 described for the first time the "polymorphic" or varied clinical picture of P.H.P. rather than "classic form" i.e. osteitis fibrosa cystica. Stressed the high serum calcium and low phosphorus in all cases. They also hypothesized the mass to secretion relationship. "It is now quite clear however, that the tumor tissue is no more potent per gram of tissue than normal tissue."

30) Ellsworth and Howard, Bull. J. Hopkins Hosp. 55:296, 1934. Demonstrated with normal human subjects that Parathormone (Collip's extract prepared by Eli Lilly) 40 units caused a natriuresis with marked alkalinization of the urine and a decrease in chloride excretion.

1935 31) Castleman and Mallory Am. J. Path. 11:1, 1935. Most classic and complete description of the gross and microscopic pathology in the American literature, with a complete review of the world's literature. They correlated the size of the gland with the clinical picture and height of serum calcium.

32) McLean and Hastings Am J. Med. Sc. 189:601, 1935. First description of the clinical use and chemistry of the protein bound and diffusable calcium in the serum and the concentration of the free Ca^{++} in circulation.

33) Albright, in a case record of the Mass. Gen. Hosp. suggested that malignant non-parathyroid tissue might produce parathyroid hormone or a parathyroid-like substance. The case was a hypernephroma simulating hyperparathyroidism. In Albright and Reifenstein, THE PARATHYROID GLANDS AND METABOLIC BONE DISEASE, P. 93, 1984.

1942 34) Patt and Luckhardt Endocrinol. 31:384, 1942. Showed in vivo that perfusion of the parathyroids with calcium deficient blood resulted in systemic hypercalcemia and suggested that the secretion of PTH was inversely proportional to the ionized calcium perfusing the glands.

1947 35) Norris, Internat. Absts. of Surg. 84:1, 1947. A remarkable review of all the reported cases of parathyroid adenoma, 322, in the world's literature from 1903 through 1947. The predominant clinical features and the gross and microscopic character of the tumor. Noted then that the female to male incidence was 3:1 for single adenomas (93.8%) and 4:1 for multiple adenomas (6.2%). The mean weight for all the adenomas was 7.0 grams.

1948 36) Barnicott, J. of Anatomy 82:233, 1948 and Barnicott and Chang Anat Rec. 111:23, 1951. Showed conclusively that grafts of parathyroid tissue adjacent to bone produced local bone resorption or a direct effect on bone.

1950 37) Crawford and Talbott. J. Clin. Invest. 24:1448, 1950. The first paper to suggest the clinical usefulness of the fractional tubular reabsorption of phosphate to assess the level of parathyroid function.

38) Mather, Lancet 2:424, 1953. Described first case of "normocalemic" primary hyperparathyroidism.

1954 39) Seldinger Acta Radiologica 42:353, 1954. First use of arteriography to localize a parathyroid adeonma.

40) Anderson, Dent, Harper, and Philpot. Lancet 2:720, 1954. Demonstrated that cortisone given as 150 mg. per day for 10 days could lower the serum calcium of non-parathyroid hypercalcemic disorders in contrast to the hypercalcemia of P.H.P. and this could be used as a differential diagnostic test.

1954, 1955 41) Talmadge and Kraintz Proc. Soc. Exp. Biol. & Med. 87:263, 1954, and

42) Talmage, Kraintz, and Buchanan, Proc. Soc. Exp. Biol. and Med. 88:600, 1955 showed in rats that parathyroidectomy caused an immediate rise in the renal clearance of calcium and that parathyroid extract caused a decrease.

1956 43) Talmage and Elliot Endocrinol. 59:27, 1956. Clearly demonstrated that parathyroid extract caused hypercalcemia when administered to nephrectomized rats dialized in the fasting state. Thus indicating a primary osseous rather than renal effect of the extract.

1957 44) Rose, Clin. Chem Acta 2:227, 1957, using the murexide chemical method for free Ca^{++} ion confirmed its reliability and applicability to clinical problems.

45) Gaillard, Schweiz. Med. Wochschr. 87: Supple 14, P 217, 1957. First to obtain bone resorption by the direct addition of purified parathyroid hormone to bone grown in tissue culture.

1958 46) Lloyd and Rose. Lancet 2:1258, 1958. Measured total and ionized Ca^{++} in PHP and non PHP hypercalcemic disorders. They found that ionized Ca^{++} was disproportionately elevated in PHP in contrast to other hypercalcemic disorders. Also in 2 cases of PHP where total calcium was normal, ionized Ca^{++} was still elevated. In 1962, Lloyd, Rose, Smeenk (Clin. Sci. 22:353, 1962) extended and confirmed these observations and concluded that plasma proteins bind relatively more calcium in PHP patients.

1959 47) Aurbach, G.D., J. Biol. Chem. 234:3179, #1959. Found that extraction of the more crude forms of parathyroid extracts with phenol instead of hot acid led to a marked degree of purification and increased potency. He recognized that cleavage of the active polypeptide took place during exposure to acid at 100° C.

1960 48) Pullman, Lavender and Aho Endocrinol. 67:570, 1960. Administered by slow infusion into one renal artery of the dog a highly purified PTH and compared with the contra lateral side. Demonstrated conclusively the direct inhibition of the tubular reabsorption of phosphorus.

1961 49) Kleeman, Bernstein, Rockney et al, Yale J. Biol. and Med. 34:1, 1961. Demonstrated in humans and dogs the effect of parathyroid hormone on the renal clearance of calcium. Hyperparathyroid states associated with decreased, and hypoparathyroid states with increased renal clearance.

1962 50) Dent, Brit. Med. J. 2:1, 95, 1962. He emphasized that the disease remains true to type in a given patient and does not change or progress from one form of PHP to the other.

1963 51) Berson, Yalow and Aurbach et al. Proc. Nat. Acad. Sc. 49:613, 1963. This is the report of the first radioimmunoassay for parathyroid hormone, both human and bovine. They utilized the highly purified hormones of Aurbach and Rasmussen as standard and for radiolabelled P.T.H.

1964 52) Wills, and McGowan Brit. Med. J. 1:1153, 1964. Based on the earlier work of Ellsworth, they studied 33 cases of PHP and 28 others with hypercalcemia to determine plasma chloride and acid-base parameters. First indication of relative or absolute hyperchloremia in PHP in contrast to other hypercalcemic disorders.

1965 53) Boonstra and Jackson Annals of Int. Med. 63:470, 1965. First study of a very large totally unselected primary care population over 5

1/2 years (26,000 consecutive patients). Specifically excluded cases referred for calcium disorder. 31 new cases of PHP, females to males 2:1, one case per 850 patients. Renal stones 20%, osseous symptoms 13%, mental and fatigue 25%.

1966 54) Berson and Yalow. Science. 154:907, 1966. First report n the clinical use of a RIA for PTH. Studied patients with adenomatous PHP, uremia and bronchiogenic carcinoma. Found cases of alter with values = PHP, even though cases were normocalcemic ones.

55) Gittes and Radde, J. Urol. 95:595, 1966. Developed a remarkable experimental model for chronic PHP. They transplanted 20, 40, 80 or more isologous functional parathyroid glands that had been removed and stored in liquid N_2 for 5-6 weeks. The glands were placed into small pockets dissected in the thigh muscle of recipients. A sustained hypercalcemia (up to 14 mg/dl) the level correlating with the number of transplants, persisted from one to three months. Thus, PHP produced by a large mass of normal tissue. Hence, it is suggested that parathyroid adenomas may involve not deranged control of hormone secretion, but only abnormal cell proliferation, with resulting excess of parathyroid tissue leading to obligatory hyperparathyroidism.

1967 56) Ross, Science 156: 1378, 1967. First report on the practical use of a calcium ion electrode for biologic fluids.

57) Ramberg, et al, Amer. J. Physiol. 213:878, 1967. In experiments in cows, found by radioimmunoassay, for the first time, that the level of PTH in the plasma was inversely related to the plasma calcium level.

58) Chase and Aurbach, Proc. Nat. Acad. of Sc., USA 58:518, 1967. Demonstated for the first time that physiologic amounts of PTH injected into rats caused a striking rise in urinary excretion of cyclic AMP that preceded the rise in phosphate clearance. Calcium infusion with hypercalcemia caused a sharp drop in urinary cyclic AMP. Proved that primary source of urinary cyclic AMP was the kidney itself.

1968 59) Berson and Yalow, J. Clin. Endo. and Metab. 28:1037, 1968. Using various preparations for purified hormones for standards, labelled hormone, and antisera demonstrated for the first time that PTH in human plasma is heterogeneous in its immunochemical composition, that removal of one or more of the peptides may depend on renal function and metabolic alteration of the hormone may account for some physiologic effects, as well as varied results with RIA tests.

60) Lloyd, Medicine 47:53, 1968. Carried out a meticulous and indepth analysis of the role of the parathyroid tumor and its relation to the clinical syndrome. Stressed that only one biologically active hormone accounted for all the manifestations. Classified 138 cases studied at Univ. Coll. Hosp., London, 1950-1965. He was able to classify a Type I and Type II syndrome, caused by 2 main types of parathyroid tumors, one growing rapidly, highly active and causing overt bone disease, and the other growing slowly, of low activity and causing kidney stones.

61) Davis, Dent and Watson. Brit. Med. J. 2:395, 1968. First in-depth discussion of the concept of "Tertiary Hyperparathyroidism" or hypercalcemia developing on the background of prolonged secondary hyperparathyroidism. Presented 12 cases. Refer to St. Goar in discussion of M.G.H. case record (N.E.J. 268: 943, 1963), coining the term for one such case.

1969 62) Reitz, Pollard, Wang et al N. Eng. J. Med. 281:348, 1969. In six patients with PHP. They carried out for the first time the localization of adenomas by selective venous catheterization and radioimmunoassay of blood from L & R jugular, L & R subclavian, L & R innominate, and superior and inferior vena cava. The "Step up" of hormone concentration correlated well with surgical localization.

63) Parfitt. Arch. Int Med. 124:269, 1969. Formulated a theoretical model for the relation between parathyroid cell mass and plasma calcium concentration. He concluded that increase tissue mass with each cell

secreting a normal amount of hormone for the level of serum calcium could explain the presence of hypercalcemia without "set point" alteration or "autonomous" secretion.

64) Bijvoet, Clin. Sc. 37:23, 1969. First extensive physiologic and analytical characterization of the tubular reabsorption of phosphate in humans. Showed that Tm PO_4/GFR, a threshold plasma concentration, probably measured a fundamental property of reabsorptive activity and by its innate characteristics is a main determinant of serum phosphate concentration. Subsequently shown to be the most sensitive and accurate measure of the effect of PTH on tubular phosphate reabsorption.

1970 65) Sherwood, Rodman and Lundberg. Proc. Natl. Acad. Sc. USA 67:631, 1970. Using in vitro methods showed that fragments of immunoreactive PTH accumulated in the culture media during incubation of PT tissue, and that these fragments had non-parallel slope of response to RIA. Arnaud, Sizemore and Oldham, Am. J. Med. 50:630, 1971, confirmed these observations.

66) Brewer and Ronan. Proc. Nat. Ac. Sc. USA. 67:1862, 1970 and Niall, Keutmann, Sauer et al Hoppe-Seyler Z Physiol. Chem 351:1586, 1970. Delineated for the first time the entire amino acid sequence of bovine, porcine and human parathyroid hormone.

67) Transbol, Hahnemann and Hornum. Acta Endocrinolgoica 65:385, 1970. Evaluated ionized, ultrafilterable, and total calcium in the serum in PHP is to determine if ionized Ca^{++} is superior to total calcium in determining hypercalcemia and the diagnosis of PHP. Patients were 47 with definite hypercalcemia and 14 with borderline hypercalcemia. This group, in a meticulous study, confirmed completely the observations of Lloyd and Rose.

1971 68) Boonstra and Jackson. Am. J. Clin. Path. 55:523, 1971. Extended their 1965 serum Ca^{++} survey of 26,000 patients to 50,000 comparable patients with increasing detection of PHP.

69) Kaplan, Katz, Ben Issac, Massry. Cancer 28:401, 1971. In a retrospective study of 166 cases of proven parathyroid adenomas (104 Surgery, 62 autopsy), plus 75 cases from literature delineated clearly the association between parathyroid adenoma and the high incidence of antecedant, concommitant or future development of nonparathyroid malignancies when compared with a control population of nonparathyroid cases.

70) Targovnick, Rodman and Sherwood. Endocrinol. 88:1477, 1971. Studying the parathyroid hormone secretion in vitro showed that it was not possible to completely inhibit the output of normal bovine glands maintained in organ culture.

71) Riggs, Arnaud, Reynolds et al. J. Clin. Invest. 50:2079, 1971. Using a guinea pig anti-porcine PTH antiserum labelled bovine PTH and human parathyroid serum as standard, compared 54 patients with PAP with 18 patients with "Hyperparathyroidism due to non-parathyroid cancer" (authors' designation). The PHP patients had lower serum calcium, 15X higher IPTH and 8 of the non PHP patients had no detectable IPTH. They saw very different RIA slopes for both groups. Concluded that PTH-like material in "ectopic" hyperparathyroidism is immunologically different from PHP patients.

72) Parsons, Neer, Potts. Endocrinol. 89:1955, 1971. Observed for the first time that the immediate effect in dogs of the acute injection of PTH was an immediate decrease in the serum calcium, followed immediately by gradual rise over next 2 hours with peak at 4 hours. In vivo observations that PTH caused an immediate shift of Ca^{++} from E.C.F. to I.C.F.

73) Habener, Binger, Deftos et al. Nature (London) New Biol 232:91, 1971, and Habener, Powell, & Murray. Proc. Nat. Acad. Sc. USA 68:2986, 1971. Showed that the parathyroid gland venous effluent contained primarily intact 1-84 hormone and that hormone in the periphery consists of one or more hormonal fragments.

1972 74) Gallagher and Nordine Lancet 1:503, 1972, and Gallagher and Wilkinson. R. Clin. Sc. & Mol. Med. 45:785, 1973. Treated postmenopausal women with PHP with estinyl estradiol and carried out metabolic studies. The treatment completely reversed the biochemical picture of PHP, decreased bone resorption rate and confirmed the in vitro experiments showing estrogen directly inhibits PTH induced bone resorption.

75) Murray, Peacock, Powell, et al. Clin. Endocrinol 1:235, 1972. Showed clearly for the first time in vivo with IPTH, parathyroid adenomas, hyperplasia and cancer, all causing PHP, were not autonomous. All were readily suppressible and stimulatable with calcium infusion and EDTA, respectively. In fact, fall of IPTH as rapid and complete as after removal of parathyroid tissue in these patients.

76) Foley, Harrison, and Arnaud. J. Pediatrics 81:1060, 1972. Described for the first time the syndrome of familial benign hypocalciuric hypercalcemia (FHH).

77) Johansson, Thoren and Werner Upsala J. Med. Sc. 77:41, 1972. Described their clinical experience with 208 cases of PHP. Prevalence 1/1000 cases in age group > 40. Number rising rapidly almost every year. Serum calcium "normal" or borderline 10% adenomas and 70% hyperplasias. 24 hour urine calcium normal in 40% adenomas and 55% hyperplasia. Stated they never had a recurrence of adenoma after cure with PTX, and no hyperplasia occurred after cure of an adenoma.

78) Murad and Pak, N. Eng. J. Med. 286:1382, 1972. Delineated in detail the use of urinary cyclic AMP excretion as a sensitive and practical tool in the diagnosis of parathyroid disorders.

79) Garabedian, Holick, DeLuca et al. Proc. Nat. Acad. Sc. 69:1673, 1972. Demonstrated for the first time that PTH stimulated the 1 hydroxylase of the kidney to cause an increased production of 1,25 $(OH)_2D_3$.

80) Keutmann, Dawson, Aurbach et al. Biochem. 11:1973, 1972 demonstrated for the first time that the biological activity of the intact PTH resides in the first 1-27 amino acids.

1973 81) Muldowney, Freaney, Spillance et al. Irish J. of Med. Sci 142:223, 1973 studies 5 cases of so-called normocalcemic PHP with hypercalciuria and multiple kidney stones with IPTH and total ionized Ca^{++} fell significantly to the normal value for each patient. Then in 1976 Muldowny, Freaney, McMullin Q.J. Med. 177:75, 1976 extended and confirmed their observations in 28 more patients. Used Slatopolsky antisera (measured both amino and carboxyl terminal).

82) Farr, Fahey and Nash. Am. J. Surg. 126:539. 1973 noted in 100 consecutive cases (1937-73) of PHP that 34 of them had other malignant tumors of various types antecedent to, coincidental with, or subsequent to, the diagnosis and treatment of the benign parathyroid adenoma. 5 were coincidental thyroid carcinoma. Of interest: first 40 cases up to 1960 had osseous: renal manifestations of PHP of 2:1, last 60 cases over 13 years had renal manifestations of stone disease: osseous manifestation now reversed at 5:1.

1974 83) Palmer, Nelson and Bacchus. Ann. Int. Med. 80:200, 1974. Studied 52 cases hypercalcemia 25 PHP to evaluate specifically for the first time the serum, chloride to phosphate ratio (Cl/P). In PHP mean chloride = 107, mean PO_4 = 2.6, ratio 32-80; non PHP hypercalcemia chloride = 98, PO_4 = 4.5 ratio = 17-32. None of their cases of no PHP hypercalcemia were H.H.M.

84) Niall, Sauer, Jacobs, et al. Proc. The Nat. Acad. of Sc. USA 71:385, 1974. Carried out the first sequencing of the amino terminal 1-34 of human parathyroid hormone.

1975 85) Arina, Yokai and Sonada. Surg. Obs. & Fun. 141:242, 1975. First published account of the use of ultrasound in the localization of parathyroid adenoma.

86) Rosen, Strawbridge and Bain. Cancer 36:1111, 1975. The first to report the probable association between therapeutic radiation to the head and neck area and the development 30 years later of parathyroid adenoma. Two later larger series Russ et al, Cancer 43:1078, 1983 and Tisell et al Acta Chir. Scand 144:7, 1978, confirmed Rosen's observation in larger series and also the association of probable radiation-induced thyroid disease and thyroid cancer.

1976 87) Mayer, Habener, and Potts. J. Clin. Invest. 57:679, 1976 studied directly hormone secretion rate in calves by analysis of hormone concentration in effluent blood. Secretion rate measured directly. In these experiments they were able to demonstrate a sustained non suppressible component of PTH secretion which continued at calcium levels from 11 to 15 mg%. They discuss major implications of this finding.

88) Hughes, Baylink & Jones. J. Clin. Invest. 58:61, 1976. Developed a sensitive, accurate radiolig and receptor assay for 25 $(OH)D_3D_3$, and 1, 25 $(OH)_2D_3$. This allowed the assessment of many physiologic and pathologic states altering vitamin D metabolism and the measurement of 1,25 $(OH)_2D_3$ in PHP.

89) Christenson, J. Clin. Lab. Invest. 36:313, 1976. Developed an extremely sensitive RIA. By extracting and concentrating fresh frozen sera, he was able to reach a detection limit of 10 pg bovine purified 12-84 PTH. Could detect 95% of normals. Mean 68 pg/ml, S.D. 18, zero in hypoparathyroidism, 96% PHP patients detected. Undetectable in 70 cases of malignant tumors with hypercalcemia.

1978 90) Drezner, and Lebovitz. Lancet 1:1004, 1978. Measured nephrogenous AMP excretion in normals, PHP, non PT hypercalcemia without malignancy, and non PT hypercalcemia with malignancy. 6, or 40% of latter had elevated c AMP consistent with increase PTH secretion. Surgical exploration found parathyroid adenoma in 4 and diffuse PT hyperlasia in 2. Normocalcemia was restored with resection in these 6. They stressed the high frequency of PHP in association with malignant disease.

91) Broadus, Deftos, and Bartter. J. Clin. Endo. and Metab. 46:477, 1978. Studied the effects of i.v. calcium on nephrogenous cyclic AMP. They found they could distincly separate hyperparathyroid patients from the normals. Their figures clearly indicate that cyclic AMP excretion fell at comparable rates in both groups, but the PHP patients started higher and ended higher. PT gland autonomy is not a feature of PHP, and the presence or absence of responsiveness is not dependent on adenoma size or adenoma vs. hyperplasia.

92) Habener, J. Clin Invest 62:436, 1978. Studied in vitro parathyroid gland slices from normal bovine and human glands, adenomas, hyperplasia and carcinoma with increases and decreases in calcium in the media. They found release of some hormone persisting with Ca^{++} at 5mM/L. Neoplastic tissue retained responsiveness to calcium. The responses were highly variable without a consistent pattern, e.g. parathryoid carcinoma response was like normal tissue. Thus no autonomy was demonstrated but also no conclusion on "set point."

93) Chambers, Dunham, Zanelli et al. Clin. Endocrinol. 9:375, 1978. Developed and discussed the application of the extremely sensitive cytochemical bioassay, the response of distal convoluted renal tubules in vitro to bovine and human PTH, to the measurement of the hormone in human plasma. Found values 10-15 times lower than those obtained with usual RIAs (e.g. 3-10 pg/ml). Chambers was first to apply this bioassay to detection of parathyroid hormone. Chambers et al J. of Endocrinol 71:87, 1976.

94) Fenton, Somers. Health Clin. Endocrinol 9:381, 1978. Using a slight modification of Chambers cytochemical bioassay for P.T.H. confirmed all of Chambers observations. Normal human values for plasma PTH ranged between 1 and 30 pg/ml. Where detectable PTH concentrations by RIA were on the average 14 fold higher than with bioassay of same sample.

95) Van Vroonhoven and Muller. Brit. J. Surgery 65:297, 1978. Studied the causes of failure in the surgical treatment of PHP through

lessons learned from 51 successful reoperations. The causes of failure: 1) inadequate anatomic knowledge and 2) inadequate surgical technique, and 3) unusual locations of the parathyroid glands.

1979 96) Mayer, Keaton, Hurst et al. Endocrinology 104:1778. 1979. Studied the effect of the plasma calcium concentration in the relative proportions of intact hormone and carboxyl fragments in parathyroid venous bl Hypercalcemia increased and hypocalcmia decreased. The proportion of fragments secreted. In all hypercalcemic states, PT and non-PT is the proportion of fragments secreted greater.

1980 97) Broadus, Horst, Llang et al. N. Eng. J. Med. 302:421, 1980. Divided a large series of their patients with PHP into those with normal and those with high levels of circulating 1,25 $(OH)_2D_3$. Two major features distinguished the two groups. Those with elevated 1,25 $(OH)_2D_3$ had significantly greater 24-hour excretion of calcium and about 4 times the rate of renal stone formation as the 1, 25 $(OH)_2D_3$ normal group. All other biochemical and clinical characteristics of the groups were similar. This study has major implications for the so called "normocalcemic" hypercalciuric stoneformer with PHP.

1980 98) Heath, Hodgson and Kennedy. N. Eng. J. Med. 302:189, 1980. Carried out a remarkable study on the incidence, morbidity and potential impact in the community of Rochester, Minn., 1965 through 1974. Incidence 7.8/100,000 population prior to introduction of routine measurement of serum calcium. Since 1974 it rose to 51/100,000 showing marked age and sex dependency: at 39 years or younger = 10/100,000, increases sharply 40 years of age or older reaching 188/100,000 in females over 60 years and 92/100,000 in males over 60 years. They predicted 35,000 to 86,000 new cases of PHP yearly in the USA.

1981 99) Broadus, in Recent Progress in Hormone Research, 37:667, 1981. Outstanding summary of all the work over a decade 1970-1980 on kidney production and excretion of cyclic AMP, primarily from the laboratories in which Broadus has worked, leading to the practical clinial adaptations of the assay to parathyroid physiology and pathophysiology, e.g. the sensitivity of the assay when properly related to functioning renal mass, i.e. G.F.R. the syndrome of PHP with "normal calcemia" hypercalciuric stone formation a consequence of the increased production of 1,25 $(OH)_2D_3$; the value of oral calcium tolerance test in the differential diagnosis of subtle cases of PHP. He also stressed the value of TmP/GFR in diagnosis. Segregation of patients with malignancy associated hypercalcemia, those with increased urinary cAMP and those with normal or suppressed excretion.

100) Scholz and Purnell. Mayo Clinic Proc. 56:473, 1981. Carried out the first major prospective study of so-called asymptomatic PHP to determine its course and need for surgery. They studied 133 patients over 10.5 years who had been referred for hypercalcemia. Criteria for immediate surgery included serum calcium greater than 11.0 mg/dl. Therefore prospective group with PHP had calciums between 10.1 and 11.0 (normal 8.1-10/1 mg/dl) 33 (23%) required surgery because of progression, and 32 died, cause of death known in 27, and 14 of these died of cardiovascular disease (ASCVD and/or hypertension) and 6 died of cancer. Ages 49-81. 23 females, 10 males. They concluded that surgery after diagnosis was still the best approach. One other study, Van'T Hogg et al (BMJ 287:16056, 1983) studied only 32 patients for only 4.2 years, concluded with the case for medical management. However 11, or 16%, with hypertension (150/100) at start increased to 16 or 50% at end of study.

101) Ontjes, Mahaffee and Wells. Metabolism 30:406, 1981. Found that with human parathyroid tissue <u>in vitro</u> normal (4), adenoma (18), and noninvolved glands (5) from adenoma patient, adenoma had reduced sensitivity to suppression of parathyroid cyclic AMP with calcium elevation, and reduced capacity for stimulatin of adenylcyclase of parthyroid cells by guany-5-yl-imidodiphosphate. They concluded that their results were consistent with "set point" alteration.

102) Brown. J. Clin. Endo. and Metab. 52:1046, 1981 Referred to the increased cirulating PTH and hypercalcemia of some patients with chronic lithium ingestion. Found in study in vitro of dispersed bovine parathyroid cells that exposure of cells for 4 hours to lithium caused

dose-related increase in the concentration of calcium required for 1/2 maximal inhibition of PTH secretion by the cells. Concluded lithium causes an acquired form of "set point" alteration.

1983 103) Lindall, Elting, Ells et al. J. Clin. Endo. and Metab. 57:1007, 1983. Developed a 2-step RIA, (1) extraction and concentration of serum PTH moieties with solid phase antiaminoterm PTH antibodies, and (2) the initial PTH immuno extract (Step 1). analyzed with a sensitive mid-region RIA intact hormone, or the substances measured by this 2-step concentrations and assay, can be detected in virtually all normal persons. Compared results with a very sensitive assay of PTH generation of cAMP in osteogenic sarcoma cell culture line which can measure as little as 2 picograms of intact PTH. Both assays = normal serum calues of about 6 pMols/L or less than 10% of carboxyl term RIA in some samples. In their assay bioactive hormone actually was 2.5 times higher in patients with PHP than in CRF patients on dialysis, many of whom had values in the normal range.

RENAL HANDLING OF PHOSPHATE

MECHANISMS OF ION TRANSPORT REGULATION BY PARATHYROID HORMONE: cAMP/Ca^{2+}/CALMODULIN AND PHOSPHOLIPID DEPENDENT PHOSPHORYLATION

K. Hruska, J. Scoble, D. Moskowitz, and M. Goligorsky

Renal Division
Washington University and The Jewish Hospital of St. Louis
St. Louis, Missouri

Since 1976, when the function of the Na^+ dependent phosphate symport of the renal brush border membrane (BBM) was first well described, numerous regulatory mechanisms of renal phosphate transport have been shown to exert effects on this carrier protein similar to their affect on proximal tubular phosphate transport (1,2). Parathyroid hormone (PTH), which decreases renal phosphate reabsorption, also causes a decrease in Na^+ dependent phosphate transport in isolated brush border membrane vesicles (BBMV) (3,4). This effect is exerted as a decrease in the Vmax of the carrier activity compatible with a mechanism acting through a decrease in the number of membrane functional units or a decrease in the overall activity of existing units. The mechanism of action of PTH has long been thought to occur through stimulation of adenylate cyclase, increasing c-AMP, and activation of a c-AMP dependent protein kinase (5,6).

In 1982, before the VIth International Workshop on Phosphate, we published studies analyzing the effects of c-AMP dependent protein kinase induced phosphorylation of renal BBMV (7). These studies revealed that the BBM exhibited two acceptor proteins which were phosphorylated by the activity of c-AMP dependent protein kinase. These two proteins exhibited molecular weights of 96 and 62,000 daltons on SDS polyacrylamide gels. We were also able to demonstrate that c-AMP dependent protein phosphorylation of BBMV was associated with a decrease in the initial rates and Vmax of the Na^+ dependent phosphate transport. Furthermore, the dose response relationship of c-AMP stimulated protein phosphorylation and inhibition of phosphate transport were identical. In subsequent studies, Hammerman et al demonstrated that the 62,000 molecular weight c-AMP dependent phosphorylated protein was dephosphorylated rapidly. This dephosphorylation was specific since the dephosphorylation of other proteins in the BBM did not occur with this rapid time course. The dephosphorylation of this protein was associated with loss of the inhibitory effects of c-AMP dependent protein phosphorylation on phosphate transport (8). Hammerman and Chase next demonstrated that the association of c-AMP dependent protein phosphorylation and dephosphorylation and effects on Na^+ dependent phosphate transport were found in another species, the mouse (9). In the HYP-Y mouse, a model of x-linked hypophosphatemia, and the normal litter mates, +Y, the mechanisms of c-AMP dependent protein phosphorylation of a 62,000 molecular weight acceptor protein and changes in Na^+ dependent phosphate transport induced by c-AMP dependent phosphorylation were observed. Thus, in both dogs and mice, changes in c-AMP dependent protein phosphorylation are associated with changes in Na^+ dependent phosphate transport at exactly the same time course of action and the same dose response to c-AMP. In our original studies, we also demonstrated that PTH, modulated c-AMP dependent protein phosphorylation and the effects of c-AMP dependent phosphorylation on Na^+ dependent phosphate transport. Thus, the above studies strongly support the hypothesis that PTH decreases renal phosphate reabsorption by inhibiting the activity of the Na^+ dependent phosphate transport carrier protein in the renal brush border membrane through stimulating the action of c-AMP dependent protein kinase. The mechanisms of inhibition through c-AMP dependent protein phosphorylation remain to be further elucidated.

In recent studies, Hammerman et al have shown that the 62,000 molecular weight protein phosphorylated by c-AMP is, in fact, the regulatory subunit of a Type I c-AMP dependent protein kinase (Hammerman, M. et al - in press). These studies utilized the effects of phosphorylation in the presence of zinc, the binding of 8 azido c-AMP, a photoaffinity reagent, and monoclonal antibodies to cardiac Type I c-AMP dependent protein kinase to identify, purify, and isolate the regulatory subunit of c-AMP dependent protein kinase in the renal BBM. These studies shed light on the controversy that has arisen regarding the function of c-AMP dependent protein phosphorylation on Na^+ dependent phosphate transport since the VIth International Workshop on Phosphate. At that meeting, and subsequently published (10,11), other investigators have been able to demonstrate c-AMP dependent phosphorylation of renal BBM proteins but have been unable to demonstrate an associated fall in Na^+ dependent phosphate transport. Since the effects of c-AMP dependent protein phosphorylation on Na^+ dependent phosphate transport have been demonstrated in two species, invoking the concept of species difference to explain the differing results is not attractive. However, examination of the reported studies (7-11) suggests that the major difference between the reports was found in the amount of the regulatory subunit of c-AMP dependent protein kinase in the BBMV. The studies of Hammerman et al exhibited a significantly greater intensity of phosphorylation of the regulatory subunit than found in the other studies. Thus, either the amount of c-AMP dependent protein kinase in the BBM, or a less intact association between the protein kinase activity and the Na^+ dependent phosphate transporter are logical explanations of the failure of some laboratories to demonstrate association between c-AMP dependent protein phosphorylation and inhibition of Na^+ dependent phosphate transport.

At the VIth International Workshop on Phosphate, we reported Ca^{2+} dependent phosphorylation of rat renal BBMV (12). Malstrom and Murer (13) have subsequently shown that Ca^{2+} dependent phosphorylation of renal BBM proteins occurs through the activity of a protein kinase C. They were not able to demonstrate the presence of a Ca^{2+}/calmodulin dependent protein kinase activity in the BBM. They also failed to show a Ca^{2+} dependent regulation of Na^+ dependent phosphate transport in vitro.

We have also undertaken studies of Ca^{2+} dependent regulatory mechanisms of cell activation through PTH. As shown in Figure 1, we have been able to demonstrate that the basolateral membrane of proximal renal tubular cells exhibits a Ca^{2+}/calmodulin dependent protein phosphorylation of two acceptor proteins. The activity of this Ca^{2+}/calmodulin dependent protein kinase is stimulated by PTH. Interestingly, the 24,000 K acceptor protein is phosphorylated both by c-AMP dependent protein kinase and Ca^{2+}/calmodulin dependent protein kinase. This represents a mechanism in the renal proximal tubular cell of amplification of activation by the two second messengers Ca^{2+} and c-AMP.

We have also instituted studies analyzing the effects of PTH on basolateral membrane phospholipid phosphorylation through the analysis of the effects of the hormone on the phosphatidylinositol cycle. As shown in Figure 2, PTH stimulates an increase in the turnover of phosphatidylinositol 4'phosphate (PhIP) and phosphatidylinositol 4'5'bisphosphate ($PhIP_2$). Since these studies were performed on BMV incubated with the hormone in vitro, they demonstrate a close association of PTH receptor

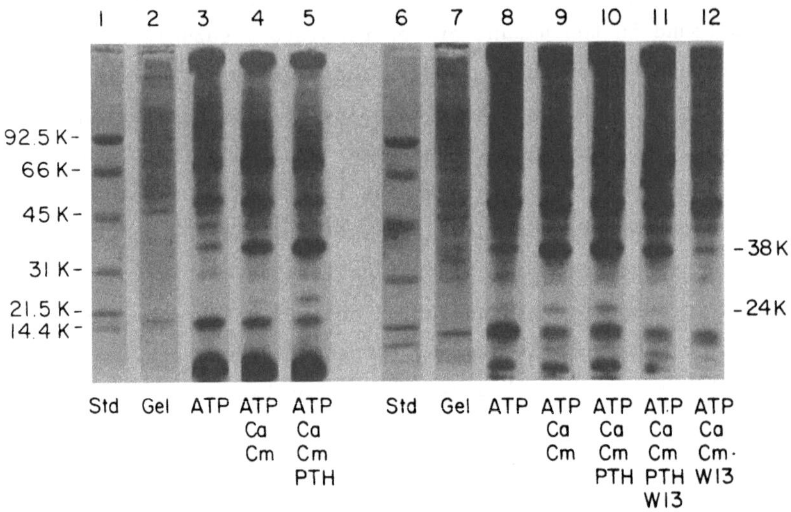

Figure 1 Stimulation of Ca^{2+}/calmodulin dependent basolateral membrane phosphorylation by PTH. Mongrel dogs were anesthetized with phenobarbital and kidneys were perfused through the renal artery with ice-cold saline. Basolateral membrane vesicles (BLMV) of renal epithelial cells were prepared from cortical homogenates by differential centrifugation and centrifugation in Percoll (8% final concentration) (17). BLMV, 0.2-0.4 mg protein, were suspended in 26 mM KCl, 130 mM sucrose, 5 mM $MgCl_2$, and 50 mM Tris HCL, pH 7.5. They were reacted with similar solutions for phosphorylation without sucrose containing Mg-ATP 10^{-3}M, containing 5-20 µC of $[\gamma^{32}Pi]$-ATP, and where appropriate: Ca^{2+}; 10^{-6}M EGTA buffer; calmodulin $3x10^{-7}$M; bovine PTH 1-84, 10^{-6}M; 3'5'c-AMP, 10^{-6}M; and W13, $2x10^{-4}$M. Reactions were stopped by addition of an ice-cold, 60 mM Tris HCl pH 5.8, containing 10% SDS (w/v) and 30% glycerol (w/v). Samples were boiled for 2 minutes, cooled and frozen at -70°C. BLMV samples containing 100-200 µg protein were added to 10% acrylamide slab gels and electrophoresis was performed for 4 hr with 6 milliamps/gel according to the method of Laemmli (18). Gels were stained with Commassie Blue, dried and exposed to radiographic film. Autoradiograms were developed after 3-7 days of exposure. The additions for each lane of two gels are listed under the respective lanes.

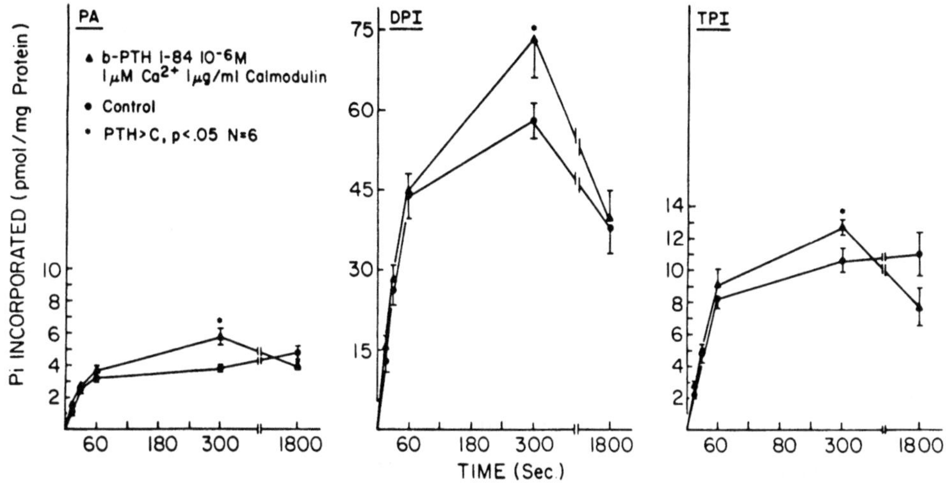

Figure 2 Stimulation of renal cortical basolateral membrane phospholipid turnover by PTH _in vitro_. BLMV were prepared as described in the legend to Figure 1. BLMV were incubated at 30°C in 130 mM sucrose, 26 mM KCl, 5 mM $MgCl_2$, 1 mM EGTA, and 50 mM Tris HCl, pH 7.5, containing 1 mM [γ^{32}Pi]ATP, and ± the following: 1 μM Ca^{2+}, 1 μg/ml calmodulin, $1x10^{-6}$M bovine PTH 1-84. Incubations were stopped by the addition of CHCL3/MEOH (1/1 v/v) containing 10 mM tetrabutylammonium sulfate at 4°C. The lipids were extracted according to Grove et al (19). Thin-layer chromatography was used to separate phospholipids and the thin-layer plates were exposed to x-ray film. Phosphorylated phospholipids were identified and removed for determination of radioactivity. The incorporation of ^{32}Pi into phosphatidic acid (PA), phosphatidylinositol 4'bisphosphate (DPI), and phosphatidylinositol 4'5'trisphosphate (TPI) were followed overtime. PTH increased the levels of ^{32}Pi-PA, DPI, and TPI significantly by 5 min of incubation. Subsequently, the level of ^{32}Pi labelled phospholipids declined compatible with stimulated phospholipid turnover through the activity of phospholipase C.

binding and stimulation of the phosphatidylinositol cycle possibly through phospholipid phosphorylation. The stimulation of the Pi cycle by PTH is associated with an increase in the release of inositol phosphates (data not shown). We have also begun studies analyzing the effects of PTH on control of proximal renal tubular cytosolic Ca^{2+} ($[Ca^{2+}]_i$). As shown in Figure 3, PTH stimulates both an immediate and sustained increase in $[Ca^{2+}]_i$ (14). The immediate increase in $[Ca^{2+}]_i$ stimulated by PTH in primary cultures of proximal renal tubular cells in suspension and on semipermeable membrane supports is compatible with cell activation through a change in $[Ca^{2+}]_i$. The increase in $[Ca^{2+}]_i$ demonstrated by PTH is compatible with recent studies by Scoble et al (15) which demonstrate a PTH stimulated increase in basolateral membrane Ca^{2+} permeability. Thus, our working hypothesis is that associated with receptor binding of PTH, an increase in the activity of a hormone activated Ca^{2+} channel is seen. This Ca^{2+} channel produces an influx of Ca^{2+} raising the $[Ca^{2+}]_i$ and activating intracellular mechanisms of Ca^{2+} sequestration principally into the endoplasmic reticulum (16). The sustained rise in $[Ca^{2+}]_i$ is important in promoting the effects of PTH on renal tubular transport function of this segment. These include inhibition of Na^+ transport through a decrease in the activity of the Na^+/H^+ exchange protein of the BBM, a decrease in apical membrane Ca^{2+} entry, and a stimulation of basolateral membrane Ca^{2+} efflux mechanisms.

Thus, in summary, PTH appears to produce its biologic effects on the proximal renal tubule by increasing the levels of two second messengers, c-AMP and $[Ca^{2+}]_i$. The exact mechanisms of receptor induced stimulation of the second messengers, their interactions, and their effects on various cell functions, including ion transport, remain to be elucidated.

REFERENCES

1. T.P. Dousa, and S.A. Kempson, Regulation of renal brush border membrane transport of phosphate, Mineral Electrolyte Metab. 7:113-121 (1982).
2. C.L. Mizgala and G.A. Quamme, Renal handling of phosphate, Physiol. Rev. 65:431-466 (1985).
3. C. Evers, H. Murer, and R. Kinne, Effect of parathyrin on the transport properties of isolated renal brush-border vesicles, Biochem. J. 172: 49-56 (1978).
4. M.R. Hammerman, I.E. Karl, and K.A. Hruska, Regulation of canine renal vesicle Pi transport by growth hormone and parathyroid hormone,

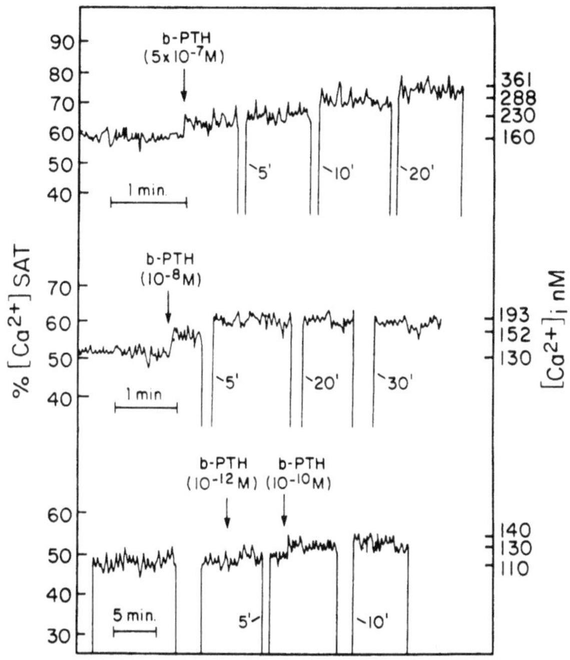

Figure 3 Effects of bovine parathyroid hormone (bPTH 1-84) on $[Ca^{2+}]_i$ and primary cultures of proximal tubular epithelial cells (14). Isolated tubular segments from Percoll gradients were seated onto cell culture flask in a media of 45% Dubelco's Modified Eagles Media, 45% HAMMS F-12, with 5% fetal calf serum, and 5% donor serum. The culture flask were maintained in a constant 95% O_2/5% CO_2 atmosphere at 37°C. After 3 days, the growth media was changed to a serum free media containing transferrin, insulin, hydrocortisone, triiodiothyronine. At confluency, the cells were suspended, loaded with quin-2 acetoxymethylester according to the methods of Tsien (20) and $[Ca^{2+}]_i$ determined from calibrated curves of quin-2 Ca^{2+} mixtures. Varying doses of bPTH 1-84, $5x10^{-7}$M (panel A), 10^{-8}M (panel B), 10^{-10}M and 10^{-12}M (panel C) on $[Ca^{2+}]_i$ of PCT cells were followed. Quin-2 loaded PCT cells were added to a cuvette and studied in a spectrofluorimeter. Calibration of the quin-2 fluorescence was conducted by permeabilization of cells to Ca^{2+} by the addition of Triton or ionomycin for maximal quin-2 signal and determination of the minimal quin-2 signal by the addition of EGTA Tris, pH 8.5. PTH produced a dose-dependent increase in $[Ca^{2+}]_i$ in both an immediate and sustained fashion.

Biochim. Biophys. Acta 603: 322-335 (1980).

5. L.J. Shlatz, I.L. Schwartz, E. Kinne-Saffran, and R. Kinne, J. Membr. Biol. 24, 131-144 (1975).
6. R. Kinne, and I.L. Schwartz, Isolated membrane vesicles in the evaluation of the nature, localization, and regulation of renal transport processes, Kidney International 14:547-556 (1978)
7. M.R. Hammerman, and K.A. Hruska, Cyclic AMP-dependent protein phosphorylation in canine renal brush-border membrane vesicles is associated with decreased phosphate transport, J. Biol. Chem. 257: 992-999, 1982.
8. M.R. Hammerman, V.A. Hansen, and J.J. Morrissey, Cyclic AMP-dependent protein phosphorylation and dephosphorylation alter phosphate transport in canine renal brush border vesicles, Biochim. Biophys. Acta 755: 10-16 (1983).
9. M.R. Hammerman, and L.R. Chase, Pi transport, phosphorylation, and dephosphorylation in renal membranes from HYP/Y mice, Am. J. Physiol. 245: F701-F706 (1983).
10. J. Biber, K. Malmstrom, V. Scalera, and H. Murer, Phosphorylation of rat kidney proximal tubular brush border membranes. Role of c-AMP dependent protein phosphorylation in the regulation of phosphate transport, Pflugers Arch 398: 221-226 (1983).
11. T. Takenawa, E. Wada, T. Tsumita, T. Masaki, C.R. Filburn, and B. Sacktor, Effect of parathyroid hormone, cyclic AMP and Ca^{2+} on the phosphorylation of brush border membranes in rabbit kidney, Mineral Electrolyte Metab. 10: 103-112 (1984).
12. K. Hruska, B. Kurnik, and M. Tsutsumi, The biochemical modifications of the brush border membrane induced by vitamin D and parathyroid hormone in their actions on phosphate, in: "Phosphate and Mineral Metabolism," S.G. Massry, G. Muschio, and E. Ritz, eds., Plenum Publishing, New York (1984).
13. K. Malmstrom, and H. Murer, Ca^{2+}-dependent protein phosphorylation in brush border membranes of rat kidney proximal tubules, Pflugers Arch. 404: 358-364 (1985).
14. K.A. Hruska, M. Goligorsky, J. Scoble, M. Tsutsumi, S. Westbrook, and D. Moskowitz, The effects of parathyroid hormone on cytosolic calcium in renal proximal tubular primary cultures. (Submitted - Am. J. Physiol.).
15. J.E. Scoble, S. Mills, and K.A. Hruska, Calcium transport in canine renal basolateral membrane vesicles. J. Clin. Invest. 75:1096-1105 (1985).

16. J.E. Scoble and K.A. Hruska, Calcium transport in the proximal tubule. (Submitted - J. Lab. Clin. Med.).

17. D.W. Windus, S. Klahr, and M.R. Hammerman, Glutamine transport in basolateral vesicles from dogs with acute respiratory acidosis, Am. J. Physiol. 247: F403-F407 (1984).

18. U.K. Laemmli, Cleavage of structural proteins during the assembly of the head of bacteriophage T_4, Nature 227: 680-685 (1970).

19. R.I. Grove, D. Fitzpatrick, and S.D. Schimmel, Effect of Ca^{++} on triphosphoinositide extraction in fusing myoblasts, Lipids 16: 691-693 (1981).

20. R.Y. Tsien, T. Pozzan, and T.J. Rink, Calcium homeostasis in intact lymphocytes: cytoplasmic free calcium monitored with a new, intracellularly trapped fluorescent indicator, J. Cell Biol. 94: 325-334 (1982).

MECHANISMS OF NAD ACTION IN REGULATION OF RENAL BRUSH BORDER MEMBRANE TRANSPORT OF PHOSPHATE

Stephen A. Kempson and Thomas P. Dousa

Department of Physiology and Biophysics, Indiana University Medical School, Indianapolis, IN

Nephrology Research Unit, Mayo Clinic and Mayo Medical School, Rochester, MN

Many different hormonal and non-hormonal stimuli change the rate of phosphate (Pi) transport across the renal brush border membrane (BBM) (1,2, 3) which probably reflects the central role of proximal tubular reabsorption in renal handling of Pi. Based on considerations of (a) the time required for the effects of these stimuli to induce a response, and (b) whether the response is dependent upon intact *de novo* protein synthesis, it appears that the various stimuli may act intracellularly through at least two general mechanisms (2).

Stimuli such as thyroid hormones (4), glucocorticoids (5), and long term dietary Pi deprivation (6) act relatively slowly, producing a change in BBM transport of Pi usually after 1-2 days, and the response is blocked by actinomycin D or cycloheximide, inhibitors of *de novo* protein synthesis (6-8). This suggests that the slow-acting intracellular mechanism involves changes in biogenesis of the Na/Pi-cotransporter (Na-Pi-COT) which becomes functional when inserted into the BBM. These changes result either in an increase or a decrease in the number of Na-Pi-COT in the BBM, or in synthesis of Na-Pi-COT with modified properties.

In contrast, stimuli such as parathyroid hormone (9,10), calcitonin (11), growth hormone (10), insulin (12), and acute dietary Pi deprivation (13,14) are fast-acting, producing a response in 1-2 hours (or less), and the response is not blocked by administration of actinomycin D (14-16). It is likely that the rapid response induced by this group of stimuli involves the action of various intracellular "second messengers" such as cyclic nucleotides, the diacylglycerol/protein kinase C and inositol triphosphate/ Ca^{2+} systems, and, as proposed recently, perhaps NAD (2,3,17,18). The role of these messengers is to modify the capacity of Na-Pi-COT in the BBM, or to change the cycling of Na-Pi-COT to and from the BBM (2,14). It is not clear why there may be multiple types of messengers, but the different stimuli may act via different messengers or different messengers may be used for the same stimulus in different subsegments of the proximal tubule, such as the early convoluted and late straight (pars recta) segments.

The evidence which led to the suggestion that NAD might function as an intracellular messenger in the modulation of BBM transport of Pi is as follows. A pharmacological dose of nicotinamide administered *in vivo* produced an increase in NAD content in the renal cortex (17). This change in NAD was very marked in the proximal tubule compared to other nephron segments (19), and this was reflected by an increase in NAD within isolated BBM vesicles (20). The increase in NAD in cortical tissue was accompanied by a specific decrease in Na-dependent Pi transport across the renal BBM, and the magnitude of these changes was inversely correlated (17). Although ATP is present in the rat kidney at a level which exceeds the NAD content (21), changes in ATP were not accompanied by changes in Pi transport (20) which illustrates the specific relationship between changes in NAD and Pi transport. Finally, the characteristic kinetic change in BBM transport of Pi, following nicotinamide treatment (and increase in renal cortical NAD), was a decrease in the Vmax with no change in the Km (20). This is exactly the same as the kinetic change which occurs following the action of most physiological stimuli such as parathyroid hormone, growth hormone, dietary Pi deprivation, and glucocorticoids (2,3). This correlation suggests that the action of at least some of the regulatory stimuli may involve NAD interaction, directly or indirectly, with the Na-Pi-COT.

The detection of intact NAD in isolated BBM fractions (20), probably trapped inside the BBM vesicles, may indicate that the cytosolic NAD pool has access to the BBM. Specific binding sites for β-NAD have been identified in the renal BBM (22). Bound radiolabelled β-NAD was displaced most effectively by addition of excess unlabelled β-NAD as compared to addition of α-NAD, NADH or possible products of NAD hydrolysis. Although the sidedness of the binding was not studied directly, the data suggest that intact cytosolic NAD may bind reversibly to specific sites on the BBM.

Although intact NAD inhibits Pi transport when added to isolated BBM vesicles, the change in Pi uptake is characterised by an increase in the Km with no change in the Vmax (20). Thus, it is unlikely that reversible non-covalent binding of intact NAD to the BBM is the single step which mediates the change in Pi transport when NAD is increased by nicotinamide administra-

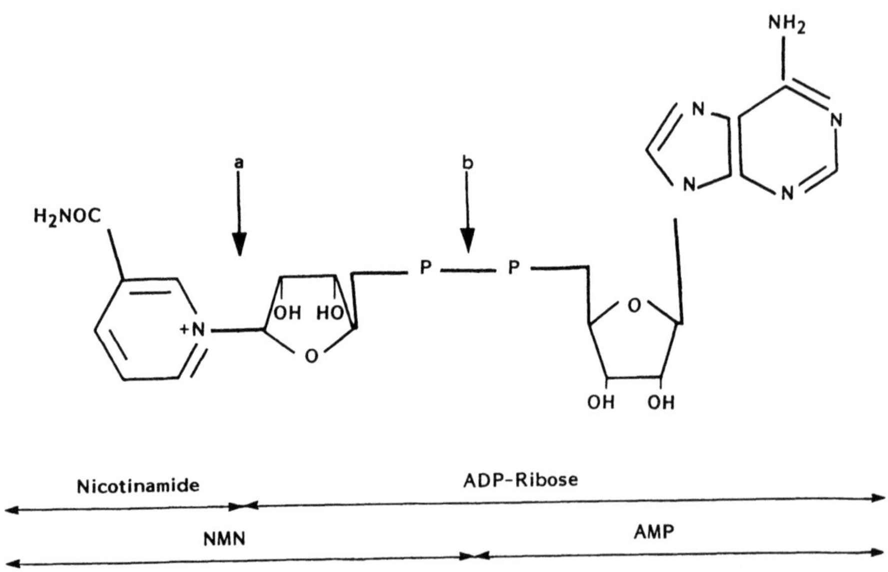

Fig. 1. Representation of NAD molecule and the products liberated by enzymatic action at a or b (P, phosphate group).

tion, since the effect of the latter is on the Vmax rather than the Km of the Pi transport system (20).

It is possible that the binding of NAD to the BBM represents the initial step in a sequence which leads to a covalent interaction between NAD and the Na-Pi-COT or related BBM proteins. There are several mechanisms by which NAD could covalently modify BBM proteins and perhaps the most likely is an ADP-ribosylation reaction. Both mono-ADP-ribosylation and poly-ADP-ribosylation are well documented mechanisms in cellular regulation (23). In theory there appear to be two reactions by which mono-ADP-ribosylation could occur, a one-step direct mechanism or a two-step indirect mechanism. Both reactions involve enzymatic splitting of NAD at point a (Fig. 1), the glycosidic bond, and transfer of ADP-ribose to side chains of amino acids of BBM proteins. These reactions are illustrated in Fig. 2, where ADPRT represents ADP-ribosyltransferase, NiAm is nicotinamide, and NGH is NAD-glycohydrolase.

The one-step ADP-ribosylation mechanism (pathway I) is a direct enzymatic transfer of ADP-ribose, with release of nicotinamide, to specific acceptor sites such as guanidino groups of arginine and other amino acids (23,24). The reaction is catalyzed by ADPRT which has been found in the renal BBM (25). It is of interest that ADPRT in various tissues has been reported to play a key role in the action of some hormones and toxins (26-30), and preliminary data indicate that parathyroid hormone may stimulate mono-ADP-ribosylation of proteins in canine renal BBM (31).

The first step in pathway II (Fig. 2) is enzymatic hydrolysis of NAD to free ADP-ribose and nicotinamide, a reaction which is catalyzed by NGH. This enzyme is one of the major NAD hydrolysing enzymes present in the renal BBM, and at least 20% of the total NGH activity of BBM is located exclusively on the cytoplasmic surface (32). The action of NGH on cytosolic NAD could produce very high local concentrations of ADP-ribose in the vicinity of the BBM. The second step in pathway II is spontaneous formation of covalent

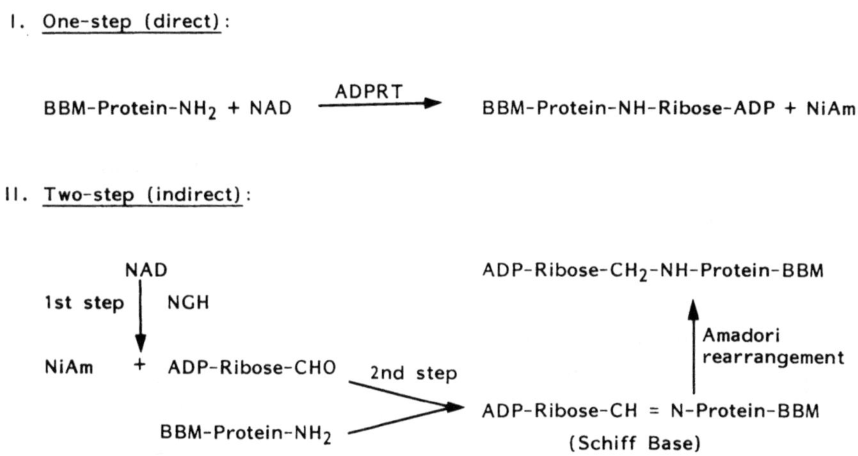

Fig. 2. Mechanisms for mono-ADP-ribosylation of BBM proteins. ADPRT, ADP-ribosyltransferase; NiAm, nicotinamide; NGH, NAD-glycohydrolase.

chemical adducts (Schiff bases) between ADP-ribose and amino groups of BBM proteins. This reaction between aldehydic groups of free sugars and amino groups of proteins is well documented, the best known example is glucosylation of the NH_2-terminus of the β-chain of hemoglobin by the non-enzymatic reaction between glucose and the amino group (33,34). The analogous reaction between free ADP-ribose and a number of different proteins has also been reported (35). The two-step reaction of pathway II, initiated by the action of NGH on NAD, appears to be the mechanism by which mono-ADP-ribosylation of mitochondrial polypeptides occurs, and remarkable specificity is possible (36).

In addition to mono-ADP-ribosylation of either type, covalent modification of BBM proteins by NAD could occur, at least in theory, by the process of adenylation (37,38). This mechanism, which has not been explored yet, would involve enzymatic splitting of NAD at point b (Fig. 1) and transfer of the AMP to protein.

If mono-ADP-ribosylation of BBM proteins by NAD is physiologically important for regulating the Na-Pi-COT, then a cellular mechanism for reversing the ADP-ribosylation process is required. Specific enzymes have been described which can remove mono-ADP-ribose units from proteins (39). These include a specific ADP-ribose-protein-lyase which removes the complete ADP-ribose moiety (40-42) as illustrated in Fig. 3 (reaction I), and enzymes which sequentially remove first AMP then ribose-5-phosphate (42) as illustrated in reaction II of Fig. 3.

The essential components for mono-ADP-ribosylation of the Na-Pi-COT in the renal BBM, and for the reversal of this process, may be present in the proximal tubule cell. Postulating that mono-ADP-ribosylation of the BBM does occur, the critical question is does this lead to specific changes in the Na-Pi-COT. Presently, this is a rather controversial issue and it is pertinent to summarize briefly the data obtained by different laboratories.

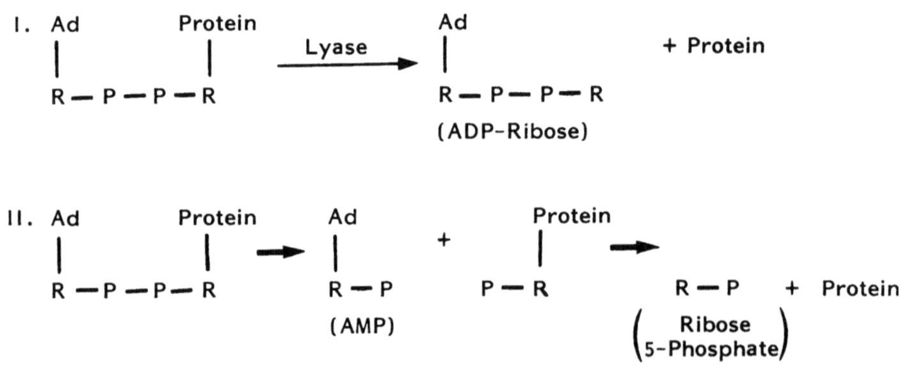

Fig. 3. Pathways of enzymatic reversal of mono-ADP-ribosylation of proteins. Ad, adenine; R, ribose; P, phosphate.

Using hypotonic shock to introduce NAD into the interior of canine renal BBM vesicles, Hammerman et al. (43) reported that mono-ADP-ribosylation of BBM proteins was accompanied by specific inhibition of Na-dependent uptake of Pi by the BBM vesicles. Kempson and Curthoys (25) characterised the presence of an ADP-ribosyltransferase in rat renal BBM, although the sidedness of the distribution of the enzyme was not addressed. Lang et al. (44) were unable to detect ADP-ribosylation of renal BBM from the rabbit. The latter study also brought to attention the important point that production of free Pi from NAD hydrolysis would produce apparent inhibition of Pi transport by diluting the specific radioactivity of the ^{32}Pi used for measurement of transport. This finding, and others (45), led some investigators to question whether NAD itself could inhibit Pi transport _in vitro_. In a recent study (20) we examined this problem in some detail and our results provide evidence that intact NAD can inhibit Pi uptake by BBM. Although BBM enzymes are able to catabolise NAD, this requires prolonged incubation. Thus, it is only under these conditions that dilution of ^{32}Pi specific radioactivity contributes to the apparent inhibition of BBM uptake of Pi. This point was addressed also by Hammerman et al. (46) who reported that under the conditions which they used to induce ADP-ribosylation, the production of Pi from hydrolysis of NAD did not account for the changes in ^{32}Pi transport. This particular issue appears to be clarified, but opinions about ADP-ribosylation are still divided. In a well controlled study (47) Gmaj and coworkers reported that encapsulation of NAD inside isolated BBM vesicles did not lead to ADP-ribosylation of BBM proteins and there was no change in Na-dependent Pi uptake. As these authors have pointed out, some of the disagreement may have arisen because slightly different procedures for preparing isolated BBM may lead in some laboratories to removal of essential yet unknown cytosolic cofactors for ADP-ribosylation.

In thinking about NAD regulation of Pi transport, the following possibility also should be considered. If one of the mechanisms for regulating BBM transport of Pi involves shuttling the Na-Pi-COT between the BBM and cytoplasmic depots (2), then NAD could regulate Pi transport indirectly by influencing the cycling process in the cytoplasm rather than by interacting directly with the Na-Pi-COT in the BBM. Membrane cycling depends on microtubules and other components of the cytoskeleton, and it is of interest to note that ADP-ribosylation of the cytoskeleton by NAD has been described recently (29,30).

In summary, the studies which have investigated the possible regulatory role of NAD in Pi transport have established a number of potentially important findings. The results of many of these studies are consistent with a role for NAD in modulating BBM transport of Pi. Some of the studies have highlighted some interpretive problems and the need for further investigations. New experimental designs, using isolated BBM vesicles, should provide NAD with access to the cytoplasmic surface of the BBM and extravesicular NAD should be carefully removed or appropriate controls will be required. Potential artifacts introduced by NAD hydrolysis should be taken into account during planning of the experiments and interpretation of the results. Finally, it should be kept in mind that, until proved otherwise, NAD may modulate Pi transport either directly by interaction with the Na-Pi-COT in the BBM, or indirectly by interaction with the cytosolic machinery involved in cycling the Na-Pi-COT.

In conclusion, considerable experimental evidence albeit still correlative points to NAD as one of the important factors in the intracellular pathways that regulate Pi transport (2,3). The challenging task ahead which should be approached with an open mind is to determine the specific intracellular mechanism of action of NAD and its internephron and intranephron site of action.

Acknowledgements. Experimental work from the authors' laboratories was supported by NIH grants AM-32148 and AM-30759. Carolyn Blankenship provided expert secretarial assistance.

REFERENCES

1. J.-P. Bonjour and J. Caverzasio, Phosphate transport in the kidney, Rev. Physiol. Biochem. Pharmacol. 100:161 (1984).
2. S. A. Kempson and T. P. Dousa, Current concepts of regulation of phosphate transport in renal proximal tubules, Biochem. Pharmacol. in press.
3. T. P. Dousa and S. A. Kempson, Regulation of renal brush border membrane transport of phosphate, Min. Electrolyte Metab. 7:113 (1982).
4. A. N. K. Yusufi, N. Murayama, M. J. Keller, and T. P. Dousa, Modulatory effect of thyroid hormones on uptake of phosphate and other solutes across luminal brush border membrane of kidney cortex. Endocrinology 116:2438 (1985).
5. S. T. Turner, G. M. Kiebzak, and T. P. Dousa, Mechanism of glucocorticoid effect on renal transport of Pi, Am. J. Physiol. 243:C227 (1982).
6. S. V. Shah, S. A. Kempson, T. E. Northrup, and T. P. Dousa, Renal adaptation to the low phosphate diet in rats; blockade by actinomycin D, J. Clin. Invest. 64:955 (1979).
7. A. N. K. Yusufi, R. Holets, and T. P. Dousa, Mechanism by which T_3 stimulates brush border membrane transport of phosphate, Abstracts 9th Int. Congress Nephrol. (Los Angeles, California) 378A (1984).
8. B. Sacktor and L. Noronha-Blob, Glucocorticoids act directly on renal cells to inhibit phosphate transport, Fed. Proc. 43:632 (1984).
9. H. Evers, H. Murer, and R. Kinne, Effect of parathyrin on the transport properties of isolated renal brush border vesicles, Biochem. J. 172: 49 (1978).
10. M. R. Hammerman, I. E. Karl, and K. A. Hruska, Regulation of canine renal vesicle Pi transport by growth hormone and parathyroid hormone, Biochim. Biophys. Acta 603:322 (1980).
11. A. Yusufi, T. Berndt, N. Murayama, F. Knox, and T. Dousa, The response of superficial and deep cortex to phosphaturic hormones in rats fed normal or low phosphate diets, Clin. Res. 32:461A (1984).
12. M. R. Hammerman, S. Rogers, V. A. Hansen, and J. R. Gavin, Insulin stimulates Pi transport in brush border vesicles from proximal tubular segments, Am. J. Physiol. 247:E616 (1984).
13. B. S. Levine, K. Ho, A. Hodsman, K. Kurokawa, and J. W. Coburn, Early renal brush border membrane adaptation to dietary phosphorus, Min. Electrolyte Metab. 10:222 (1984).
14. J. Caverzasio, C. D. A. Brown, J. Biber, J.-P. Bonjour, and H. Murer, Adaptation of phosphate transport in phosphate-deprived $LLC-PK_1$ cells, Am. J. Physiol. 248:F122 (1985).
15. H. C. Rasmussen, C. Arnaud, and C. Hawker, Actinomycin D and the response to parathyroid hormone, Science 144:1019 (1964).
16. B. S. Levine, K. Kurokawa, and J. W. Coburn, Renal adaptation to diet phosphorus; early events, Abstracts 9th Int. Congress Nephrol. (Los Angeles, California) 55A (1984).
17. S. A. Kempson, G. Colon-Otero, S.-Y. L. Ou, S. T. Turner, and T. P. Dousa, Possible role of nicotinamide adenine dinucleotide as an intracellular regulator of renal transport of phosphate in the rat, J. Clin. Invest. 67:1347 (1981).
18. R. J. Lefkowitz and M. G. Caron, Adrenergic receptors; molecular mechanisms of clinically relevant regulation, Clin. Res. 33:395 (1985).
19. T. P. Dousa, A. N. K. Yusufi, E. Kusano, and J. L. Braun-Werness, Effect of nicotinamide administration on NAD content in proximal tubules, Kidney Int. 23:222 (1983).
20. S. A. Kempson, S. T. Turner, A. N. K. Yusufi, and T. P. Dousa, Actions

of NAD on renal brush border transport of phosphate in vivo and in vivo, Am. J. Physiol. 249:in press (1985).
21. D. H. Williamson and J. T. Brosnan, Concentrations of metabolites in animal tissues, in: Methods of Enzymatic Analysis, H. U. Bergmeyer, ed., Academic Press, New York, vol. 4:2266 (1974).
22. J. L. Braun-Werness, B. A. Jackson, P. G. Werness, and T. P. Dousa, Binding of nicotinamide adenine dinucleotide by the renal brush border membrane from rat kidney cortex, Biochim. Biophys. Acta 732: 553 (1983).
23. K. Ueda and O. Hayaishi, ADP-Ribosylation, Ann. Rev. Biochem. 54:73 (1985).
24. D. A. Yost and J. Moss, Amino acid-specific ADP-ribosylation, J. Biol. Chem. 258:4926 (1983).
25. S. A. Kempson and N. P. Curthoys, NAD-dependent ADP-ribosyltransferase in renal brush border membranes, Am. J. Physiol. 245:C449 (1983).
26. S. Filetti and B. Rapoport, Hormonal stimulation of eucaryotic cell ADP-ribosylation. Effect of thyrotropin on thyroid cells, J. Clin. Invest. 68:461 (1981).
27. M. J. S. DeWolf, P. Vitti, F. S. Ambesi-Impiombato, and L. D. Kohn, Thyroid membrane ADP ribosyltransferase activity, J. Biol. Chem. 256: 12287 (1981).
28. D. M. Gill and R. Meren, ADP-ribosylation of membrane proteins catalysed by cholera toxin; basis of the activation of adenylate cyclase, Proc. Nat. Acad. Sci. USA 75:3050 (1978).
29. H. R. Kaslow, V. E. Groppi, M. E. Abood, and H. E. Bourne, Cholera toxin can catalyze ADP-ribosylation of cytoskeletal proteins, J. Cell Biol. 91:410 (1981).
30. D. J. Hawkins and E.T. Browning, Tubulin adenosine diphosphate ribosylation is catalysed by cholera toxin, Biochemistry 21:4474 (1982).
31. M. R. Hammerman, D. E. Cohn, J. Tamayo, and K. J. Martin, PTH increases ADP-ribosylation of canine renal brush border membrane proteins, Kidney Int. 23:100 (1983).
32. S. A. Kempson, NAD-Glycohydrolase in renal brush border membranes, Am. J. Physiol. 249:F366 (1985).
33. H. F. Bunn, R. Shapiro, M. McManus, L. Garrick, M. J. McDonald, P. M. Gallop, and K. H. Gabbay, Structural heterogeneity of human hemoglobin A due to non-enzymatic glycosylation, J. Biol. Chem. 254:3892 (1979).
34. O. H. Wieland, Protein modification by non-enzymatic glucosylation; possible role in the development of diabetic complications, Mol. Cell. Endocrinol. 29:125 (1983).
35. E. Kun, A. C. Y. Chang, M. L. Sharma, A. M. Ferro, and D. Nitecki, Covalent modification of proteins by metabolites of NAD^+, Proc. Nat. Acad. Sci. USA 73:3131 (1976).
36. H. Hilz, R. Koch, W. Fanick, K. Klapproth, and P. Adamietz, Nonenzymic ADP-ribosylation of specific mitochondrial polypeptides, Proc. Nat. Acad. Sci. USA 81:3929 (1984).
37. B. M. Olivera, Z. W. Hall, Y. Anraku, J. R. Chien, and I. R. Lehman, On the mechanism of the polynucleotide joining reaction, Cold Spring Harbor Symp. Quant. Biol. 33:27 (1968).
38. R. I. Gumport and I. R. Lehman, Structure of the DNA ligase-adenylate intermediate; lysine (ε-amino)-linked adenosine monophosphoramidate, Proc. Nat. Acad. Sci. USA 68:2559 (1971).
39. S. C. B. Yan, An unexpected twist in the reversal of ADP-ribosylation, Trends Biochem. Sci. 9:331, (1984).
40. H. Okayama, M. Honda, and O. Hayaishi, Novel enzyme from rat liver that cleaves an ADP-ribosyl histone linkage, Proc. Nat. Acad. Sci. USA 75:2254 (1978).
41. J. Oka, K. Ueda, O. Hayaishi, H. Komura, and K. Nakanishi, ADP-ribosyl protein lyase. Purification, properties and identification of the product, J. Biol. Chem. 259:986 (1984).

42. K. P. Smith, R. C. Benjamin, J. Moss, and M. K. Jacobson, Identification of enzymatic activities which process protein bound mono(ADP-ribose), Biochem. Biophys. Res. Commun. 126:136 (1985).
43. M. R. Hammerman, V. A. Hansen, and J. J. Morrissey, ADP-ribosylation of canine renal brush border membrane vesicle proteins is associated with decreased phosphate transport, J. Biol. Chem. 257:12380 (1982).
44. R. P. Lang, N. Yanagawa, E. P. Nord, L. Sakhrani, S. H. Lee, and L. G. Fine, Nucleotide inhibition of phosphate transport in the renal proximal tubule, Am. J. Physiol. 245:F263 (1983).
45. H. S. Tenenhouse and Y. L. Chu, Hydrolysis of nicotinamide-adenine dinucleotide by purified renal brush border membranes, Biochem. J. 204: 635 (1982).
46. M. R. Hammerman, V. M. Corpus, and J. J. Morrissey, NAD^+-induced inhibition of phosphate transport in canine renal brush border membranes. Mediation through a process other than or in addition to NAD^+ hydrolysis, Biochim. Biophys. Acta 732:110 (1983).
47. P. Gmaj, J. Biber, S. Angielski, G. Stange, and H. Murer, Intravesicular NAD has no effect on sodium-dependent phosphate transport in isolated renal brush border membrane vesicles, Pfluegers Arch. 400: 60 (1984).

INTERACTIONS BETWEEN INORGANIC PHOSPHATE AND ENERGY METABOLISM IN RENAL CORTICAL TUBULES

Vincent W. Dennis and Peter C. Brazy

Division of Nephrology
Duke University Medical Center
Durham, N.C.

Inorganic phosphate interacts with essentially all major metabolic pathways, primarily through the phosphorylation of intermediates via ATP or other nucleotides. In this regard, oxidative phosphorylation has an absolute requirement for inorganic phosphate (1) but it is not certain whether phosphate under certain conditions may regulate or control mitochondrial respiration. It is generally believed that mitochondrial respiration is determined by the extramitochondrial concentration ratio of [ADP]/[ATP]. On the other hand, Wilson et al (2) have advanced the concept that mitochondrial respiration may be governed by the extra-mitochondrial activities of inorganic phosphate as well as ADP and ATP. It is clear that inorganic phosphate is important for mitochondrial anion transport processes that facilitate the transfer of reducing equivalents between cytosolic and mitochondrial compartments (3). In view of these considerations, we performed a series of studies that examine the interactions between phosphate and respiration in renal cortical tubules.

METHODS

Transport studies were performed in segments of proximal convoluted tubules from rabbit kidney. In general, the tubules were perfused with an intraluminal fluid that resembles glomerular ultrafiltrate and bathed in a medium made directly from the corresponding intraluminal fluids by adding defatted albumin and additional calcium chloride to a final calcium concentration of 3 mM. Net sodium transport was measured as the fluid absorption rate (Jv) using ^{125}I-iothalmate. Phosphate transport (J-Phos) was measured as the lumen-to-bath flux of ^{33}P-phosphate (4).

Metabolic measurements were made in suspensions of cortical tubules prepared from the cortex of rabbit kidney. Kidneys from female New Zealand White rabbits were flushed in situ for 15 minutes with a solution containing 80 mg/dl collagenase. The collagenase solution was subsequently washed out, the kidneys were excised, and the cortex was dissected. Cortical tubules were then dispersed, washed three times in cold medium and centrifuged on a cushion of Ficoll (400,000 M wt.) to remove non-vital cells and cellular debris. The tubular pellet was washed three times in the experimental medium and then resuspended to give a concentration of 3-5 mg tubular protein/ml. The experimental media were similar to our intraluminal perfusion fluids except that they contained dextran rather than albumin.

After preparation of tubules, aliquots of the tubule suspensions were transferred directly into a closed chamber for measurement of oxygen consumption rates (QO_2). Oxygen consumption rates were measured polarographically in a closed chamber using a Clark-type electrode (4).

RESULTS AND DISCUSSION

Interactions between phosphate transport and metabolism. Insight into possible special relationships between phosphate transport and cellular metabolism were derived initially from studies involving arsenate (4). Arsenate competes with phosphate and inhibits its action at a number of biochemical sites. We therefore viewed the actions of arsenate as tracing the metabolic course of inorganic phosphate. We demonstrated that arsenate in the luminal fluid of proximal convoluted tubules causes proportional reductions in phosphate and fluid absorption. These effects did not occur with 5 mM arsenate in the bath. This action of arsenate on phosphate transport is expected from studies of renal brush border membrane vesicles showing that arsenate competitively inhibits phosphate uptake (5). Surprisingly, however, arsenate in the perfused proximal tubule caused reductions in net sodium transport proportional to the reductions in phosphate transport. If arsenate acts solely by blocking the luminal uptake of phosphate, one would expect only a slight reduction in net sodium transport since phosphate transport from lumen-to-bath represents only about 5 percent of net sodium transport.

In related studies, we demonstrated that arsenate increases oxygen consumption by about 27 percent and also decreases NADH fluorescence (4). The simultaneous increase in oxygen consumption and decrease in NADH fluorescence are consistent with the uncoupling of oxygen consumption from the phosphorylation of ADP to ATP, or with the stimulation of cellular ATPase activity. To distinguish these two possibilities, we repeated the studies in the presence of 0.5 mM ouabain to inhibit sodium-potassium ATPase activity totally, and thereby decrease the respiratory rate of the tissue. In the presence of ouabain, arsenate still increased oxygen consumption and decreased NADH fluorescence. From these data, arsenate appears to act by uncoupling oxidative phosphorylation rather than by stimulating ATPase activity.

We also tested other, more potent uncouplers of oxidative metabolism. The additon of m-chlorocarbonyl cyanide phenylhydrazone (CCCP) to a final concentration of 1 uM caused proportional reductions in Jv and phosphate absorption of 64 and 72 percent, respectively. In a suspension of cortical tubules, 1 uM CCCP increased oxygen consumption by 51%, consistent with the action of CCCP to uncouple mitochondrial respiration from oxidative phosphorylation. Again, the striking observation was the proportional reduction in both sodium and phosphate absorption in spite of lesser reductions in glucose transport. Collectively, these data indicate that inhibitors or oxidative phosphorylation such as arsenate and more specifically CCCP have striking effects on phosphate transport. These results suggest that a close relationship exists between phosphate transport and oxidative metabolism in the proximal tubule.

The next series of observations tested whether stimulation rather than inhibition of oxidative metabolism was associated with increases in phosphate transport. We examined the effects of various short-chain fatty acids and tricarboxylic acid cycle intermediates on transport and oxygen consumption in renal cortical tubules (6). The addition of 1 mM valerate, butyrate, or succinate to rabbit cortical tubules increased QO_2 by 38, 24, or 48 percent, respectively, without changing the ATP content. Under similar conditions, valerate increased phosphate transport in the proximal convoluted tubule from 3.61 ± 0.52 to 7.82 ± 1.62 pmol/mm·min ($P<0.005$);

butyrate increased it from 5.64 ± 0.77 to 7.86 ± 1.02 ($P<0.02$); and succinate increased it from 4.31 ± 1.40 to 5.69 ± 1.45 ($P<0.005$). There were no detectable changes in net sodium transport as estimated by Jv. Thus, stimulations of oxidative metabolism increased phosphate transport without detectable changes in net sodium transport.

Substrates such as valerate, butyrate, and succinate also stimulate gluconeogenesis in renal cortical tubules. Accordingly, we used 3-mercaptopicolinic acid, an inhibitor of phosphoenolpyruvate carboxykinase, to inhibit gluconeogenesis. In the presence of 3-MPA, valerate continued to stimulate phosphate transport and oxygen consumption in spite of the elimination of changes in glucose production (6,7). Thus, the action of valerate on phosphate transport and oxidative metabolism is not dependent on changes on glucose production.

We also considered the possibility that the increased phosphate transport associated with these metabolic substrates may be caused by increased incorporation of radioisotopic phosphate into an expanded pool of intracellular phosphates. To test this possibility, ^{32}P-phosphate uptake was measured as a function of time in suspensions of proximal renal tubules (6). Phosphate uptake was a curvilinear of time. There was an early rapid phase that appeared to saturate at about one minute and a later slower phase. The rapid early uptake was linear for about 45 seconds. Under control conditions, this early rate of phosphate uptake was 5.2 ± 0.2 nmol/mg protein·min and was increased by valerate to 8.7 ± 0.2 nmol/mg protein·min ($P<0.001$). Valerate had no effect on the later slower uptake phase. This suggests that the increased transport of phosphate observed in perfused tubules was not associated with phosphate accumulation by the cells. The valerate-induced increases in the early rapid phase of phosphate uptake may represent the initial membrane-dependent step in phosphate transport and suggests that valerate may increase the apparent permeability of the luminal membrane to phosphate entry.

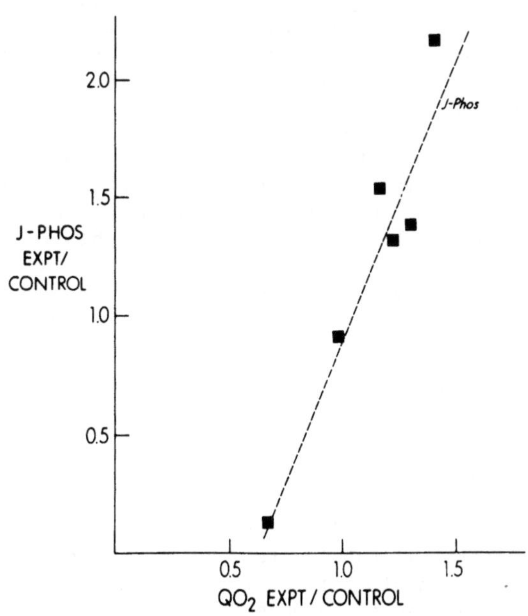

Figure 1.

Figure 1 represents a summary of the observations relating changes in phosphate transport to changes in oxygen consumption. The line correlating these observations indicates a striking relationship between phosphate transport and changes in oxidative metabolism in the range from inhibited to basal to stimulated rates of oxygen consumption.

Metabolic requirement for inorganic phosphate. One implication of the preceding studies is that the proximal convoluted tubule under certain conditions might be substrate-limited with regard to inorganic phosphate entering via the luminal brush border. To test this hypothesis, we examined whether or not the selective removal of inorganic phosphate from the intraluminal perfusate might affect energy metabolism as monitored by changes in Jv or net sodium transport, an index of metabolic rate. We demonstrated that if proximal convoluted tubules are dissected and perfused initially with phosphate-free fluids, Jv averaged -0.01 ± 0.08 nl/mm·min and was only partially restored by perfusion with phosphate-containing fluid for up to one hour. During all these studies, the bathing medium was normal rabbit serum containing 2.3 mM phosphate. We interpret these studies as indicating that intracellular phosphate may be depleted during the periods of warm and cold ischemia associated with the isolation of single tubules and that this depletion cannot be corrected by the limited or absent influx of phosphate from the basolateral surface. That is, under these conditions, oxidative metabolism necessary for transport is limited by the absence of phosphate influx from the lumen (8).

One mechanism by which phosphate limitation can affect mitochondrial respiration may involve an inhibition of the dicarboxylate anion transporter. This anion transport mechanism uses intramitochondrial inorganic phosphate to facilitate the mitochondrial uptake of divalent anions, especially the dicarboxylic acids of the tricarboxylic acid cycle. To test this possibility, we examined the effects of n-butylmalonate on Jv in isolated perfused proximal convoluted tubules and on QO_2 and ATP content in suspensions of cortical tubules (9). N-butylmalonate is a competitive inhibitor of the mitochondrial transport mechanism. Addition of 5 mM n-butylmalonate to the intraluminal and bathing fluids reduced Jv in isolated perfused proximal tubules from 0.83 ± 0.13 to 0.36 ± 0.07 nl/mm·min ($P<0.001$). These effects were reversed by the addition of 1 mM succinate, a result consistent with competitive inhibition of mitochondrial substrate availability. In suspensions of cortical tubules with phosphate-containing media, the addition of 5 mM n-butylmalonate reduced both coupled and uncoupled QO_2, and ATP content without affecting the ouabain-insensitive or non-transport related QO_2. In phosphate-free media the addition of n-butylmalonate to suspensions of cortical tubules had no further effect on any of the components of QO_2 or ATP content. These results indicate that the inhibition of dicarboxylate anion exchange across the mitochondrial membranes can significantly effect transport and metabolic functions of proximal convoluted tubules in a pattern similar to that seen with phosphate limitation.

Quantitative estimates of phosphate requirements. The preceding data indicate that oxidative metabolism in renal cortical tubules can be limited by inadequate influx of inorganic phosphate through the luminal surface. This presumably reflects reductions in intracellular and cytosolic inorganic phosphate concentration. In this regard, the concentration range of intracellular phosphate that may limit oxidative phosphorylation by cortical tubules is unknown. We therefore determined the quantitative relationship between the rate of ATP production and the intracellular Pi concentration by measuring the stimulation of QO_2 in response to added ADP at several concentrations of inorganic phosphate (9). In these studies, cortical tubules were prepared in phosphate-free media. Digitonin (0.3 mg/ml) was added to permeabilize the plasma membrane and provide access for ADP added to a concentration of 0.375 mM. This concentration gives maximal stimulation of oxidative metabolism. In the presence of ADP, phosphate concentration in the initially phosphate-free medium was increased sequentially and resulted in step-wise increases in QO_2. An Eadi-Hofstee plot of these data indicated an apparent K_m of about 0.8 mM phosphate. That is, under conditions of adequate ADP availability, oxidative metabolism in renal cortical tubules may be phosphate-limited at ambient phosphate concentrations less than about 1 mM. The full interpretation of this observation is dependent on knowledge of the cytosolic inorganic phosphate concentration which may be in this range (10).

CONCLUSION

Proximal convoluted tubules from rabbit kidney appear to require phosphate entry across luminal membranes to meet the demands of cellular metabolism. When phosphate availability is severely limited, all components of oxidative metabolism are reduced, ATP content is reduced, and fluid absorption is eliminated. The present study suggests that phosphate availability under these conditions limits oxidative phosphorylation and that this state may occur when cytosolic concentration is less than about 1 mM. Conversely, the provision of metabolic substrates that stimulate oxidative metabolism is associated with increased phosphate transport. These relationships between oxidative metabolism and the flux of inorganic phosphate represent mutual interdependence, but the mechanisms by which these relationships take place are still undefined.

REFERENCES

1. H. A. Lardy and H. Wellman, Oxidative phosphorylations: Role of inorganic phosphate and acceptor systems in control of metabolic rates, J. Biol. Chem. 195:215 (1952).

2. D. F. Wilson, M. Stubbs, N. Oshino, and M. Ericinska, Thermodynamic relationships between the mitochondrial oxidation-reduction reactions and cellular ATP levels in ascites tumor cells and perfused rat liver, Biochem. 13:5305 (1974).

3. L. Sestoft, and P. D. Bartels, Regulation of metabolism by inorganic phosphate, in: "Short-Term Regulation of Liver Metabolism," L. Hue, and G. Van de Werve, eds., Amsterdam: North-Holland (1981).

4. P. C. Brazy, R. S. Balaban, S. R. Gullans, L. J. Mandel, and V. W. Dennis, Inhibition of metabolism: Relative effects of arsenate on sodium, phosphate and glucose transport by the rabbit proximal tubule, J. Clin. Invest. 66:1211 (1980).

5. N. Hoffman, M. Thees, and R. Kinne, Phosphate transport by isolated renal brush border vesicles, Pfugers Archiv. 362:147 (1976).

6. S. R. Gullans, P. C. Brazy, L. J. Mandel, and V. W. Dennis, Stimulation of phosphate transport in the proximal tubule by metabolic substrates, Am. J. Physiol. 247:F582 (1984).

7. S. R. Gullans, P. C. Brazy, V. W. Dennis, and L. J. Mandel, Interactions between gluconeogenesis and sodium transport in rabbit proximal tubule, Am. J. Physiol. 246:F859 (1984).

8. P. C. Brazy, S. R. Gullans, L. J. Mandel, and V. W. Dennis, Metabolic requirement for inorganic phosphate by the rabbit proximal tubule: Evidence for a Crabtree effect, J. Clin. Invest. 70:53 (1982).

9. P. C. Brazy, L. J. Mandel, S. R. Gullans, and S. P. Soltoff, Interactions between phosphate and oxidative metabolism in proximal renal tubules, Am. J. Physiol. 247:F575 (1984).

10. D. Freeman, S. Bartlett, G. Radda, and B. Ross, Energetics of sodium transport in the kidney, Saturation transfer 31P-NMR, Biochim. Biophys. Acta 762:325 (1983).

REGULATION OF Na-PHOSPHATE COTRANSPORT IN CULTURED RENAL EPITHELIAL CELLS: PROTEIN-SYNTHESIS-DEPENDENT AND PROTEIN-SYNTHESIS-INDEPENDENT PATHWAYS

Heini Murer, Jürg Biber, Kerstin Malmström, Ingvild Mohrmann, and Michael Coady

Institute of Physiology
University of Zurich
8057 Zurich, Switzerland

INTRODUCTION

The regulation of Na-phosphate cotransport across the apical membrane can, in analogy to classical enzymology, be divided into: short-term and long-term regulation. Short-term regulation involves: (i) allosteric and (ii) other activation-inactivation processes, e.g. covalent modifications such as phosphorylation or ADP-ribosylation. Long-term regulation usually involves de novo synthesis of the cotransporter. An additional regulatory mechanism can be considered for plasma membrane-bound systems: regulation via insertion or withdrawal of transporter from the membranes (endo-/exo-cytosis; for review see: Gmaj and Murer, 1985; Kempson and Dousa, 1985).

Allosteric events are most likely important in the early phases of phosphate transport regulation, for example during acid/base changes and in the parathyroid hormone action (mediated by e.g. changes of intracellular pH and/or intracellular free phosphate as a consequence of altered gluconeogenesis). Covalent modifications may also be of importance in the parathyroid hormone dependent inhibition of Na-phosphate cotransport. Protein-synthesis is involved e.g. in phosphate deprivation. Short-term regulation may precede long-term regulation, e.g. early effect in deprivation/overload. On the other hand, long-term regulation may also occur as a secondary event, in situations where the "physiological" conditions normally lead to a short-term regulation (acute effect) but under maintained or extreme conditions, e.g. chronic acid/base changes, prolonged infusions or injections of high concentrations of parathyroid hormone; (for review see: Gmaj and Murer, 1985).

Studies with brush border membrane vesicles have documented that the changes in renal phosphate transport are - at least in part - retained in isolated membranes and always involve changes in the apparent V_{max} (for review see: Kempson and Dousa, 1985; Gmaj and Murer, 1985; Dousa and Kempson, 1982). Most likely these alterations are the product of long-term regulation as short-term regulation involves fast reversible mechanisms and would not be retained during lengthy procedures of membrane isolation. Cellular events (regulatory cascade) which lead to altered transport at the membrane level can therefore only be explored in a intact cellular system.

THE USE OF ESTABLISHED CELL LINES TO STUDY RENAL Na-PHOSPHATE COTRANSPORT AND ITS REGULATION

In order to study the cellular events in regulation of Na-phosphate cotransport we have used three continuous epithelial cell lines. $LLC-PK_1$ derived from pig kidney (Hull et al., 1976), $JTC-12.P_3$, derived from a monkey kidney (Katsuta et al., 1960), and OK-cells, derived from an American Opossum kidney (Koyama et al., 1978). All three cell lines transport phosphate in a Na-dependent way (Biber and Murer, 1985; Biber et al., 1983; Caverzasio et al., 1985 a; 1985 b; Malmström and Murer, 1985 b; Noronha-Blob et al., 1984; Rabito, 1983). Na-phosphate cotransport has been shown to be present in apical membrane vesicles isolated from $LLC-PK_1$ and OK-cells (Caverzasio et al., 1985 a; Malmström and Murer, 1985 b; Biber and Murer, 1985; Brown et al., 1984).

The basic properties of phosphate transport into monolayer of $LLC-PK_1$, $JTC-12.P_3$ and OK-cells are shown in Figure 1 and in Table 1 (Malmström and Murer, 1985 b). In all three cell lines, phosphate uptake is strictly Na-dependent (Figure 1); the apparent K_m is between 80 and 230 µM whereas the apparent V_{max} varies by a factor of 10 with the highest rates in the OK-cells (Table 1).

The $JTC-12.P_3$ cells and the OK-cells respond to parathyroid hormone (PTH) with large increases in cAMP whereas in $LLC-PK_1$ cells only a minimal increase was observed (Ishizuka et al., 1978; Noronha-Blob et al., 1984; Teitelbaum and Strewler, 1984; Malmström and Murer, 1985 b; Caverzasio et al., 1985 b). Na-phosphate cotransport was found to be inhibited by increased cAMP in OK-cells but not in $JTC-12.P_3$ and $LLC-PK_1$ cells (Malmström and Murer, 1985 b).

Extensive studies on the regulation of Na-phosphate cotransport by phosphate deprivation/overload have been performed in $LLC-PK_1$ cells. Alterations of the phosphate content in the growth media provoke corresponding alterations in Na-phosphate cotransport (Caverzasio et al., 1985 a; Biber and Murer, 1985). Preliminary experiments have shown that the adaptive response is also expressed in the OK-cells (Malmström and Murer, unpublished observations).

We have also studied whether activation of protein kinase C is involved in the regulation of phosphate transport (Mohrmann et al., to be published). 12-O-tetradecanoylphorbol-13-acetate (TPA); a phorbolester induces a time dependent increase of Na-phosphate cotransport in $LLC-PK_1$ cells.

The effect of PTH (cAMP) and phosphate deprivation/overload are specific for Na-phosphate cotransport. In contrast, the TPA-induced increase of phosphate transport is parallelled by alterations in Na-alanine cotransport. Alterations of phosphate transport in the monolayer were related to changes of phosphate transport in the isolated apical membrane vesicles, implying that the changes are not related to changes of the driving forces.

THE INFLUENCE OF PHOSPHATE DEPRIVATION/OVERLOAD ON Na-PHOSPHATE CO-TRANSPORT IN $LLC-PK_1$

Monolayers of $LLC-PK_1$ cells exposed to a phosphate free medium showed a marked increase of Na-dependent phosphate transport (Figure 2), extrapo-

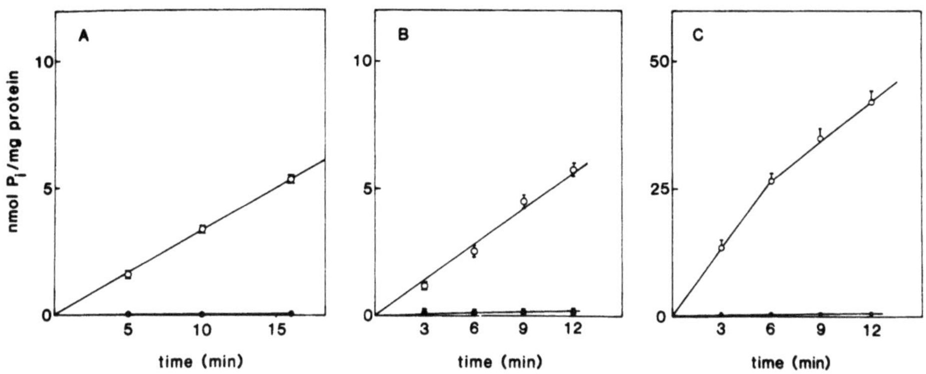

Figure 1. Time course of phosphate uptake by $LLC-PK_1$ (A), $JTC-12.P_3$ (B) and OK (C) cells: Uptake of 0.1 mM phosphate was determined as described by Malmström and Murer (1985 b) with either 137 mM NaCl (o) or N-methyl-D-glucamine (•) in the incubation medium. Note that the scale for OK cells is 5 times smaller. This figure is taken from Malmström and Murer (1985 b).

Table 1. Kinetic parameters for Na^+-dependent phosphate transport in different cell lines

Cell type	V_{max} (nmol mg^{-1} · 3 min^{-1})	K_m (µM)
$LLC-PK_1$	3.5 ± 1.6	194 ± 30
OK	34.0 ± 4.5	86 ± 8
$JTC-12.P_3$	3.4 ± 0.2	228 ± 5

The phosphate transport was followed at 25 °C for $LLC-PK_1$ and at 37 °C for $JTC-12.P_3$ and OK. The values are means ± SD from at least two independent experiments performed in triplicates. The phosphate transport was analyzed at 7 different concentrations between 8 µM and 1 mM phosphate. The data are derived from Lineweaver Burk plots. This table is derived from Malmström and Murer (1985 b).

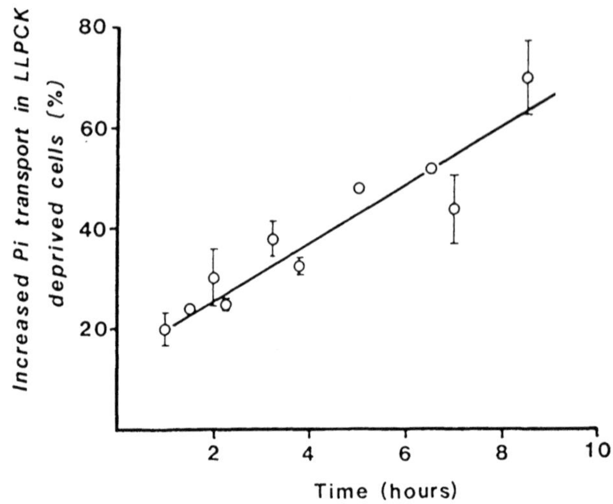

Figure 2. Time course of the increase in phosphate transport in phosphate-deprived LLC-PK_1 cells: Cells were incubated for the times indicated in a phosphate free medium. Na-dependent transport of phosphate was determined with 0.1 mM phosphate after 5 min. As a reference (0 % increase) cells were incubated for the same times in a medium containing 2 mM phosphate. Each point represents the mean ± SD of five determinations. This figure is taken from Caverzasio et al. (1985 a).

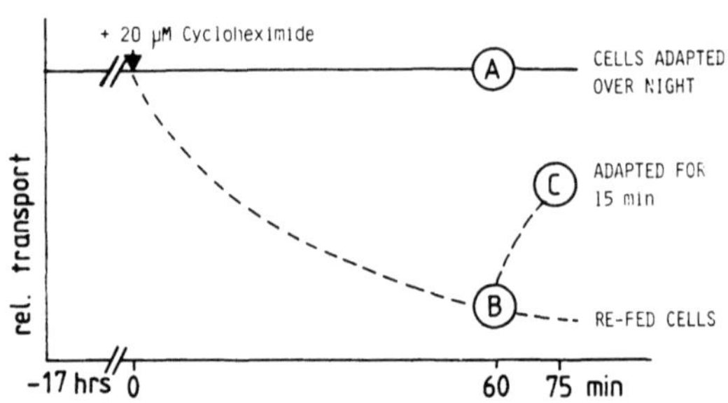

Na⁺-dependent transport of phosphate measured at:	control	+ 20 μM Cycloheximide
	(pmoles $P_i \cdot mg^{-1} \cdot 5\ min^{-1}$)	
A	3532 ± 182	2850 ± 78
B	1665 ± 61	1760 ± 89
C	2546 ± 146	2449 ± 114

Figure 3. Effect of cycloheximide on the rapid "up-regulation" of Na-phosphate cotransport: Cells were deprived of phosphate over night and at time zero refed with 2 mM phosphate either in the presence or absence of 20 μM cycloheximide (B). (A) represents over night adapted cells. The data are given as means ± SD of four determinations. This figure is taken from Murer and Biber (1985).

lation of the exposure time close to zero suggests a fast and a slow component in the adaptive response (Caverzasio et al., 1985 a). Subsequent experiments revealed that only the slow component was prevented by inhibitors of protein-synthesis (Caverzasio et al., 1985 a).

The fast protein-synthesis-independent adaptive response is further illustrated in Figure 3. Cells were exposed overnight to phosphate free medium and then refed with phosphate, at refeeding, protein synthesis was blocked by cycloheximide. Refeeding was followed by a second phosphate deprivation (Biber and Murer, 1985). The data shown in Figure 3 illustrate that the "down regulation" of phosphate transport and the fast re-activation are protein synthesis-independent.

The threshold phosphate concentration for the initiation of the fast and the slow regulatory response to phosphate deprivation are different: The fast response starts below 0.8 mM and the slow response starts below 0.1 mM, i.e. only at severe depletion (Biber and Murer, 1985).

Little is known about the cellular mechanism involved in these events. The slow response involves protein synthesis. The rapid inactivation when refeeding or the rapid re-activation after a second deprivation are independent of protein synthesis and can either be achieved by a covalent modification reaction (i.e. phosphorylation) of for example the transporter itself or by an endo/exocytotic process.

In an attempt to identify the carrier involved in Na-phosphate cotransport we have incubated the cells with ^{35}S-L-methionine in growth media containing normal phosphate or no phosphate, i.e. under conditions where the cells show long-term adaptation of phosphate transport. Apical membranes were isolated from control and adapted cells and the proteins were separated on SDS-polyacrylamide gels. The autoradiograms of the gels are shown in Figure 4.

The labelling pattern of the "adapted" and "non-adapted" brush border membrane proteins reveal no apparent differences. Preliminary experiments separating the peptides on two-dimensional gels have also failed to detect any changes in the pattern of labelled proteins (Biber and Murer, unpublished observations). Several parameters may contribute to these negative findings, e.g. the molecule involved in transport is a minor component of the brush border membrane.

THE PARATHYROID HORMONE (PTH) DEPENDENT INHIBITION OF PHOSPHATE TRANSPORT IN OK-CELLS

OK-cells respond to PTH at concentrations between 10^{-10} and 10^{-7} M with an increase of intracellular cAMP (Malmström and Murer, 1985 b). Phosphate transport is inhibited in OK cells which have been exposed to as little as 10^{-10} M PTH; the reduction of phosphate transport is potentiated by increasing PTH-concentrations and by prolonged exposure to PTH (Figure 5).

Interestingly, at the time when cAMP reaches its maximum (2-5 minutes after exposure to PTH) phosphate transport is not significantly altered. The effect of parathyroid hormone is specific for phosphate transport, Na-dependent alanine transport is not altered (Malmström and Murer, 1985 b).

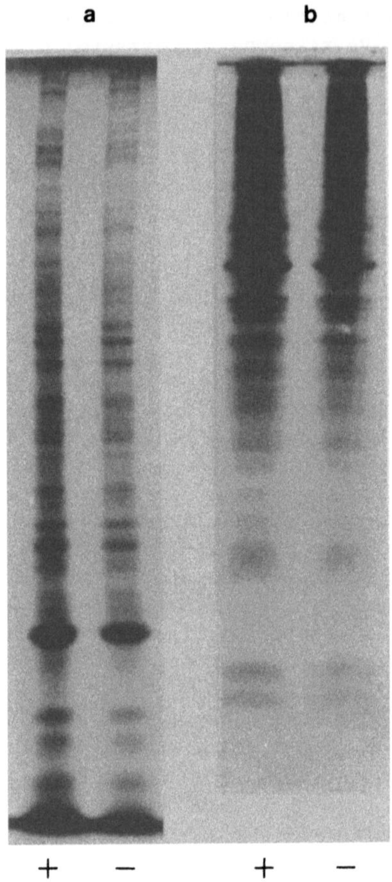

Figure 4. Effect of phosphate deprivation on ^{35}S-L-methionine incorporation into proteins of apical vesicles isolated from LLC-PK_1. The cells were exposed for 16 hours to growth media containing 2 mM phosphate (+) or to phosphate free media (-) in the presence of 0.1 mM ^{35}S-L-methionine. The conditions for adaptation and membrane isolation were taken from Caverzasio et al. (1985 a) and from Biber and Murer (1985). The membrane proteins were separated either on 8.0 % (a) or on 15 % (b) SDS polyacrylamid gels (Laemmli, 1970) and processed for autoradiography (Malmström and Murer, 1985 a).

The inhibition by PTH is related to an effect on the V_{max} of Na-phosphate cotransport and is released after removal of PTH (Table 2) concomitant to a decline of the cellular cAMP level (Malmström and Murer, 1985 b). Vesicles isolated from cells exposed to 10^{-10} M PTH for 4 hours transport phosphate also with reduced V_{max} (Murer and Malmström, 1985 b).

At present only speculations can be made concerning the cellular mechanisms involved in the PTH-dependent inhibition of phosphate transport in OK cells. The inhibition is a protein-synthesis-independent event (Caverzasio et al., 1985 b; Malmström and Murer, unpublished observations). The involvement of the cAMP system is evident as PTH action can be mimicked by other activators of the adenylate cyclase (Malmström and Murer, 1985 b).

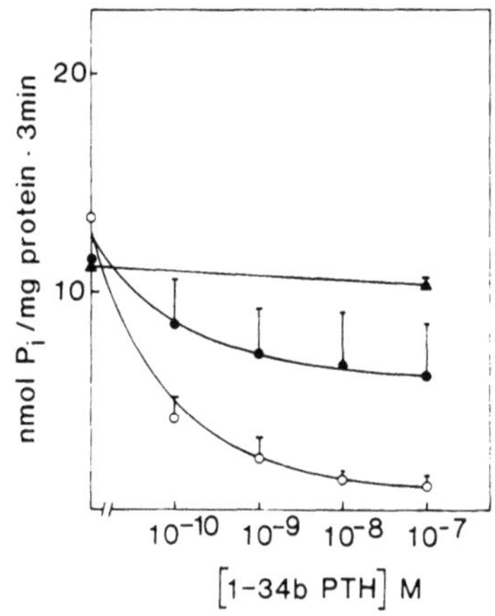

Figure 5. Inhibition of Na^+-dependent phosphate transport in OK cells as a function of PTH concentration. Confluent cell monolayers were exposed to various concentration of PTH either for 5 min (▲), 1 h (●) and for 4 h (o). The uptake of 0.1 mM phosphate into the cells was determined as described by Malmström and Murer (1985 b). The values are means ± SD of three independent experiments made in triplicates. This figure is taken from Malmström and Murer (1985 b).

Table 2. Kinetic parameters of phosphate transport in OK cells; a time dependence of the PTH effect.

	V_{max} (nmol/mg prot.•3 min)	K_m (µM)
CONTROL	30.8 ± 5.3	86.5 ± 8.2
1-34b PTH (10^{-10} M)		
1 h	21.4 ± 0.1	89.5 ± 2.6
2 h	11.2 ± 5.2	77.1 ± 22
3 h	9.2 ± 2.6	93.0 ± 2.2
4 h	10.2 ± 1.1	105.9 ± 36
REVERSED		
4 h PTH + 5 h no PTH	24.8 ± 0.9	86.3 ± 18.4

The Na^+-dependent phosphate transport was studied for 3 min at 37 °C. Concentration of 1-34b PTH was 10^{-10} M. The values are derived from Lineveaver Burk plots and means ± SD of three independent experiments made in triplicates. Phosphate transport was analyzed at 7 different concentrations between 8 µM and 1 mM. These data are in part derived from Malmström and Murer (1985 b).

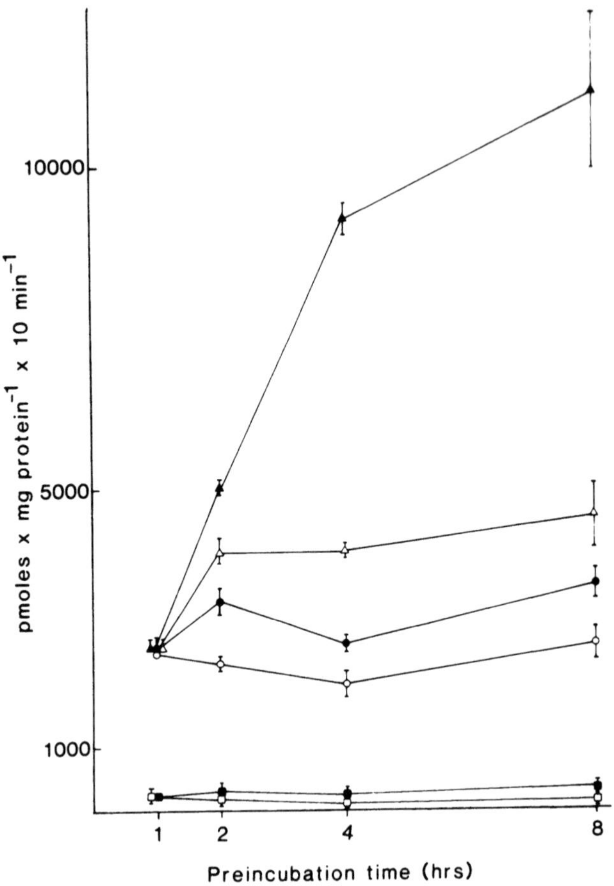

Figure 6. Influence of TPA on phosphate uptake by confluent LLC-PK_1 monolayers: Monolayers were incubated with 10 ng/ml TPA (▲, △ , x) for different times as indicated in the figure. Monolayers exposed to the carrier vehicle only (0.01 %o DMSO, ●, o ■) served as controls. Incubation was performed either in the absence (△ , ●, ■ , x) or in the presence of 20 μM cycloheximide added 1 hour prior to TPA-addition. The uptake of 0.1 mM phosphate was measured either in the presence of 137 mM NaCl (▲ , △ , ●, o) or 137 mM N-methyl-D-glucamine (■, x). These data are taken from Mohrmann et al. (to be published).

The specific regulatory mechanism in the PTH-action is beyond the generation of cAMP and is a rather slow process. This conclusion is warranted by (i) the time lap between maximal increase in cAMP and maximal reduction of phosphate transport and (ii) by the observation that in two other cell lines LLC-PK_1 and JTC-12.P_3 cAMP can be increased in a PTH-dependent as well as PTH-independent manner without reducing the phosphate transport (Malmström and Murer, 1985 b). cAMP-dependent protein kinase(s) is(are) most likely involved in the PTH action and may either induce a direct phosphorylation of the Na-phosphate cotransporter or a phosphorylation which initiates a regulatory cascade whose final response is reduction of Na-dependent phosphate transport.

STIMULATION OF PHOSPHATE TRANSPORT IN LLC-PK_1 CELLS BY TPA (PHORBOLESTER)

Recent experiments with brush border vesicles from rat and rabbit kidney cortex have provided evidence for the presence of protein kinase C as well as for target peptides for protein kinase C (Barrett et al., 1985; Malmström and Murer, 1985 a). To elucidate a possible role of protein kinase C in the regulation of phosphate transport we have investigated the effect of TPA - known to activate the protein kinase C - on phosphate transport in LLC-PK_1 cells (Mohrmann et al., to be published).

Incubation of LLC-PK_1 cells with TPA (1 ng/ml) produced a time-dependent increase in phosphate transport (Figure 6). Two separate events are involved in TPA-dependent alteration of phosphate transport: (i) a fast protein-synthesis-independent increase and (ii) a slow protein-synthesis-dependent one. An identical response to TPA is observed for Na-alanine cotransport (Mohrmann et al., to be published). The fast response and the slow response involves changes in the apparent V_{max} of the Na-phosphate cotransport mechanism and are retained in isolated vesicles (Mohrmann et al., to be published). The stimulation by TPA of Na-phosphate and Na-alanine cotransport is not related to a general proliferative effect while TPA-administration did not increase thymidine incorporation into DNA (Mohrmann et al., to be published).

This data suggest that activation of protein kinase C regulates Na-phosphate cotransport in parallel to other Na-dependent transport systems, e.g. Na-alanine cotransport system (see also: Amsler et al., 1983). This regulation involves a protein-synthesis-independent fast process and a protein-synthesis-dependent slow process similar to the alterations in phosphate transport in response to phosphate derivation.

CONCLUSIONS

The data on phosphate transport in established renal epithelial cell lines and in apical membrane vesicles isolated thereof illustrate that Na-phosphate cotransport is regulated by cellular pathways involving protein-synthesis and by pathways independent of protein-synthesis. At present the studies are on a phenomenological level, i.e. we have been able to show different events but cannot offer a detailed view of the regulatory "mechanisms". It is the task of future experiments to analyze the cellular mechanisms (regulatory cascade) involved in alterations in Na-phosphate cotransport activity. Certainly a major effort must be directed to the molecular identification of the transport unit. Only then it will be possible to follow all the cellular events involved in regulation of Na-phosphate cotransport, in cultured cells and hopefully also in the proximal tubular epithelium.

ACKNOWLEDGEMENT

The work discussed from the authors laboratory was supported by the Swiss National Foundation.

REFERENCES

Amsler, K., Schaffer, C., and Cook, J.S., 1983, Growth dependent AJB and me-AJB in LLC-PK_1 cells:Effects of differentiation inducers and of TPA, J. Cell. Physiol., 114:184.

Barrett, P.Q., Zwailich, K., and Rasmussen, H., 1985, Protein kinase C activity in renal microvillus membranes, Biochem. Biophys. Res. Commun., 128:494.

Biber, J., and Murer, H., 1985, Na/P_i cotransport in LLC-PK_1 cells: Fast adaptive response to P_i deprivation, Amer. J. Physiol., in press.

Biber, J., Brown, C.D.A., and Murer, H., 1983, Sodium-dependent transport of phosphate in LLC-PK_1 cells, Biochim. Biophys. Acta, 735:325.

Brown, C.D.A., Bodmer, M., Biber, J., and Murer, H., 1984, Sodium-dependent phosphate transport by apical membrane vesicles from a cultured renal epithelial cell line (LLC-PK_1), Biochim. Biophys. Acta., 769:471.

Caverzasio, J., Brown, C.D.A., Biber, J., Bonjour, J-P., and Murer, H., 1985, Na-dependent phosphate transport in phosphate deprived LLC-PK_1 cells. Amer. J. Physiol., 248:F122.

Caverzasio, J., Rizzoli, R., Bonjour, J-P., 1985, Sodium-dependent phosphate transport is inhibited by parathyroid hormone and cyclic AMP stimulation in an opossum kidney cell line, J. Biol. Chem., submitted.

Dousa, T.P., and Kempson, S.A., 1982, Regulation of renal brush border membrane transport of phosphate, Mineral. Electrolyte Metab., 7:113.

Gmaj, P., and Murer, H., 1985, Cellular mechanisms in inorganic phosphate transport in the kidney. Physiol. Rev., in press.

Hull, R.N., Cherry, W.R., and Weaver, G.W., 1976, The origin and characteristics of a pig kidney cell strain, LLC-PK_1, In Vitro, 12:670.

Ishizuka, J., Tadano, K., Natato, N., Niimura, Y., and Nagai, Y., 1978, Hormone specific responses and biosynthesis of sulpholipids in cell lines derived from mammalian kidney, Biochim. Biophys. Acta., 541:467.

Katsuta, H., Takaoka, T., Tagaya, I., and Kikuchi, K., 1960, Cultivation of monkey kidney cells in simplified replicate tissue culture, Jap. Exp. J. Med., 30:485.

Kempson, S.A., and Dousa, T.P., 1985, Current concepts of regulation of phosphate transport in renal proximal tubules, Biochem. Pharmacol., in press.

Koyama, H., Goodpasture, C., Miller, M.M., Teplitz, R.L., and Riggs, A.D., 1978, Establishment and characterization of a cell line from the American opossum (Didelphys Virginiana), In Vitro, 14:239.

Laemmli, U.K., 1970, Cleavage of structural protein during the assembly of the head of bacteriophage T_4, Nature, 227:680.

Malmström, K., and Murer, H., 1985a, Ca^{2+}-dependent phosphorylation in brush border membranes of rat kidney proximal tubules, Pflügers Arch., 404:358.

Malmström, K., and Murer, H., 1985b, PTH inhibits phosphate transport in OK cells but not in LLC-PK_1 and JTC-12.P3 cells, Amer. J. Physiol. submitted.

Noronha-Blob, L., Filburn, C., and Sacktor, B., 1984, Phosphate uptake by kidney epithelial (LLC-PK_1) cells, Arch. Biochem. Biophys., 234:265.

Rabito, C.A., Phosphate uptake by a kidney cell line (LLC-PK_1), Amer. J. Physiol. 245; F22.

Teitelbaum, A.P., and Strewler, G.J., Parathyroid hormone receptors coupled to cyclic adenosine monophosphate formation in an established renal cell line, Endocrinology, 114:980.

PHOSPHORIN, A PHOSPHATE-BINDING HYDROPHOBIC PROTEIN ISOLATED FROM RENAL BRUSH BORDER MEMBRANES

Ralph J. Kessler, Duke A. Vaughn,
Christian Schäli*, and Darrell D. Fanestil

Department of Medicine, Division of Nephrology, M-023B
University of California, San Diego
La Jolla, CA 92093

INTRODUCTION

Reabsorption of inorganic phosphate (Pi) by the renal proximal tubule occurs from the tubular fluid across the brush border membrane into the cell interior by a Na^+-dependent co-transport system. A number of the features of this Na^+-Pi symport system have been documented by earlier workers(1,2). These include a description of the kinetic parameters and of the alterations produced by in vivo manipulation of dietary phosphate(3) or vitamin D(4) or by parathyroid hormone status(5). However, the molecular constituent or constituents of this Na^+-Pi symporter have not been identified. We summarize here some features of a putative constituent of the Na^+-Pi symporter, a proteolipid that can be extracted from renal brush borders. We have called this proteolipid "phosphorin" because it binds Pi with selectivity and high affinity and because it is located in brush borders that conduct Na^+-Pi symport. Although our working hypothesis is that phosphorin might be a constituent of the Na^+-Pi symporter, firm evidence for such a role for phosphorin remains to be advanced.

SPECIAL METHODOLOGICAL ASPECTS

Proteolipids are, as originally described by Folch(6), proteins with the solubility properties of lipids - i.e., they are soluble in the organic solvents commonly employed to extract lipoidic material from tissues. We obtain crude phosphorin from renal brush border membrane vesicles (BBMV) that have been prepared by a divalent metal ion-dependent differential centrifugation method(7). Crude phosphorin is extracted from BBMV (we usually use rabbit kidney as a source for BBMV) into chloroform:methanol (1:2, V/V), washed as described in detail by Kessler et al.(8), and concentrated under reduced pressure before being suspended in chloroform:methanol (2:1, V/V). A more purified preparation can be prepared as outlined in Table I.

Binding of ligands (e.g., ^{32}Pi or $^{45}Ca^{2+}$) is quantitated in a two phase partition assay in which phosphorin (typically, about 1μg protein) is placed in 0.5ml 1-butanol:chloroform:methanol (150:50:25, V/V). Radioactive ligand

* Current address: Institute of Pharmacology, University of Lausanne, Rue du Bugnon 21, CH-1005 Lausanne, Switzerland

TABLE I
Ether Precipitation of Phosphorin

* Dissolve crude proteolipid in small volume of chloroform:methanol (2:1)
* Add 5 volumes diethyl ether and store at -20°C overnight
* Collect precipitated proteolipid by centrifugation

The ether precipitation step removes unbound lipoidal material

* Repeat ether precipitation 2 more times

and any reagents being tested are added in less than 20μl of buffer and mixed to form a one phase system. After 10 min. at room temperature 0.6ml of 2.0M sucrose is added and the samples are shaken vigorously. Brief centrifugation displaces the dense aqueous phase to the bottom of the tube and, thereby enables direct sampling of the organic phase for quantitation of radioactivity that has been bound by molecules that are highly soluble in the organic phase.

BINDING OF Pi BY PHOSPHORIN

Crude phosphorin binds Pi with high affinity ($K_{0.5} \simeq 8$ M), as shown in Fig. 1. Moreover, the sigmoidal relationship between binding of Pi and the concentration of Pi is characteristic of that found with site-site interaction features of positive cooperativity. Analysis of the data in Fig. 1 by the Hill equation yields a Hill coefficient of nearly 2.0 (Fig. 2) - again, this is characteristic of positive cooperativity.

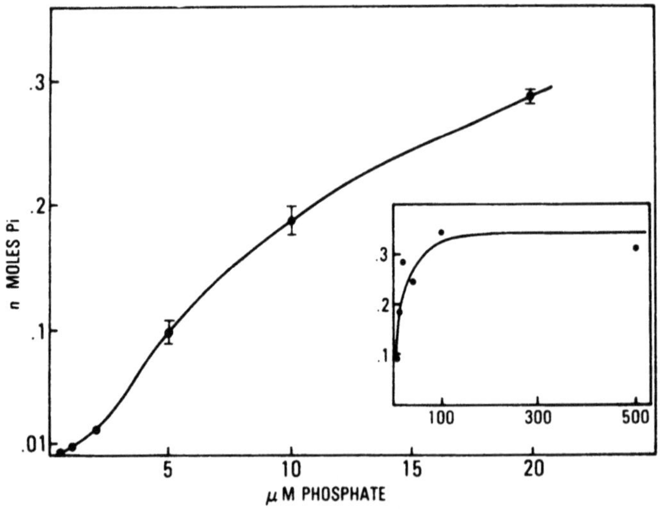

Fig. 1. The effect of phosphate concentration on phosphate binding. Dependence of [^{32}P]phosphate binding by the proteolipid on phosphate concentration. The phosphate solutions were buffered with 2mM MES, pH 6.5. The concentration of the proteolipid was 5.8 μg/assay. Values are mean ± S.E. of duplicate determinations. At low phosphate concentrations, the standard error is within the size of the circle (●). Inset, measurement of phosphate binding in the concentration range of 5 to 500μM [^{32}P]phosphate.

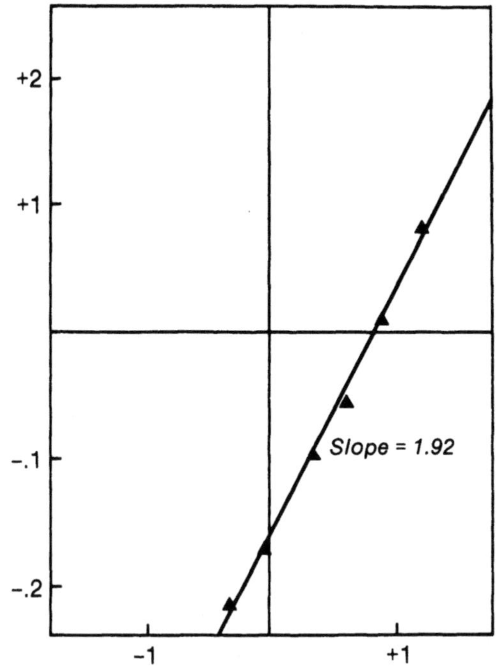

Fig. 2.

Table II

Inhibition of Binding of Phosphate by Various Anions

Addition	^{32}Pi Bound-Percent of Control
None	100
Lactate	85
SO_3^-	60
Cl^-	51
NO_3^-	60
Acetate	85

Data from Kessler et al.[8]

Table II. Inhibition of [^{32}P]phosphate binding by various anions. The ^{32}Pi concentration was 1μM; the competing anions were present at 250μM; buffer was 2mM MES, pH 6.5. The concentrations are calculated for the single phase before phase separation. All assays were performed in triplicate. Values are means ± S.E. Salts are present at a concentration 250 times greater than phosphate.

The binding of Pi is selective for phosphate and arsenate, the biologically toxic competitor of phosphate. As shown in Table II, a 250 fold molar excess of several other anions produces only minor inhibition of binding of Pi. This is indicative of the selectivity of phosphorin for Pi.

Binding of Pi by phosphorin is not inhibited by 30 min. pre-exposure to (100μM) tetraphenylarsonium, pentachlorophenol, N-ethylmaleimide or (20μM) phenylmercuric acetate. This insensitivity to reagents that react with sulfhydral residues clearly distinguishes the Pi-binding activity of phosphorin from that of the mitochondrial phosphate-binding protein.

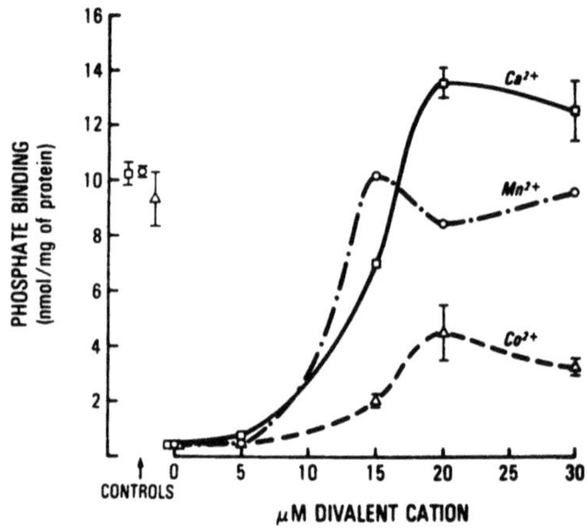

Fig. 3. Restoration of phosphate binding to the proteolipid phosphorin by divalent metals after inhibition by the chelator EGTA. The [^{32}P]phosphate concentration was 10μM buffered with 1mM MES at pH 6.5. The proteolipid concentration was 2.1μg/assay. The proteolipid was inhibited by preincubation with 5μM EGTA followed by a 10-min incubation ± the indicated divalent metal ion. The Pi binding to the proteolipid was determined by the binding assay described previously. Values are mean ± S.E. of triplicate determinations. "Controls" indicates binding by phosphorin not exposed to EGTA.

RELEVANCE OF DIVALENT METAL IONS

The unexpected observation that binding of Pi by phosphorin is inhibited by chelators of divalent metals (including EDTA, EGTA, oxalate and citrate) suggested the possibility that binding of Pi by phosphorin might involve divalent metals (Me^{2+}). Kessler and Vaughn(9) described several features of the dependence of the binding of Pi upon the presence of Me^{2+}. Phosphorin which had been partially purified by ether precipitation (see Table I) lost much of its ability to bind Pi. When micromolar quantities of Me^{2+} were added back in the presence of 5μM EGTA, the added Ca^{2+}, Mn^{2+} or Co^{2+} restored the ability of the phosphorin to bind Pi, once the concentration of the Me^{2+} exceeded that of the EGTA that was present in the assay (Fig. 3.).

At first glance the finding that the binding of Pi by phosphorin required Me^{2+} seemed to indicate that phosphorin could not be a component of the Na^+-Pi symporter in BBMV, since treatment of BBMV with EDTA did not inhibit Na^+-Pi symport activity. However, a more detailed examination of the issue by Kessler and Vaughn(9) paradoxically provided indirect support for the possibility that phosphorin may be a constituent of the Na^+-Pi symporter. Kessler and Vaughn found that treatment of BBMV with EDTA plus the lipophilic Me^{2+} ionophore A-23187 resulted in a decrease in the initial rate of Na^+ gradient-dependent uptake of Pi by the BBMV (Fig. 4). This

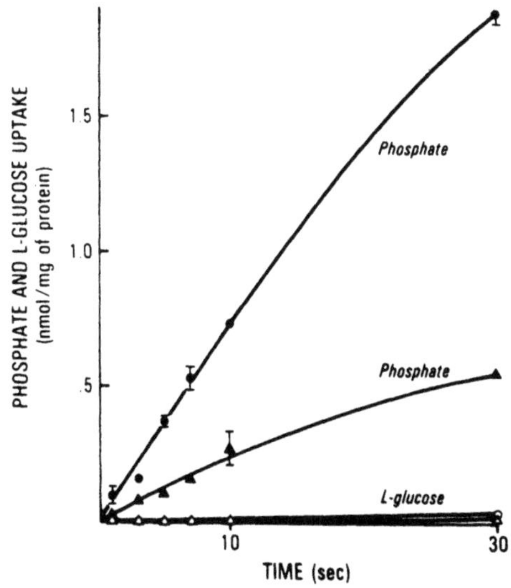

Fig. 4. The time course of Na^+-dependent phosphate uptake and the permeability of the BBMV to L-glucose. The BBMV were preincubated for 15 min at 30°C in suspension medium containing 300mM mannitol buffered at pH 7.0 with 20mM Hepes/Tris ± EDTA (5mM) and A-23187 (6μM). The uptake was initiated by vortex mixing in a 10-μl drop of the brush-border suspension with a 10-μl drop of uptake medium containing 100mM mannitol, 200mM NaCl, 0.2mM $[^{32}P]K_2HPO_4$ buffered at pH 7.0 with 20mM Hepes/Tris and 20μM L-$[^3H]$glucose. The mixture was vortexed for the indicated time, at which point the uptake was terminated by the addition of 1 ml of ice-cold stop solution. Values are mean ± S.E. of triplicate uptake determinations. Where not shown, the standard error is within the size of the symbol. ▲, △, represent EDTA-treated BBMV.

inhibition was due to a decrease in the maximal rate of transport of Pi. EDTA plus A-23187 also inhibited Na^+-facilitated transport of Pi (i.e., transport of Pi in the presence of Na^+ but in the absence of a concentration gradient for Na^+), indicating that the effect of the chelator plus ionophore was not solely due to any alteration in the driving force supplied by the Na^+ gradient. There was no effect of treatment with EDTA plus A-23187 on the Km for Pi. Finally, Na^+-Pi symport activity that had been inhibited by EDTA plus A-23187 could be restored toward normal by re-addition of Mn^{2+} (Table III). This is analogous to the restoration of binding of Pi by re-addition of Me^{2+} to phosphorin that had been depleted on Me^{2+} (Fig. 3).

Table III
Effect of divalent metal chelators on Na^+-dependent phosphate uptake in BBMV

The BBMV were suspended in 300mM mannitol buffered at pH 7.0 with 20mM Hepes/Tris. Aliquots of the BBMV were preincubated at 30°C for 15 min ± chelator and ionophore, where indicated. The phosphate uptake was initiated with the addition of the extravesicular medium containing 100mM mannitol, 100mM NaCl, and 0.1mM [^{32}P]phosphate in 20mM Hepes/Tris, pH 7.0. Values are expressed as a per cent of maximal (peak) uptake of Pi in the absence of both chelators and ionophore.

Chelator	A23187 (6μM)	Maximum Pi uptake
		% of control
None	-	100
	+	96
EDTA 6.25mM	-	67
	+	38
EGTA 5mM	-	101
	+	97
Oxalate 5mM	-	85
	+	61
Citrate 5mM	-	72
	+	58

Note that the inhibition of Na^+-Pi symport in BBMV by EDTA required the presence of the lipophilic ionophore. This suggests that the requisite Me^{2+} is in a hydrophobic phase of the BBMV where EDTA by itself fails to gain access, but from where the lipophilic ionophore can "shuttle" the Me^{2+} to the EDTA in the aqueous phase of the medium. Thus, the Me^{2+} that is required for Na^+-Pi symport activity appears to be in a very lipophilic environment. This is the same type of environment in which the Me^{2+} bound by the extremely hydrophobic phosphorin must reside.

INHIBITORS OF THE BINDING OF Me^{2+} OR OF Pi BY PHOSPHORIN

The binding of Pi by phosphorin can also be inhibited by pre-treatment of the proteolipid with either amine-reactive chemicals (DIDS, SITS, dinitrofluorobenzene) or by cyclohexandione, an arginine reactive chemical. Unpublished studies indicate that the amine-reactive reagents, but not cyclohexandione, also inhibit the binding of Me^{2+} by phosphorin. These observations suggest that it is a complex of phosphorin and Me^{2+} which is the active form of the proteolipid for the binding of Pi.

PHOSPHORIN IS NOT PART OF ALKALINE PHOSPHATASE

Alkaline phosphatase binds Pi. The possibility that phosphorin was a part of alkaline phosphatase was examined by Schäli, et al.[10] These studies capitalized on the finding by Yusufi et al.[11] that phosphatidylinositol-specific phospholipase C could release alkaline phosphatase from BBMV without affecting Na^+-dependent symport of Pi by the BBMV. Treatment of BBMV with the phospholipase released about 85% of the brush border membrane-linked alkaline phosphatase activity into the supernatant, where the activity was recovered. This treatment did not appreciably alter Na^+-Pi symport activity of the BBMV, a finding in agreement with those of Yusufi et al. Importantly, the phosphate binding activity characteristic of phosphorin that could be extracted from the BBMV and assayed for Pi-binding in either the absence or presence of added Me^{2+} was not altered by the phospholipase treatment (Fig. 5). These results clearly dissociate alkaline phosphatase activity, but not Na^+-Pi symport activity, from phosphorin.

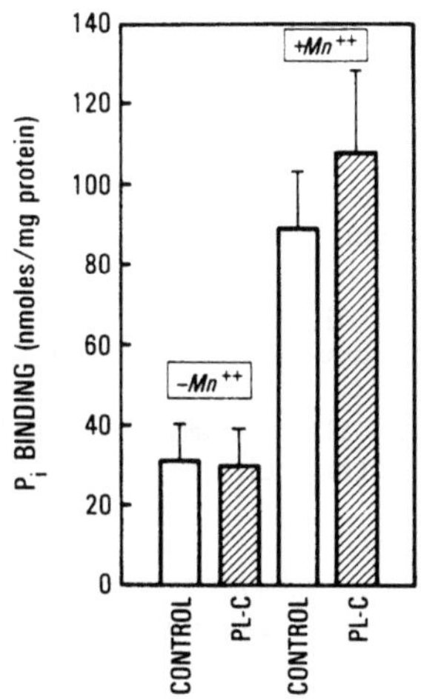

Fig. 5. Phosphate binding to phosphorin isolated from control and alkaline phosphatase depleted brush border membrane vesicles. The phosphate binding was conducted using the standard binding assay described in Fig. 3. The final concentration of Mn^{2+} and Pi was 10μM buffered with MES at pH 7.0. The proteolipid concentration was 1μg per assay. Values are mean ± S.E. of duplicate determinations.

STUDIES ON THE PURIFICATION OF PHOSPHORIN

The purification of phosphorin from all other components in the chloroform:methanol extract of BBMV has proven to be a difficult task. The following is a brief tabulation of efforts we have made to purify phosphorin further.

i) LH-20 chromatography was conducted using increasing methanol concentrations as the elutant. Separation of protein was poor and only ~60% of the applied protein was recovered from the column before the flow rate went to zero. Several other solvents were used without appreciable improvement in results.

ii) LH-60 chromatography. Results were not even as "satisfactory" as with LH-20.

iii) Cellulose Phosphate. Phosphorin always eluted in the solvent front without appreciable purification. Many solvent combinations were tried, because Pi binding was fully recovered (in contrast to many other methods - see below).

iv) Silica chromatography. There's good news and bad news. The good news was that a batch of old silica that had been on the shelf in the lab for 10± years gave good (60-90%) recovery and a several fold purification. However we ran out of the old silica and purchased new silica. The bad news is that every new batch of silica failed to produce purification! Consultation with the library provided information that the chromatographic characteristics of silica vary with the degree of hydration. We tried hydrating new silica to varying degrees, but were never able to achieve a degree of purification approaching that produced by the "old" silica.

v) Florisil chromatography. Only 2% of applied protein could be recovered.

vi) Basic alumina columns. There was nice separation of protein into 5 peaks. But there was no binding of Pi (in the absence or presence of added Me^{2+}) in any of the peaks.

vii) Biobeads SM-2 chromatography in chloroform:methanol (varied ratios). No Pi-binding activity could be found in any eluted fraction.

viii) Fractogel TSK HW-40. The column material was not stable in the presence of chloroform (and chloroform is needed to keep the protein in soluble form).

ix) C-18 reverse phase HPLC. Almost no protein stuck to the column. All protein and Pi-binding came through with the solvent front - i.e., no purification.

IS PHOSPHORIN A COMPONENT OF THE Na^+-Pi SYMPORTER?

This question remains unanswered at this time. Several strategies are being pursued in an attempt to obtain a definitive answer to this question. One strategy has been reported on in preliminary form by Schäli et al.[12] In these studies BBMV were treated with non-ionic detergents. The resulting solubilized membrane proteins were then separated into three fractions, according to their isoelectric points, by chromatofocusing. One fraction contained about 25% of the applied protein and all of the recovered Pi-binding activity. This Pi-binding activity, which was characteristic of that of phosphorin, was enriched three fold. When the various chromatofocusing fractions were reconstituted into liposomes and assayed for Na^+-Pi symport activity, Na^+-Pi symport was found only in the proteoliposomes that had been re-constituted from the protein fraction which had contained the Pi-binding activity of phosphorin. Moreover, the Na^+-Pi symport activity was enriched 2.7 fold in this fraction. This initial, quantitative co-purification of the Pi-binding activity of phosphorin and of Na^+-Pi symport activity lends support to the hypothesis that phosphorin might be a constituent of a more complex renal Na^+-Pi symporter. However, much more rigorous tests of this hypothesis are needed. Never-the-less, a listing of evidence compatible with a role for phosphorin in renal phosphate transport is presented in Table IV.

Table IV
EVIDENCE COMPATIBLE WITH A ROLE FOR
PHOSPHORIN IN RENAL PHOSPHATE TRANSPORT

1. Found in brush border of kidney
2. Binds phosphate with high affinity and specificity
3. Arsenate is a competitive inhibitor
4. Both binding and transport of phosphate are inhibited by:
   A) Divalent ion chelators
   B) Amine-reactive reagents
5. A divalent metal is required for both phosphate transport and phosphate binding to phosphorin
6. Initial studies show co-purification of binding and transport

ACKNOWLEDGEMENTS

This work has been supported by Research Grant AM-32168 from the National Institutes of Health. C. Schäli was a recipient of research fellowships from the Swiss National Foundation for Scientific Research and from the National Kidney Foundation. The expert word processing provided by Mrs. Patricia Spindler is gratefully acknowledged.

REFERENCES

1. N. Hoffman, M. Theis and R. Kinne. Phosphate Transport by Isolated Renal Brush Border Vesicles. Eur. J. Physiol. 362:147-156 (1976).
2. L. Cheng and B. Sacktor. Sodium Gradient-Dependent Phosphate Transport in Renal Brush Border Membrane Vesicles. J. Biol. Chem. 256:1556 (1981)
3. S.A. Kempson, S.V. Shah, P.G. Wenness, T. Berndt, P.H. Lee, L.H. Smith, F.G. Knox and T.P. Dousa. Renal Brush Border Membranes Adaption to Phosphorus Deprivation: Effects of Fasting Versus Low-Phosphorus Diet. Kidney Int. 18:36 (1980)
4. D.B. N. Lee, M.W. Walling, J.G. Siliss and J.W. Coburn. Calcium and Inorganic Phosphate Transport in Rat Colon: Dissociated Response to 1,25-Dihydroxyvitamin D_3. J. Clin. Invest. 65:1326 (1980)
5. R. Stoll, R. Kinne, H. Murer, H. Fleisch and J.P. Bonjour. Phosphate Transport by Rat Renal Brush Border Membrane Vesicles: Influence of Dietary Phosphate, Thyroparathyroidectomy and 1,25-Dihydroxyvitamin D_3. Pfluegers Arch. Eur. J. Physiol. 380:47-52 (1979)
6. J. Folch-Pi and M.B. Lees. Proteolipids, a New Type of Tissue Lipoproteins: Their Isolation from Brain. J. Biol. Chem. 191:807 (1951)
7. A.G. Booth and A.J. Kenny. A Rapid Method for the Preparation of Microvilli from Rabbit Kidney. Biochem. J. 142:575 (1974)
8. R.J. Kessler, D.A. Vaughn and D.D. Fanestil. Phosphate-binding Proteolipid from Brush Border. J. Biol. Chem. 257:14311 (1982)
9. R.J. Kessler and D.A. Vaughn. Divalent Metal is Required for Both Phosphate Transport and Phosphate Binding to Phosphorin, a Proteolipid Isolated from Brush Border Membrane Vesicles. J. Biol. Chem. 259: 9059 (1984)

10. C. Schäli, D.A. Vaughn and D.D. Fanestil. Enzymatic Removal of Alkaline Phosphatase from Renal Brush-Border Membranes. Effect on Phosphate Transport and on Phosphate Binding. Biochim. Biophys. Acta 769:277 (1984)
11. A.N. K. Yusufi, M.G. Low, S.T. Turner and T.P. Dousa. Selective Removal of Alkaline Phosphatase from Renal Brush-Border Membrane and Sodium-Dependent Brush-Border Membrane Transport. J. Biol. Chem. 258:5695 (1983)
12. C. Schäli and D.D. Fanestil. Solubilization and Reconstitution of the Renal Phosphate Transporter. Biochim. Biophys. Acta (in press)

INTRARENAL PHOSPHATE REABSORPTION: ROLE OF NEPHRON HETEROGENEITY

Aviad Haramati

Department of Physiology and Biophysics
Georgetown University School of Medicine
Washington, D.C. 20007

INTRODUCTION

The possibility that phosphate handling may differ between superficial and deep nephrons has been suggested by several investigators over the past 15 years, to explain the discrepancy between phosphate delivery to the distal tubule of superficial nephrons and phosphate excretion in the urine (1,2). Subsequent micropuncture studies have confirmed this notion. During phosphate conservation (by TPTX), phosphate reabsorption was enhanced in deep nephron proximal tubules (3), whereas, when the excretion of phosphate was high, phosphate reabsorption was relatively greater in superficial nephrons (4,5). However, the role of nephron heterogeneity in the regulation of phosphate homeostasis was not clear (6). Moreover, in contrast to the in vivo micropuncture experiments, in vitro studies using isolated, perfused tubule segments failed to find any differences in phosphate transport between superficial and deep nephrons (7,8). To help resolve some of the controversy, this review will highlight recent findings from our laboratory regarding phosphate transport in superficial and deep nephrons and the response to regulators of phosphate reabsorption, which will point to the importance of nephron heterogeneity in the renal handling of phosphate.

TUBULAR CAPACITY FOR PHOSPHATE REABSORPTION

As mentioned earlier, previous in vitro microperfusion studies reported phosphate transport rates for segments of rabbit proximal tubules that were not different between superficial and deep nephrons. However, since the in vivo evidence for nephron heterogeneity of phosphate reabsorption is based on micropuncture studies in the rat, cat and psammomys (3-5,9), it is possible that the rabbit is different from other species in its intrarenal handling of phosphate. Indeed, the renal response to parathyroid hormone and calcitonin differs between the rabbit and the rat (10).

Another explanation may relate to the anatomical differences between superficial and deep nephrons. Phosphate transport is threefold greater in the convoluted than in the straight segment of the proximal tubule

(7,8). Since the ratio of convoluted to straight portions of the proximal tubule is more than 6 times greater in the deep nephron than in the superficial nephron (11), the presence of a longer proximal convoluted segment may represent enhanced reabsorptive capacity. Accordingly, measurements of the phosphate transport capacity for the entire proximal tubule, in vivo, may reveal differences between superficial and deep nephrons that would not be detected in tubule segments in vitro.

To test this hypothesis, the tubular capacity for phosphate reabsorption was measured in the proximal tubule of superficial and deep nephrons (12). Experiments were performed in 20 acutely TPTX Munich-Wistar rats fed a normal phoshate diet (0.7%). Four groups were infused with differing amounts of phosphate (0,2,4 or 6 umol Pi/min) to increase the filtered phosphate load. The sites selected for micropuncture were the superficial early distal tubule and the deep nephron loop of Henle, which reflect delivery from superficial and deep nephron proximal tubules, respectively. Single nephron glomerular filtration rates (SNGFR) in deep nephrons (47 ± 5 nl/min) were significantly higher than in superficial nephrons (28 ± 2 nl/min, $p < .05$). Therefore, the results are expressed as absolute phosphate reabsorption per nl SNGFR.

As shown in figure 1, with increases in plasma phosphate concentration, phosphate reabsorption in the superficial proximal tubule remained stable, whereas phosphate reabsorption in the deep nephron proximal tubule continued to progressively increase. The maximum rate of phosphate transport was 3.2 ± 0.4 pmol/nl in the superficial proximal tubule, which was significantly less than the rate of 5.1 ± 0.1 pmol/nl observed for the deep nephron proximal tubule. Thus, when measured in vivo, the maximum capacity for phosphate reabsorption was greater in deep than in superficial nephrons in TPTX rats.

RESPONSE TO REGULATORS OF PHOSPHATE REABSORPTION

In the following series of studies phosphate reabsorption in superficial and deep nephrons was evaluated during changes in the primary regulators of phosphate transport, namely: parathyroid hormone and dietary phosphate intake.

Parathyroid Hormone

The effect of parathyroid hormone on phosphate reabsorption in superficial and deep nephrons was evaluated in nine Munich-Wistar rats fed a normal phosphate diet (0.7%) (13). All rats were acutely TPTX to control for the effects of endogenous parathyroid hormone and calcitonin. Comparisons were made before and during infusion of parathyroid hormone (synthetic, 1-34 Beckman Instrument, Palo Alto, CA). The hormone was administered with a bolus injection (33 U/kg) and maintained with a sustaining infusion (1 U/kg . min).

During the control period, the fractional phosphate reabsorption was relatively greater from deep than from superficial nephrons ($94.3 \pm 1.1\%$ vs. $84.3 \pm 3.8\%$, $p < .05$). With infusion of parathyroid hormone, phosphate reabsorption declined in both groups to similar levels (72.0 ± 2.1 vs. 70.3 ± 5.2, NS). Since the enhanced reabsorption of phosphate by deep nephrons during the control period was absent in the presence of parathyroid hormone, the studies suggested that the inhibition of phosphate reabsorption by parathyroid hormone tends to be greater in deep nephrons.

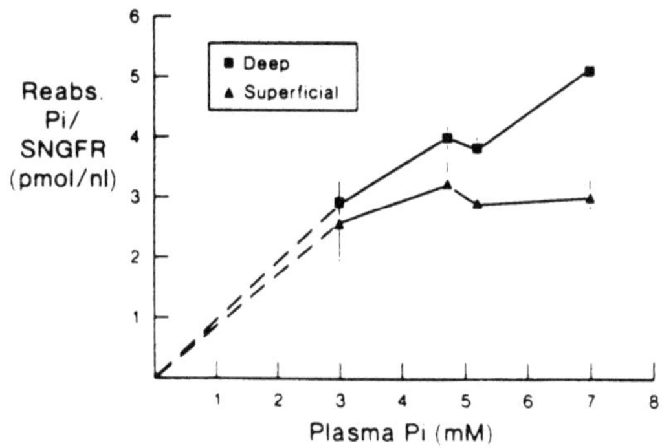

Fig. 1. Relationship between phosphate reabsorption and filtered load in superficial and deep nephron proximal tubules. Both phosphate reabsorption and filtered load are normalized by the SNGFR. Dashes connect lines to the origin. (Reprinted from reference 12 with permission.)

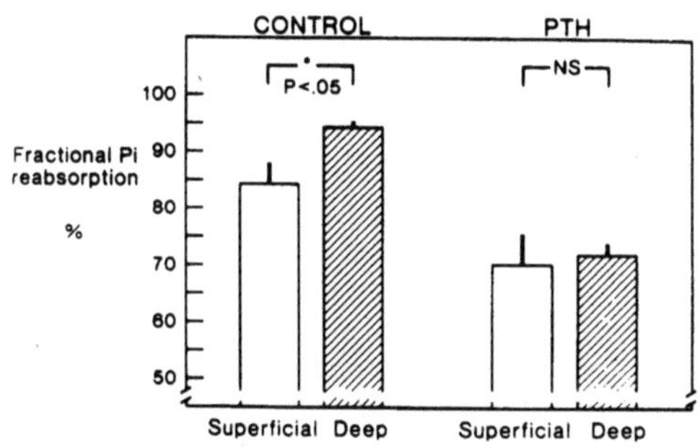

Fig. 2. Fractional phosphate reabsorption in superficial and deep nephron proximal tubules during control and infusion of parathyroid hormone (PTH) in TPTX rats. (Reprinted from reference 13 with permission.)

The effect of parathyroid hormone was further explored in another group of seven TPTX rats that were infused with phosphate (12). Comparisons were made during infusion of phosphate alone (2-6 umol/min) and after an hour during which parathyroid hormone (same preparation and dose as before) was added to the phosphate infusion solutions. The results are illustrated in the figure below.

During the period of phosphate infusion alone, fractional phosphate reabsorption was greater in deep than in superficial proximal tubules (76.9 ± 3.6% vs. 56.7 ± 7.0%, $p < .05$). When parathyroid hormone was added to the infusion of phosphate, fractional phosphate reabsorption was significantly lower in deep than in superficial proximal tubules (8.5 ± 9.5% vs. 30.6 ± 6.8%, $p < .05$). Thus, during administration of hormone alone (previous study), a predominant effect of parathyroid hormone in deep nephrons was suggested but not obvious. However, in the presence of phosphate infusions, the effect of parathyroid hormone to inhibit phosphate reabsorption to a greater extent in deep nephrons is clearly evident.

Dietary Phosphate

This study tested the hypothesis that superficial and deep nephrons respond differently to changes in dietary phosphate intake (14). Comparisons were made between superficial and deep nephrons in rats fed either low (0.07%), normal (0.7%) or high (1.8%) phosphate diet for four days. On the day of the experiment the rats were acutely TPTX. The results are depicted in figure 4.

In rats fed low phosphate diet, fractional phosphate deliveries were similar from superficial and deep nephron proximal tubules (7.5 ± 2.5% vs. 9.1 ± 2.2%, NS). With increases in dietary phosphate intake, fractional delivery of phosphate increased to a greater extent from superficial than from deep nephron proximal tubules (7.5 ± 2.5% to 33.9 ± 4.6%, $p < .05$ in superficial nephrons vs. 9.1 ± 2.2% to 14.0 ± 2.7%, NS in deep nephrons). Thus, changes in dietary phosphate intake primarily affect phosphate reabsorption in the superficial nephron.

CONCLUSIONS

It is evident that nephron heterogeneity exists with regard to the intrarenal reabsorption of phosphate. The intrinsic capacity to reabsorb phosphate is greater in deep nephrons, as is the inhibition of phosphate transport by parathyroid hormone. On the other hand, the adaptation to changes in dietary phosphate intake occurs to a larger extent in superficial nephrons. The anatomic differences between superficial and deep nephron proximal tubules provide a reasonable explanation for some of the functional differences. However, since the precise distribution of deep and superficial nephrons in regard to phosphate transport is unknown, it is difficult to state with certainty to what extent the heterogeneous handling of phosphate contributes to the final urinary excretion of phosphate. Nevertheless, there are instances where nephron heterogeneity could have important implications; for example, during the maturation of renal function. Nephrogenesis begins in the juxtamedullary region and continues with the development of outer cortical nephrons (15). Since the immature animal is in positive phosphate balance and avidly retains phos-

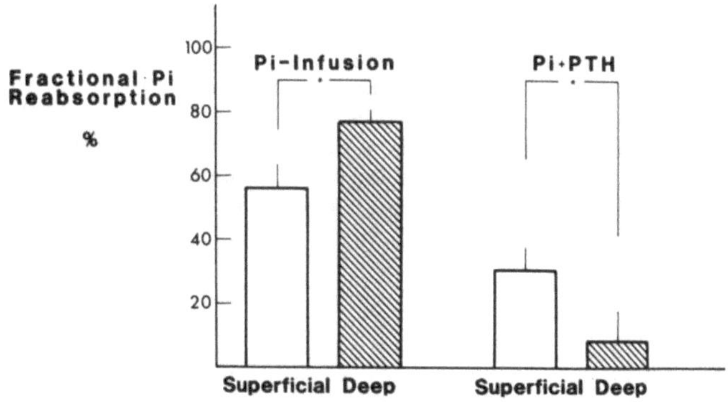

Fig. 3. Effect of parathyroid hormone (PTH) on fractional phosphate reabsorption in superficial and deep nephron proximal tubules in phosphate-infused TPTX rats. (Data from reference 12.)

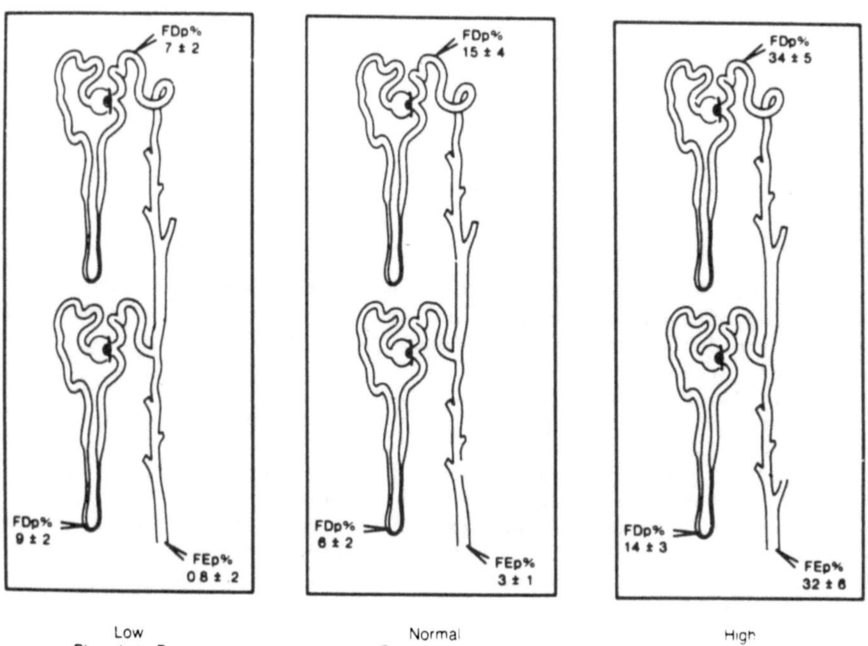

Fig. 4. Fractional delivery of phosphate (FDPi%) from superficial and deep nephron proximal tubules in TPTX rats fed low-, normal-, or high- phosphate diet. (Data from reference 14.)

phate (16), the relative predominance of deep nephrons (with a higher reabsorptive capacity) may contribute to the enhanced reabsorption of phosphate. Furthermore, in view of the minimal change that occurs in deep nephron phosphate reabsorption during an increase in dietary phosphate intake, it is possible that an immature animal would display a blunted adaptive response to high phosphate diet. In a related issue, the phosphaturic response to parathyroid hormone in immature rats has been shown to be significantly blunted compared to adult rats (17). To what extent that is a consequence of changes in deep nephrons is yet to be investigated. Clearly, although the presence of nephron heterogeneity of phosphate reabsorption has been established, our understanding of its significance for phosphate homeostasis is just emerging.

ACKNOWLEDGEMENTS

The work described was performed during the author's tenure at the Department of Physiology, Mayo Medical School, Rochester, MN, and was supported by National Science Foundation Grant PCM-83-08724 and by a grant from the Reimann Foundation.

The author gratefully acknowledges Terri Berndt, John Haas, Susan Mulroney and Marcy Onsgard for providing technical assistance, and Holly Hargreaves for typing the manuscript.

Address reprint requests to Dr. Aviad Haramati, Department of Physiology and Biophysics, Georgetown University School of Medicine, 3900 Reservoir Road, N.W., Washington, D.C. 20007.

REFERENCES

1. Amiel, C., H. Kuntziger, and G. Richet. Micropuncture study of handling of phosphate by proximal and distal nephron in normal and parathyroidectomized rat: Evidence for distal reabsorption. Pflugers Arch. 317:93-109, 1970.
2. Boudry, J., U. Troehler, M. Touabi, H. Fleisch, and J. Bonjour. Secretion of inorganic phosphate in the rat nephron. Clin. Sci. Mol. Med. 48:475-489, 1975.
3. Haas, J., T. Berndt, and F. Knox. Nephron heterogeneity of phosphate reabsorption. Am. J. Physiol. 234:F287-F300, 1978.
4. Goldfarb, S. Juxtamedullary and superficial nephron phosphate reabsorption in the cat. Am. J. Physiol. 239:F336-F342, 1980.
5. Knox, F., J. Haas, T. Berndt, G. Marchand, and S. Youngberg. Phosphate transport in superficial and deep nephrons in phosphate-loaded rats. Am. J. Physiol. 233:F150-F153, 1977.
6. Haramati, A. and F. Knox. Is phosphate reabsorbed by the distal nephron? Min. Electrolyte Metab. 6:165-173, 1981.
7. McKeown, J., P. Brazy, and V. Dennis. Intrarenal heterogeneity for fluid, phosphate, and glucose absorption in the rabbit. Am. J. Physiol. 237:F312-F318, 1979.
8. Brazy, P., J. McKeown, R. Harris, and V. Dennis. Comparative effects of dietary phosphate, unilateral nephrectomy, and parathyroid hormone on phosphate transport by the rabbit proximal tubule. Kidney Int. 17:788-800, 1980.

9. deRouffignac, C., F. Morel, and N. Roinel. Micropuncture study of water and electrolyte movement along the loop of Henle in psammomys with special reference to magnesium, calcium and phosphorus. Pflugers Arch. 344:309-326, 1973.
10. Berndt, T. and F. Knox. Effects of parathyroid hormone and calcitonin on electrolyte excretion in the rabbit. Kidney Int. 17:473-478, 1980.
11. Woodhall, P., C. Tisher, C. Simonton, R. Robinson. Relationship between para-aminohippurate secretion and cellular morphology in rabbit proximal tubules. J. Clin. Invest. 61:1320-1329, 1978.
12. Haramati, A. Tubular capacity for phosphate reabsorption in superficial and deep nephrons. Am. J. Physiol. 248:F729-F733, 1985.
13. Haramati, A., J. Haas, and F. Knox. Nephron heterogeneity of phosphate reabsorption. Effect of parathyroid hormone. Am. J. Physiol. 246:F155-F158,1984.
14. Haramati, A., J. Haas, and F. Knox. Adaptation of deep and superficial nephrons to changes in dietary phosphate intake. Am. J. Physiol. 244:F265-F269, 1983.
15. Oliver, J. A Quantitative Study Of Developmental And Evolutionary Mammalian Renal Architectonics. Hoeber Medical Division, Harper and Row, New York, 1968.
16. Spitzer, A. F. Kaskel, L. Feld, H. Trachtman, V. Johnson, Y. Choi and A. Kumar. Regulation of phosphate homeostasis during growth. Sem. Nephrol. 3:87-93, 1983.
17. Webster, S. and A. Haramati. Developmental changes in the phosphaturic response to parathyroid hormone in the rat. Am. J. Physiol. (in press), 1985.

RESISTANCE TO THE PHOSPHATURIC EFFECT OF PARATHYROID HORMONE DURING PHOSPHATE CONSERVATION IS DUE TO ENHANCED PHOSPHATE REABSORPTION BY THE PARS RECTA

Franklyn G. Knox, Theresa J. Berndt and John A. Haas

Department of Physiology and Biophysics
Mayo Medical School, Rochester, Minnesota

INTRODUCTION

Studies of the renal adaptation to changes in the dietary intake of phosphate have provided important insights into the mechanisms of regulation of phosphate transport by the renal tubule. Tröhler, Bonjour and Fleisch (1) and Steele and DeLuca (2) showed that phosphate deprivation resulted in marked increases in phosphate reabsorption which were independent of plasma phosphate and parathyroid hormone (PTH). Studies utilizing several different approaches support the concept that adaptive changes occur in the proximal tubule. Studies utilizing microperfusion, micropuncture, brush border membrane vesicles and isolated tubules all illustrate enhanced phosphate transport by proximal tubules in animals fed a low phosphate diet (3-6). Further, rats fed a low phosphate diet provide a model of resistance to the phosphaturic responses to hormones (7). In the case of PTH, where the expected increases in cAMP excretion and hypocalciuria occur, there is little or no increase in phosphate excretion.

Most investigators thought it was likely that the resistance to the phosphaturic effects of PTH occurred in the proximal convoluted tubule. This is the major site of phosphate reabsorption and a predominant site of enhanced transport in response to dietary phosphate deprivation. However, recent micropuncture results suggest that the proximal convoluted tubule may respond to PTH in a normal fashion, and that the resistance to the phosphaturic effects of the hormone may be due to enhanced reabsorption of phosphate by the pars recta portion of the proximal tubule (8, 9). Phosphate transport by brush border membrane vesicles is also inhibited by PTH but without a phosphaturia in phosphate-deprived animals (10). The possibility of a relative hypo-responsiveness of the pars recta to the phosphaturic effect of PTH in phosphate deprivation is supported by in vitro microperfusion studies of rabbit tubules (11).

Although much less well studied, hypocapnic rats display similar characteristics. The phosphaturic effect of parathyroid hormone is markedly blunted by respiratory alkalosis (12).

In a recent study, Quamme and Wong demonstrated that an alkaline luminal pH enhanced proximal phosphate reabsorption and this enhanced phosphate reabsorption was inhibited by parathyroid hormone (13). The observation that parathyroid hormone inhibits proximal phosphate reabsorption at an alkaline pH is somewhat surprising since respiratory alkalosis blunts the phosphaturic response to parathyroid hormone (12) and a blunted response might also be expected in the proximal tubule. Therefore, the objective of the present study was to compare the effects of parathyroid hormone on segmental phosphate reabsorption during respiratory alkalosis with the effects of feeding a low phosphate diet.

METHODS

Male Sprague-Dawley rats weighing 250-400g body weight were fed either a normal phosphate diet (0.7% P), or a low phosphate diet (0.07% P) (ICN Pharmaceuticals) for three days prior to the experiment. Animals were allowed food and water ad libitum. Rats were anesthetized with Inactin, 100 mg/kg body weight. All rats were acutely TPTX by heat cautery, and a tracheostomy was performed. Catheters were inserted into the jugular veins for infusions, and into the carotid artery for blood sampling and blood pressure monitoring. Body temperature was maintained between 36°C and 38°C with a heated table. Rats were infused with isoncotic albumin at a rate equal to 1% of their body weight for 1 hour. This infusion was then replaced by 0.9% NaCl at a rate equal to 4% body weight per hour. One hour after initiation of this infusion, the rate was decreased to equal the urine flow rate of approximately 4 ml/hr. The animals were also infused with 6% inulin at a rate of 2 ml/hr. All rats were then mechanically hyperventilated at a rate of approximately 300 cc/min. Blood P_{CO_2} was controlled by changing the composition of inspired air. After a 90 minute equilibration period, a flank incision was made, the kidney was placed in a holder, the left ureter was catheterized with PE 50 tubing, and the kidney was prepared for micropuncture. Lissamine green dye (5%) was injected to identify late proximal and early distal tubules. Three protocols were performed:

<u>Group 1. Normocapnia + PTH (n = 11)</u>

These rats were fed a normal phosphate diet and were respired with 10% CO_2 and 90% oxygen to achieve normal blood acid-base parameters. Two hours after TPTX, a control urine collection was initiated and free flow micropuncture collections were taken from superficial late proximal and early distal tubules. Parathyroid hormone (Synthetic 1-34, Beckman Bioproducts) was then administered with a priming dose of 33 U/kg and a sustaining dose of 1 U/kg·min. Recollection micropuncture was performed one hour after initiating the hormone infusion.

<u>Group 2. Respiratory Alkalosis + PTH (n = 9)</u>

Rats were fed normal phosphate diet and were respired with 100% oxygen to induce respiratory alkalosis. The protocol was identical to Group 1.

<u>Group 3. Low Phosphate Diet + PTH (n = 11)</u>

These rats were fed a low phosphate diet. The protocol was identical to Group 1.

Inulin concentrations in tubular fluid, urine, and plasma were determined by the microfluorometric method of Vurek and Pegram (14). The volume of tubule fluid was measured with a 1 µl constant bore capillary.

Phosphate concentrations in plasma, urine, and tubule fluid were determined by the Chen method (15). Sodium and potassium were measured by flame photometry (Instrumentation Laboratory, Inc., Lexington, MA). Calcium was determined by atomic absorption spectrometry. Blood pH was determined using a blood gas analyzer (Instrumentation Laboratory, Inc., Lexington, MA). Statistical comparisons were made using the student's paired or group t-test as appropriate. All results are presented as means $\pm$ 1 SE. Significance was designated as $p<0.05$.

RESULTS

Table 1 summarizes the clearance data for the micropunctured kidney comparing the normal and respiratory alkalotic groups, and the rats fed a low phosphate diet.

Table 1. Effect of Parathyroid Hormone on Clearance Parameters

	NPD, Normocapnic[1] (n=11)		NPD, Respiratory[1] Alkalosis (n=9)		LPD, Normocapnic[2] (n=11)	
	C	PTH	C	PTH	C	PTH
URINE						
GFR ml/min	1.4	1.3	1.3	1.3	0.9	1.1
$\pm$ SE	$\pm$.2	$\pm$.1	$\pm$.2	$\pm$.3	$\pm$.1	$\pm$.1
FEp%	2.1	36.6*	1.4	5.9	0.8	4.8*
$\pm$SE	$\pm$0.5	$\pm$5.0	$\pm$0.9	$\pm$2.2	$\pm$0.1	$\pm$1.3
BLOOD						
P_p mM	2.68	2.40*	1.82	1.99	2.27	2.84*
$\pm$SE	$\pm$.12	$\pm$.11	$\pm$.13	$\pm$.08	$\pm$0.15	$\pm$0.15
P_{Ca} mM	2.09	2.11	2.32	2.24	1.93	1.82
$\pm$SE	$\pm$.13	$\pm$.10	$\pm$0.24	$\pm$0.25	$\pm$.04	$\pm$.10
pH	7.38	7.39	7.45	7.52*	7.40	7.36
$\pm$SE	$\pm$.02	$\pm$.02	$\pm$.01	$\pm$.01	$\pm$0.03	$\pm$.03
P_{CO_2} mmHg	37	38	28	24*	45	44
$\pm$SE	$\pm$2	$\pm$2	$\pm$1	$\pm$1	$\pm$2	$\pm$2
BP mmHg	140	144	148	152	136	126
$\pm$SE	$\pm$4	$\pm$5	$\pm$5	$\pm$5	$\pm$5	$\pm$6

* Significant difference as compared to control animals - paired t-test; ( ) = number of animals; NPD, normal phosphate diet; LPD, low phosphate diet; C, mean of control collections; GFR, glomerular filtration rate; FEp%, fractional excretion of phosphate; P_p, plasma phosphate concentration; P_{Ca}, plasma calcium concentration; pH, plasma pH; P_{CO_2}, plasma CO_2; BP, mean arterial blood pressure.

[1]data from reference 8; [2]data from reference 4

In normocapnic rats fed a normal phosphate diet, parathyroid hormone administration significantly increased the FEp% from 2.1 $\pm$ 0.5 to 36.6 $\pm$ 5.0%. In respiratory alkalotic rats, parathyroid hormone tended to increase the FEp% from 1.4 $\pm$ 0.9 to 5.9 $\pm$ 2.2%, but the change was not significant. In the phosphate deprived rats, PTH significantly increased the FEp% from 0.8 $\pm$ 1% to 4.8 $\pm$ 1.3% (Fig. 1).

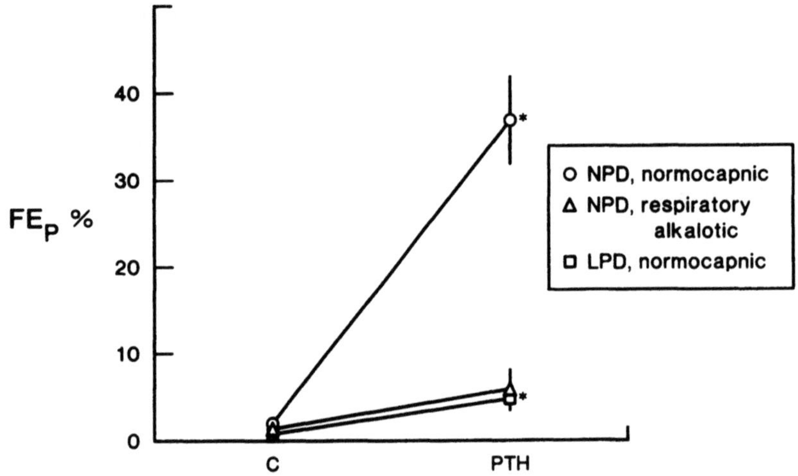

Fig. 1. Effect of PTH on the fractional excretion of phosphate (FE_P%) during phosphate conservation. An asterisk (*) indicates significance, $p<0.05$.

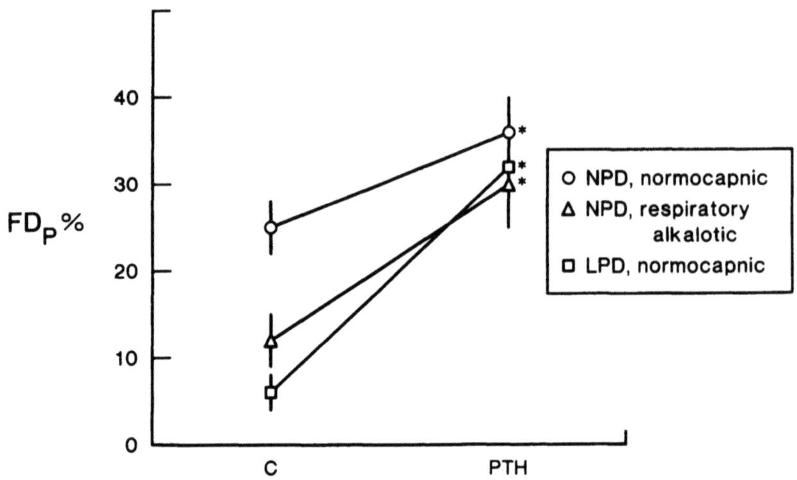

Fig. 2. Effect of PTH on the fractional delivery of phosphate (FD_P%) from the proximal convoluted tubule during phosphate conservation. An asterisk (*) indicates significance, $p<0.05$

Figure 2 presents the fractional delivery of phosphate (FDp%) to the superficial late proximal tubule in the three groups. Parathyroid hormone administration in normocapnic rats fed a normal phosphate diet significantly increased the FDp% from 25.3 ± 3.0% to 36.2 ± 3.8% (16 tubules). In respiratory alkalotic rats, parathyroid hormone significantly increased the FDp% from 12.2 ± 3.1% to 30.3 ± 4.9% (15 tubules). Parathyroid hormone also significantly increased the FDp% from 6.2 ± 1.8% to 32.8 ± 6.7% (8 tubules) in rats fed a low phosphate diet.

Figure 3 presents the FDp% to the superficial early distal tubule in the three groups. Parathyroid hormone administration in normal rats significantly increased FDp% at the early distal tubule from 9.3 ± 3.9% to 38.7 ± 7.4% (11 tubules). In contrast to the normal rats, parathyroid hormone administration in respiratory alkalotic rats increased FDp% from 4.5 ± 1.7% to 12.9 ± 3.4%, $p<0.05$ (13 tubules). Parathyroid hormone significantly increased the FDp% at the early distal tubule in the proximal tubule group from 3.4 ± 0.8% to 9.6 ± 2.2% (12 tubules) in the rats fed a low phosphate diet.

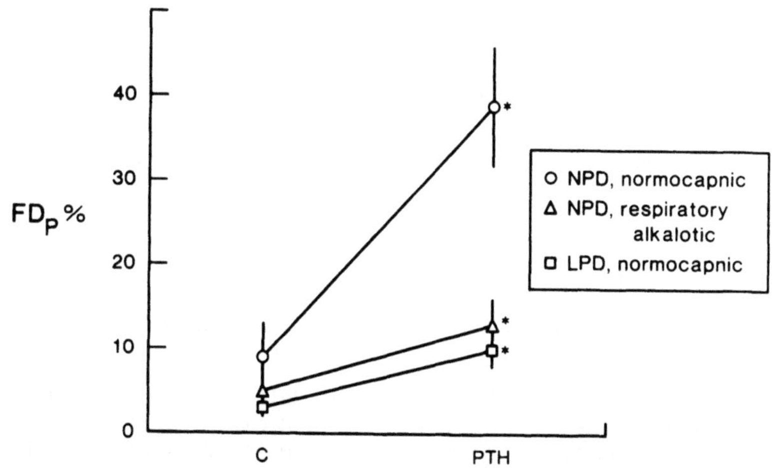

Fig. 3. Effect of PTH on the fractional delivery of phosphate (FDp%) to the early distal tubule during phosphate conservation.

DISCUSSION

Parathyroid hormone inhibited phosphate reabsorption by the superficial proximal tubules in normal, respiratory alkalotic, and phosphate deprived rats so that delivery of phosphate from the late proximal tubule was similar in all of the groups. While the delivery of phosphate from the late proximal tubule was similar in all three groups (Fig. 4), the delivery of phosphate remaining at the early distal tubule puncture site was markedly different between the normal and either the respiratory alkalotic or the phosphate deprived group. In normal rats, parathyroid hormone also inhibited phosphate reabsorption between the late proximal and early distal tubule, while in the respiratory alkalotic and the phosphate deprived groups the fractional delivery of phosphate significantly decreased between the late proximal and early distal tubule. Avid phosphate reabsorption along the pars recta is apparent in the respiratory alkalotic and phosphate restricted rats in the presence of parathyroid hormone. We interpret the enhanced phosphate reabsorption between the late proximal tubule and early distal tubule to be due to reabsorption by the proximal straight tubule since no reabsorption has been demonstrated in other segments of the loop of Henle (16). This enhanced phosphate reabsorption along the pars recta in the presence of PTH is similar in both the respiratory alkalotic and phosphate deprived rats in that the response to parathyroid hormone along the proximal tubule is intact; however, enhanced phosphate reabsorption by the pars recta blunts the phosphaturic response to parathyroid hormone (8, 9). These studies suggest that the pars recta may play an important regulatory role during conditions of phosphate conservation.

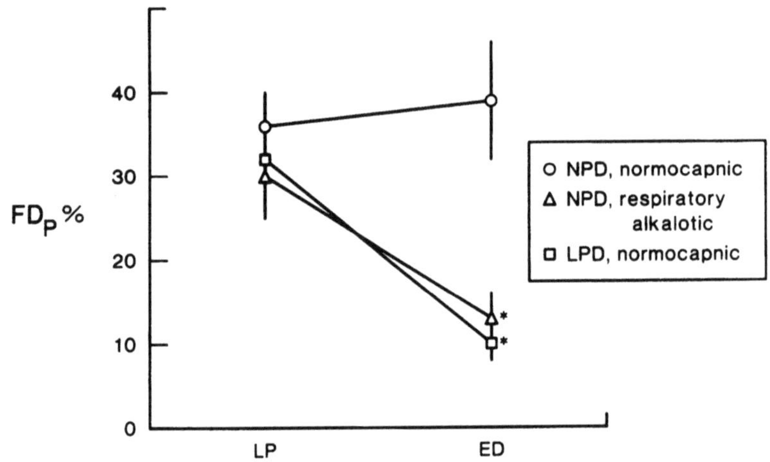

Fig. 4. Effect of PTH on the fractional delivery of phosphate (FD_P%) between the late proximal (LP) and early distal (ED) tubules during phosphate conservation. An asterisk (*) indicates significance, $p<0.05$.

Previous studies concluded that the blunted urinary phosphaturic response to parathyroid hormone during respiratory alkalosis or phosphate deprivation was not due to a defect in the adenylate cyclase-cAMP system since they exhibit a normal increase in urinary cAMP in response to parathyroid hormone (7, 10). While urinary cAMP reflects the whole kidney response to parathyroid hormone administration, the response of discrete nephron segments to adenylate cyclase stimulation during respiratory alkalosis is unknown. The inhibition of phosphate reabsorption along the proximal convoluted tubule by parathyroid hormone during respiratory alkalosis and phosphate deprivation suggests that the adenylate cyclase-cAMP system is intact in the proximal convoluted tubule. However, the enhanced phosphate reabsorption along the pars recta in the presence of parathyroid hormone raises the possibility that respiratory alkalosis and phosphate deprivation may interfere with the adenylate cyclase-cAMP system in the pars recta.

We conclude that the blunted phosphaturic effect of parathyroid hormone during respiratory alkalosis or phosphate deprivation is not due to a blunted inhibition of phosphate reabsorption by the proximal convoluted tubule, but is primarily due to enhanced reabsorption by the pars recta.

ACKNOWLEDGMENT

The authors gratefully acknowledge the technical assistance of Marcy Onsgard and the secretarial assistance of June M. Hanke.

This study was supported by the National Institute of Arthritis, Diabetes, Digestive and Kidney Diseases, AM 19715, and by the Mayo Foundation.

REFERENCES

1. H. Tröhler, J.-P. Bonjour, and H.Fleisch, Inorganic phosphate homeostasis: Renal adaptation to the dietary intake in intact and thyroparathyroidectomized rats. J. Clin. Invest. 57:264 (1976).
2. J. H. Steele and H. F. DeLuca, Influence of dietary phosphorus on renal phosphate reabsorption in the parathyroidectomized rat. J. Clin. Invest. 57:867 (1976).
3. K. J. Ullrich, G. Rumrich, and S. Kloss, Phosphate transport in the proximal convolution of the rat kidney. I. Tubular heterogeneity, effect of parathyroid hormone in acute and chronic parathyroidectomized animals and effect of phosphate. Pflügers Arch. 372:269 (1977).
4. J. A. Haas, T. J. Berndt, A. Haramati, and F. G. Knox, Nephron sites of action of nicotinamide on phosphate reabsorption. Am. J. Physiol. 246:F27 (1984).
5. R. Stoll, R. Kinne, and H. Murer, Effect of dietary phosphate intake on phosphate transport by isolated rat renal brush border vesicles. Biochem. J. 180:465 (1979).
6. S. A. Kempson, and T. P. Dousa, Phosphate transport across renal cortical brush border membrane vesicles from rats stabilized on a normal, high or low phosphate diet. Life Sci. 24:881 (1979).
7. T. H. Steele, Renal resistance to parathyroid hormone during phosphorus deprivation. J. Clin. Invest. 58:1461 (1976).
8. J. A. Haas, T.J. Berndt, A. Haramati, and F.G. Knox, Nephron sites of action of nicotinamide on phosphate reabsorption. Am. J. Physiol. 246:F27 (1984).
9. E. Pastoriza-Munoz, D. R. Mishler and C. Lechene, Effect of phosphate deprivation on phosphate reabsorption in rat nephron: Role of PTH. Am. J. Physiol. 244:F140 (1983).
10. K. Hruska and M. Hammerman, Parathyroid hormone inhibition of phosphate transport in renal brush border vesicles from phosphate-depleted dogs. Biochim. Biophys. Acta. 645:351)1981).
11. P. C. Brazy, J. W. McKeown, R. H. Harris, and V. W. Dennis, Comparative effects of dietary phosphate, unilateral nephrectomy, and parathyroid hormone on phosphate transport by the rabbit proximal tubule. Kidney Int. 17:788 (1980).
12. A. Hoppe, M. Metler, T. J. Berndt, F. G. Knox, and S. Angielski. Effect of respiratory alkalosis on renal phosphate excretion. Am. J. Physiol. 243:F471 (1982).

13. A. A. Quamme and N. L. M. Wong. Phosphate transport in the proximal convoluted tubule: Effect of intraluminal pH. Am. J. Physiol. 246:F323 (1984).
14. C. G. Vurek and S. E. Pegram. Fluorometric method for the determination of nanogram quantities of inulin. Anal. Biochem. 10:409 (1966).
15. P. Chen, T. Toribara, and H. Warner, Microdetermination of phosphorus. Anal. Chem. 28:1756 (1956).
16. A. S. Rocha, J. B. Magaldi and J. P. Kokko, Calcium and phosphate transport in isolated segment of rabbit Henle's loop. J. Clin. Invest. 59:975 (1977).
17. T. J. Berndt and F. G. Knox, Nephron site of resistance to the phosphaturic effect of parathyroid hormone during respiratory alkalosis. Am. J. Physiol. (in press, 1985).

BRUSH BORDER MEMBRANE PHOSPHATE

TRANSPORT EFFECTS OF VOLUME EXPANSION

Jules B. Puschett, Joseph Liput and Christopher Galya

Renal-Electrolyte Division, Department of Medicine
University of Pittsburgh, School of Medicine, and the
Presbyterian-University and Veterans Administration
Hospitals, Pittsburgh, PA.

Volume expansion reproducibly results in the development of phosphaturia both in the experimental animal[1-4] and in man.[5,6] Although there is evidence that the expansion procedure reduces phosphate reabsorption at more than one site within the nephron,[7] the proximal tubule is considered to be a major source of the excreted phosphate, based largely upon micropuncture experiments.[3,8-10] This would be expected, since the majority of filtered phosphate appears to be reabsorbed in the proximal nephron,[11] at least in animals with intact parathyroid glands.[12,13] When the parathyroid glands are removed however, phosphate transport in later nephron segments appears to be enhanced such that volume expansion is less phosphaturic in parathyroidectomized than in intact animals.[1,13,14]

The initial and potentially rate-limiting step in the renal tubular reabsorption of phosphate is its uptake across the luminal brush border of the renal tubular cell.[15] The current studies were performed to verify proximal tubular participation in the phosphaturia of volume expansion, to determine if the in vivo infusion of saline could result in measurable in vitro alterations in brush border membrane (BBM) vesicle phosphate transport, and to determine the relationship between brush border and whole kidney phosphate transport.

Methods

Experiments were performed in female Sprague-Dawley rats weighing 150-250 grams. The animals were anesthetized with sodium pentobarbital, a bladder catheter was placed and carotid cannulation was performed for blood sampling purposes. An experimental and a control rat were studied simultaneously on each experimental day. An inulin infusion was begun in saline at .04 ml/min and two control urine collections were obtained. Blood was then drawn, following which volume expansion with saline was performed over a thirty minute period in the experimental animal. A second blood sample was then taken and three thirty minute post-expansion urine collections were obtained following which a final blood was taken. Blood and urine specimens were analyzed for inulin, sodium, phosphate and calcium by methods previously described.[16,17] The animals were then sacrificed and the kidneys were removed for the performance of BBM vesicle transport studies. Vesicles were prepared with sodium-dependent phosphate uptake determined by a calcium precipitation technique previously reported from this laboratory.[16,17] Purity of the BBM preparation was determined utilizing an assay for gamma glutamyl transpeptidase. Experiments in which enrichment was less than 8:1 (vesicles:homogenate) were excluded.

Results

Shown in figure 1 is a representative experiment in which the measurement of the BBM transport of phosphate in an expanded rat is compared to that in a control rat which was studied simultaneously. It can be seen from this diagram that phosphate uptake was substantially inhibited in the vesicles obtained from the expanded animal as compared to the control rat. This is especially evident at the 0.5 and 1 minute time points. Furthermore, phosphate transport was quantitatively very similar at the 120 minute equilibrium time point suggesting equality of vesicle size.

In figure 2 are provided the clearance data for those rats whose BBM studies were presented in figure 1. There were no changes in inulin clearance (C_{In}) in either animal. Percentage sodium excretion rose from less than 1% to better than 20% of filtered load in the expanded rat, while it remained less than 1% in the control rat. Furthermore, percentage phosphate excretion rose by four-fold in the expanded animal whereas it varied little in the control rat. In addition, the plasma calcium value was unchanged in the expanded animal (it was 4 mEq/L in both the control and experimental phases of the experiment). Similar findings have been obtained in several additional experiments.

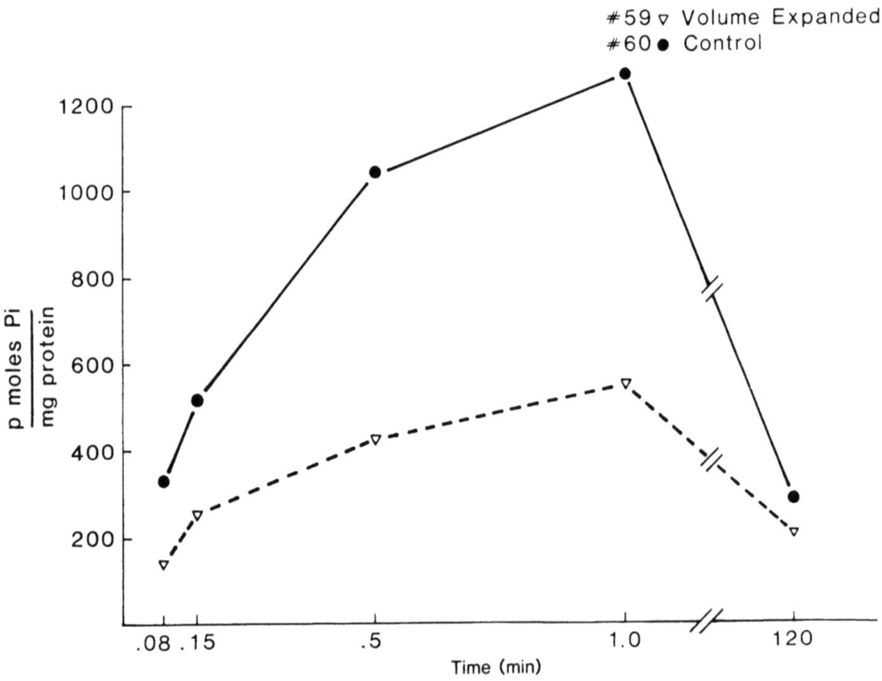

Fig. 1: Brush border vesicle uptake of inorganic phosphate (plotted on the ordinate) at the time periods represented on the abscissa. The closed circles depict the data in the control rat and the open triangles represent the observations obtained in the expanded animal. Each data point represents the mean of triplicate determinations.

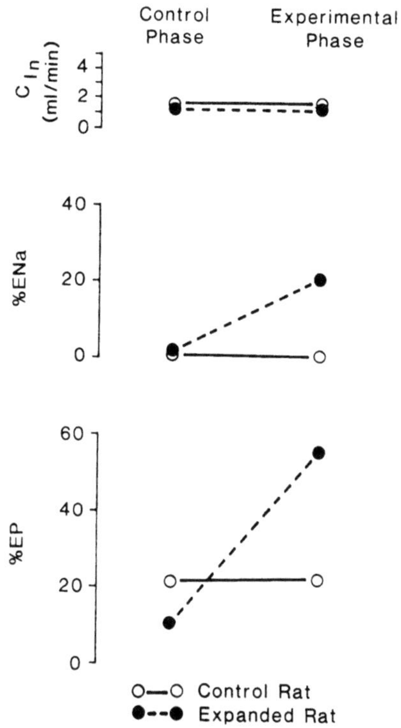

Fig. 2: Clearance data obtained in paired volume expanded and control rats. The expansion procedure resulted in an increase in percentage sodium (%ENa) and phosphate excretion (%EP), but no change in glomerular filtration rate (C_{In}). There were no changes in these parameters in the control rat.

Discussion

According to current concepts, reabsorbed phosphate is first transported across the luminal brush border membrane by a process that probably involves one or more membrane proteins which act as transporters.[15] Given the evidence in support of the view that phosphate is rebsorbed in a major way proximally,[11] and that volume expansion inhibits proximal phosphate reabsorption,[3,4,7,10] it seemed reasonable to test the hypothesis that luminal phosphate uptake in the proximal nephron would be altered by volume expansion. The BBM vesicle technique provided the opportunity to evaluate this hypothesis and the preliminary studies reported here indicate that this is indeed correct. In the data presented from a representative study (figures 1 and 2), 10% (BW) volume expansion not only causes a major phosphaturia but inhibited phosphate uptake in BBM vesicles prepared from an expanded rat compared to the values obtained in its paired control.

We have previously performed studies in the dog in which either parathyroid hormone (PTH) or cyclic AMP was infused in vivo, followed by the harvesting of the kidneys of these animals for measurement of BBM vesicle phosphate transport.[16] In these studies, the left kidney was removed to be utilized to obtain control observations. The hormone or nucleotide was then administered and the right (experimental) kidney was then harvested. Both following the injection of PTH and 8-bromo cyclic AMP, phosphate uptake in the vesicles obtained from the experimental kidney was inhibited when compared to the values determined in the control kidney.[16] In the current studies, the protocol involved determining, instead, the transport of phosphate in kidneys removed from expanded animals compared to that measured in control rats. The data demonstrate that the changes produced in proximal tubule cellular transport by volume expansion persist after the kidneys have been removed and are consistent with other data from this laboratory.[18] In a series of studies performed by Pitts et al in rabbits, proximal straight tubular segments removed from expanded animals demonstrated reduced fluid and phosphate transport compared to those of paired controls.[18] Thus, the data presented in this report and the observations of Pitts et al, suggest that the expansion procedure results in a change in membrane transport which persists even after the removal of the kidney from its normal milieux.

Acknowledgement

The authors thank Diane Puschett and Suzanne Manning for their invaluable technical assistance. These studies were supported, in part, by the Veterans Administration.

References

1. S. G. Massry, J. W. Coburn and C.R. Kleeman. The influence of extracellular volume expansion on renal phosphate reabsorption in the dog. J. Clin. Invest. 48:1237 (1969).
2. W. N. Suki, M. Martinez-Maldonado, D. Rouse and A. Terry. Effect of expansion of extracellular fluid volume on renal phosphate handling. J. Clin. Invest. 48:1888 (1969).
3. J. B. Puschett, Z. S. Agus, D. Senesky and M. Goldberg. Effects of saline loading and aortic obstruction on proximal phosphate transport. Am. J. Physiol. 223:851 (1972).
4. J. B. Puschett, H. A. Feldman and Diane B. Sylk. Dissociative effects of volume expansion on renal tubular ionic transport. Mineral Electrolyte Metabol. 2:7-19 (1979).
5. J. B. Puschett, M. Goldberg. The relationship between the renal handling of phosphate and bicarbonate. J. Lab. Clin. Med. 73:956 (1969).
6. T. H. Steele. Increased urinary phosphate excretion following volume expansion in normal man. Metabolism 19:129 (1970).
7. C. Amiel. Sites of renal tubular reabsorption of phosphate, in: "Renal Handling of Phosphate," S. G. Massry and H. Fleisch, eds, Plenum Press, New York, pp. 39-57, (1980).
8. H. Kuntziger, C. Amiel and C. Gaudebout. Phosphate handling by the rat nephron during saline diuresis. Kidney International 2:318 (1972).
9. P. Poujeol, D. Chabardes, N. Roinel and C. DeRouffignac. Influence of extracellular fluid volume expansion on magnesium, calcium and phosphate handling along the rat nephron. Pflügers Arch 365:203 (1976).
10. A. Frick. Proximal tubular reabsorption of inorganic phosphate during saline infusion in the rat. Am. J. Physiol. 223:1034 (1972).
11. C. Amiel, H. Kuntziger and G. Richet. Micropuncture study of handling of phosphate by proximal nephron in normal and parathyroidectomized rat. Evidence for distal reabsorption. Pflügers Arch 317:93 (1970).
12. P. C. Fernandez and J. B. Puschett. Proximal tubular actions of metolazone and chlorothiazide. Am. J. Physiol. 225:954 (1973).
13. L. H. Beck and M. Goldberg. Mechanism of the blunted phosphaturia in saline-loaded thyroparathyroidectomized dogs. Kidney International. 6:18 (1974).

14. A. Frick and I. Durasin. Effect of expansion of extracellular fluid volume on the maximal reabsorptive capacity for inorganic phosphate in parathyroidectomized and intact rats. Pflügers Arch 370:115 (1977).
15. F. G. Knox, A. Hoppe, S. A. Kempson, S. V. Shah and T. P. Dousa. Cellular mechanisms of phosphate transport, in: "Renal Handling of Phosphate" S. G. Massry and H. Fleisch, eds, Plenum Press, New York, pp. 79-114, (1980).
16. S. A. Kempson, J. C. Kowalski and J. B. Puschett. Inhibition of renal brush border phosphate transport and stimulation of renal gluconeogenesis by cyclic AMP and parathyroid hormone. Biochemical Pharm. 32:1533 (1983).
17. S. A. Kempson, J. C. Kowalski and J. B. Puschett. Direct effect of metolazone on sodium-dependent transport across the renal brush border membrane. J. Lab. Clin. Med. 101:308 (1983).
18. T. O. Pitts, T. Buckenheimer, T. Brown, F. Shuler and J. B. Puschett. Volume expansion - induced phosphaturia in the rabbit - participation of the proximal straight tubule. Abstracts of the 17th Annual Meeting of the American Society of Nephrology, Washington, D.C., September, 1984, p. 25A.

THE DIRECT EFFECT OF 1,25-DIHYDROXYCHOLECALCIFEROL ON MEMBRANE PHOSPHOLIPID COMPOSITION AND PHOSPHATE TRANSPORT

B.R.C. Kurnik, M. Huskey, and K.A. Hruska

Renal Division
Washington University and The Jewish Hospital of St. Louis
St. Louis, Missouri

INTRODUCTION

Current evidence suggests that the mechanism of action of several hormones may be mediated through alterations in membrane lipid composition[1-4]. Both the content and the degree of fatty acid unsaturation of phosphatidylcholine (PC) in renal proximal tubular and duodenal brush border membranes is increased after 1,25-dihydroxycholecalciferol (1,25$(OH)_2D_3$) administration to animals[5,6]. In both the intestine[5,7] and the kidney[8] indirect evidence suggests that these lipid changes may be important in stimulating calcium and phosphate transport. In recent studies performed in this laboratory fluorescent phospholipid derivatives have been used to demonstrate that 1,25$(OH)_2D_3$ rapidly and directly stimulates the transfer of PC from liposomes to renal brush border membrane vesicles (BBMV)[9]. The following studies were performed to investigate the effects of 1,25$(OH)_2D_3$ induced membrane phospholipid alterations on phosphate transport.

MATERIALS AND METHODS

Renal BBMV were prepared from either normal or vitamin D deficient rats by a technique of divalent cation precipitation and differential centrifugation[10]. The final vesicle preparation was suspended in 150 mM KCl, 2 mM Mes-Tris pH 7.5.

Small unilamellar liposomes, 25-75 nm in diameter, were prepared by a method of bath sonication and ultracentrifugation[11]. The liposomes were composed of dioleoylphosphatidylcholine (DOPC) or ß-linoleic phosphatidylcholine (ß-linoleic-PC) with or without fluorescent phospholipid derivatives. The liposomes also contained $1,25(OH)_2D_3$ at specificied concentrations. The fluorescent phospholipid derivatives contained fluorescent groups in either the #2 fatty acid (NBD-PC) or head group (N-rhodamine-PE) positions. The concentrations were chosen so as to "self-quench", that is, the liposomes alone had a minimal relative fluorescence (.015, NBD-PC; .04, N-Rh-PE). Self-quenching concentrations were determined to be 30 mol% for NBD-PC and 20 mol% for N-Rh-PE. A decrease in the fluorescent phospholipid concentration was measurable as an increase in fluorescence. This could occur through either a direct transfer of the fluorescent phospholipid molecules from the liposome to the vesicle membranes or fusion of liposomes with the BBMV. NBD-PC has previously been shown to transfer spontaneously between membranes while N-Rh-PE does not[12]. Electron micrographs and chromatographic ellusion patterns of the liposomes were the same and independent of hormone or phospholipid composition.

Following their preparation, liposomes (50 µg of lipid/ml) and BBMV (1 mg vesicle protein/ml), were incubated together and subsequently used for three types of studies (Figure 1).

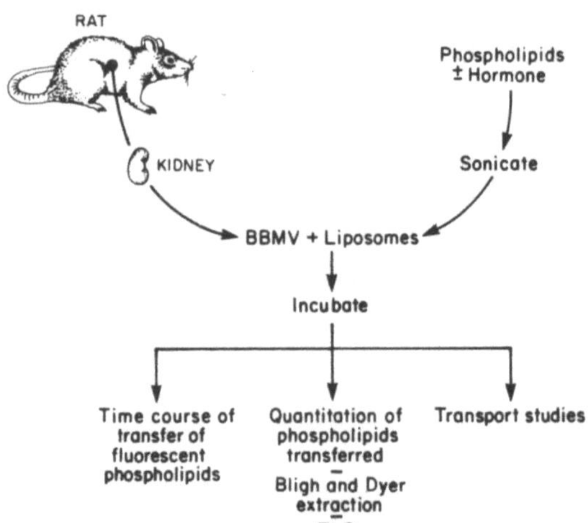

Figure 1 Rat renal BBMV and liposomes of various composition were prepared as described in the Methods. Three types of studies were performed: 1) the time course of transfer of fluorescent phospholipid, 2) the vesicle phospholipids were extracted and separated for quantitation, 3) the vesicles were separated from the liposomes and tranport measured.

In the first type of study, the transfer of fluorescent phospholipid from various liposomes to the BBMV was recorded as an increase in relative fluorescence over time. Fluorescence was measured using an SLM Spectrofluorometer (Urbana, IL) set at the appropriate wavelengths. Fluorescent intensity was continuously monitored at the rate of 1 measurement every second, relayed to a computer and recorded.

In a second set of experiments, vesicle phospholipid content was evaluated following incubation with various liposomes. To quantitate the fluorescent phospholipid transferred, vesicles were incubated with liposomes composed of 30 mol% NBD-PC, 2 mol% N-Rh-PE, 68 mol% DOPC with or without 1,25$(OH)_2D_3$ for 60 min at 0° C. The vesicles were then pelleted by centrifugation (35,000 x g), the supernatant (free liposomes) discarded, and the pellet washed two more times. The pellets containing vesicle lipid were extracted by the method of Bligh and Dyer,[13] using chloroform:methanol:.1 N HCl (1:2:.8). Lipid extracts were analyzed by chromatography on Silica Gel 60 thin-layer plates (E. Merck, Darmstadt, Germany) in chloroform:methanol:28% ammonium hydroxide (65:35:5) along with standards which included N-Rh-PE and NBD-PC. The spots containing N-Rh-PE and NBD-PC were scraped and re-extracted with chloroform:-methanol:water (1:2:.8). The amount of fluoresent lipid retrieved was quantitated by comparing the amount of fluorescence in a sample to the fluorescence of a standard curve generated using known concentrations of either N-Rh-PE or NBD-PC. Non-fluorescent vesicle phospholipid was quantitated following incubations with liposomes composed of either DOPC or ß-linoleic-PC with or without 1,25$(OH)_2D_3$. After separation, the vescile phospholipids were measured by phosphate assay[14].

In a third set of experiments phosphate and glucose transport were studied. Following incubation with various liposomes for 30 min at 0° C, the vesicles were separated by centrifugation and resuspended in 150 mM KCl, 2 mM Mes-Tris pH 7.5, to achieve a final protein concentration of 100-200 μg/10 λ. The preparation was aliquoted into individual test tubes and used for transport studies as previously described[10].

Fluorescent phospholipids were obtained from Avanti Polar Lipids (Birmingham, Alabama), and periodically monitored for purity by thin-layer chromatography. Vitamin D metabolites were a gift from Hoffman LaRoche (Nutley, New Jersey). Other chemicals were of the highest purity available for commerical sources. All solutions were filtered through a .45 μm Millipore filter on the day of the experiment.

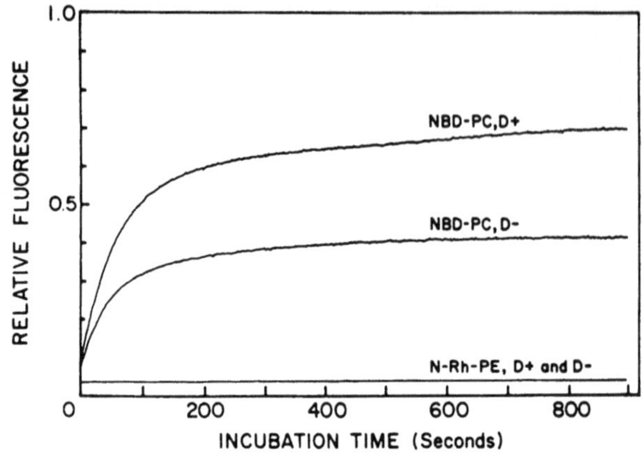

Figure 2 Liposomes containing self-quenching concentrations of either NBD-PC (30 mol%) or N-Rh-PE (20 mol%) and DOPC ± 10^{-7} M 1,25$(OH)_2D_3$ were incubated with BBMV derived from vitamin D deficient rats for 15 min at 0° C. The increase in fluorescence was monitored at the appropriate wavelength (NBD-PC, excitation 467 nm, emission 528 nm; N-Rh-PE, excitation 553 nm, emission 583 nm). Representative example, N=6.

Table 1 Quantitation of Transfer of NBD-PC, and N-Rh-PE from Liposomes to BBMV Prepared From From Vitamin D Deficient Rats

	D-	1,25$(OH)_2D_3$ (10^{-7}M)
NBD-PC, ng/hr (N=6)	1385 ± 265	*2068 ± 263
N-Rh-PE, ng/hr (N=3)	5 ± 3	4 ± 4

Liposomes composed of 30 mol% NBD-PC, 2 mol% N-Rh-PE and 68 mol% DOPC with or without 10^{-7} M 1,25$(OH)_2D_3$ were incubated with BBMV from vitamin D deficient rats at 0° C for one hour. The vesicle phospholipids were extracted and quantitated as described in the "Methods". *D+>D-, p<.05

RESULTS

When liposomes containing 30 mol% NBD-PC were incubated with BBMV prepared from vitamin D deficient rats, an increase in relative fluorescence was immediately detectable which reached a level of .41 in 15 min. Liposomes containing 10^{-7} M 1,25$(OH)_2D_3$ stimulated the initial rate and amount of NBD fluorescence to .70 at 15 min. Minimal fluorescence in the presence or absence of 1,25$(OH)_2D_3$ was detected when BBMV were incubated with liposomes containing 20 mol% N-Rh-PE (Figure 2).

The results of incubating liposomes composed of 30 mol% NBD-PC, 2 mol% N-Rh-PE, and 68 mol% DOPC with BBMV from vitamin D deficient rats and quantitating the vesicle lipids show that 1,25$(OH)_2D_3$ (10^{-7} M) stimulated the transfer of NBD-PC from 1385 ± 265 to 2068 ± 263 ng (Table 1, $p<.05$). There was minimal transfer of N-Rh-PE in the presence or absence of hormone. Following incubation of BBMV with liposomes composed of 10^{-7} M 1,25$(OH)_2D_3$ and DOPC, the vesicle PC content increased from 49.5 ± 2.3 µg/mg protein to 56.9 ± 2.3 µg/mg protein (Table 2). When the liposomes were composed of ß-linoleic-PC and 1,25$(OH)_2D_3$ (10^{-7} M), vesicle PC content increased similarly to 58.5 ± 3.2 µg/mg protein. Concentrations of 1,25$(OH)_2D_3$ of 10^{-11} M and below did not affect the vesicle PC content regardless of the liposome lipid composition.

Table 2 Phosphatidylcholine Content of BBMV Prepared from Vitamin D Deficient Rats Following Incubation with Liposomes

	Vesicles PC Content µg/mg protein/hr
Control (N=10)	49.5 ± 2.3
DOPC + 1,25$(OH)_2D_3$ 10^{-7}M (N=7)	*56.9 ± 2.3
DOPC + 1,25$(OH)_2D_3$ 10^{-11}M (N=3)	49.5 ± 5.2
ß-linoleic-PC + 1,25$(OH)_2D_3$ 10^{-7}M (N=3)	*58.5 ± 3.2
ß-linoleic-PC + 1,25$(OH)_2D_3$ 10^{-11}M (N=5)	48.8 ± 3.4
ß-linoleic-PC + 1,25$(OH)_2D_3$ 10^{-14}M (N=3)	45.5 ± 11

Liposomes composed of either DOPC or ß-linoleic-PC with or without 1,25$(OH)_2D_3$ at the specified concentrations were incubated with BBMV prepared from vitamin D deficient rats for 1 hour at 0° C. The vesicle phospholipids were extracted and quantitated as described in the Methods. *D+>D-, $p<.05$

Table 3 Quantitation of Transfer of NBD-PC from Liposomes to Renal BBMV Prepared from Normal Rats

Group	ng of NBD-PC/hr	N
D-	2119 ± 18	12
$1,25(OH)_2D_3$		
10^{-7}M	*2599 ± 225	7
10^{-11}M	*2630 ± 62	4
10^{-14}M	*2984 ± 146	3
10^{-17}M	*2437 ± 174	4

Liposomes composed of 30 mol% NBD-PC, 70 mol% DOPC ± $1,25(OH)_2D_3$, at specified concentrations, were incubated with BBMV from normal rats at 0° C for one hour. The vesicle phospholipids were extracted and quantitated as described in the "Methods". *D+>D-, $p<.02$

When liposomes containing 30 mol% NBD-PC and 10^{-7} M $1,25(OH)_2D_3$ were added to BBMV prepared from normal rats, the NBD-PC content increased from 2119 ± 18 ng to 2599 ± 225 ng ($p<.02$) (Table 3). Baseline transfer of NBD-PC to BBMV derived from normal rats was significantly greater than to BBMV derived from vitamin D deficient rats ($p<.05$). A dose-response demonstrated that $1,25(OH)_2D_3$ stimulated transfer of NBD-PC at concentrations ranging from 10^{-7} M to 10^{-17} M, with the maximum effect occurring at 10^{-14} M (Table 3).

Phosphate transport was studied into BBMV prepared from vitamin D deficient rats following incubation with liposomes of various compositions. Figure 3 demonstrates that phosphate transport was stimulated from 287 ± 30 to 431 ± 41 pmol/mg protein/15 sec. when the liposomes were composed of 10^{-7} M $1,25(OH)_2D_3$ and ß-linoleic-PC. This effect was not seen if the liposomes were composed of DOPC in the presence or absence of $1,25(OH)_2D_3$. Phosphate uptake at 90 min, representing the steady state, were not different between groups. Phosphate transport was similarly stimulated when the BBMV were prepared from normal rats from 443 ± 33 to 601 ± 42 pmol/mg protein/15 sec (Figure 4). Liposomes containing 10^{-11} M or 10^{-14} M $1,25(OH)_2D_3$ did not stimulate phosphate transport.

Sodium-dependent glucose transport was studied into BBMV from both normal and vitamin D deficient rats. Uptake at 15 sec of incubation was 114 ± 23 pmol/mg protein using BBMV derived from vitamin D deficient

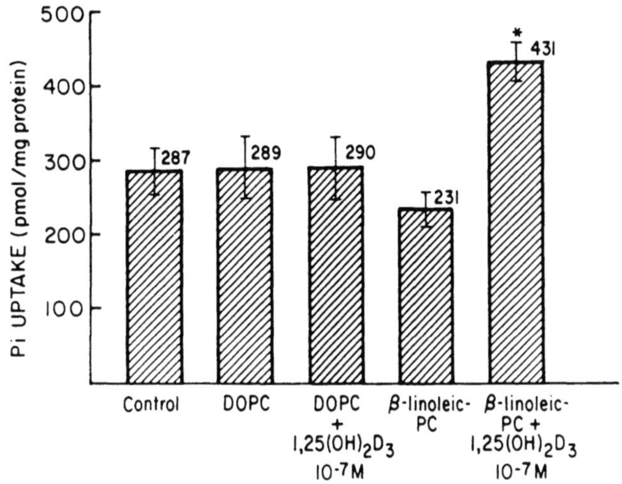

Figure 3 Liposomes composed of either DOPC or β-linoleic-PC with or without 10^{-7} M $1,25(OH)_2D_3$ were prepared and incubated with BBMV from vitamin D deficient rats for 30 min at 0° C. Phosphate transport was measured as described in the Methods. Na-dependent phosphate uptakes at steady state (90 min) were: 183 ± 25, control; 271 ± 29, DOPC; 225 ± 26, DOPC + $1,25(OH)_2D_3$;180 ± 19, β-linoleic-PC; 217 ± 50, β-linoleic-PC + $1,25(OH)_2D_3$. Mean ± SE, N=6 in each group. *Greater than other groups, p<.05.

animals, and 115 ± 13 pmol/mg protein when the vesicles were pre-incubated with liposomes composed of β-linoleic-PC and $1,25(OH)_2D_3$ (10^{-7} M). Uptakes at 90 min of incubation (steady state) were also unchanged between groups (74 ± 8 pmol/mg protein, controls; 56 ± 11 pmol/mg protein, β-linoleic-PC + 10^{-7} M $1,25(OH)_2D_3$). Glucose uptakes were similar when the BBMV were derived from normal animals and were not affected by incubations with liposomes.

DISCUSSION

These studies suggest that the increase in the PC content of both the renal and duodenal brush border membranes, which has been previously reported following in-vivo $1,25(OH)_2D_3$ treatment, may be mediated through a facilitation of the direct transfer of PC to the BBM.

Membrane phospholipid biosynthesis occurs primarily in the endoplasmic reticulum with possible further modifications in the Golgi apparatus. The lipids are then incorporated into the plasma membrane, either by direct transfer via phospholipid exchange or transfer proteins, or by fusion of intracellular lipidic particles with the plasma membrane. The data presented here suggests that fusion is not occurring since only the NBD-PC and not the N-Rh-PE was transferred. Because these fluorescent phospholipid derivatives and their property of spontaneous transfer is not physiologic, the experiments were also performed using liposomes composed of DOPC or β-linoleic-PC with or without $1,25(OH)_2D_3$. The results

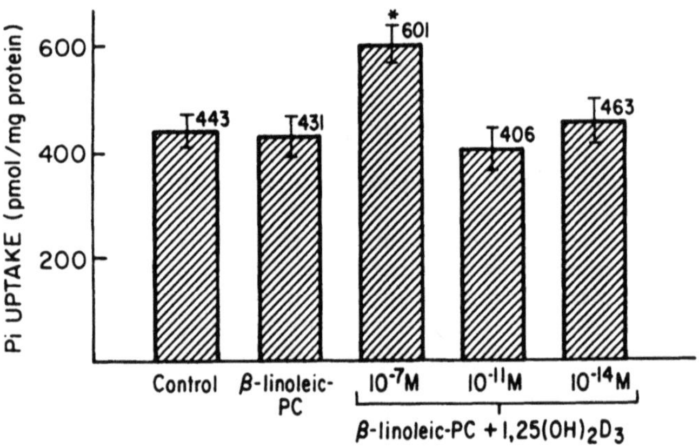

Figure 4 Liposomes composed of β-linoleic-PC with or without $1,25(OH)_2D_3$ at the specified concentrations were incubated with BBMV derived from normal rats for 30 min at 0° C. Phosphate transport was measured as described in the Methods. Phosphate uptakes at steady state (90 min) were: control, 249 ± 45; β-linoleic-PC, 234 ± 63; β-linoleic-PC + 10^{-7} M $1,25(OH)_2D_3$, 317 ± 86; β-linoleic-PC + 10^{-11} M $1,25(OH)_2D_3$, 178 ± 22; and β-linoleic-PC + 10^{-14} M $1,25(OH)_2D_3$ 193 ± 37 pmol/mg protein/90 min. Data are expressed as mean ± SE, N=6 in each group. *Greater than other groups, p<.02.

demonstrate that the $1,25(OH)_2D_3$ stimulated phospholipid transfer is not unique to the fluorescent PC derivative. In addition, a quantitatively similar increase in total phosphatidylcholine content occurs whether the liposomes are composed of DOPC or β-linoleic-PC and 10^{-7} M $1,25(OH)_2D_3$. Concentrations of $1,25(OH)_2D_3$ of 10^{-11} M and below have no effect on vesicle PC content.

The transfer of NBD-PC is significantly greater when the BBMV are derived from normal animals than vitamin D-deficient animals in the absence of $1,25(OH)_2D_3$. One possibility for this difference is that a protein, possibly a phosphatidycholine transfer protein, is being stimulated by $1,25(OH)_2D_3$. One could speculate that this protein is present at greater concentrations in BBMV prepared from normal rats than in BBMV prepared from vitamin D deficient rats. Differences in the amount of PC transfer between normal and vitamin D deficient vesicles could also be dependent on differences in the membrane phospholipid composition.

When the liposomes were composed of NBD-PC the peak effect on NBD-PC transfer occurred at 10^{-14} M $1,25(OH)_2D_3$, while the maximum effect using liposomes composed of DOPC or ß-linoleic-PC occurred at 10^{-7}M $1,25(OH)_2D_3$. This difference in dose response could reflect a different specificity of a phospholipid transfer protein which could be dependent on fatty acid chain length[15] rather than on fatty acid desaturation. This may be related to the structure of NBD-PC which contains only a six carbon chain in the beta position. The NBD group, also in the ß position, may itself react differently with the transfer protein. Alternatively, NBD-PC could alter liposome fluidity, maximizing the effects of sub-physiologic $1,25(OH)_2D_3$ concentrations.

In order to determine the functional significance of these membrane phospholipid alterations, phosphate and glucose transport were studied. The results show that phosphate transport is stimulated into BBMV from either normal or vitamin D deficient rats following incubation with liposomes composed of ß-linoleic-PC and $1,25(OH)_2D_3$ (10^{-7} M). Transport was not stimulated if the liposomes were composed of a more saturated PC, despite a quantitatively similar transfer of this species. Transport was not stimulated at concentrations of $1,25(OH)_2D_3$ of 10^{-11} M and below, consistent with the dose response for hormone stimulated non-fluorescent PC transfer. These results suggest that a specific increase in ß-linoleic-PC directly stimulates phosphate transport. This effect could be mediated through a change in the lipid micro-domain of the phosphate carrier protein resulting in an increase in its activity. Specificity for this effect on phosphate transport is provided by the data demonstrating that glucose transport is not altered.

The effects of $1,25(OH)_2D_3$ on non-fluorescent phospholipid transfer and phosphate transport occur at concentration of 10^{-7} M but not 10^{-11} M. Further studies will be needed to determine the dose response at intermediate concentrations of $1,25(OH)_2D_3$. Intracellular $1,25(OH)_2D_3$ concentrations have not been measured and it is possible that they may be higher than serum levels in view of the fact that $1,25(OH)_2D_3$ is produced in renal proximal tubular cells.

In conclusion, these studies suggest that $1,25(OH)_2D_3$ rapidly and directly modifies the membrane phospholipid composition and that a specific increase in membrane β-linoleic-PC content results in a stimulation of sodium-dependent phosphate transport.

REFERENCES

1. R.V. Farese, D.E. Barnes, J.S. Davis, M.L. Standaert, and R.J. Pollet, Effects of insulin and protein synthesis inhibitors on phospholipid metabolism, diaclyglycerol levles, and pyruvate dehydrogenase activity in BC3H-1 cultured myocytes, J. Biol. Chem. 259:7094 (1984).
2. D. Marver, and J.P. Kokko, Renal target sites and the mechanism of action of aldosterone, Miner. & Electrolyte Metab. 9:1 (1983).
3. I. Litosch, S.H. Lin, and J.N. Fain, Rapid changes in hepatocyte phosphoinositides induced by vasopressin, J. Biol. Chem. 258:13727 (1983).
4. V. Meltzer, S. Weinreb, E. Bellorin-Font, and K.A. Hruska, Parathyroid hormone stimulation of renal phosphoinositide metabolism is a cyclic nucleotide-independent effect. Biochim. Biophys. Acta 712:258 (1982).
5. M. Tsutsumi, U. Alvarez, L.V. Avioli, and K.A. Hruska, The effect of $1,25(OH)_2D_3$ on phospholipid composition of rat renal brush border membrane, Am. J. Physiol. 249:___ (1985) in press.
6. T. Matsumoto, O. Fontaine, and H. Rasmussen, Effect of 1,25-dihydroxy vitamin D_3 on phospholipid metabolism in chick duodenal mucosal cell, J. Biol. Che. 256:3354 (1981).
7. T. Matsumoto, O. Fontaine, and H. Rasmussen, Effect of 1,25-dihydroxy vitamin D_3 on phosphate uptake into chick intestinal brush border membrane vesicles, Biochim. Biophys. Acta 599:13 (1980).
8. B.R.C. Kurnik, and K.A. Hruska, Mechanisms of stimulation of renal phosphate transport by 1,25-dihydroxycholecalciferol (Biochim. Biophys. Acta - in press).
9. B.R.C. Kurnik, and K.A. Hruska, Effects of $1,25(OH)_2D_3$ directly stimulates phosphatidylcholine (PC) transfer, Kidney Int. 27:119 (1985).
10. B.R.C. Kurnik, and K.A. Hruska, Effects of 1,25-dihydroxy-cholecalciferol on phosphate transport in vitamin D-deprived rats, Am. J. Physiol. 247:F177 (1984).
11. B.R.C. Kurnik, M. Huskey, D. Hagerty, and K.A. Hruska, Vitamin D metabolites stimulate phosphatidylcholine transfer to renal brush border membranes. Submitted to J. Biol. Chem., May 1985.
12. D.K. Struck, and R.E. Pagano, Insertion of fluorescent phospholipids into the plasma membrane of a mammalian cell, J. Biol. Chem. 255:5404 (1980).
13. E.G. Bligh, and W.J. Dyer, A rapid method of total lipid extraction and purification, Can. J. Biochem. Physiol. 37:911 (1959).
14. B.N. Ames, and D.T. Dubin, The role of polyamines in the neutralization of bacteriophage deoxyribonucleic acid, J. Biol. Chem. 235:769 (1960).
15. R. Welti, and G.M. Helmkamp, Jr., Acyl chain specificity of phosphatidylcholine transfer protein from bovine liver, J. Biol. Chem. 259:6937 (1984).

POSSIBLE ROLE OF CALCIUM IN PARATHYROID HORMONE ACTION ON PHOSPHATE TRANSPORT IN RABBIT RENAL PROXIMAL TUBULES

Norimoto Yanagawa and Ok D. Jo

Nephrology Division, Medical and Research Services, Veterans Administration Medical Center, Sepulveda, California 91343 and Department of Medicine, UCLA School of Medicine, Los Angeles, California

In the renal proximal tubule, parathyroid hormone (PTH) stimulates gluconeogenesis (GNG) and suppresses fluid and phosphate reabsorption. Since PTH stimulates cAMP formation in this nephron segment, and the administration of cAMP reproduces hormonal actions, it is generally accepted that PTH exerts its biological effects through cAMP as the intracellular messenger.

In the rabbit kidney, PTH-dependent adenylate cyclase activity has been shown to be present in both proximal convoluted tubule (PCT) and proximal straight tubule (PST) (1). However, in the <u>in vitro</u> isolated perfused rabbit renal proximal tubules, <u>Dennis et al</u> found that while PTH inhibited fluid absorption in both PCT and PST, the phosphate absorption rate was suppressed only in PST but not in PCT (2). The mechanism of the resistance to the effect of PTH on phosphate absorption rate in rabbit PCT is not clear.

To test if there is also a similar heterogeneous response to PTH effect on GNG in rabbit renal proximal tubules, we have recently studied the PTH effect on GNG in PCT and PST isolated from rabbit kidney (3). It was found that PTH, as well as 8-bromo-cAMP, stimulated GNG only in PST but not in PCT. On the other hand, in PST, it was found that GNG was stimulated by an increase, while suppressed by a decrease, in the incubation medium calcium concentration. At either increased or decreased calcium concentration, PTH, as well as 8-bromo-cAMP, failed to stimulate GNG in PST. These results are consistent with the contention that PTH stimulates renal GNG by its effect to increase the intracellular calcium in the proximal tubular cells (4). In the isolated rabbit PCT, however, it was found that an increase in the incubation medium calcium concentration was able to stimulate GNG. If indeed that an increase in the intracellular calcium is required for PTH to stimulate GNG, the above finding that GNG in rabbit PCT can be stimulated by an increase in the ambient calcium concentration while it cannot be stimulated by PTH, would suggest that the resistance to the stimulatory effect of PTH on GNG in rabbit PCT may be due the inability of the hormone to increase the intracellular calcium in this nephron segment.

In contrast to the effect of PTH on GNG where the involvement of calcium is well demonstrated, whether or not calcium is also involved in the inhibitory effect of PTH on phosphate transport rate in renal proximal tubule is less clear. Such possibility, however, has been suggested from previous observations made under different pathological and experimental conditions (5-15). With the aforementioned results from GNG studies, we therefore postulated that the resistance to the inhibitory effect of PTH on phosphate absorption rate in rabbit PCT may also be due to the altered response of the intracellular calcium to the hormone. We have tested such postulation in the <u>in vitro</u> isolated perfused rabbit renal proximal tubules. It was expected that an increase in the ambient calcium concentration--a maneuver which may increase the intracellular calcium--may allow PTH to inhibit the phosphate absorption rate in rabbit PCT.

In the isolated perfused PCT, when bathed in an artificial medium containig 1 mM ionized calcium concentration (Ca^{++}), 1,34-bPTH (1 U/ml), as well as 8-bromo-cAMP (Ø.1 mM), did not affect the lumen to bath phosphate transport rate (Jp.) (Figure 1). When the Ca++ in the bathing medium was increased to 2 mM, alone did not affect Jp, but allowed 1,34-bPTH (1 U/ml), as well as 8-bromo-cAMP (Ø.1 mM), to inhibit Jp significantly by aprroximately 3Ø% (Fig. 2). These results are thus consistent with the postulations that intracellular calcium and hence calcium-dependent intracellular processes are involved in the action of PTH on Jp, and that the resistance to such hormonal action in rabbit PCT may be due to the altered response of the intracellular calcium to PTH in this nephron segment. However, the exact changes in the intracellular calcium concentration were not measured in these studies, and therefore any definitive conclusion awaits for further studies.

In order to further delineate the involvement of calcium-dependent intracellular processes in the effect of PTH on Jp, the effects of two structurally different intracellular calcium antagonists, trifluoperazine (TFP) and N-(6-aminohexyl)-5-chloro-1-naphthlenesulfonamide (W-7), were studied. As is shown in Table 1, the addition of Ø.1 mM TFP or W-7 to the bath medium, alone did not affect Jp significantly, but prevented the anticipated inhibitory effects of PTH and cAMP on Jp in both PST and PCT. Various studies have suggested the possibility that PTH may inhibit the phosphate absorption rate in renal proximal tubule through, at least in part, its effect on the phosphate uptake function of the brush border membrane (BBM). It has been demonstrated that treatment with PTH, either <u>in vivo</u> (16,17) or <u>in vitro</u> (18,19), suppresses the phosphate uptake by the <u>subsequently</u> isolated renal brush border membrane (BBM) vesicles. In addition, it has been shown that cAMP-dependent protein kinase is present on BBM (2Ø) and that cAMP-induced protein phosphorylation in BBM is associated with a decrease in phosphate uptake (21,22). These findings led to the proposition that protein phosphorylation by cAMP-dependent protein kinase on BBM represents the actual final

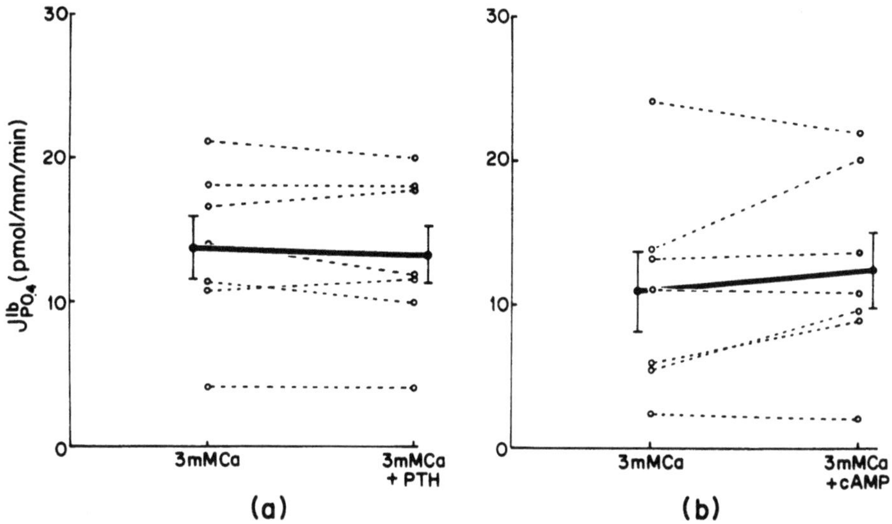

Figure 1. Effects of 1,34-bPTH (1 U/ml) and 8-bromo-cAMP (10^{-4} M) on Jp across the isolated perfused PCT. With 1 mM Ca^{++} bath medium (i.e. total Ca concentration of 3 mM in the presence of 6% bovine serum albumin), both PTH and cAMP did not affect Jp significantly. Broken lines indicate the results from each experiment. Means with standard errors are shown by the solid lines.

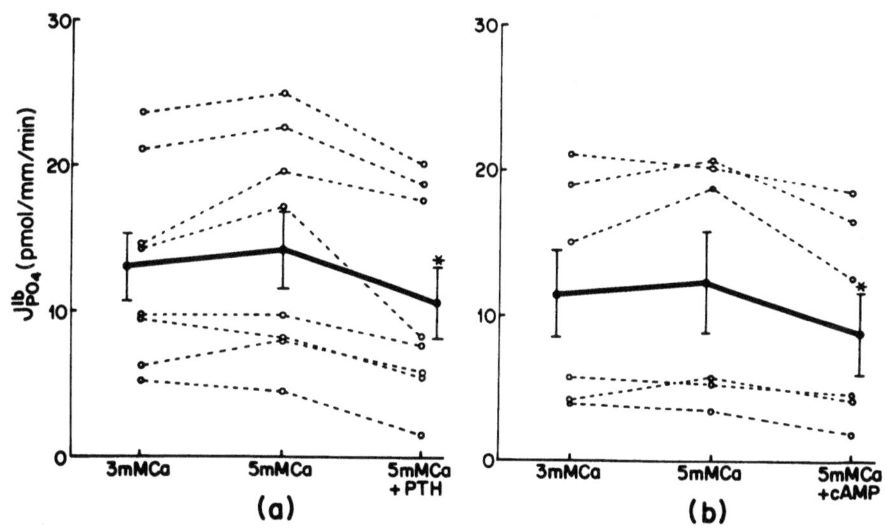

Figure 2. Effects of Ca, 1,34-bPTH (1 U/ml) and 8-bromo-cAMP (10^{-4} M) on Jp across the isolated perfused PCT. An increase in the Ca^{++} in the bath medium from 1 to 2 mM (i.e. total Ca concentration from 3 to 5 mM in the presence of 6% bovine serum albumin), alone did not affect Jp, but allowed both PTH and cAMP to inhibit Jp significantly. Broken lines indicate the results from each experiment. Means with standard errors are shown by the solid line. Asterisk (*) indicates significant difference ($p < 0.05$).

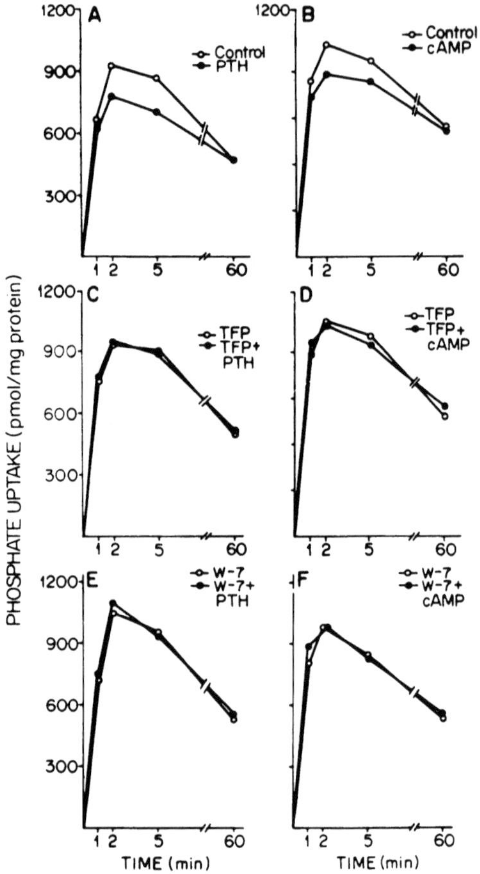

Figure 3. Effects of 1-34-bPTH (1 U/ml), 8-bromo-cAMP (10^{-4} M), TFP (10^{-4} M) and W-7 (10^{-4} M) on phosphate uptake by BBM vesicles. Preincubation of proximal tubule suspension with PTH (left upper panel) or cAMP (right upper panel) induced a significant inhibition on the phosphate uptake by the BBM vesicles isolated therefrom. The presence of either TFP (middle panels) or W-7 (lower panels) in the preincubation medium, however, abolished the inhibitory effects of both PTH and cAMP. Representative studies are shown in the figure.

Table 1. Intracellular calcium antagonists, TFP and W-7 (10^{-4} M), prevented the inhibitory effects of 1,34-bPTH (1 U/ml) and 8-bromo-cAMP (10^{-4} M) on Jp (pmol/mm/min.) in isolated perfused rabbit PCT and PST. The bath Ca^{++} was 1 or 2 mM for PST and PCT, respectively. Within each group, none of these results is statistically different from its immediate previous period (p 0.05). (Mean ± S.E.)

	Control	TFP	TFP +PTH	Control	TFP	TFP +cAMP
PCT	13.28 ±2.90	11.18 ±2.29 (n=7)	11.05 ±2.65	10.51 ±2.00	9.05 ±2.14 (n=6)	8.31 ±3.66
PST	2.60 ±0.46	2.11 ±0.47 (n=9)	1.96 ±0.33	2.20 ±0.29	1.67 ±0.40 (n=5)	2.54 ±0.36

	Control	W-7	W-7 +PTH	Control	W-7	W-7 +cAMP
PCT	9.79 ±1.84	7.68 ±1.29 (n=5)	7.53 ±0.80	10.92 ±2.54	8,85 ±1.88 (n=4)	8.30 ±2.16
PST	1.57 ±0.42	1.38 ±0.31 (n=6)	1.22 ±0.21	1.17 ±0.18	1.13 ±0.22 (n=4)	1.21 ±0.40

step of the inhibitory effect of PTH on phosphate transport in renal proximal tubule. However, inconsistent results on the in vitro effect of cAMP on protein phosphorylation and phosphate uptake in isolated BBM have been reported recently (23).

Since the addition of intracellular calcium antagonists, TFP and W-7, prevented the effects of PTH and cAMP on Jp across the proximal tubule, we have further examined the effects of TFP and W-7 on the inhibitory effects of PTH and cAMP on the phosphate uptake by BBM vesicles. Proximal tubular suspension was first prepared from rabbit renal cortex by homogenization and sieving processes similar to those described previously (24). After these proximal tubules were incubated with or without various testing agents, including 1,34-bPTH (1 U/ml), 8-bromo-cAMP (0.1 mM), TFP (0.1 mM) or W-7 (0.1 mM) for 20 minutes, BBM vesicles were then isolated thereof by the conventional calcium precipitation method, and the phosphate uptake by the BBM vesicles was measured using a Millipore rapid filtration procedure under an inward sodium gradient of 100 mM. Figure 3 shows the results of representative experiments. Preincubation of proximal tubules with PTH or cAMP induced a significant inhibition on phosphate uptake by the BBM vesicles isolated therefrom. The addition of TFP or W-7 to the preincubation medium, alone was without effect, but abolished the inhibitory effects of PTH and cAMP.

In summary, the present studies provide evidences in support of the role of calcium-dependent intracellular processes in the inhibitory effect of PTH on phosphate transport by either the proximal tubule as a whole or by the BBM isolated from the proximal tubules. The resistance to such hormonal effect in isolated rabbit PCT may be due an altered response of the intracellular calcium to PTH. Reservation must be maintained, however, in interpreting these results due to the inability of monitoring the intracellular calcium in the isolated perfused proximal tubules and the rather nonspecific actions of TFP and W-7. The definitive conclusion awaits for further studies.

REFERENCES

1. D. Chabardes, M. Imbert, A. Clique, M. Montegut, F. Morel., PTH sensitive adenyl cyclase activity in different segments of the rabbit nephron, Pfluegers Arch. 345:229-239 (1975).
2. V.W. Dennis, E. Bello-Reuss, R.R. Robinson, Response of phosphate transport to parathyroid hormone in segments of rabbit nephron, Am. J. Physiol. 233:F29-F38 (1977).
3. Ok Jo, N. Yanagawa, Absence of parathyroid hormone effect on gluconeogenesis in rabbit proximal convoluted tubules, Kidney Int. 25:266 (1984)
4. N. Nagata, H. Rasmussen, Parathyroid hormone, 3'5'AMP, Ca^{++}, and renal gluconeogenesis, Proc. Natl. Acad. Sci. USA. 65:368-374 (1970).
5. F. Glorieux, C.R. Scriver, Loss of a parathyroid hormone-sensitive component of phosphate transport in X-linked hypophosphatemia, Science 175:997-1000 (1972).

6. M. Drezner, F.A. Nelson, H.E. Lebovitz, Pseudohypoparathyroidism type II: A possible defect in the reception of the cAMP signal, New Engl. J. Med. 289:1056-1060 (1972).
7. T.H. Steele, Renal rsistance to parathyroid hormone during phosphate deprivation, J. Clin. Invest. 58:1461-1464 (1976).
8. D. Aarskog, L. Aksnes, T. Markestad, Effect of parathyroid hormone on cAMP and 1,25-dihydroxyvitamin D formation and renal handling of phosphate in vitamin D-dependent rickets, Pediatr. 71:59-63 (1983)
9. M. Drezner, F.A. Nelson, M. Haussler, H.T. McPherson, H.E. Lebovitz, 1,25-Dihydroxycholecalciferol deficiency: The probable cause of hypocalcemia and metabolic bone disease in pseudohypoparathyroidism. J. Clin. Endocrinol. Metab. 42:621-628 (1976).
10. H.J. Gloor, J.P. Bonjour, J. Caverzasio, H. Fleisch, Resistance to the phosphaturic and calcemic actions of parathyroid hormone during phosphate depletion. Prevention by 1,25-dihydroxyvitamin D_3, J. Clin. Invest. 63:371-377 (1979).
11. H.J. Rodriquez, H. Villarreal, S. Klahr, E. Slatopolsky, Pseudohypoparathyroidism type II: Resistance of normal renal responsiveness to parathyroid hormone by calcium administration, J. Clin. Endocrinol. Metab. 39:639-701 (1974).
12. W. Stogmann, J.A. Fischer, Pseudohypoparathyroidism: Disappereance of resistance to parathyroid extract during treatment with vitamin D. Am. J. Med. 59:140-144 (1975).
13. S.M. Suh, D. Fraser, S.W. Kooh, Pseudohypoparathyroidism responsiveness to parathyroid extract induced by vitamin D_2 therapy, J. Clin. Endocrinol. Metab. 30: 609-614 (1970)
14. J.L. Cuche, C.E. Ott, G.R. Marchand, J.A. Diaz-Buro, F.G. Knox, Intrarenal calcium in phosphate handling, Am. J. Physiol. 230:790-796 (1976).
15. J. Winaver, J. Fragola, T.C. Chen, D.B. Sylk, J.S. Robertson, J.B. Puschett, Potentiation by calcium of the proximal tubular transport effects of parathyroid hormone. Min.Electrol. Metab. 8:275-288 (1982).
16. C. Evers, H. Murer, R. Kinne, Effect of parathyrin on the transport properties of isolated renal brush border vesicles, Biochem. J. 172:49-56 (1978).
17. M.R. Hammerman, I.E. Karl, K.A. Hruska, Regulation of canine renal vesicle Pi transport by growth hormone and parathryoid hormone. Biochim. Biophy. Acta 603:322-335 (1980).
18. L. Chaeng, C.T. Liang, B. Sacktor, In vitro effects of parathyroid hormone on kidney cortical slices: cAMP responses and concomitant inhibition of the Na gradient-dependent uptake of phosphate by brush border membrane vesicles isolated from the renal slices. Endocrinol. Res.Commun. 8(2):97-110 (1981).
19. T. Takenawa, E. Wada, T. Tsumita, T. Masaki, C.R. Filburn, B. Sacktor, Effect of parathyroid hormone, cyclic AMP and Ca on the phosphorylation of brush border membranes in rabbit kidney. Min. Electrol. Metab. 10:103-112 (1984).
20. R. Kinne, L.J. Shlatz, E. Kinne-Saffran, I.L. Schwartz, Distribution of membrane-bound cyclic AMP dependent

protein kinase in plasma membrane of cells of kidney cortex. J.Membr. Biol. 24:145-159 (1975).

21. M.R. Hammerman, K.A.Hruska, Cyclic AMP-dependent protein phosphorylation in canine renal brush border membrane vesicles is associated with decreased phosphate transport, J. Biol. Chem. 257:992-999 (1982).
22. M.R. Hammerman, V.A. Hansen, J.J. Morrisey, Cyclic AMP-dependent protein phosphorylation and dephosphorylation alter phosphate transport in canine renal brush border vesicles, Biochim. Biophy. Acta 755:10-16 (1983).
23. J. Biber, K. Malmstrom, V. Scalera, H. Murer, Phosphorylation of rat kidney proximal tubular brush border membranes. Role of cAMP dependent protein phosphorylation in the regulation of phosphate transport, Pflugers Arch. 398:221-226 (1983).
24. S.D. Chung, N. Alavi, D. Livingston, S. Hiller, M. Taub, Characterization of primary rabbit kidney cultures that express proximal tubule functions in a hormonally defined medium, J. Cell Biol. 95:118-126 (1982).

CALCIUM STIMULATES SODIUM-DEPENDENT PHOSPHATE ABSORPTION IN RABBIT PROXIMAL CONVOLUTED TUBULES PERFUSED IN VITRO

Diane Rouse and Wadi N. Suki

Baylor College of Medicine and The Methodist Hospital

Houston, Texas

INTRODUCTION

There has been considerable evidence for an effect of calcium on the absorption of phosphate by the kidney. Two micropuncture studies have agreed that increasing serum calcium enhances distal nephron phosphate absorption.[1, 2] However, in the proximal tubule, one study found enhancement of phosphate absorption,[1] the other inhibition.[2] Because the various indirect effects of increasing serum calcium may have played a role in the outcome of in vivo studies, an in vitro study using the technique of isolated tubule microperfusion was designed.

Results, presented previously, showed that in proximal convoluted tubule, S_2, segments as bath and perfusate calcium are simultaneously increased, lumen-to-bath phosphate flux increases, and as ambient calcium is decreased, phosphate flux decreases. Increasing ambient calcium in proximal straight tubules had no effect on phosphate flux.

With the effect defined, the mechanism by which calcium modulates phosphate absorption was investigated. Increasing cytosolic calcium with ionophore, inhibiting the calcium-calmodulin complex with trifluoperazine, and exposure to the calcium channel blocker, nifedipine, did not alter lumen-to-bath phosphate flux. Additional studies showed that increasing only bath calcium had no effect on lumen-to-bath phosphate flux. However, increasing only perfusate calcium enhanced phosphate flux to a similar degree attained when bath and perfusate calcium were increased simultaneously. Finally, studies designed to determine if calcium enhances phosphate absorption by increasing H^+ secretion proved negative.

Although these results suggest that luminal calcium may alter the kinetics of the phosphate transporter, there are at least two other possible explanations. Firstly, increasing calcium may lead to the increased formation of more permeant phosphate species which enhances bidirectional phosphate transport without changing net flux. Secondly, increasing luminal calcium may increase the concentration of poorly soluable calcium-phosphate salts which precipitate in the lumen. Both of these two physicochemical reactions would artificially increase lumen-to-bath phosphate flux. In order to rule out the two alternate explanations net phosphate flux studies were designed.

MATERIALS AND METHODS

General Procedure

Proximal convoluted tubule, S_2, segments were dissected from New Zealand white rabbit kidneys and perfused according to the method of Burg and colleagues.[3] An artificial solution resembling rabbit serum ultrafiltrate was used to perfuse and bathe the tubule segments. The specific activity of ^{32}P was identical in bath and perfusate for the measurement of net phosphate flux. ^3H inulin was added to the perfusate to measure fluid absorption, Jv, and 0.35 g/dl defatted bovine albumin was added to the bath. Transepithelial potential difference, PD, was measured using agarose-Ringer's bridges and calomel electrodes as described by Rocha and Kokko.[4] The bath was gassed with 5%CO_2, 95%O_2, exchanged continuously at the rate of 0.5 ml/min, and maintained at either 37°C or 25°C. Net phosphate flux and Jv were calculated as described previously.[5]

Experimental Protocols

Group I. In five S_2 segments, net phosphate flux was measured in the presence of 1.8 mM ambient calcium and then 3.6 mM ambient calcium. Bath temperature was maintained at 37°C.

Group II. The protocol above was repeated in six additional S_2 segments but instead bath temperature was maintained at 25°C.

Data Analysis

The means of 3-4 collections for each control and experimental period were compared using the t-test for paired samples.

RESULTS

Group I. At a bath temperature of 37°C, Jv in this group was 1.2 $\pm$ 0.3 nl/mm.min using 1.8 mM ambient calcium, and 1.0 $\pm$ 0.2 using 3.6 mM ambient calcium ($P>0.05$). PD was not altered by increasing bath and perfusate calcium being -2.1 $\pm$ 0.5 mV with 1.8 mM calcium and -1.5 $\pm$ 0.3 with 3.6 mM calcium ($P>0.05$). Net phosphate flux increased from 3.9 $\pm$ 0.8 pmol/mm.min with 1.8 mM calcium to 6.5 $\pm$ 1.3 with 3.6 mM calcium ($P<0.05$).

Group II. In this group studied at a bath temperature of 25°C neither Jv nor PD were significantly different from 0 using either 1.8 or 3.6 mM bath and perfusate calcium. Jv was 0.1 $\pm$ 0.1 nl/mm.min with 1.8 mM calcium, and 0.2 $\pm$ 0.1 with 3.6 mM calcium; and PD was -0.3 $\pm$ 0.1 mV before and -0.3 $\pm$ 0.8 after raising ambient calcium. Net phosphate flux was low, being 1.2 $\pm$ 0.3 pmol/mm.min with 1.8 mM calcium, and failed to increase as calcium was increased to 3.6 mM, being 1.2 $\pm$ 0.4 pmol/mm.min.

DISCUSSION

Our previous results have suggested that luminal calcium directly affects phosphate absorption in proximal convoluted tubule segments. The present net phosphate flux data provide further evidence for a physiological mechanism.

The studies, performed at 37°C, rule out the possibility that increasing calcium promotes the formation of more permeant phosphate species and increases influx as well as efflux with no change in net flux. Thus, lumen-to-bath phosphate flux is representative of net phosphate flux. However, there remained the possibility that as calcium is increased, poorly soluable Ca-PO_4 salts form and precipitate in the tubular lumen, artificially raising phosphate absorption. In order to distinguish between a physicochemical phenomenon and the stimulation of a true reabsorptive process, net phosphate flux studies were performed at 25°C. Using the lower temperature could yield two pieces of information. Firstly, if precipitation of poorly soluable Ca-PO_4 salts is occurring, the lower temperature would enhance this process and further increase net phosphate absorption. Secondly, carrying out studies at room temperature should determine if the effect of calcium is exerted on an energy-requiring phosphate transport process. Because net phosphate absorption was reduced when bath temperature was maintained at 25°C and failed to increase as bath and perfusate calcium was increased, it is not likely that precipitation of Ca-PO_4 salts is artificially increasing phosphate absorption. These studies also suggest that calcium stimulates an energy-requiring component of phosphate absorption.

We suggest that calcium modulates the translocation of phosphate by altering the kinetics of its transporter at the luminal membrane.

REFERENCES

1. C. Amiel, H. Kuntziger, S. Couette, C. Coureau, and N. Bergounioux. Evidence for a parathyroid hormone-independent calcium modulation of phosphate transport along the nephron. J. Clin. Invest. 57:256 (1976).

2. S. Goldfarb, P. Bosanac, M. Goldberg, and Z.S. Agus. Effects of calcium on renal tubular phosphate reabsorption. Am. J. Physiol. 234:F22 (1978).

3. M. Burg, J. Grantham, M. Abramow, and J. Orloff. Preparation and study of fragments for single rabbit nephrons. Am. J. Physiol. 210:1293 (1966).

4. A.S. Rocha and J.P. Kokko. Sodium chloride and water transport in the medullary thick ascending limb of Henle. J. Clin. Invest. 52:612 (1973).

5. D. Rouse, R.C.K. Ng, and W.N. Suki. Calcium transport in the pars recta and thin descending limb of Henle of the rabbit, perfused in vitro. J. Clin. Invest. 65:37 (1980).

CHARACTERISTICS OF PHOSPHATE TRANSPORT ACROSS THE BASOLATERAL MEMBRANE OF THE PROXIMAL TUBULE: AN IN VIVO STUDY

Tim S. Yan and Gary A. Quamme

Department of Medicine
University of British Columbia
Vancouver, B.C., Canada

INTRODUCTION

The final event in renal epithelial phosphate absorption is the movement of phosphate from the cytosol across the basolateral membrane into the peritubule capillary. The transmembrane potential difference is about -65 mV and the chemical concentration gradient is in the order of 1-5 mM inside-to-3mM outside. Accordingly, the transfer of phosphate may be passive down its electrical gradient. This is thought to be facilitated by a phosphate carrier within the basolateral membrane [1,2,3,4].

Some evidence has been given which suggests that phosphate may share a common anion exchange system on the basolateral membrane [3,4]. In this system, phosphate moving from the cytosol across the basolateral membrane is exchanged for another inorganic anion such as sulfate, chloride, bicarbonate (hydroxyl ions), or another organic anion, for instance acetate, lactate, pyruvate, or oxalate [4]. In addition to accelerating the efflux of phosphate by their presence in the peritubular capillary, these organic anions may also inhibit phosphate exit by competing for the common carrier from the cis side or the cytosolic side of the basolateral membrane. Accordingly, phosphate movement across the basolateral membrane may not be accurately controlled but dependent on the presence of many other anions. This system is analogous to the anion exchange process observed in the red blood cell [5]. Events at the basolateral membrane may modulate transcellular phosphate reabsorption although at the present time it is not thought to be the rate-limiting step.

The present study was designed to investigate the effects of phosphate on the basolateral membrane. Unidirectional phosphate transport was measured in the early proximal tubule of rats maintained on relatively low dietary phosphate content. Plasma phosphate was progressively elevated to determine the effects of peritubular phosphate concentration on net and unidirectional phosphate flux.

METHODS

Studies were performed on male Wistar rats, weighing 180-200g. Rats were maintained for 5-7 days on either a normal or a low phosphate diet prior to experimentation. The low phosphate diet contained: phosphate 0.01%, calcium 0.6% and magnesium 0.7% (diet #902206, Nutritional Biochemicals Div. ICN, Life Services Group, Cleveland, O). The normal phosphate diet was this diet to which phosphate was added in appropriate amounts to provide 0.44%.

Proximal tubules were perfused with solutions of constant phosphate concentration as the plasma phosphate was progressively elevated. Accordingly, the effect of phosphate concentration gradients across the basolateral membrane could be evaluated independent of the filtered phosphate concentration.

Thyroparathyroidectomy (TPTX) was performed by thermocautery two hours prior to experimentation. The left jugular vein was cannulated for infusion of a modified Ringer solution [6]. The rats were volume expanded to 3% body weight with 3 ml infused over 30 minutes and 60-80 ul/min over the duration of the experiment to replace urinary losses. Plasma and urine were sampled periodically throughout the experiment to enable the determination of plasma electrolyte clearances.

Plasma phosphate concentration was elevated by infusing solutions containing variable amounts of phosphate, 50-200mM in place of the sustaining Ringer infusate at 60 ul/min. As plasma calcium may directly affect proximal phosphate absorption, calcium was infused in the femoral vein in appropriate amounts to avoid changes in plasma calcium concentration [6].

The details of the microperfusion technique have previously been given [6,7,8]. Briefly, early proximal tubules were perfused with a thermally insulated microperfusion pump. Surface nephrons were perfused at a constant rate of 15 nl/min. Timed total collections were obtained from the proximal tubule with the aid of an oil block. Early proximal convoluted tubules were localized with Microfil (Canton Bio-Medical Products, Boulder, CO) and the length of the perfused segment determined by microdissection of the tubule filled with Microfil and comparison of the cast with an object micrometer. Data obtained from tubule segments located beyond 2-2.5mm from the glomerulus was not included in the present study. The composition of the perfusion solution was designed to prevent net sodium and water transport. The alkaline perfusion solution contained in mM: Na, 112±4; Cl, 80±10; K, 3.51±0.25; Ca, 1.35±0.04; Mg, 0.27±0.02; alanine, 5; acetate, 7.5; glucose, 5; and raffinose, 30-50. Na_2HPO_4/NaH_2PO_4 was added (0-20mM) to the perfusion solution in isosmotic substitution for NaCl. In addition, the perfusion solution contained 25 mM HEPES (N-2-hydroxethylpiperazine-N 2-ethanesulfonic acidic, pKa 7.55, Calbiochem, La Jolla, CA) buffered to pH 7.65 when the solution was equilibrated with 5% CO_2-95% O_2. The acidic perfusion solution contained in mM: Na, 109±6; Cl, 102±2; K, 4.22±0.12; Ca, 1.75±0.06; Mg, 0.38±0.02; phosphate, 0-20; alanine, 5; acetate, 7.5: glucose, 5 and raffinose, 30-50. The acidic perfusate was buffered with MES (2-N-morpholinoethanesulfonic acid, pKa 6.15, Calbiochem, La Jolla, CA) to pH 6.5 when equilibrated with 5% CO_2-95% O_2. Unless otherwise stated, the bicarbonate concentrations of the alkaline and acidic perfusion solutions were 36.47±0.79mM and 4.10±0.26mM, respectively. The osmolality of the perfusion solutions was made equivalent to plasma by addition of appropriate amounts of raffinose. In addition, 15-20 uCi/ml [^{3}H] inulin (New England Nuclear, Montreal) was included to determine water movement and 5-10 uCi/ml [^{32}P] phosphate (New England Nuclear, Montreal) to quantitate unidirectional phosphate transport.

Reabsorption rates were calculated by standard formulae previously given [6], and statistics include comparison of observations from different groups of tubules using the nonpaired Student's t test. Apparent carrier kinetics were determined by altering the mean luminal phosphate concentration and comparing the dependence of unidirectional lumen-to-peritubular capillary phosphate flux to the average luminal phosphate concentration. The apparent Km and Jmax values determined by least-squares linear regression lines from a Lineweaver Burk kinetic transformation of the mean values. Statistical comparisons of these parameters were determined with the use of multiple regression analysis.

RESULTS

Table 1 summarizes the mean plasma and urinary excretion data for the acutely TPTX rats fed normal or low phosphate diets. Fig. 1 illustrates the clearance data. Fractional phosphate excretion was less than 0.05%, in the rats maintained on the phosphate-restricted diet prior to TPTX, compared to 3-5% for the intact rats maintained on a normal phosphate diet. The rats on a phosphate deficient diet had significantly

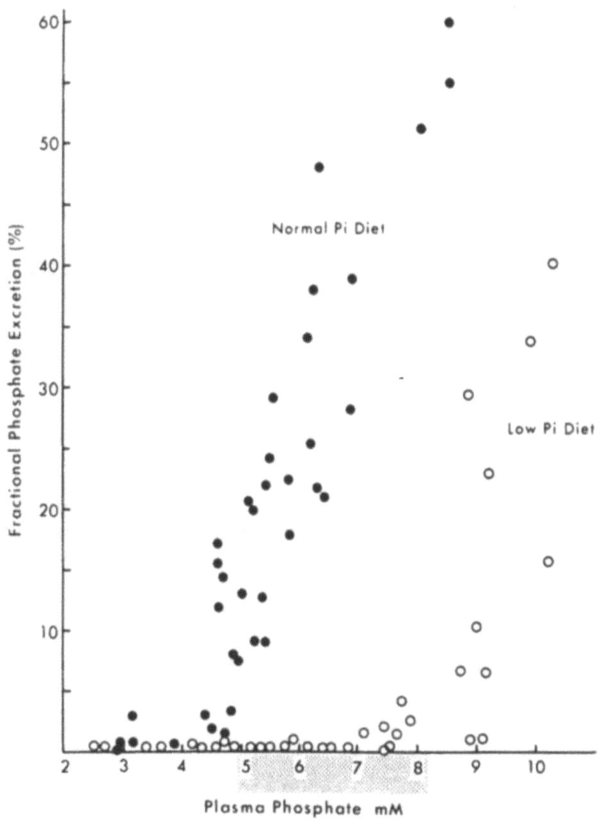

Fig. 1 Effect of plasma phosphate concentration on urinary phosphate excretion in rats maintained on normal and low dietary phosphate contents. Plasma calcium concentration was normalized during the hyperphosphatemic periods.

lower GFR which may be related to the lower body weight or to the phosphate depletion per se. In addition, fractional calcium excretion was markedly elevated (17% vs 5%) in rats fed phosphate deficient diets compared to those on normal diets. Plasma and urinary phosphate excretion were significantly less following removal of the parathyroid glands in phosphate-depleted rats compared to TPTX animals on normal phosphate intakes (table 1).

Table 1. Summary of plasma and urinary excretion data for rats maintained on phosphate restricted diets.

	Normal Plasma Phosphate	Elevated Peritubular Phosphate
Hematocrit %	45.1 ± 0.5	46.0 ±0.9
Pprotein g%	4.34 ± 0.10	4.75 ± 0.07
GFR ml/min	0.53 ± 0.04	0.50 ± 0.04
P_{pH}	7.41 ± 0.02	7.44 ± 0.02
$P_{HCO_3^-}$ mM	24.2 ± 1.3	20.5 ± 1.5
FE_{HCO_3} %	0.41 ± 0.12*	0.13 ± 0.08
P_{Na} mM	141 ± 1	142 ± 1
FE_{Na} %	2.77 ± 0.05	0.37 ± 0.09
P_K mM	4.09 ± 0.16*	
FE_K %	24.5 ± 2.3*	
P_{Ca} mM	2.66 ± 0.08*	2.33 ± 0.06
FE_{Ca} %	17.15 ± 0.92*	3.01 ± 0.73
P_{Pi} mM	2.41 ± 0.05*	6.38 ± 0.45
FE_{Pi} %	<.01 *	0.63 ± 0.31
	(28)	(12)

The values for the normal TPTX rats were taken from a previous study which was done about the same time as the present investigation (12). The * indicates the significance of the difference, $p<.001$, in the respective parameters between the animals on low and high dietary phosphate compared to normal dietary phosphate, as assesseds by nonpaired Student's t test.

Fig. 2 illustrates the microperfusion data for unidirectional phosphate absorption of the rats on a phosphate-restricted diet. The reabsorptive rate was increased from solutions of both pH values at all levels of intraluminal phosphate. The apparent kinetic parameters at pH 6.5 were Jmax 42.73±5.48 $pmol.min^{-1}.mm^{-1}$ and Km 1.65±0.10 mM in phosphate-depleted rats. The apparent Jmax and Km values for the alkaline studies were, 53.47±3.71 $pmol.min^{-1}.mm^{-1}$ and 2.45±0.06 mM, respectively.

Fig. 2 demonstrates the effect of elevated peritubule phosphate concentration on phosphate absorption in rats adapted to low dietary phosphate. In contrast to normal animals, modest elevation of plasma phosphate from normal concentrations of 2.4 to values of 6.4 mM enhanced phosphate absorption at both intraluminal pH values. The apparent kinetic parameters for the intraluminal alkaline, pH 7.5, studies were: Jmax, 86.21±12.63 $pmol.min.^{-1}.mm^{-1}$ and Km, 1.79±0.14 mM with a correlation coefficient of r = 0.99 for the Lineweaver-Burk transformation of this data. Both the apparent Jmax and Km values were significantly different, F<0.01, from rats with normal plasma phosphate concentrations maintained on low dietary phosphate content. The respective parameters determined with acidic lumen pH values were: Jmax, 140.84±86.76 $pmol.min.^{-1}.mm^{-1}$ and and; Km, 6.07±0.56 mM. The Jmax was significantly greater than normophosphatemic rats. Accordingly, modest elevations of plasma phosphate concentration increase unidirectional and net phosphate absorption in animals maintained on diets deficient in phosphate.

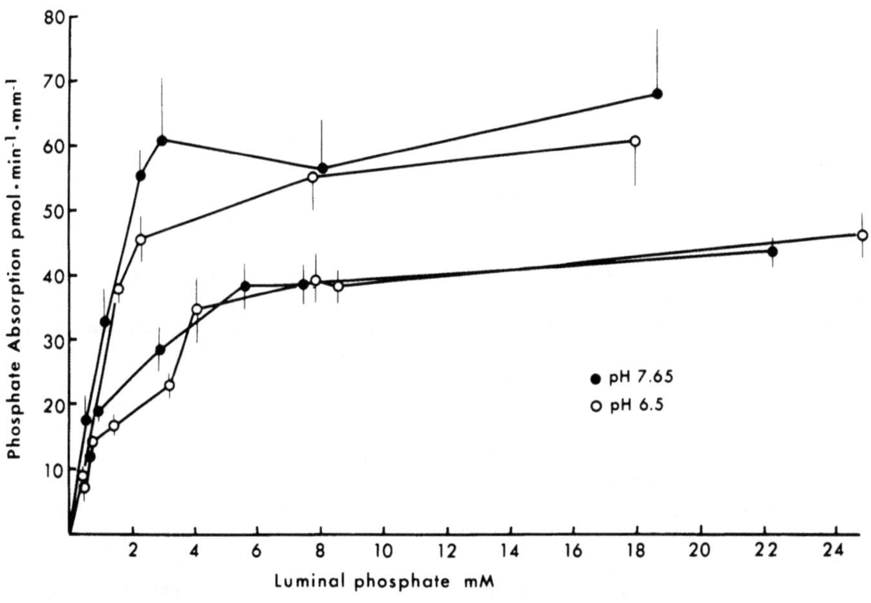

Fig. 2 Effect of peritubule phosphate concentration on phosphate absorption in the early proximal convoluted tubule of rats maintained on phosphate-restricted diets.

In summary, elevation of plasma phosphate concentration and the peritubule concentration decreases phosphate reabsorption in animals maintained on normal dietary intake [6,8]. The magnitude of the inhibition is dependent on the levels of plasma phosphate concentration. In contrast, in rats adapted to low dietary content, modest elevation enhances phosphate transport. Accordingly, when the renal cell is adapted to low available phosphate, the presence of phosphate on the basolateral membrane enhances unidirectional phosphate transport. This data supports the notion of a phosphate exchange process located on the basolateral membrane which can modulate net phosphate transport under certain circumstances. These conclusions are in line with the observation that phosphate-deprived animals retain filtered phosphate in the face of hyperphosphatemia. This is evident from the data presented in figure 1 and has previously been reported by others [9,10,11]. However, greatly elevated plasma phosphate concentrations diminish phosphate transport despite the adaptation to the phosphate-restricted diets.

DISCUSSION

Although sodium-phosphate cotransport across the brush border membrane is thought to be the rate-limiting step in phosphate absorption, evidence has been reported that the basolateral membrane may affect transport under certain circumstances (16). In order to shed some light on the exit step across the basolateral membrane, we elevated the plasma phosphate concentration at constant intraluminal concentrations. Elevation of peritubular phosphate inhibited net flux and unidirectional efflux in normal animals. Accordingly, the evidence given here suggests that relatively small acute changes in peritubule phosphate may modulate net phosphate transport through the action on the basolateral membrane. These actions are independent of the intraluminal phosphate concentration. Interestingly, elevation of plasma phosphate when the brush border transport is optimal ie. animals maintained on a low dietary phosphate, results in enhanced unidirectional phosphate absorption. This may be accounted for by a phosphate exchange system on the basolateral membrane. Phosphate may exchange across the basolateral membrane leading to an increase in transport across the brush border membrane and thus enhance net phosphate absorption. Recent evidence lends credence to this postulate [4,12,13,14,15]. For instance, Brazy and colleagues have reported that elevation of peritubule or bath sulfate may enhance sulfate absorption by the isolated rabbit proximal tubule. Others, have demonstrated anion exchange systems [4,15] in isolated basolateral membranes which may be specific for phosphate [13,14]. Accordingly, net phosphate absorption within the early proximal tubule may be a balance, in part, between events occuring at the brush border membrane and the basolateral membrane. Elevation of plasma phosphate may increase net transport in situations when the system at the brush border membrane is operating maximally. The events occuring at the two membranes act in concert to retain phosphate in this condition. Enhancement of net phosphate absorption is not normally observed in rats on normal dietary intake because brush border uptake is set too low to demonstrate stimulation of net transport.

ACKNOWLEDGEMENTS

The skilled technical assistance of Don Huysmans and the secretarial expertise of Ginette Vogel are gratefully acknowledged.

This investigation was supported by grants to G.A. Quamme from the British Columbia Health Care Foundation and the Medical Research Council of Canada (MA 5793).

REFERENCES

1. S. Grinstein, R. J. Turner, M. Silverman, and A. Rothstein, Inorganic anion transport in kidney and intestinal brush border and basolateral membranes. Am. J. Physiol. 238 (Renal Fluid Electrolyte Physiol.):F452 (1980).
2. K. J. Ullrich, and H. Murer, Sulfate and phosphate transport in the renal proximal tubule. Philos. Trans. R. Soc. Lond. Ser. B. 229:549 (1982).
3. H. Murer, and G. Burckhardt, Membrane transport of anions across epithelia of mammalian small intestine and kidney proximal tubule. Rev. Physiol. Biochem. Pharmacol. 96:1-51 (1983).
4. I. Loew, T. Friedrich, and G. Burckhardt, Properties of an anion exchanger in rat renal basolateral membrane vesicles. Am. J. Physiol. 246 (Renal Fluid Electrolyte Physiol. 15):1,F334-F342 (1984).
5. P. A. G. Fortes, Anion movements in red blood cells. Ann. NY Acad. Sci. 341:175 (1980).
6. G. A. Quamme, and N. L. M. Wong, Phosphate transport in the proximal convoluted tubule: effect of intraluminal pH. Am. J. Physiol., 246 (Renal Fluid Electrolyte Physiol. 15):F323-F333 (1984).
7. G. A. Quamme, and N. L. M. Wong, Effects of metabolic acidosis, alkalosis and dietary hydrogen ion inake on phosphate transport in the proximal convoluted tubule. Am. J. Physiol., in press (1985).
8. G. A. Quamme, C. L. Mizgala, S. J. Whiting, and N. L. M. Wong, Effect of intralumnal pH and dietary phosphate on phosphate transport in the proximal convoluted tubule. Am. J. Physiol., in press (1985).
9. T. H. Steele, and H. G. Deluca, Influence of dietary phosphorus on renal phosphate reabsorption in the parathyroidectomized rat. J. Clin. Invest. 57: 867 (1976).
10. R. Stoll, R. Kinne, and H. Murer. Effect of dietary phosphate intake on phosphate transport by isolated rat renal brush-border vesicles. Biochem. J. 180:564-570 (1979).
11. H. Trohler, J.-P. Bonjour, and H. Fleisch, Inorganic phosphate homeostasis. Renal adaptation to the dietary intake in intact and thyroparathyroidectomized rats. J. Clin. Invest. 57:264 (1976).
12. P. C. Brazy, and V. W. Dennis, Sulfate transport in rabbit proximal convoluted tubules: presence of anion exchange. Am. J. Physiol. 241 (Renal Fluid Electrolyte Physiol. 10):F300-F307 (1981).
13. G. A. Quamme, Characteristics of phosphate transport across the basolateral membrane of the proximal tubule. Kidney Inter. 26:26A (1984).
14. S. J. Schwab, and M. R. Hammerman, Sodium independent phosphate transport in basolateral membrane vesicles from dog kidney. Kidney Inter. 26:28A (1984).
15. K. J. Ullrich, G. Rumrich, and S. Kloess, Contraluminal sulfate transport in the proximal tubule of the rat. I Kinetics, effects of K^+, Na^+, Ca^{2+}, H^+, and anions. Pfluegers Arch. 402:264 (1984).

RENAL PROXIMAL TUBULAR BRUSH-BORDER MEMBRANE TRANSPORT OF PHOSPHATE WITH ACUTE ACIDOSIS

Barton S. Levine, Jeffrey A. Kraut, and Delta R. Mishler

The Medical and Research Services, Veterans Administration Medical Center West Los Angeles, and the Department of Medicine UCLA School of Medicine, Los Angeles, CA 90073

Introduction

Recent studies from our laboratory (1) and that of other investigators (2) have shown that metabolic acidosis of several days duration is associated with suppression of phosphate transport by isolated brush-border membrane vesicles (BBMV) of the proximal tubule. The suppression is specific for the Pi transporter, since other Na-dependent transport processes are not altered.

The signal to the BBM to reduce Pi transport with acidosis is unclear. Possible signals include reduction in extracellular or intracellular pH. If a fall in extracellular pH is the primary signal, then acidemia produced by a fall in serum HCO_3, or by an increase in CO_2 concentration, should have comparable effects on Pi transport. By contrast, if the signal is primarily a fall in intracellular pH, the effect of metabolic acidosis and respiratory acidosis on renal Pi transport might differ. To examine these possibilities, Pi transport by isolated BBMV was measured after acidosis of 1-3 hours duration, a time when the effects of metabolic and respiratory acidosis on intracellular pH are thought to be most disparate (3). The results of these studies indicate that acute respiratory acidosis, but not acute metabolic acidosis, depresses Na-dependent Pi transport by the BBMV of the proximal tubule. These data indicate that a fall in systemic pH, per se, is not the signal for inhibiting Pi transport by the BBM with acidosis of short duration.

Methods

Sprague-Dawley rats weighing 175 to 200 grams and thyroparathyroidectomized 2 days prior to study were utilized in all studies.

Protocol I: To examine the effect of hypercapnia on Pi transport, 4-6 rats were anesthetized with pentobarbital (50mg/ml), 0.1 ml/100 gm BWT., a tracheostomy performed and the rats intubated with PE 200 tubing. The femoral artery was catheterized for sampling of blood and an external jugular vein catheterized for the infusion of solutions. A bolus of 140 mM NaCl, 0.2 ml/100 gm BWT was given followed by a continuous infusion of 140 mM NaCl at a rate of 2 ml/hr. The animals were placed on Harvard respirators and given tubocurare, 0.12 mg intravenously, to paralyze respiratory muscles. All rats were initially ventilated with room air during a 15 minute equilibriation period. Half the rats were then switched to 10% CO2, 20% O2, and 70% N2 for 60 minutes (**RA**) while the other half were ventilated with room air (**RAC**). Bloods for measurement of pH, pCO2, plasma P and Ca, were obtained at 0, 15 and 60 minutes of acidosis. The rats were then killed and the renal cortex obtained for preparation of BBMV.

Protocol II: Respiratory acidosis causes a rise in plasma P. To examine the impact of this rise in plasma P on Pi transport, rats with RA were compared to controls infused with Pi. Following the 15 minute equilibration period, the control rats were infused with a 45mM mixture of Na_2HPO_4 and NaH_2PO_4 (4:1 ratio) at 1.5 ml/100 gm BWT/hour (**RAC + P**) while rats with respiratory acidosis were infused with an equivalent amount of NaCl (**RA + Na**).

Protocol III: Acute metabolic acidosis was induced by the infusion of NH_4Cl for 3 hours, a time when previous experiments have demonstrated a phosphaturia (4). On the morning of study the rats were anesthetized with pentobarbital 0.1 ml/100 gm BWT, and the femoral artery and jugular vein cannulated using PE 50 tubing for arterial blood collections and intravenous infusions, respectively. A bolus of 140 mM NaCl 0.12 ml/100 gm BWT was given followed by continuous infusion of 140 mM NaCl at 0.85 ml/hour/100 gm BWT. A tracheosteomy was performed and PE 200 tubing was inserted and the rats were allowed to equilibrate for a 30 minute period. The experimental group was then infused with 140 mM NH_4Cl at 0.85 ml/hr/100 gm BWT (**MA**) while controls (**MA-C**) continued to receive 140 mM NaCl at the same rate. The infusions were continued for 3 hours and arterial blood collections were obtained for measurement of plasma P, Ca, and blood pH and pCO2. The rats were then killed and renal cortex obtained for BBMV preparation.

Analytical methods: Plasma Ca, and P and arterial pH and pCO_2 were measured as previously described (1,5). Plasma HCO_3 was calculated utilizing the Henderson-Hasselbalch equation with a solubility coefficient of 0.0309 and pKa of 6.1.

The method used to prepare BBMV was similar to that of Beck and Sacktor (6) using a $CaCl_2$ precipitation technique. Transport studies were done using standard millipore filtration techniques (1,5). Transport conditions for Pi and glucose are described in the legends for each figure and table. In each experiment BBMV were prepared from control and experimental groups and studied in parallel. As reported by others (7,8), although absolute values for Pi transport showed marked variations from day to day, relative changes resulting from experimental manipulations were extremely reproducible. Therefore, data from representative experiments are depicted in each figure.

Activities of various marker enzymes were assayed and as previously described (1,5), showed similar enrichments between groups. Fractional recoveries of protein and alkaline phosphatase were also similar. Differences between groups were assessed using paired and group t-tests where indicated (9).

Results

Acid-base parameters and plasma chemistries from RA-C and RA at killing are shown in the top of Table 1. As expected, CO_2 and HCO_3 concentrations were higher while arterial pH was lower in RA than RAC. Plasma P rose within 15 minutes in RA, and reached a value of 12.2 ± 1.9 by 60 minutes. By contrast, in RA-C plasma P remained stable. Plasma Ca did not change significantly in either group.

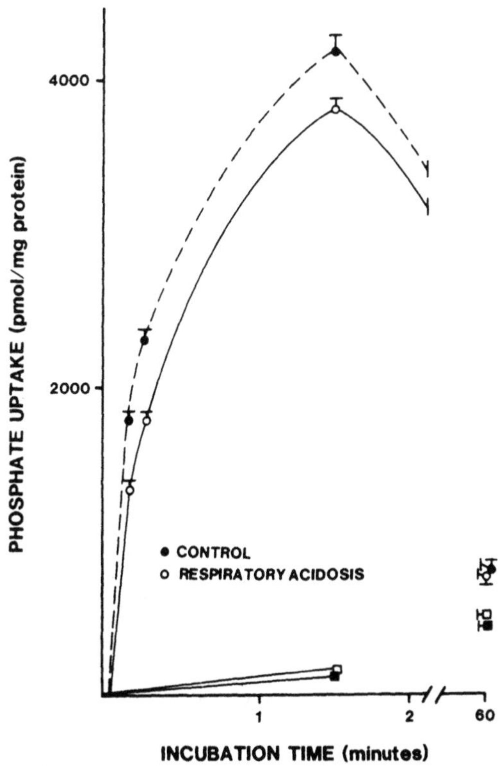

Figure 1. Time course for Pi uptake by BBMV from controls and rats with respiratory acidosis of 1 hour duration. Each point represents the mean ±SEM of 4 replicate incubations. BBMV were suspended in 320mM Sorbitol, 5 mM Hepes titrated to pH 7.5 and incubated at 22°C in a solution with a final concentration of 100 mM NaCl, 120 mM Sorbitol, 0.1 mM $K_2H^{32}PO_4$ (1 uCi/tube) 5 mM Hepes titrated to pH 7.5. In the presence of a 100mM Na-gradient (circles), Pi uptake was lower with respiratory acidosis at 10 and 15 seconds incubation, $p < 0.02$, n=4-5 (unpaired t-test). With a 100 mM K-gradient (squares) there was no difference in Pi uptake between groups.

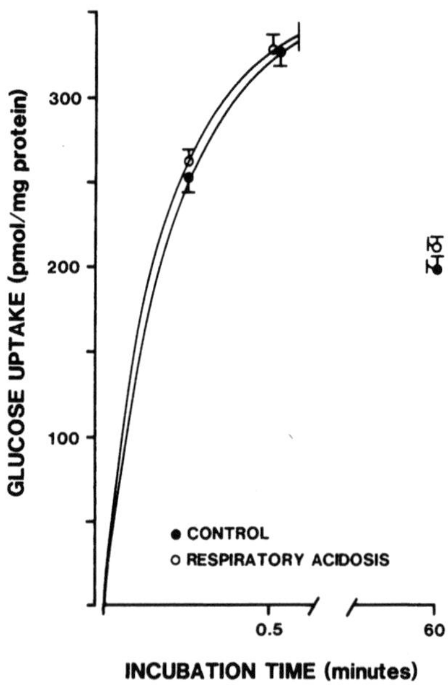

Figure 2. Time course for glucose uptake by BBMV from controls and rats with respiratory acidosis of 1 hour duration. Each point represents the mean ± SEM of three replicate incubations. BBMV were suspended in 320 mM sorbitol, 5 mM Hepes, titrated to pH 7.5 with tris base, and incubated at 22°C in uptake solution with a final concentration of 100 mM NaCl, 121 mM Sorbitol, 0.1 mM U-^{14}-D-glucose (1 uCi/tube) and 5 mM Hepes titrated to pH 7.5 with tris base. Glucose uptake was not different between groups at 0.25, 0.5, or 60 minutes of incubation.

Typical time courses for Pi and glucose transport by BBMV from RA-C and RA are shown in Figures 1 and 2. The uptake of Pi, in the presence of a 100 mM Na-gradient, increased rapidly and reached a peak within 1-3 minutes of incubation, followed by a gradual decline toward equilibrium values (60 minutes). Na-dependent Pi transport by BBMV at 10 and 15 seconds of incubation were lower in RA than in RA-C; 60 minute values were not different. Similar results were obtained in 8 additional experiments. In contrast to Pi transport, Na-dependent glucose uptake at 0.25, 0.50 and 60 minutes of incubation were not different between groups. Na-dependent Pi uptake, as the Pi concentration in the incubation medium was varied from 0.05 to 1 mM demonstrated typical saturation kinetics in both groups. When the data were plotted as Eadie-Hofstee plot the Vmax for Pi uptake, calculated from the y-intercept, was lower in RA than in RAC, 1.6 and 2.2 nmol/mg protein/10 seconds, respectively. Kt calculated from the slope of the lines was not different between groups. Similar results were obtained in 3 additional kinetic studies.

Acid-base parameters for RAC-P and RA-Na at killing are shown at the bottom of Table 1. Blood pH, PCO_2 and HCO_3 in both groups were similar to that found in Protocol I in the absence of a Pi infusion. With a Pi infusion, plasma P rose in RAC and reached levels similar to that found in RA-Na; 13.19±0.57 mg/dl in RAC-P and 12.98±.034 in RA-Na. Plasma Ca was significantly lower in RAC-P than in RA-Na; 4.92±0.26 mg/dl in RAC-P and 6.07±0.41 in RA-Na, $P<0.05$.

Na-dependent Pi transport at 0.08, 0.25 and 60 minutes of incubation for RAC-P and RA-Na are shown in Table 2. In contrast to the previous study, with equalization of plasma P, Pi uptake in the groups was not different.

Table 1. Acid-base parameters and plasma Ca and P for controls (RAC) and rats with respiratory acidosis (RA) at killing.

	Arterial			Plasma	
	pH Units	PCO_2 mm Hg	HCO_3 meq/l	Ca	P mg/dl
RAC	7.38 ±0.07	39.2 ±7.3	21.8 ±0.6	5.9 ±0.5	9.4 ±2.0
RA	7.06** ±0.05	97.9** ±17.1	26.7** ±1.2 (12)	6.0 ±0.5	12.2** ±1.9
RAC-P[a]	7.33 ±0.038	40.2 ±6.8	19.9 ±0.9	4.9 ±0.3	13.2 ±0.6
RA-Na	7.08** ±0.058	91.0** ±5.5	25.5** ±0.8 (8)	6.07* ±0.4	13.0 ±0.3

Data are mean ±SEM. a = A 45 mM phosphate solution was infused into controls at 1.5 ml/hour/100 gm BWT (RAC-P). Rats with respiratory acidosis were infused with equivalent amount of NaCl (RA-Na). * = $p<0.05$ compared to control, ** = $p<0.01$ compared to control, group t-test.

Acid-base parameters, plasma Ca, and P obtained at killing in MAC and MA are shown in Table 3. There were no significant differences in plasma Ca or P between groups. As expected, arterial pH and HCO_3 were lower in MA than in MAC.

Transport by BBMV from MAC and MA are shown in Table 4. In contrast to respiratory acidosis, both Na-dependent Pi and glucose uptakes in MAC and MA were not different.

Table 2. Pi and glucose transport by BBMV from controls infused with Pi (RAC-P) and rats with respiratory acidosis infused with NaCl (RA-Na).

	Phosphate pmol/mg protein			Glucose pmol/mg protein	
	.08[a]	.25	60	.25	60
RAC-P	240 ±33	501 ±80	340 ±53	120 ±11	100 ±9
RA-Na	253 ±51	532 ±119	357 ±32	114 ±9	113 ±16

Data are mean ± sem, n=5 with 2-3 rats per group per experiment. a= incubation in minutes. Each incubation time for each experiment was done in triplicate. BBMV were suspended in 320 mM Sorbitol, 5mM Hepes titrated to pH 7.5 and incubated at 22°C in a solution with a final concentration of 100 mM NaCl, 120 mM Sorbitol, 0.1 mM Pi or glucose, titrated to pH 7.5.

Table 3. Acid-base parameters and plasma P and Calcium in controls (NaCl) and acidotic (NH_4Cl) rats.

	Serum		Arterial blood		
	P mg/dl	Ca mg/dl	pH Units	PCO_2 mm Hg	HCO_3 mM/l
NaCl	9.67 ±0.62	3.90 ±0.21	7.36 ±0.016	35.9 ±1.9	19.7 ±0.8
NH_4Cl	10.01 ±0.87	3.98 ±0.67	7.10* ±0.036	34.5 ±2.1	10.6* ±0.9

Data are mean± SEM, n=10. The 140 mM NaCl or NH_4Cl solutions were given for 3 hours at 0.85 ml/100 gm BWT/hour. * = p<0.01 compared to the NaCl group, group t-test.

Discussion

The present studies indicate that acute respiratory acidosis inhibits Na-dependent Pi transport by the renal proximal tubular BBM. This effect of respiratory acidosis appeared to be specific for Na-dependent Pi transport, since: 1) suppression of Pi transport was absent when the Na-gradient was eliminated by substitution with 100 mM KCl and 2) Na-dependent glucose uptake was not altered.

Kinetic studies showed that acute respiratory acidosis resulted in a reduction in the maximal transport rate (Vmax) of the Pi carrier but did not alter the affinity of the carrier (Kt) for Pi. This effect could be due either to a decrease in the number of available carriers or a decrease in the rate of carrier movement.

The reduction in Pi transport by the BBM with respiratory acidosis does not appear to be due to a fall in extracellular pH, per se, since metabolic acidosis of comparable severity did not affect Pi uptake. These data are consistent with the studies of Webb et al (10) and Guntapalli et al (11) which showed that normalization of systemic pH by the infusion of HCO_3 did not prevent the phosphaturia of respiratory acidosis.

A rise in serum P is seen in respiratory acidosis, whereas serum P does not rise appreciably with metabolic acidosis generated by an NH_4Cl infusion. Recent studies by Cheng et al (12) demonstrated that a rise in serum P produced by a Pi infusion suppressed Na-dependent Pi transport at the BBM. These observations suggest that the inhibition of Pi transport observed with respiratory acidosis might be due, in part, to an increase in plasma P. To examine this possibility, we measured Pi uptake in controls when serum Pi was raised to levels similar to those found with respiratory acidosis. Under these conditions, Pi transport with respiratory acidosis was not different than controls. These data suggest that the rise in serum P noted with respiratory acidosis contributes to the suppression of Pi transport observed.

Table 4. Transport by BBMV prepared from controls (NaCl) or acidotic (NH_4Cl) rats.

	Phosphate pmol/mgprotein			Glucose pmol/mg protein	
	.25[a]	1.5	60	.25	60
NaCl	471 ±91	1026 ±164	416 ±81	268 ±88	114 ±31
NH_4Cl	578 ±175	1182 ±336	378 ±104	227 ±55	116 ±23

Data are mean ± SEM, n=4, with 2-3 rats per group per experiment. a = incubation time in minutes. The solutions were given for 3 hours at a rate of 0.85 ml/100 gm BWT/hour. Each incubation time for each experiment was done in triplicate. BBMV were suspended in 320 mM Sorbitol, 5mM Hepes titrated to pH 7.5 and incubated at 22°C in a solution with a final concentration of 100 mM NaCl, 120 mM Sorbitol, 0.1 mM Pi or glucose, titrated to pH 7.5. There were no significant differences in transport at any incubation period.

Alterations in serum Ca may also affect renal Pi transport (13,14,15). In the present study, total plasma Ca did not change. It is possible, however, that ionized Ca may have increased with acidemia as reported by others (4,16). An elevation in ionized Ca under hypocalcemic conditions has been reported to enhance renal Pi reabsorption and increase Pi efflux from perfused proximal tubules (15). Thus, even if ionized Ca was higher in the acidotic animals than in controls, an increase and not a reduction in Pi transport would be expected.

A fall in intracellular pH presumably accompanies the fall in extracellular pH with respiratory acidosis (3). Several studies have indicated that alterations in intracellular pH may modulate the activity of transport systems at the luminal membrane of the proximal tubule including Pi transport (17,18). It has been suggested that acidosis induced by an increase in CO_2 concentration is associated with a more rapid reduction in intracellular pH than that produced by a reduction in HCO_3 concentrations. If a fall in intracellular pH was the signal to the transporter to reduce Pi transport, than the achievement of a lower intracellular pH with respiratory acidosis than with metabolic acidosis within the time frame of the present study might explain the different effects of respiratory and metabolic acidosis on BBM Pi transport. The results of the present study cannot exclude a role for changes in intracellular pH in mediating the phosphaturia.

In summary, respiratory acidosis of 1 hour duration specifically suppressed Pi transport by the renal proximal tubular BBM while metabolic acidosis of 3 hours duration was without effect. A possible signal for this effect is the increase in plasma P occurring with respiratory acidosis. However, a fall in intracellular pH cannot be excluded as a contributory factor. By contrast, a change in systemic pH alone does not appear to play an important role in this process.

References

1. B.S. Levine, K. Ho, J.A. Kraut, J.W. Coburn and K. Kurokawa. Effect of metabolic acidosis on phosphate transport by the renal brush border membrane. Biochem. Biophys. Acta. 727:7 (1983).
2. S.A. Kempson. Effect of metabolic acidosis on renal brushborder membrane adaptation to low phosphorus diet. Kidney Internat. 22:225 (1982).
3. A. Roos and W.F. Boron. Intracellular pH. Phys. Rev. 61(2):296 (1981).
4. J. Guntupalli, B. Eby and K.Lau. Mechanism for the phosphaturia of NH_4Cl: dependence on acidemia but not on diet PO_4 or PTH. Am. J. Physiol. 242(11):F552 (1982).
5. B.S. Levine, K. Ho, A. Hodsman, K. Kurokawa and J.W. Coburn. Early renal brush border membrane adaptation to dietary phosphorus. Min. Elec. Metab. 10:222 (1984).
6. J.C. Beck and B. Sacktor: The sodium electrochemical potential mediated uphill transport of D-glucose in renal brush border membrane vesicles. J. Biol. Chem. 253:5531 (1978).
7. P.W. Barrett, J.M. Gertner and H. Rasmussen. Effect of dietary phosphate on transport of pig microvillus vesicles. Am. J. Physiol. 8:F352 (1980).
8. C. Evers, H. Murer and R. Kinne. Effect of parathyrin on the transport properties of isolated renal brush-border vesicles. Biochem J. 172:49 (1978).
9. G.W. Snedcor and W.G. Cochran. Statistical methods. Ames, Iowa: Iowa State University Press pp. 91 (1967).

10. R.K. Webb, P.B. Woodhall, C.C. Tisher, G. Glaubiger, F.A. Neelon and R.R. Robinson. Relationship between phosphaturia and acute hypercapnia in the rat. J Clin. Invest. 60:829 (1977).
11. J. Guntupalli, and E. Bourke. Evidence that the phophaturic effect of acute hypercapnia is not due to the low systemic pH of hypercapnia. Kidney Internat. (in press).
12. L. Cheng, C. Dersch, E. Kraus, D. Spector and B. Sacktor. Renal adaptation to phosphate load in the acutely thyroparathyroidectomized rat: rapid alteration in brush border membrane phosphate transport. Am. J. Physiol. 246(15):F488 (1984).
13. H. Oberleithner, R. Greger and F. Lang. Role of calcium in the decline of phosphate reabsorption during phosphate loading in acutely thyroparathyroidectomized rats. Pflugers Arch. 374:249 (1978).
14. A. Frick and I. Durasin. Maximal reabsorptive capacity for inorganic phosphate (Tmpi) in the absence of parathyroid hormone: Decrease on the Tmpi during prolonged administration of phosphate and the role of calcium. Pflugers Arch. 377:9 (1978).
15. W.N. Suki. Effect of serum calcium on phosphate transport. Adv. Exp. Med. Biol. 178:25 (1984).
16. B. Hoffken, D.K. Parkinson, P. Storms and I.C. Raddle. Effects of alterations of blood pH on calcium ion activity in rat plasma. Clin. Orthop. 78:30 (1971).
17. P.S. Aronson, J. Nee and M.A. Suhm. Modifier role of internal H^+ in activating the Na^+-H^+ exchanger in renal microvillus membrane vesicles. Nature 299:161 (1982).
18. J.L. Kinsella and P.S. Aronson. Properties of the Na^+ -H^+ exchanger in renal microvillus membrane vesicles. Am. J. Physiol. 238:F461 (1980).

EFFECTS OF GLUCAGON ON RENAL AND EXTRARENAL HANDLING OF Pi IN RODENTS: EVIDENCE FOR Pi MOBILIZING ACTIVITY

R.C. Mühlbauer, J.-P. Bonjour and H. Fleisch

Department of Pathophysiology, University of Bern
Murtenstrasse 35,
CH-3010 Bern, Switzerland

INTRODUCTION

Glucagon is known to be released during fasting (1) and between meals (2), and it is generally accepted that the presence of this hormone is important in the prevention of hypoglycemia. Furthermore, glucagon has also been found to be phosphaturic in man (3) and animals (4,5) fed diets containing normal amounts of inorganic phosphate (Pi). Since these phosphaturic effects are associated with decrements in Plasma Pi concentration ($[Pi]_{Pl.}$) one has to conclude that glucagon had acted through a change in the renal handling of Pi.

It has been shown that phosphaturia induced by other agents such as parathyroid hormone (PTH) is blunted in rats fed a low Pi (LP) diet (6,7). The renal adaption mechanism (8,9), by which the LP diet increases the threshold of Pi excretion into urine, is considered to be responsible for this blunting. Whether glucagon induced phosphaturia would likewise become blunted in the LP diet condition is presently not known.

In a first step we show that, as already demonstrated with PTH, phosphaturia induced by glucagon is also blunted in rats fed LP diet. In these studies it appeared that while in rats fed high Pi diet, $[Pi]_{Pl.}$ decreased during glucagon infusion, an increment of $[Pi]_{Pl.}$ was observed during glucagon administration in rats fed LP diet. This finding suggests that glucagon might act by another mechanism, i.e. to mobilize Pi from body stores.

This question was further examined in a separate experiment using the mouse. The animals were again fed a LP diet because in this condition the Pi fluxes from gut into blood and from blood into urine are virtually nil, provided that the amount of Pi filtered by the kidney does not exceed the maximal tubular reabsorptive capacity. It was found that in mice fed LP diet glucagon indeed increases $[Pi]_{Pl.}$ indicating that this hormone may be a potent Pi mobilizing hormone.

MATERIALS AND METHODS

Diets

Until the experimental phase, all animals were reared in standard animal facilities which comply the Swiss and NIH guidelines for care and use of laboratory animals. They had free access to tap water, and were fed ad libitum a standard laboratory chow (Altromin 1314, Klingentalmühle AG, Kaiseraugst, Switzerland) containing 0.9 g calcium, 0.7 g phosphorus and 80 I.U. vitamin D_3 per 100 g dry weight (DW).

Then the animals were fed for 9 days diets (Sodi 34, Klingentalmühle) containing either low phosphorus (LP), 0.25 g/100 g DW, or high phosphorus (HP) 1,2 g/100 g DW. Both diets contained a similar amount of calcium, sodium and vitamin D. This was achieved by addition of appropriate amounts of neutral sodium phosphate, sodium chloride and Ca gluconate to the basic diet low in phosphate, calcium and vitamin D_3. This special food was always given between 4 and 6 p.m., including the night before the experiment. The animals had free access to distilled water. To allow complete food consumption by the animals of both dietary groups, appropriate food portions were given (group feeding).

Clearance Experiments in Rats

Male Wistar rats weighing 182 ± 4.4 g ($\bar{x}$ ± SD) (from our own colony) were group-fed the above described LP and HP diets for 9 days. On the tenth day after onset of the group-feeding, they were anesthetized by intraperitoneal injection of Inactin (Byk-Gulden, Konstanz, FRG) at a dose of about 100 mg/kg BW. The rats were then placed on an operating table set at 37°C.

After tracheotomy, PP_{50} catheters (Portex Lt. Hythe Kent, England) were inserted into the right carotid artery, the right jugular vein, the left femoral vein and a PP 110 catheter into the bladder. After the surgical preparation all animals received a priming injection (0.5 ml/100 g BW) of a 0.15 M NaCl solution containing inulin at a concentration of 40 g/l through the jugular vein catheter. This was followed by a constant infusion of the same solution described above throughout the experiment at a rate of 3 ml/hr. During the first 105 minutes, a 0.15 M NaCl solution was also infused into the femoral vein at a rate of 3 ml/hr. Then a priming dose (2.5 µg/rat) of porcine Glucagon (Novo Industri A/S, Copenhagen, Denmark) immediately followed by a sustaining dose of 25 ng/min/100 g rat was infused by this route for the next 105 minutes. This was followed by 0.15 M NaCl solution for an additional hour at the same rate.

After an equilibration period of 45 min, the first urine was collected for 60 min (control, pre-glucagon period). Then after a 15 min equilibration period, three urine collections of 30 min each (glucagon period) were made. During the next 15 min the urine was discarded and a last urine collection was made during 60 min (post-glucagon period). At the midtime of each urine collection, a blood sample of about 80 µl was obtained from the carotid artery.

Assessment of $[Pi]_{Pl.}$ in Mice

In a second experiment, the effect of zinc-protamine-glucagon Novo (a generous gift from Aldepha AG, Zürich, representing Novo Industri in Switzerland) on $[Pi]_{Pl.}$ was measured in two months old female NMRJ mice from our colony fed the above described LP diet for nine days. Glucagon was given at 16 µg/g BW and the controls received the same volume (5µl/g BW) of the zinc-protamine solvent. Blood specimens were obtained by retroorbital puncture under ether anesthesia using microhematocritcapillaries.

Analytical Procedures

Plasma samples for Pi and/or for Inulin determination were deproteinized with 10 % trichloracetic acid (TCA) on the day of the experiment. Pi in plasma and urine was determined using a molybdate malachite green reagent as previously described (10). Inulin was determined using diphenylamine (10).

The content of phosphorus in the diet was determined after incineration using the above mentioned method. Ca in the diet was measured by complexometric titration with EGTA.

Statistical Analysis

The experimental results are expressed as mean values $\pm$ SEM. The significance of differences between groups was evaluated by the Student t-test for unpaired samples and if appropriate for paired samples.

RESULTS

Effect of Glucagon Infusion on the Fractional Excretion of Pi, (FE_{Pi}) and on $[Pi]_{Pl}$. in Rats fed HP and LP Diets

The FE_{Pi} before, during and after glucagon infusion in rats fed HP or LP diets is shown in Fig. 1. In rats fed HP diet glucagon infusion elicited a prompt phosphaturic response (increase of FE_{Pi}). This effect was not reversible within 45 minutes after stopping the hormone infusion. In rats fed LP diet, FE_{Pi} was, as expected, significantly lower in the pre-glucagon control period as compared to HP diet fed rats. In contrast to HP diet fed rats, glucagon infusion in LP diet fed animals did not elicit a phosphaturic response. In the post-glucagon control period some increase of FE_{Pi} could be observed, which, however, did not reach significance level.

The upper panel of Fig. 1 shows the corresponding values for $[Pi]_{Pl}$, which was, as expected, significantly lower during the pre-glucagon period in rats fed LP as compared to HP diet. In the group fed HP diet, $[Pi]_{Pl}$. decreased already after 15 minutes of glucagon infusion with a further decrease up to 90 minutes during the hormone infusion and a rebound of $[Pi]_{Pl}$. after stopping the treatment. In rats fed LP diet, after 90 minutes of glucagon infusion, however, a significant increment of $[Pi]_{Pl}$. was observed. $[Pi]_{Pl}$. of rats fed LP diet actually exceeded the value observed in HP diet fed animals.

Table 1 shows the inulin clearance (C_{In}), filtered load of Pi and urine volume data from the animal groups reported above.

C_{In} of the rats fed HP diet increased during glucagon infusion and decreased promptly to control levels after stopping the hormone infusion. C_{In} appeared lower, however not significantly, in LP diet fed rats during the pre-glucagon period as compared to HP diet fed animals. Glucagon infusion increased C_{In} to about the same level as in HP diet fed counterparts.

The changes in urine volume throughout the experiment was very similar in both dietary groups. A nearly 3-fold increase was observed during glucagon infusion and this persisted during the post-glucagon period.

The filtered load of Pi(FL_{Pi}) in rats fed LP diet was, as expected, significantly lower during the pre-glucagon control period as compared to HP diet fed animals. Glucagon infusion in HP diet rats increased FL_{Pi}

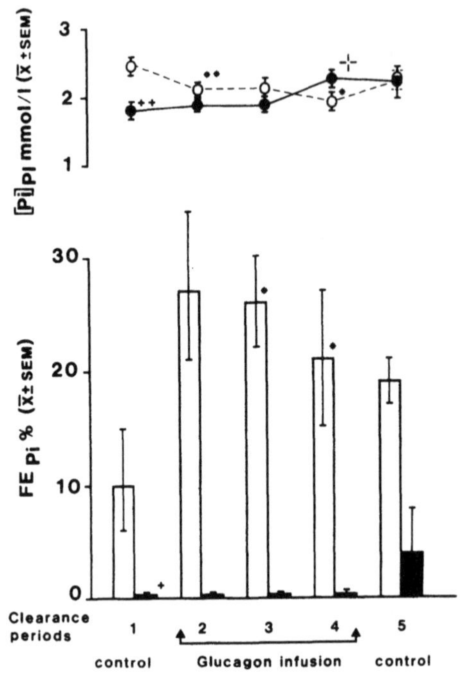

Figure 1

Effect of glucagon infusion on plasma Pi concentration (upper panel) and fractional excretion of Pi in the urine (lower panel) in rats fed a low Pi (LP ●, ■) or a high Pi (HP ○, □) diet for 9 days. Number of animals = 6 in each dietary group. Values are means ± SEM. * $p < 0.05$; ** $p < 0.01$ as compared with the initial value of the HP diet group. + $p < 0.05$; ++ $p < 0.01$ as compared with the HP diet group. -¦- $p < 0.05$ as compared with the value obtained during the previous (period 3) clearance period.

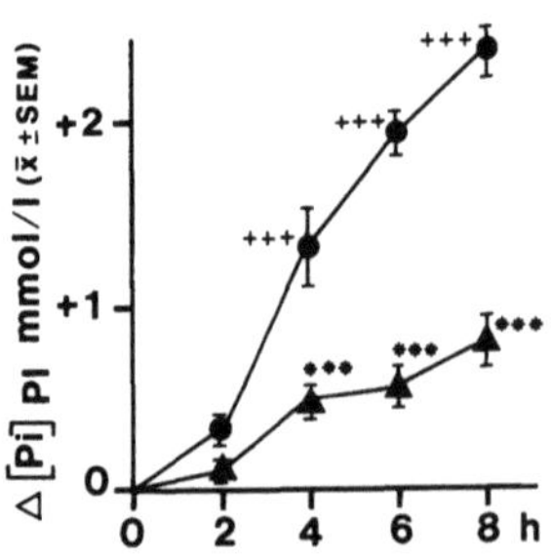

Figure 2

Changes of plasma Pi concentration following a single injection of solvent (▲) or zinc protamin glucagon at the dose 16 (●) µg/g BW subcutaneously in mice fed low Pi diet. Number of mice = 13 for control and 5 for glucagon. Values are means ± SEM.
*** $p < 0.001$ as compared to the initial value. +++ $p < 0.001$ as compared to the solvent group.

Table 1

	Control	Glucagon			Control
Urine collection	1	2	3	4	5
C_{In} ml/min HP diet	1.69 ± 0.36	2.12 ± 0.39	1.94 ± 0.34	2.46 ± 0.52*	1.70 ± 0.30
LP diet	0.90 ± 0.10	1.95 ± 0.34**	1.97 ± 0.46*	2.10 ± 0.50*	1.99 ± 0.53
Urine volume ml/30' HP diet	1.01 ± 0.28	2.72 ± 0.59**	3.17 ± 0.69*	3.22 ± 0.70*	3.54 ± 0.48**
LP diet	1.10 ± 0.53	2.71 ± 0.82*	2.76 ± 0.78*	2.82 ± 0.75*	2.77 ± 0.31***
Filtered load of Pi umoles/min HP diet	4.07 ± 0.82	4.57 ± 1.00	4.20 ± 0.95	4.85 ± 1.24	3.69 ± 0.60
LP diet	1.65 ± 0.24+	3.60 ± 0.60*	3.56 ± 0.76*	4.57 ± 1.04*	4.28 ± 1.03*

* $p < 0.05$; ** $p < 0.01$; *** $p < 0.005$ significantly different from the first urine collection of the same dietary group. + $p < 0.05$ significantly lower as compared to the HP diet group.

only slightly, whereas in LP diet fed animals it increased FL_{Pi} significantly to levels not significantly different from HP diet fed counterparts. Thus, glucagon did not elicit a phosphaturic response in LP diet fed rats despite a similar FL_{Pi}.

Effects of Glucagon in Mice

Before conducting the experiment described below, three pilot experiments were performed (results not given). Glucagon was shown to be also phosphaturic in mice fed HP diet. Furthermore the blunting of glucagon induced phosphaturia was confirmed in the mouse fed LP diet. Finally it was shown that a single injection of zinc-protamin glucagon (slow glucagon) caused similar effects as crystalline pure glucagon.

Time Course of the Response of $[Pi]_{Pl}$ to a Single Injection of Zinc-Protamin Glucagon to Mice Fed LP Diet

Fig. 2 shows the time course of $[Pi]_{Pl}$ during 8 hours fasting in mice fed LP diet injected either with solvent or with glucagon. The spontaneous increase of $[Pi]_{Pl}$ in control mice became significant after 4 hours. Glucagon injection was followed by an important increment of $[Pi]_{Pl}$ above the control values, which reached significance by 4 hours.

DISCUSSION

The present work has confirmed the phosphaturic and concomitant hypophosphatemic effect of glucagon in rats fed HP diet (4,5). It also demonstrated that the phosphaturic effect of glucagon infusion was blunted in rats previously fed LP diet. This resistance to the phosphaturic effect cannot be explained by a different vascular effect of glucagon in LP diet fed rats since the elevation of glomerular filtration rate and urinary volume, a salient feature of glucagon treatment (11,12,13,14), was similar in both dietary groups, thus leading to comparable filtered load of Pi in the two.

This blunted phosphaturia in LP diet fed animals was similar to what has been shown with parathyroid hormone infusion (6,7) in the rat. It shows that the powerful dietary adaptation mechanism (8,9), which increases the tubular reabsorptive capacity in animals fed LP diet, could not be overcome within 8 hrs of glucagon treatment.

Furthermore, this study has shown that glucagon is a potent Pi mobilizing hormone. This became apparent first by the fact that the glucagon induced increase of Pi found in urine in HP fed rats by far exeeded the decrease of Pi in plasma and ECF. It was then also recognized by the elevation of $[Pi]_{Pl}$ observed in both rats and mice fed LP diets. In animals fed HP diet, this mobilization of Pi could not be detected in plasma, probably because in view of the lower renal transport threshold, the Pi was immediately excreted in the urine. One might argue, that part of the Pi found in blood and urine could have originated from food remnants that were still present in the gut at the beginning of the experiment. However, an effect of glucagon on the Pi absorption in the gut has to our knowledge not been described. Furthermore, inspection of the gastrointestinal tract in mice group-fed HP and LP diets at 8.00 a.m., corresponding to the time at which the experiments were started, showed that the stomach, duodenum and jejunum were practically empty. The ileum contained only a few food remnants and in the colon only small amounts of faeces were found, excluding therefore that Pi contained in food remnants in the gut as an important source of Pi. Finally, in a previous experiment (15) in mice fed very LP diet (0.03 g P/100 g DW) 8 hour fasting induced a similar increment of $[Pi]_{Pl}$. 0.71 $\pm$ 0.17 mmol/l (n=8) as compared to 0.82 $\pm$ 0.14 mmol/l (n=13) obtained in the present study. It may therefore be concluded that, in LP diet fed mice, most if not all of the Pi appearing in plasma stems from other sources than gut. The source from which Pi is mobilized by glucagon is not yet known. Bone (16) and liver (17) are candidates since they have been proposed as sources of the Pi mobilized during fasting.

The phosphatemic response to fasting and glucagon is similar. Since glucagon is released during the intermeal periods to protect the organism from hypoglycemia (2) it is conceivable that the spontaneous mobilisation of Pi from body stores, readily observed during this time as an increment of $[Pi]_{Pl}$ in fasting animals adapted to LP diet may be due to the endogenously secreted glucagon. However, at present we cannot yet say whether the spontaneous or by glucagon injection induced increase of $[Pi]_{Pl}$ were direct effects of glucagon or are mediated by other hormones or factors.

It should be mentioned that the doses of glucagon used in the present experiments were high and probably not physiological. However, the physiological amounts of glucagon released between meals are unknown in mice. Furthermore, in our experiments the hormone was given to intact animals during the intermeal period, a time where one has to expect already high levels of endogenous hormone. Thus, a high sensitivity to exogenous glucagon cannot be expected in this condition. A similar problem is encoutered with parathyroid hormone where much higher doses are needed to elicit a phosphaturic response in intact animals as compared to the doses used in parathyroidectomized animals. Furthermore, if the liver would be the target organ for the Pi mobilizing effect of glucagon described here, it would make sense that high systemic levels of glucagon are needed to elicit an effect, because the level of glucagon in portal blood can be very much higher than the one normally found in plasma (18).

In conclusion, the present work has confirmed that glucagon is a potent phosphaturic agent. It can, however, not overcome the dietary adaptation mechanism. Furthermore, glucagon appears to be a potent mobiliser of Pi from body stores. Therefore glucagon might well be a hormone important in Pi homeostasis.

ACKNOWLEDGEMENTS

The expert technical assistance of Mrs. A. Moosavi-Wyler, Mrs. R. Rubli-Bolinger and Mr. D. Pulver is gratefully acknowledged. We thank Dr. Marco Cecchini for stimulating discussions, Mrs. C. Stieger for drawing the Figures and Mrs. B. Meier and Mrs. C. Pantaleo for preparing the manuscript.

This work has been supported by the Swiss National Science Foundation (3.937.82) and the Ausbildungs- und Förderungsfonds der Arbeitsgemeinschaft für Osteosynthese (AO), Chur, Switzerland.

REFERENCES

1. Saudek, C.D., P.R. Boulter, and R.A. Arky. The natriuretic effect of glucagon and its role in starvation. J. Clin. Endocrinol. Metab. 36:761-765, 1973.
2. Unger, R.H. Insulin-glucagon relationships in the defense against hypoglycemia. Diabetes 32:575-583, 1983.
3. Elrick, H., E.R. Huffman, C.J. Hlad jr., N. Whipple, and A. Staub. Effects of glucagon on renal function in man. J. Clin. Endocrinol. and Metab. 18:813-824, 1958.
4. Bailly, C., and C. Amiel. Effect of glucagon on magnesium renal reabsorption in the rat. Pflügers Arch. 392: 360-365, 1982.
5. Butlen, D., and S. Jard. Renal handling of 3'-5'-cyclig AMP in the rat. The possible role of luminal 3'-5'-cyclic AMP in the tubular reabsorption of phosphate. Pflügers Arch. 331:172-190, 1972.
6. Gloor, H.J., J.-P. Bonjour, J. Caverzasio, and H. Fleisch. Resistance to the phosphaturic and calcemic actions of parathyroid hormone during phosphate depletion. Prevention by 1,25-dihydroxyvitamin D_3. J. Clin. Invest. 63:371-377, 1979.
7. Steele, T.H., J.L. Underwood, B.A. Stromberg, and C.A. Larmore. Renal resistance to parathyroid hormone during phosphorus deprivation. J. Clin. Invest. 58:1461-1464, 1976.
8. Steele, T.H., and H.F. DeLuca. Influence of dietary phosphorus on renal phosphate reabsorption in the parathyroidectomized rat. J. Clin. Invest. 57:867-874, 1976.
9. Tröhler, U., J.-P. Bonjour, and H. Fleisch. Inorganic phosphate homeostasis. Renal adaptation to the dietary intake in intact and thyroparathyroidectomized rats. J. Clin. Invest. 57:264-273, 1976.
10. Mühlbauer, R.C., J.-P. Bonjour, and H. Fleisch. Tubular localization of adaptation to dietary phosphate in rats. Am. J. Physiol. 233 (Renal Fluid Electrolyte Physiol. 2):F342-F348, 1977.
11. Franke, H., G. Gronow, and K. Petersen. Glucagon induced funtional changes of isolated perfused rat kidney. Curr. Probl. Clin. Biochem. 8:424-434, 1977.
12. Levy, M. The effect of glucagon on glomerular filtration rate in dogs during reduction of renal blood flow. Can. J. Physiol. Pharmacol. 53:660-668, 1975.
13. Parving, H.-H., J. Noer, H. Kehlet, C.E. Morgensen, P. Aa. Svendsen, and L. Heding. The effect of short-term glucagon infusion on kidney function in normal man. Diabetologia 13:323-325, 1977.
14. Uranga, J., R. Fuenzalida, A.L.Rapoport, and E. del Castillo. Effect of glucagon and glomerulopressin on the renal function of the dog. Horm. Metab. Res. 11:275-279, 1979.
15. Mühlbauer, R.C., J.-P. Bonjour, and H. Fleisch. Abnormal hyper-phosphatemic response to fasting in X-linked hypophosphatemic mice. Min. Electrol. Metab. 10:362-365, 1984.
16. Talmage, R.V., S.H. Doppelt, and J.H. Postma jr. Observations on the relationship of parathryoid hormone and calcitonin to plasma and liver phosphate. Proc. Soc. Exp. Biol. Med. 153:131-137, 1976.
17. Talmage, R.V., C.J. VanderWiel, and D.B. Raneri. Factors affecting parathyroid hormone-induced hypophosphatemia and ^{32}P specific activity in thyroparathyroidectomized rats. Calcif. Tissue Res. 27:239-246, 1979.
18. Blackard, W.G., N.C. Nelson, and S.S. Andrews. Portal and peripheral vein immunoreactive glucagon concentrations after arginine or glucose infusions. Diabetes 23:199-202, 1974.

RENAL HANDLING OF CALCIUM AND MAGNESIUM

CALCIUM TRANSPORT BY THE PROXIMAL TUBULE

Fred S. Wright
Yale University
VA Medical Center
New Haven, CT

Karol Bomsztyk
University of Washington
Nephrology Division
Seattle, WA

The proximal tubule contributes to renal handling of calcium by reabsorbing approximately 60% of the filtered quantity (Lassiter et al., 1963). Studies employing in vivo and in vitro perfusion of single proximal tubules have begun to provide some information about the transport mechanisms that contribute to recovery of filtered calcium by the proximal tubule.

RELATION TO SODIUM TRANSPORT

One feature of renal calcium handling that has been considered in discussions of transport mechanisms is that fractional rates of calcium absorption and sodium absorption generally change in parallel. This parallelism has been observed in both whole kidney clearance experiments (Walser, 1961) and in micropuncture studies of proximal tubule function (Lassiter et al., 1963; Agus et al., 1973; Beck and Goldberg, 1973; Edwards et al., 1973; Poujeol et al., 1976). In view of this relation several authors have concluded that proximal calcium absorption is "secondary to the absorption of sodium" (Suki, 1979) or is "intimately linked" to transport of sodium (Poujeol et al., 1976).

In the course of a recent series of studies involving in vivo microperfusion of proximal tubules in exposed kidneys of anesthetized rats, we were able to examine relative rates of sodium and calcium transport under conditions not previously tested (Bomsztyk et al., 1984). We perfused single proximal tubules with solutions having constant total sodium and calcium concentrations, but different free ionic calcium concentrations. Mannitol was added to perfusion solutions to reduce net fluid absorption to zero. In one series of experiments a control perfusion solution, resembling fluid reaching the late proximal tubule, was modified by adding 0.1-1.6 mM EGTA. In another series, perfusion fluid chloride was replaced with either nitrate, thiocyanate, sulfate or citrate. These non-chloride anions complexed varying amounts of the total calcium. Total calcium concentration in nanoliter volumes of tubule fluid was measured by atomic absorption spectrometry (Bomsztyk et al., 1984). In contrast to previous studies in which a number of experimental maneuvers resulted in generally parallel changes in sodium and calcium fluxes, the changes in anion composition created conditions in which changes in sodium and calcium fluxes were dissociated (Bomsztyk et al., 1984). Adding EGTA caused calcium transport to decrease and then reverse direction, whereas sodium transport rates were not affected. When chloride was replaced with nitrate, calcium absorption was not affected, but sodium absorption was decreased. When chloride was replaced with thiocyanate, calcium absorption was increased while sodium absorption was not changed. Finally, replacement of chloride with either sulfate or citrate resulted in reductions in calcium absorption approximately two times larger than the reductions in sodium absorption.

ACTIVE TRANSPORT

In order to assess driving forces affecting calcium transport we measured both transepithelial voltage and the activity of calcium ions in capillary plasma and tubule fluid (Bomsztyk et al., 1984). Voltage was measured with conventional KCl-filled microelectrodes. Calcium activity was measured with liquid ion-exchanger microelectrodes. The ion-selective electrodes exhibited near Nernstian behavior in the concentration range between 10^{-6} to 10^{-2} molar.

By measuring calcium activity in perfusion fluid and in fluid reaching the end of the proximal tubule we determined mean luminal calcium activity and the average transepithelial difference in calcium activity. Using the additional information from voltage measurements with conventional microelectrodes, we calculated the electrochemical activity difference as the driving force term in the Goldman-Hodgkin-Katz equation. In experiments with the control perfusion solution the average calcium activity was slightly lower in the tubule lumen than in capillary plasma, and the transepithelial voltage was slightly lumen positive. These oppositely directed forces cancelled, and the transepithelial electrochemical driving force was not different from zero. Despite this lack of external driving force we measured a significant net absorptive flux of calcium, 3.5 pmol/min·mm $\pm$ 0.3 SE (Bomsztyk et al., 1984). These results suggest that an active transport component contributes to proximal calcium reabsorption. Ullrich et al. (1976) reported results of experiments employing radioactive tracer flux measurements indicating that net calcium absorption can proceed against an electrochemical gradient. Their experiments also employed in vivo microperfusion of rat proximal tubules. Contrasting results were reported recently by Ng et al. (1984). In this study bidirectional fluxes of ^{45}Ca were used to assess calcium transport by isolated rabbit proximal tubules perfused in vitro. With perfusion and bathing solutions chosen to eliminate calcium concentration differences, transepithelial voltage, and net fluid movement, net calcium flux was not different from zero. At present it is not clear whether the absence of an active transport component in rabbit tubules indicates a species difference, an effect of in vitro perfusion, or some other difference in the treatment of the perfused tubules.

DIFFUSION

We have also obtained evidence that proximal calcium transport has an important passive component (Bomsztyk et al., 1984). By modifying the anion content of our perfusion solutions, and thus varying the electrochemical driving force for calcium transport, we were able to relate changes in net calcium transport to changes in the transepithelial electrochemical concentration difference. Net absorption was decreased, and then converted to net secretion, when luminal calcium activity was decreased and when the transepithelial voltage became lumen negative. The least squares linear regression line, that best fits the relation between net flux of calcium and the transepithelial electrochemical concentration difference, has a slope of 18.9 nl/min. Dividing by the average perfused length in these experiments, 1.95 mm, the apparent permeability for calcium is 16.1 x 10^{-7} cm^2/s. Assuming a tubule radius of 15 um the apparent permeability can be expressed as a function of tubule surface area, and is 1.7 um/s. This permeability value is similar to values reported by Murayama et al. (1972) and by Ullrich et al. (1976), and is 50-fold higher than the value recently reported by Ng et al. (1984) for rabbit proximal tubules perfused in vitro. The higher values for calcium permeability in rat proximal tubules are nearly equal to previously reported values for Na, K and Cl permeability determined in rat tubules from electrophysiological measurements (Ullrich, 1973).

SOLUTE-SOLVENT COUPLING

Although we added mannitol to perfusion fluids in an effort to eliminate net fluid movement, a small but significant rate of fluid absorption persisted in some experiments. Therefore, to assess the possible contribution that net volume transport could make to proximal calcium transport, we undertook another series of experiments in which a wider range of perfusion fluid osmolalities was tested (Bomsztyk and Wright, 1982). Transepithelial fluid transport was varied by increasing the osmolality of luminal perfusion fluid. When tubules were perfused with a solution similar in composition to tubule fluid normally present in the distal portion of the proximal tubule, we observed net fluid absorption; adding mannitol (30, 60 and 75 mM) to this solution first reduced fluid absorption to zero, and then reversed the direction of fluid movement. Associated with these changes in net fluid transport, net calcium transport also decreased and then reversed. With 30 mM mannitol added to the perfusion solution (total osmolality 322 mOsm/kg) proximal fluid transport was 0.04 nl/min·mm ± 0.09 SE and was not different from zero. Calcium was absorbed from this solution at 1.7 pmol/min·mm ± 0.3 SE. This significant transport rate in the absence of net fluid movement indicates that a solute-solvent coupling mechanism is not required for calcium transport. With perfusion solutions both more and less concentrated it was possible to vary net volume transport between 2.3 nl/min·mm (absorption) and -2.2 nl/min·mm (secretion). Calcium flux measurements with these solutions showed that calcium transport is affected by osmotically driven water flow. This result contrasts with observations in four recent studies employing rabbit proximal tubules perfused in vitro (Hierholzer et al., 1980; Jacobson et al., 1982; Corman and DiStephano, 1983; and Ng et al., 1984). It is possible that our study, employing in vivo perfusion of rat proximal tubules, succeeded in demonstrating solute-solvent coupling because we were able to produce larger changes in net fluid transport than were effected in the studies with isolated tubules. It is also possible that sieving coefficients are lower for rabbit proximal tubules than for rat tubules and, thus, are not as responsive to changes in volume flux as rat tubules are.

SODIUM-CALCIUM INTERACTION

In view of the wide variety of experimental circumstances in which sodium and calcium absorption by the proximal tubule change in parallel, even though our anion substitution experiments had shown that changes in sodium and calcium absorption could be dissociated, we also wanted to examine the effect on calcium absorption of reducing luminal sodium concentration. In this series of experiments sodium in perfusion fluid was replaced with either tetramethylammonium (TMA) or lithium (Bomsztyk and Wright, 1983). A control solution similar to fluid normally present in the late proximal tubule was again used for comparison. The 145 mM Na in this solution was partially replaced by either 35 or 85 mM TMA or by 135 mM Li. For each solution mannitol was added to reduce net volume flux to zero. Total sodium and calcium concentrations were measured by atomic absorption spectrometry. Ionic calcium was measured with liquid ion-exchanger microelectrodes. Transepithelial voltage was measured with KCl-filled microelectrodes.

As expected, reducing sodium concentration in perfusion fluid reduced net sodium absorption. Sodium was absorbed from the control solution; sodium entered the tubules perfused with the TMA and Li solutions. Calcium concentrations in collected tubule fluid samples were uniformly lower than in perfusion fluids, a reflection of net calcium absorption in each case. Two observations regarding net calcium absorption are noteworthy. First, reduction of luminal Na concentration stimulated net calcium absorption. Second, a greater reduction in luminal Na concentration was required for the same effect when Li was the replacement cation. Measurements of transepithelial voltage provide a partial explanation for these results. With all four perfusion solutions the lumen was positive with respect to the capillary. With the control solution the voltage averaged 0.8 mV. When tubules were perfused with the TMA solutions, voltage increased to 3.2 and 6.6 mV. With the lithium solution the voltage was not different from control values.

With data for lumen and capillary calcium activity, and transepithelial voltage it was possible to relate net calcium fluxes to the transepithelial electrochemical gradient. Recall that in our experiments involving anion substitutions (Bomsztyk et al., 1984) the slope of the linear relation between net calcium flux and transepithelial electrochemical gradient, the apparent permeability for calcium, was approximately 20×10^{-7} cm^2/s. In the present experiments (Bomsztyk and Wright, 1983) the results with the control and the TMA solutions followed this same relation. The increase in calcium absorption from the solutions in which Na was replaced with TMA appear to be fully accounted for by the increase in transepithelial driving force resulting from the increase in lumen positive transepithelial voltage. The results with the lithium solution, however, do not conform to this relation. Replacement of most sodium by lithium stimulated calcium absorption even though the transepithelial electrochemical potential difference was reduced to zero. Thus, with lithium replacement the increase in calcium absorption was not the result of a change in passive driving force. Rather, it appears that an increase in an active transport component is responsible. One possible mechanism that might be involved is a basolateral calcium-sodium exchanger that links downhill movement of sodium from blood to cell with uphill extrusion of calcium from cell to blood.

SUMMARY

Taken together the results of these in vivo microperfusion experiments indicate that calcium absorption by the proximal tubule depends on more than one transport mechanism. We have observed that net calcium flux is affected by changes in calcium ion activity (even with constant total calcium concentration) and in transepithelial voltage. This sensitivity of calcium flux to changes in electrochemical driving force points to a diffusional component of calcium transport. Ng et al. (1984) have recently concluded that simple diffusion accounts for the majority of calcium absorption by superficial proximal convoluted tubules of the rabbit. The pathway for this diffusional component may involve paracellular channels. The permeability of this pathway appears to be as high for calcium as it is for sodium, potassium and chloride.

Calcium flux is also affected by changes in osmotic water flow. The effect of changes in volume flow on calcium transport occurs even in the absence of concentration changes in bulk solutions. Thus, it does not appear to be the result of changes in passive driving forces secondary to dilution or concentration of tubule fluid. At present we are not able to distinguish between two other possible mechanisms: solute polarization in a microscopic unstirred fluid layer adjacent to the cell membrane, or true entrainment of calcium in the stream of osmotically driven water flow (solvent drag). Either mechanism could provide an additional component of total calcium transport independent of changes in bulk phase ion concentrations and electrical driving forces.

A third component of total calcium absorption appears to involve active transport. In the absence of net volume movement, and with no transepithelial driving force, we have observed a significant absorptive net calcium flux. This component of calcium transport likely proceeds through proximal tubule cells. It may involve downhill movement of calcium ions from lumen to cell and uphill extrusion of calcium across the basolateral membrane in exchange for sodium. The stimulation of net calcium absorption caused by replacing luminal sodium with lithium may be the result of activation of this basolateral sodium-calcium exchange.

REFERENCES

Agus, A. Z., Gardner, L. B., and Beck, L. H., 1973, Effect of parathyroid hormone on renal tubular reabsorption of calcium, sodium, and phosphate, Am. J. Physiol., 224:1143.

Beck, L. H., and Goldberg, M., 1973, Effects of acetazolamide and parathyroidectomy on renal transport of sodium, calcium, and phosphate, Am. J. Physiol., 224:1136.

Bomsztyk, K., and Wright, F.S., 1982, Effects of transepithelial fluid flux on transepithelial voltage and transport of calcium, sodium, chloride, and potassium by renal proximal tubule, Kidney Int., 21:269.

Bomsztyk, K., and Wright, F. S., 1983, Calcium absorption by proximal tubule stimulated by reducing luminal sodium concentration, Kidney Int., 23:95.

Bomsztyk, K., George, J. P., and Wright, F. S., 1984, Effects of luminal fluid anions on calcium transport by proximal tubule, Am. J. Physiol., 246 (Renal Fluid Electrolyte Physiol. 15):F600.

Corman, B., and DiStephano, A., 1983, Evidence for movement of water without solutes through kidney proximal tubule, Kidney Int., 23:252A.

Edwards, B. R., Baer, P. G., Sutton, R. A. L., and Dirks, J. H., 1973, Micropuncture study of diuretic effects on sodium and calcium reabsorption in the dog nephron, J. Clin. Invest., 52:2418.

Hierholzer, K., Kawamura, S., Seldein, D. W., Kokko, J. P., and Jacobson, H.R., 1980, Reflection coefficients of various substrates across superficial and juxtamedullary proximal convoluted segments of rabbit nephrons, Mineral Electrolyte Metab., 3:172.

Jacobson, H. R., Kokko, J. P., Seldin, D. W., and Holmberg, C., 1982, Lack of solvent drag of NaCl and $NaHCO_3$ in rabbit proximal tubules, Am. J. Physiol., 243 (Renal Fluid Electrolyte Physiol. 12):F342.

Lassiter, W. E., Gottschalk, C. W., and Mylle, M., 1963, Micropuncture study of renal tubular reabsorption of calcium in normal rodents, Am. J. Physiol., 204:771.

Murayama, Y., Morel, F., and LeGrimellec, C., 1972, Phosphate, calcium and magnesium transfers in proximal tubules and loops of Henle, as measured by single nephron microperfusion experiments in the rat, Pflugers Arch., 333:1.

Ng, R.C.K., Rouse, D., and Suki, W.N., 1984, Calcium transport in the rabbit superficial proximal convoluted tubule, J. Clin. Invest., 74:834.

Poujeol, P., Chabardes, G, Roinel, N., and de Rouffignac, C., 1976, Influence of extracellular fluid volume expansion on magnesium, calcium, and phosphate handling along the rat nephron, Pflugers Arch., 365:203.

Suki, W. N., 1979, Calcium transport in the nephron, Am. J. Physiol., 237 (Renal Fluid Electrolyte Physiol. 6):F1.

Ullrich, K. J., 1973, Permeability characteristics of the mammalian nephron, and F. Sauer, Appendix: Nonequilibrium thermodynamics of kidney tubule transport, in: "Handbook of Physiology, Renal Physiology", J. Orloff, and R.W. Berliner, eds., Am. Physiol. Soc., Washington, D.C..

Ullrich, K. J., Rumrich, G., and Kloss, S., 1976, Active Ca^{++} reabsorption in the proximal tubule of the rat kidney, Pflugers Arch., 364:223.

Walser, M., 1961, Calcium clearance as a function of sodium clearance in the dog, Am. J. Physiol., 200:1099.

MECHANISMS OF CALCIUM TRANSPORT IN THE THICK ASCENDING LIMB OF THE LOOP OF HENLE

Zalman S. Agus, Fuad N. Ziyadeh, and Ellie Kelepouris

Renal Electrolyte Section, Department of Medicine
University of Pennsylvania School of Medicine
Philadelphia, PA

In the micropuncture segment between late proximal convoluted tubule and early distal tubule, approximately 20-25% of the filtered load of calcium is transported[1]. An undetermined fraction of the transport may occur in the proximal straight tubule where studies by Rouse, et al[2] in the isolated perfused superficial straight tubule of the rabbit have suggested a substantial calcium efflux from the lumen. Beyond this segment, the thick ascending limb would appear to account for the bulk of calcium transport which occurs in the loop of Henle. Thus, studies by Rocha, et al,[3] and Rouse, et al[2] have shown a very low bidirectional permeability to calcium and little if any net flux in the thin descending and ascending loop segments.

The mechanism of calcium transport in the cortical thick ascending limb has been studied by a number of laboratories and remains a source of controversy. In contrast, the medullary thick ascending limb has been studied by only one group and, therefore, there is little disagreement in the literature. In these studies by Suki, et al[4], rabbit medullary thick ascending limbs were perfused with equal sodium concentration in perfusate and bath exhibiting a mean transepithelial voltage of 5.5 mV, lumen positive. There was significant net calcium absorption with a flux ratio consistent with a passive, voltage driven mechanism. Reduction in transepithelial voltage with intraluminal furosemide was associated with a proportional reduction in calcium transport again consistent with simple passive movement. Addition of cyclic AMP analogues or calcitonin (but not PTH), however, produced a marked stimulation of calcium transport (from 1.3 to 2.3 pEq/mm/min) without a detectable change in transepithelial voltage. Thus, either cyclic AMP induces or uncovers an active transport mechanism or it induces rectification of the paracellular pathway allowing greater calcium efflux at the same driving voltage. Studies have not been performed evaluating the effects of calcitonin or cAMP on calcium efflux in the absence of a driving voltage to settle this issue. The situation is not as clear in the cortical segment of the thick ascending limb. Basically, two types of data have been obtained with perfusion studies in the unstimulated state. One group of investigators[5,6,7] have found results which qualitatively are very similar to those described in the medullary segment, and they propose, therefore, that the lumen positive transepithelial voltage drives passive calcium diffusion across the tubule. The major evidence for this hypothesis is the close correlation between voltage and calcium flux. In the study by Bourdeau and Burg[5], calcium absorption occurred at positive voltages and an increasing voltage was associated with an increasing net calcium flux. Conversely, calcium secretion occurred when the voltage was made lumen negative by addition of furosemide and manipulation of the bath to lumen chemical gradients for sodium and chloride. Similar studies were reported by Shareghi and Stoner[6], and we in our laboratory[7] were also able to corroborate a linear relationship between calcium transport and transepithelial voltage with calcium secretion occurring at negative voltages.

In direct contrast, three other laboratories[3,4,8] were unable to document a direct correlation between calcium transport and voltage, and concluded, therefore, that active transcellular transport must play an important role in this segment. The major finding in these studies was that, despite reduction of voltage with ouabain[3,8] or chemical voltage clamping[8], there remained a significant net calcium flux. There was some disagreement concerning the effects of furosemide. Suki, et al found no effect of 10^{-5} molar furosemide (although transepithelial voltage was not reduced to zero) while Imai demonstrated a decrease in both voltage and calcium efflux. I should point out parenthetically that Bordeau, et al have subsequently demonstrated a dose response to furosemide and complete inhibition of net calcium flux by furosemide even in the presence of PTH when furosemide (10^{-4} M) reduces the voltage to zero[8]. A variety of explanations have been offered for these discrepancies[10], and these have been enumerated elsewhere relating principally to methodologic problems in flux studies with calcium[45 5,10,11]. I shall offer no new insights and prefer instead to focus upon the effects of hormonal stimulation.

All investigators agree that PTH increases calcium absorption across the cortical thick ascending limb, that this effect is mimicked by cyclic AMP analogues and that there is no detectable change in transepithelial voltage[4,7,12,13,14]. This would seem to imply that PTH stimulates active calcium transport, presumably by increasing apical cell membrane permeability to calcium and/or increasing calcium extrusion at the basolateral membrane. However, as shown by Bourdeau and Burg[12,15], when the transepithelial voltage was reduced to near zero, neither PTH nor cAMP had any effect on calcium transport. They, therefore, suggested that the PTH stimulated component of calcium transport was also passive and voltage dependent, and that the effect of PTH and cAMP was to increase the permeability of the paracellular pathway allowing rectification of the voltage-driven transport. This hypothesis would suggest that when the voltage gradient was reversed that PTH would result in an increase in calcium secretion. This was not observed, however, and in fact there was even a decrease in secretion. If the rectification hypothesis cannot explain these observations, stimulation of active transport by PTH cannot explain the lack of effect of PTH in the absence of a voltage. I would like to suggest the possibility of a third hypothesis. In the studies by Bourdeau and Burg[12] as well as our own[7], two manipulations were performed to reduce voltage to zero - fluid composition was altered and intraluminal furosemide was added. Bourdeau, et al have clearly shown in separate experiments that furosemide in sufficient dose can completely block the calcium transport response to PTH[9]. Furosemide lowers the transepithelial voltage and inhibits chloride transport. This raises the possibility that it is chloride transport per se rather than voltage which is necessary for the effects of PTH. To evaluate this possibility requires the ability to reduce voltage to zero without interfering with chloride transport. As this is technically difficult to do in the perfused tubule, we have evaluated this issue with short circuiting techniques in another chloride transporting epithelium - frog skin glands[16].

In all of these studies, whole thickness frog skin is mounted in standard Ussing chambers and continuously short circuited. Unidirectional Ca^{45} flux measurements are made in paired hemiskins studied simultaneously. In the presence of amiloride, which completely abolishes sodium absorption by the surface epithelium, both cyclic AMP and isoproterenol stimulate an amiloride-resistant short-circuit current which others have shown represents electrogenic chloride secretion by glands within the skin preparation. In the baseline condition, there is no net calcium transport but, in response to cyclic AMP or isoproterenol, there is a marked increase in the S to M calcium-45 flux, little change in the M to S flux and, therefore, a large increase in net secretory flux. This secretion occurs in the absence of concentration gradients for calcium and in the absence of a voltage gradient as the tissue is short-circuited, and is presumably, therefore, active calcium transport. Similar to studies in other chloride transporting epithelia, prior administration of furosemide completely abolishes the short-circuit current response. Furosemide also completely inhibits the calcium transport response to cyclic AMP. Thus, we have reproduced the observation reported in the ascending limb studies. There is net calcium transport, stimulated by cAMP in the absence of a change in voltage, which is inhibited by furosemide. Stimulated calcium transport in this system is chloride and not voltage dependent and, in other studies, we were able to abolish the calcium transport response to cAMP by replacing serosal chloride with nitrate or sulfate. The calcium chloride linkage is not absolute as the chloride transport pathway can be stimulated without altering calcium transport with either PGE_2 or a calcium ionophore[17]. Thus, administration of 10^{-5} M PGE_2 is associated with a marked stimulation of chloride secretion as represented by the short circuit current while there is no change in net calcium flux.

We have chosen to interpret these data as suggesting two chloride pathways as shown in Figure 1. The first is an electrogenic chloride secretory pathway as described in other chloride secreting epithelia consisting of a Na-K-2 Cl co-transporter at the serosal membrane and a conductive chloride channel at the mucosal membrane of frog skin gland cells. This pathway is responsible for the stimulation of chloride secretion observed with a variety of secretagogues including isoproterenol, cyclic AMP, prostaglandins and calcium ionophore. This system seems to account for the bulk of transported chloride.

In addition, since cyclic AMP induces net calcium secretion, a mechanism should be available which would provide sufficient energy to allow uphill calcium exit at the mucosal membrane. Such a process does not appear to utilize a mucosal Na/Ca exchanger because we have found persistence of calcium secretion with removal of sodium from the mucosal solution. However, to explain the dependence of calcium secretion upon chloride transport, we propose the existence of a calcium-chloride co-transporter at the mucosal membrane which is inducible by cyclic AMP. The calculated apparent stoichiometry for this co-transporter is 5-chloride for each calcium to allow calcium exit from the cell driven by the favorable electrochemical gradient for chloride.

The existence and relevance of such a co-transport system to that component of calcium transport which is stimulated by cyclic AMP in the cortical and medullary ascending limbs of Henle's loop remain to be evaluated. To demonstrate the chloride rather than voltage dependence will require perfusion studies combining furosemide and cAMP in the presence of a positive voltage produced by chemical gradients and/or the demonstration of a stimulation by cAMP in the absence of furosemide while voltage has been abolished by chemical gradients.

In summary, approximately 20-25% of the filtered load of calcium is absorbed in the cortical and medullary thick ascending limbs of the loop of Henle. The mechanism of this transport remains controversial with some laboratories providing evidence for passive paracellular diffusion driven by the lumen positive transepithelial voltage. Other laboratories have provided evidence for voltage independence of calcium transport in the cortical segment implying active transcellular movement, the mechanism of which is unclear. Calcium transport in this segment is

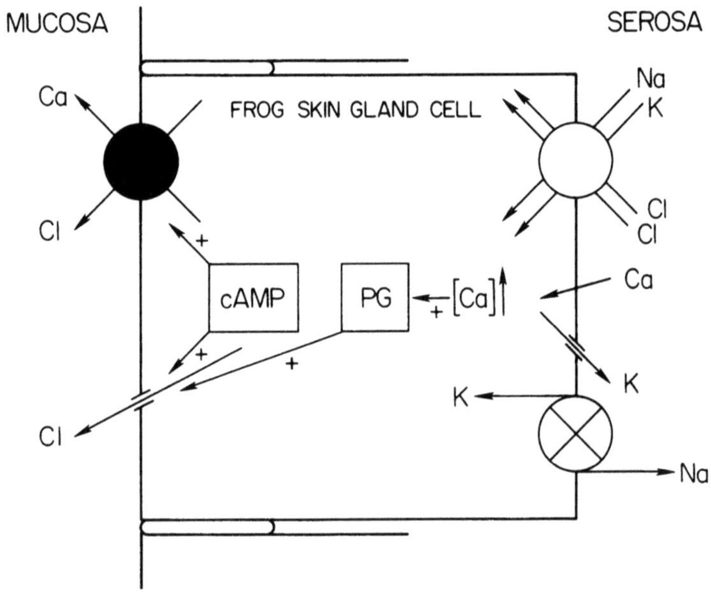

Fig. 1. Hypothesis concerning calcium and chloride secretion in the glands of frog skin. The figure depicts two pathways for chloride exit across the mucosal membrane. A conductive pathway is stimulated by prostaglandins and cyclic AMP. A calcium-chloride co-transporter is stimulated only by cyclic AMP and is driven by the electrochemical gradient for chloride.

stimulated by calcitonin in the medullary component and PTH in the cortical segment acting presumably via stimulation of adenylcyclase. The mechanism by which cAMP stimulates calcium transport is unknown but it is inhibitable by furosemide and/or elimination of the transepithelial voltage. Possible hypotheses to be evaluated include changes in the paracellular conductance allowing an increase in calcium movement in response to a voltage gradient or the induction of a calcium chloride co-transporter by cAMP with energy provided by the electrochemical gradient for chloride across the basolateral membrane.

REFERENCES

1. W. E. Lassiter, C. W. Gottschalk, and M. Mylle, Micropuncture study of renal tubular reabsorption of calcium in normal rodents, Am. J. Physiol. 204:771 (1963).
2. D. Rouse, R. C. K. Ng, and W. N. Suki, Calcium transport in the pars recta and thin descending limb of Henle of the rabbit, perfused in vitro, J. Clin. Inves. 65:37 (1980).
3. A. S. Rocha, J. B. Magaldi, and J. P. Kokko, Calcium and phosphate transport in isolated segments of rabbit Henle's loop, J. Clin. Invest. 59:975 (1977).
4. W. N. Suki, D. Rouse, R. C. K. Ng, and J. P. Kokko, Calcium transport in the thick ascending limb of Henle: Heterogeneity of function in the medullary and cortical segments, J. Clin. Invest. 66:1004 (1980).
5. J. E. Bourdeau and M. B. Burg, Effects of PTH on calcium transport across the cortical thick ascending limb of Henle's loop, Amer. J. Physiol. 238:F350 (1979).
6. G. R. Shareghi and L. C. Stoner, Calcium transport across segments of the rabbit distal nephron in vitro, Amer. J. Physiol. 235:F367 (1978).
7. G. R. Shareghi and Z. S. Agus, Magnesium transport in the cortical thick ascending limb of Henle's loop of the rabbit, J. Clin. Invest. 69:759 (1982).
8. M. Imai, Calcium transport across the rabbit thick ascending limb of Henle's loop perfused in vitro, Pflug. Archiv. 374:255 (1978).
9. J. E. Bourdeau, S. L. Buss, and G. G. Turck, Inhibition of calcium absorption in the cortical thick ascending limb of Henle's loop by furosemide, J. Pharm. and Exp. Therap. 211:815 (1982).
10. M. B. Burg, Thick ascending limb of Henle's loop, Kid. Int. 22:454 (1982).
11. W. N. Suki, Calcium transport in the nephron, Amer. J. Physiol. 237:F1 (1979).
12. J. E. Bourdeau and M. B. Burg, Effect of PTH on calcium transport across the cortical thick ascending limb of Henle's loop, Amer. J. Physiol. 239:F121 (1980).
13. M. Imai, Effects of parathyroid hormone and N^6m O^2-dibutyryl cyclic AMP on Ca^{2+} transport across the rabbit distal nephron segments perfused in vitro, Pflug. Archiv. 390:145 (1981).
14. W. N. Suki and D. Rouse, Hormonal regulation of calcium transport in thick ascending limb renal tubules, Amer. J. Physiol. 241:F171 (1981).
15. J. E. Bourdeau, Calcium transport across the cortical thick ascending limb of Henle's loop, in: "Calcium and Phosphate Transport Across Biomembranes", F. Bronner and M. Peterlik, eds., Academic Press, New York (1981).
16. F. N. Ziyadeh, E. Kelepouris, M. M. Civan, and Z. S. Agus, Cyclic AMP and -adrenergic stimulated chloride-dependent calcium secretion in frog skin, Amer. J. Physiol., In Press (1985).
17. F. N. Ziyadeh, E. Kelepouris, and Z. S. Agus, Possible role for Ca/Cl co-transporter in Ca secretion by glands of frog skin, Clin. Res. 33:590A (1985).

CALCIUM TRANSPORT IN THE DISTAL CONVOLUTED TUBULE

Linda S. Costanzo

Department of Physiology and Biophysics
Medical College of Virginia
Richmond, Virginia 23298

In 1963, Lassiter and coworkers, using micropuncture techniques, provided the first quantitative assessment of calcium transport along the nephron (1). Compiling their data and those of subsequent investigators, it is apparent that about 65% of the filtered calcium load is reabsorbed in the proximal convoluted tubule, another 25% in the loop of Henle and about 8% in the distal tubule. While the quantitative contribution of the distal convoluted tubule seems small, it should be noted that comparably small fractions of sodium and water are reabsorbed distally (2). At the same time, the distal nephron is a major regulatory site for these substances (2). Thus it seemed worthwhile to characterize distal tubular calcium transport and to study the potential hormonal regulation and effects of diuretic agents.

The kinetics of distal calcium transport were first described using techniques of free flow micropuncture (3). When the luminal calcium concentration profile was mapped along the rat distal tubule the tubular concentration fell as a function of length to values that were less than one-third those in plasma ultrafiltrate. Fractional reabsorption by the entire rat distal tubule was 8% of the filtered load, a value similar to that reported by Lassiter et al (1). Since this reabsorptive process proceeded against a large chemical (3) and electrical (4) gradient, these results constituted direct proof that calcium reabsorption across the distal tubular epithelium was an active transport process.

Since the distal tubule operates in a load-dependent fashion when transporting sodium (3), the relationship between distal calcium reabsorption and delivered load was examined using the technique of *in vivo* microperfusion (Figure 1). One millimeter segments of distal tubules with two surface loops were perfused at 6 nl/min with a solution simulating early distal tubular fluid and containing C-14 inulin as a volume marker. Collection of the fluid was made at the downstream site with a second micropipette and analyzed for calcium and sodium by flameless atomic absorption spectrophotometry and for inulin by liquid scintillation counting. The arithmetic difference between delivered and collected rates for fluid and electrolytes constituted the rate of reabsorption. Calcium reabsorption was a linear function of delivered load when the calcium load was varied by changing the perfusion rate (3). Reabsorption was constant at about 70% the delivered load without evidence of saturation, even at loads which were three-times normal.

This load-dependent reabsorption was the result of a concentration-dependent efflux from lumen to blood and a small, almost undetectable backflux from blood to lumen (3). The negligible backflux suggests that the calcium permeability of the distal epithelium is very low.

It has been increasingly apparent from the work of anatomists that the distal convoluted tubule is not a single, homogeneous structure. Rather, as described by Kaissling (5) and others, there are at least four cell types in the rat distal tubule. The "true" distal convoluted tubule cell occupies the first third of the distal tubule. The latter two-thirds consists of intercalated cells, connecting tubule cells and finally, principal cells. In the rat, particularly, there are not distinct transitions from one cell type to another.

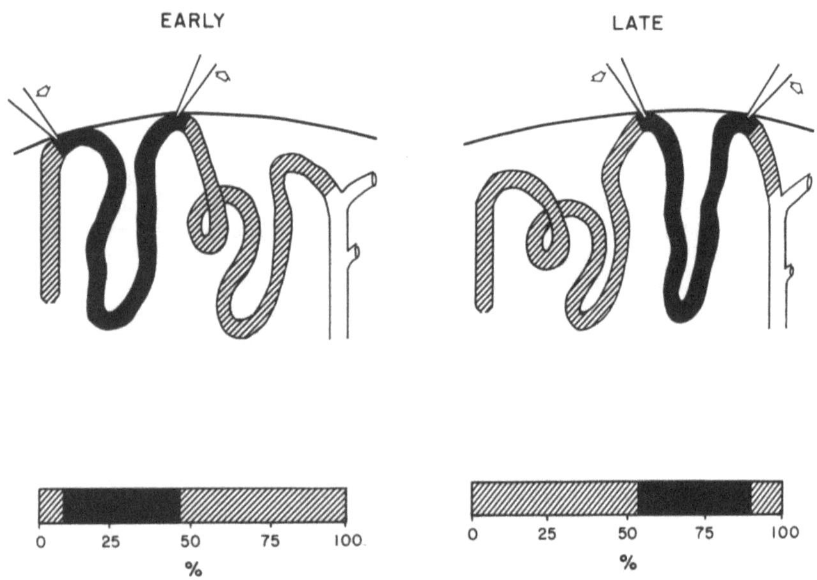

Figure 1. Illustration of technique for measuring lengths of perfused distal tubules and localization of segments in early and late distal tubule. Solid area plus shaded area shows entire length of distal tubule from macula densa to convergence with another distal tubule.

In order to deal with the structural heterogeneity of the rat distal tubule, we modified the in vivo microperfusion technique so that the function of the "early" and "late" distal tubule could be studied separately (Figure 1). Perfused segments were classified retrospectively as either belonging to the first or last 50% of the distal tubule. The positions of the "early" or "late" distal segments were referred to the macula densa (0%) and to the confluence of two distal tubules to form a collecting tubule (100%). Using this modified technique, the transport functions of early and late tubules were compared (6). Notably, calcium and sodium transport rates, when corrected for tubular length were not different. The segments could be distinquished, however, on the basis of their water reabsorptive rates. A much higher rate of osmotic water flux was observed across late than across early segments, probably a result of the localized anti-diuretic hormone sensitivity of the latter portion of the rat distal tubule (7).

It has become apparent that, as for sodium and water, the final physiologic control over the urinary excretion of calcium is exerted at the level of the distal tubule. Some of the factors or agents which alter distal tubule calcium transport are listed in Table 1.

Table 1. Influences on Distal Tubular Calcium Reabsorption

Agent	Calcium Transport
Parathyroid hormone	↑
Alkalosis	↑
Acidosis	↓
Phosphate depletion	↓
Phosphate repletion	↑

The most important hormonal regulation of renal calcium transport is exerted by parathyroid hormone. Agus and coworkers (8) were the first to assign the hormone's calcium-retaining property to the distal tubule. Shareghi and Stoner subsequently demonstrated that parathyroid hormone directly increases calcium transport in isolated perfused segments of rabbit distal tubule and connecting tubule (9). We confirmed these results in *in vivo* perfused rat distal tubules and further showed that potent cyclic AMP analogues could duplicate the hormone's action perfectly, suggesting that the calcium-retaining action is mediated by stimulation of a parathyroid hormone-sensitive adenylate cyclase (10). In the latter study, enhanced distal calcium reabsorption was entirely specific for parathyroid hormone and not a feature of the distal action of anti-diuretic hormone (10).

Two groups of diuretics have the unusual property of causing renal calcium retention: the thiazides and the potassium-sparing diuretics such as amiloride. The hypocalciuric effects of both diuretics have been localized to the distal convoluted tubule.

The thiazide diuretics act primarily in the early portion of the rat distal tubule (Figure 2) (11). Early and late tubules were perfused with a solution containing chlorothiazide, and their transport rates were compared to the pre-drug value. The thiazide promoted calcium reabsorption only in the early distal segments, whereas late segments were unaffected. Although not shown in a figure, the drug inhibited sodium transport in the same early distal segments. This dissociation of calcium and sodium transport is a hallmark of the renal thiazide action.

Although less well-known, amiloride is also a calcium-retaining diuretic (12) whose action resides in the distal tubule. In distinction to thiazides, however, it's action is confined to the late distal segment (Figure 3) (6). When early and late distal tubules were perfused with control and amiloride-containing solutions, the drug enhanced calcium reabsorption in the late but not in the early segments. Like thiazides, Na transport was inhibited simultaneously, but only in the late distal tubule. Thus, thiazides and amiloride produce qualitatively similar hypocalciuria, but at different locations of the distal tubule.

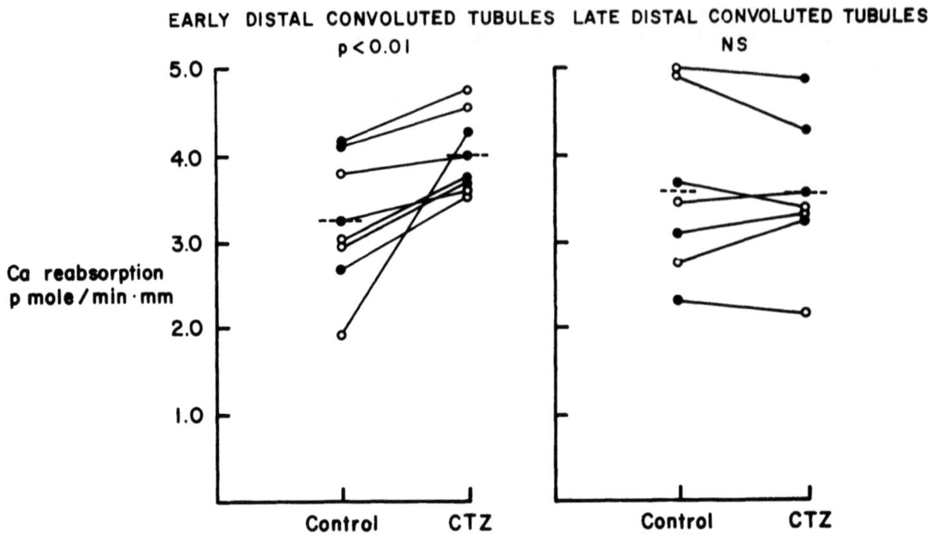

Figure 2. Effect of chlorothiazide (CTZ) on calcium reabsorption in early (left) and late (right) distal tubules. Each tubule received control and CTZ ($5x10^{-4}$M) solutions. Open circles indicate solution perfused first. Horizontal lines give means for control and CTZ.

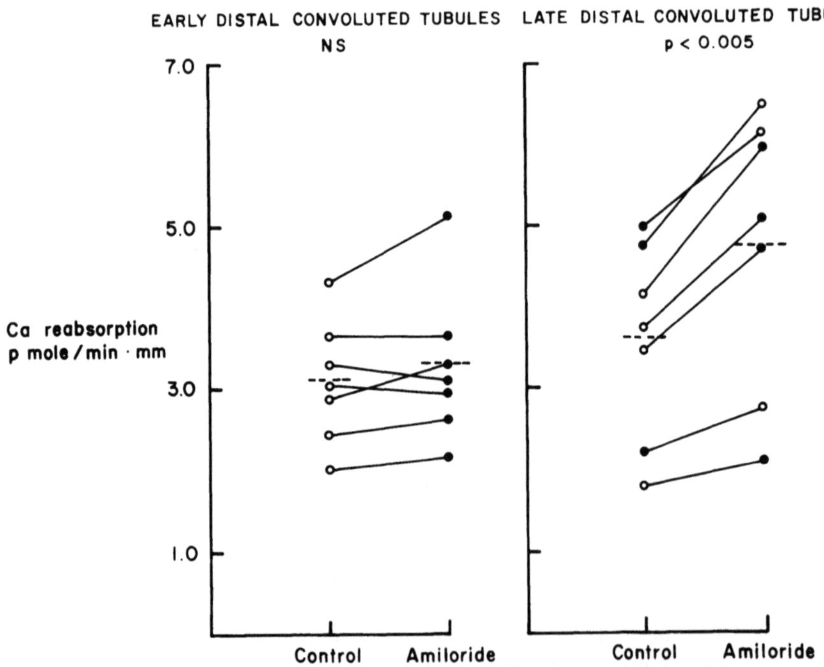

Figure 3. Effect of amiloride on calcium reabsorption in early (left) and late (right) distal tubules. Each tubule received control and amiloride (10^{-5}M) solutions. Symbols are as in Figure 2.

A mechanism for the simultaneous natriuretic and hypocalciuric actions of amiloride has been proposed (6). Both calcium and sodium can enter distal epithelial cells down their respective electrochemical potential gradients. Sodium can exit the cell by an active, ATP-driven Na-K pump. Calcium can be pumped uphill via a Ca-ATPase (13) or can exit the cell via a Ca-Na exchanger which energized by the downhill movement of sodium ions across the peritubular membrane. When amiloride is present, we suggest that luminal sodium entry is inhibited, consistent with the drug's established mechanism of action (6). Intracellular sodium ion activity would be expected to fall and as a result, transepithelial sodium reabsorption would be reduced, as is observed experimentally. In addition, the rate of Ca-Na exchange across the peritubular membrane would be enhanced as a result of the new, more favorable sodium gradient. Thus, transepithelial calcium transport would be enhanced, also observed experimentally. A similar mechanism of action has been suggested for the thiazide diuretics.

Thiazides and amiloride have different sites of action in the distal convoluted tubule. The thiazide diuretics interact with a cell type in the early distal tubule, probably the true distal convoluted tubule cell. Amiloride interacts with one of the cell types in the late distal tubule. Years ago, experiments suggested that these two diuretics could produce additive hypocalciuria (12), an observation, which is entirely consistent with the separate sites of diuretic action. We recently investigated the possibility that the additive hypocalciuria might have some implications for the treatment of children with Vitamin D-induced hypercalciuria (14). In fact, addition of amiloride to the standard hydrochlorothiazide regimen produced a significantly greater reduction in urinary calcium excretion than did the thiazide alone. Since amiloride can prevent thiazide-induced potassium wasting (14), this drug combination may provide sensible management of certain hypercalciurias.

In conclusion, the distal convoluted tubule appears to be the site where renal calcium reabsorption can be regulated, separate from sodium reabsorption. In contrast to the parallel behavior of calcium and sodium in the proximal nephron and loop of Henle, transport of these ions in the distal tubule is frequently dissociated. Examples of this dissociation include: during changes in circulating levels of parathyroid hormone, alterations in acid-base status, phosphate depletion or repletion and under the influence of diuretics having their site of action in the distal tubule.

References

1. W.E. Lassiter, C.W. Gottschalk and M. Mylle. Micropuncture study of renal tubular reabsorption of calcium in normal rodents. Am. J. Physiol. 204:771 (1963).
2. G. Giebisch and E.E. Windhager. Electrolyte transport across renal tubular membranes. In: Handbook of Physiology. Renal Physiology. Ed, J. Orloff and R.W. Berliner. Am. Physiol. Soc., Washington, DC (1973).
3. L.S. Costanzo and E.E. Windhager. Calcium and sodium transport by the distal convoluted tubule of the rat. Am. J. Physiol. 235:F492 (1978).
4. F.S. Wright. Increasing magnitude of electrical potential along the renal distal tubule. Am. J. Physiol. 220:624 (1971).
5. B. Kaissling. Structural aspects of adaptive changes in renal electrolyte excretion. Am. J. Physiol. 243:F211 (1982).

6. L.S. Costanzo. Comparison of calcium and sodium transport in early and late rat distal tubules: effect of amiloride. Am. J. Physiol. 246:F937 (1984).
7. F. Morel. Sites of hormone action in the mammalian nephron. Am. J. Physiol. 240:F159 (1981).
8. Z.S. Agus, P.J.S. Chiu and M. Goldberg. Regulation of urinary calcium excretion in the rat. Am. J. Physiol. 232:F545 (1977).
9. G.R. Shareghi and L.C. Stoner. Calcium transport across segments of the rabbit distal nephron in vitro. Am. J. Physiol. 235:F367 (1978).
10. L.S. Costanzo and E.E. Windhager. Effects of PTH, ADH and cyclic AMP on distal tubular Ca and Na reabsorption. Am. J. Physiol. 239:F478 (1980).
11. L.S. Costanzo. Localization of diuretic action in microperfused rat distal tubules: Ca and Na transport. Am. J. Physiol. 248:F527 (1985).
12. L.S. Costanzo and I.M. Weiner. Relationship between clearances of Ca and Na: effect of distal diuretics and PTH. Am. J. Physiol. 230:67 (1976).
13. A. Doucet and A.I. Katz. High-affinity Ca-Mg-ATPase along the rabbit nephron. Am. J. Physiol. 242:F346 (1982).
14. U. Alon, L.S. Costanzo and J.C.M. Chan. Additive hypocalciuric effects of amiloride and hydrochlorothiazide in patients treated with calcitriol. Min. Elec. Metab. 10:379 (1984).

RENAL MEMBRANE TRANSPORT OF CALCIUM

Yusuke Tsukamoto and Wadi N. Suki

Department of Medicine, Kitasato University School of Medicine, Japan; Renal Section, Baylor College of Medicine Houston, Texas

The development of techniques for the isolation of highly purified basolateral membrane vesicles (BLMV) from the renal tubular epithelial cells enabled us to investigate the mechanism of calcium extrusion at this barrier of the cell. Two mechanisms of calcium transport known to exist in other tissues have been identified in the renal BLMV. These mechanisms are ATP-dependent Ca^{2+} transport and Na^{+}/Ca^{2+} exchange. Yet, several aspects of these transporters are quite unique in the renal tubular cells.

ATP-dependent Ca^{2+} transport

ATP-dependent Ca^{2+} uptake was first investigated by using a crude preparation of plasma membrane from rat kidney (1). In this experiment, an apparent Km for free Ca^{2+} of the Mg^{2+}-ATP dependent Ca^{2+} uptake was 20 μM in the plasma membrane fraction, which was the same as in the microsomal fraction. Since this value is far beyond cytosolic free Ca^{2+} concentration, it should be considered that this Ca uptake was due to contamination with other cell components (e.g. mitochondria and endoplasmic reticulum, etc.). In order to avoid this contamination, Gmaj et al (2) purified the BLMV by free flow electrophoresis. According to their study, an apparent Km for free Ca^{2+} of the ATP-dependent Ca^{2+} uptake was 0.5 μM, which agreed well with the reported cytosolic free Ca^{2+} concentration (less than 0.5 μM). This ATP-dependent Ca^{2+} uptake requires Mg^{2+} for its activity. Since the distribution of this activity in the fractions of electrophoresis was similar to that of the (Na + K)-ATPase activity, this Ca^{2+} uptake should not be due to contamination of other cell components but to basolateral membrane. They also reported that this Ca^{2+} uptake was calmodulin dependent and was inhibited by a low concentration of vanadate.

Preparation of BLMV by Percoll R gradient and free flow electrophoresis methods presents critical problems for the study of Ca transport. First, this preparation of BLMV consists of a heterogeneous population of vesicles. We estimated the orientation of BLMV and the leakiness to ouabain by the method of Forbush (3); 35% of the BLMV were found to be leaky to ouabain and ATP and were considered to be a seat membrane and leaky vesicle. Of the tight vesicles, 29% were right-side out and 71% were inside-out. Since Ca^{2+}-ATPase of BLMV can be speculated to have an ATP binding site at the cytosolic face of the membrane, Ca^{2+} uptake only by tight inside-out vesicles, which is only 46% of all BLMV, may be considered to reflect the activity of Ca^{2+}-ATPase. This may be one of the reasons why the velocity of ATP-dependent Ca^{2+} uptake was much smaller than the velocity of Ca^{2+}-ATPase (4).

Second, when both brush-border (BBMV) and basolateral membranes from rat kidney cortex were used for Ca^{2+} uptake in the absence of ATP, only 20% of Ca^{2+} taken up accounted for free Ca^{2+} inside vesicles, while 60% was bound to the inner side of vesicles and 20% was bound to the outer side of vesicles (5). Mg^{2+}-ATP stimulated the Ca^{2+} binding to both BBMV and BLMV. Since Ca^{2+} bound to the BBMV was not inhibited by calcium ionophores, the Ca^{2+} binding at this side of the plasma membrane could not be operated by Ca^{2+}-ATPase (6). In the BLMV, the Ca^{2+} binding might be the transitory stage of the transmembrane Ca^{2+} movement and it could be operated by Ca^{2+}-ATPase. However, we cannot deny the possibility that other calcium binding proteins might be involved in the ATP dependent Ca^{2+} uptake.

In order to reconcile these problems, one ought to characterize the Ca^{2+}-ATPase of the basolateral membrane of the kidney. In other tissues such as erythrocyte membrane (7), and cardiac sarcolemma (8), isolation and reconstitution studies have revealed that Mg^{2+}-ATP dependent Ca^{2+} uptake was operated by high affinity (Ca^{2+} + Mg^{2+})-ATPase. We measured Ca^{2+} dependent ATP hydrolysis by the basolateral membrane from rat kidney cortex (9). This membrane preparation derives mainly from the proximal tubules (10). The assay was performed by measuring liberated ^{32}Pi from the (γ-^{32}P) ATP by the leaky basolateral membrane preparation. The results demonstrated the presence of a high affinity and low capacity Ca^{2+}-ATPase (Km for free Ca^{2+}=172 nM, Vmax=144 nmol Pi/min/mg protein) and a low affinity and high capacity Ca^{2+}-ATPase (Km for free Ca^{2+}=25 μM, Vmax=896 nmol Pi/min/mg protein). Since Mg^{2+} caused a huge amount of background by Mg^{2+} ATPase, Mg^{2+} was omitted from the assay mixture. As a matter of fact, Mg^{2+} inhibited high affinity Ca^{2+} dependent ATP hydrolysis competitively and inhibited low affinity ATP hydrolysis non-competitively. Since only a small amount of additional $MgCl_2$ (10 μM) could compete with Ca binding to the ATPase, it is very hard to assume that an endogenous Mg^{2+} in the membrane activates Ca^{2+}-dependent ATP hydrolysis. Gmaj et al previously reported the high affinity (Ca^{2+} + Mg^{2+}) ATPase by using a coupled enzyme assay (11). Mg^{2+} dependency cannot be evaluated by this assay method because Mg^{2+} is required for the pyruvate kinase in the assay mixture. Recently, they also reported the presence of Ca^{2+} stimulated Mg^{2+} independent ATPase (12). These findings raise the question whether two different high affinity Ca^{2+}-ATPases exist in the basolateral membrane of kidney cortex. We know that ATP dependent Ca^{2+} uptake by BLMV requires Mg^{2+}. This uptake was reported to be calmodulin dependent and was inhibited by a low concentration of vanadate (2,4). These are common characteristics among the Ca^{2+} transporting ATPases in other tissues. However, even in the presence of Mg^{2+} in the assay mixture, we failed to show either vanadate sensitivity or calmodulin dependency of high affinity Ca^{2+}-ATPase activity (9). We also failed to demonstrate the involvement of two enzyme kinetics of ATP hydrolysis in the presence of both Mg^{2+} and high affinity range of free Ca^{2+}. Only purification of the Ca^{2+}-ATPase in this membrane can solve the question whether two high affinity Ca^{2+}-ATPases co-exist.

We also studied the regulation of high and low affinity Ca^{2+}-ATPase activity. Vitamin D deficient rats showed significantly lower activity of high and low affinity enzymes than those of the $1,25(OH)_2D_3$ repleted D deficient rats (9). Since D deficient rats showed high PTH levels, the effect of thyroparathyroidectomy (TPTX) with thyroxin repletion was also investigated (13). The results showed that TPTX rats had a significantly decreased activity of high and low affinity Ca^{2+}-ATPases 48 hours after the surgery, suggesting a stimulatory effect of PTH on these enzymes. However, the enzyme activities returned to normal levels 2 weeks after the surgery. Since serum $1,25(OH)_2D_3$ level was lower in the TPTX rats than in the sham-operated rats, another unknown factor(s) might have stimulated the activity. PTH has been demonstrated to also stimulate Ca^{2+} binding to BBMV (14).

Na^+/Ca^{2+} exchange

In many tissues, a Na^+/Ca^{2+} exchange system plays a key role in the extrusion of cellular Ca^{2+} and the regulation of the cytosolic Ca^{2+} concentration. In the kidney, the presence of a Na^+/Ca^{2+} exchange was first suggested from the findings that in microperfused tubules removal of Na^+ from, or addition of ouabain to, the peritubular fluid inhibits active Ca^{2+} transport (15). Gmaj et al established the presence of Na^+/Ca^{2+} exchanger in the BLMV of rat kidney cortex (2). Jayakumar et al (10) characterized this transporter further and demonstrated that the enhancement of Ca^{2+} uptake by the intravesicular > extravesicular Na^+ gradient was inhibited by ionophores that dissipated the gradient, and was increased by an outside negative membrane potential. Kinetic studies estimated a $[Ca^{2+}]$ 0.5 of 8 μM. This value is higher than reported cytosolic free Ca concentration, but agrees well with the $[Ca^{2+}]$ 0.5 for the Na^+/Ca^{2+} exchanger in cardiac sarcolemma (16). The relationship between Na^+ concentration and rate of Ca^{2+} efflux was sigmoidal: a $[Na^+]$ 0.5 of 15 mM and a Hill coefficient of 2.5 were estimated (10).

Removal of parathyroid glands from the rat resulted in a 40% decrease in the Na^+ gradient-dependent Ca^{2+} uptake 48 to 72 hours after TPTX (10). The stimulatory effect of PTH was also supported by Scoble et al using the canine kidney (17). The canine kidney shares the kinetics of Na^+/Ca^{2+} exchange with the rat kidney. The $[Ca^{2+}]$ 0.5 was 10 μM in the BLMV from the canine kidney. PTH also has a stimulatory effect on this Ca^{2+} transport.

Conclusion

We do not know how these two transporters share the role of Ca^{2+} extrusion in the renal tubular epithelial cells. Simple comparison of velocity between two transporters is not valid because of mixed populations of vesicles. It is possible to assert, however, that these Ca^{2+} transporters must play a key role in regulating cytosolic free Ca^{2+} concentration which in turn regulates many cell activities as a calcium messenger system (18).

References

1. L. Moore, D.F. Fitzpatrick, T.S. Chen, and E.J. Landon, Calcium pump activity of the renal plasma membrane and renal microsomes, Bioch. Biophys. Acta 345:405 (1974).
2. P. Gmaj, H. Murer, and R. Kinne, Calcium ion transport across plasma membranes isolated from rat kidney cortex, Biochem. J. 178:549 (1979).
3. B. Forbush, III, Characterization of right-side-out membrane vesicles rich in (Na,K)-ATPase and isolated from dog kidney outer medulla, J. Biol. Chem. 257:12678 (1982).
4. P. Gmaj, M. Zurini, H. Murer, and E. Carafoli, A high-affinity, calmodulin-dependent Ca^{2+} pump in the basal-lateral plasma membranes of kidney cortex, Eur. J. Biochem. 136:71 (1983).
5. Z. Taylor, G. Richison, and J.A.L. Arruda, High-affinity calcium binding sites in luminal and basolateral renal membranes, Am. J. Physiol. 248:F472 (1985).
6. K.A. Hruska, S.C. Mills, S. Khalifa, and M.R. Hammerman, Phosphorylation of renal brush-border membrane vesicles, J. Biol. Chem. 258:2501 (1983).
7. V. Niggli, E.S. Adunyah, J.T. Penniston, and E. Carafoli, Purified (Ca^{2+}-Mg^{2+})-ATPase of the erythrocyte membrane, J. Biol. Chem. 256:395 (1981).
8. P. Caroni, and E. Carafoli, The Ca^{2+} pumping ATPase of heart sarcolemma, J. Biol. Chem. 256:3263 (1981).
9. Y. Tsukamoto, and B. Sacktor, Characterization of Ca^{2+}-dependent ATPase in basolateral membrane of rat renal cortex: Regulatory role of $1,25(OH)_2D_3$, Kidney Int. 25:156 (1984).

10. A. Jayakumar, L. Cheng, C.T. Liang, and B. Sacktor, Sodium gradient-dependent calcium uptake in renal basolateral membrane vesicles, J. Biol. Chem. 259:10827 (1984).
11. P. Gmaj, H. Murer, and E. Carafoli, Localization and properties of a high-affinity (Ca^{2+} + Mg^{2+})-ATPase in isolated kidney cortex plasma membranes, FEBS letters 144:226 (1982).
12. W. Ghijsen, P. Gmaj, and H. Murer, Ca^{2+}-stimulated Mg^{2+}-pumping ATPase, Biochem. Biophys. Acta 778:481 (1984).
13. Y. Tsukamoto, and W.N. Suki, Regulation of Ca^{2+}-ATPase in the basolateral membrane of rat kidney cortex, Kidney Int. 27:128 (1985).
14. S. Khalifa, S. Mills, and K.A. Hruska, Stimulation of calcium uptake by parathyroid hormone in renal brush-border membrane vesicles, J. Biol. Chem. 258:14400 (1983).
15. K.J. Ullrich, G. Rumrich, and S. Kloss, Active Ca^{2+} reabsorption in the proximal tubule of the rat kidney. Dependence on sodium and buffer transport, Pflugers Arch. Eur. J. Physiol. 364:223 (1976).
16. K.D. Philipson, and A.Y. Nishimoto, Na^{+}-Ca^{2+} exchange in inside-out cardiac sarcolemmal vesicles, J. Biol. Chem. 257:5111 (1982).
17. J.E. Scoble, S. Mills, and K.A. Hruska, Calcium transport in canine renal basolateral membrane vesicles. Effect of parathyroid hormones, J. Clin. Invest. 75:1096 (1985).
18. H.S. Chase, Jr., Does calcium couple the apical and basolateral membrane permeabilities in epithelia? Am. J. Physiol. 247:F869 (1984).

RENAL TUBULAR DISORDERS ASSOCIATED WITH HYPOCALCIURIA

Roger A.L. Sutton

Department of Medicine, University of British Columbia

Vancouver, B.C. Canada

INTRODUCTION

Hypocalciuria (an abnormally low urinary calcium excretion) may be absolute and relative. Robertson and Morgan[1] reported that the 24-hour urinary calcium excretion of normal subjects showed a non-Gaussian distribution with a distortion at the lower end due to the build-up of values over a range of lower limits. Less than 15% of normal male subjects had urinary calcium excretions below 100mg per 24 hours. Relative hypocalciuria may be defined as a fractional or absolute urinary calcium excretion lower than would be expected on the basis of the patients' filtered calcium load and urinary excretion of sodium. For example, in mild renal impairment (creatinine clearance 106-40ml/min.), Popovtzer et al.[2] observed a significantly lower fractional calcium excretion, but a significantly higher fractional sodium excretion than normal, implying preferential renal calcium retention, which they attributed to the presence of secondary hyperparathyroidism. Hypocalciuria may result from the actions of several factors known to enhance renal calcium reabsorption, including parathyroid hormone (PTH)[3], metabolic alkalosis[4], thiazide diuretics[5], hypocalcemia[6], and perhaps hypomagnesia[7] and vitamin D deficiency. Hypocalciuria cannot therefore be confidently attributed to a primary renal tubular defect unless these factors have been excluded as contributing causes. Absolute or relative hypocalciuria may be observed in patients with low, normal or high serum calcium levels.

CLASSIFICATION

Hypocalciuria with hypocalcemia

A reduction in the filtered calcium load would be expected to be associated with hypocalciuria. Furthermore, hypocalcemia may itself enhance tubular reabsorption of calcium[6]. However, where hypocalcemia results from hypoparathyroidism, absence of the calcium retaining effect of PTH may result in a higher urinary calcium excretion than in other hypocalcemic conditions, such as vitamin D deficiency, in which secondary hyperparathyroidism presumably contributes to the usually striking hypocalciuria[8].

Hypocalciuria with hypercalcemia

Primary hyperparathyroidism is usually accompanied by hypercalciuria, though the urinary calcium excretion may be lower than in other hypercalcemic disorders in which the parathyroid glands are suppressed, such as sarcoidosis, vitamin D intoxication and "humoral hypercalcemia of malignancy." In familial hypocalciuric hypercalcemia (FHH), urinary calcium excretion is relatively low, being about one third of that of patients with a similar degree of hypercalcemia due to primary hyperparathyroidism[9]. Attie et al.[10] have shown that the characteristic hypocalciuria in FHH is present even after total parathyroidectomy, and is therefore apparently not due to renal hypersensitivity to PTH. Both Watanabe and Sutton[11,12] and Attie et al.[10] showed that diuretic agents acting predominantly in the thick ascending limb of Henle's loop (furosemide and ethacrynic acid) tend to abolish the relative hypocalciuria in FHH. For example, in the study of Watanabe and Sutton, the urinary calcium:sodium ratio was significantly lower in FHH than in primary hyperparathyroidism, both under baseline conditions and after the administration of chlorothiazide (which acts mainly in the distal convoluted tubule[13]), but did not differ in the two conditions after furosemide. These results suggest that the hypocalciuria in FHH may be due to enhanced calcium reabsorption in the thick ascending limb of Henle's loop. Hypercalcemia has been shown to inhibit calcium transport in the thick ascending limb[7]. It is possible that in FHH there is an impairment of the tubular sensitivity to this effect of hypercalcemia, just as there must be an insensitivity to the normal inhibitory effect of hypercalcemia on PTH release in this disorder, in which hypercalcemia is associated with normal or elevated PTH levels. There is also a tendency to hypermagnesemia in FHH[9] (in contrast to primary hyperparathyroidism, in which hypomagnesemia may occur[14]), and this may reflect enhanced reabsorption of magnesium, as well as calcium, in the thick ascending limb.

Hypocalciuria with normocalcemia

An abnormally low urinary calcium excretion in a normocalcemic patient may be attributable to secondary hyperparathyroidism (as in vitamin D deficiency), to metabolic alkalosis[4] or to the administration of thiazide diuretics[5]. Recently, hypocalciuria has been reported in other clinical conditions in which there is other evidence suggesting the presence of a renal tubular disorder, for example, in cisplatin nephropathy and in Bartter's syndrome.

Cisplatin nephropathy. Up to one third of patients who have received a full course of cisplatin for the treatment of a malignancy are left with a residual renal lesion characterized by a modest reduction in glomerular filtration rate and by hypomagnesemia due to renal magnesium wasting[15]. We have studied a small group of these chronically hypomagnesemic patients[16] (mean serum magnesium 1.3mg/dl) and have found that despite normocalcemia, normal PTH levels, and absence of metabolic alkalosis, the mean urinary calcium excretion was only 82mg per 24 hours, significantly lower than that of matched normal control subjects (206mg, $p < 0.01$). The baseline urinary magnesium excretion was not different in patients and controls (116 vs. 114mg/day). The serum magnesium level has been shown to influence calcium transport (reciprocally) in the loop of Henle in the rat[7]. It seemed possible, therefore, that the hypocalciuria in this condition might be secondary to hypomagnesemia. Oral magnesium loading, sufficient to elevate the mean serum magnesium level from 1.3mg/dl to a low normal level (mean 1.7mg/dl) did not correct the hypocalciuria (mean 24-hour urine calcium 80mg). However, an intravenous infusion of magnesium chloride sufficient to raise the serum magnesium concentration to above 3.5mg/dl increased both fractional urinary magnesium and calcium excretion markedly, though to a lesser extent than in normal control subjects. These data suggested that, in cisplatin nephropathy, the effect of the serum magnesium level on divalent cation reabsorption may be abnormal, such that hypomagnesemia fails to enhance magnesium reabsorption normally, and hypermagnesemia fails to inhibit both calcium and magnesium reabsorption normally. Such a defect may reside in the ascending limb of Henle's loop, the site at which extracellular fluid calcium and magnesium levels are known to reciprocally affect reabsorption of calcium and magnesium[7,17]. An alternative location for such a defect might be the distal convoluted tubule, a site at which the reabsorption of calcium and magnesium can be dissociated for example by thiazide diuretics[18]. However, it seems unlikely that a primary enhancement of active calcium reabsorption (such as is induced by thiazide diuretics and amiloride in the distal convoluted tubule[19]) would result from a drug-induced nephrotoxicity.

Bartter's syndrome and Congenital "Magnesium Losing Kidney". Bartter's syndrome has been reported to be associated with high[20], normal[21] or low[22,23] urinary calcium excretions. Likewise, in some cases reported as congenital "magnesium losing kidney" urinary calcium excretion has been reported to be high[24,25] while in others hypocalciuria has been noted[26]. The distinction between Bartter's syndrome and congenital "magnesium losing kidney" is not always clear, and it is possible that these conditions are variants of the same disorder. We have studied three siblings[27] with renal magnesium wasting, normotension, hyperreninemia, hyperaldosteronism and hypokalemia in whom magnesium supplementation did not correct the hyperreninemia and who might therefore be classified as having Bartter's syndrome[25]. In many respects, our patients were similar to others described as having congenital "magnesium losing kidney"[26]. These patients had striking hypocalciuria (mean 28mg/day), normocalcemia, intermittent mild metabolic alkalosis and normal PTH levels. Their distal fractional chloride reabsorption [$C_{H_2O}/(C_{H_2O} + C_{Cl})$], determined during hypotonic saline diuresis, was subnormal (mean 62%, normal >80%) as has been reported in Bartter's syndrome[28] and interpreted as evidence of a defect in sodium chloride transport in the thick ascending limb of Henle's loop[28]. Although such a defect at a major site of magnesium reabsorption could certainly explain the occurrence of renal magnesium wasting in some patients with Bartter's syndrome, it would not explain why magnesium wasting is an inconsistent feature of the syndrome. We reasoned that, if a defect in sodium chloride transport in the thick ascending limb were the primary renal tubular lesion in our patients, and if furosemide acts at this site, the patients would be expected to show a blunted response to furosemide. Furthermore, if calcium reabsorption were also enhanced at this site, the hypocalciuria might be abolished by furosemide. In the event, however, following intravenous furosemide these patients showed sodium and chloride responses equal to or greater than those of normal control subjects, and the striking hypocalciuria (relative to controls) was not abolished. These data suggest that, in this variant of Bartter's syndrome, the presumed abnormalities of sodium, chloride and calcium transport are not located at the furosemide-sensitive site. However, there is evidence of heterogeneity within the thick ascending limb[29] and the defects could be in a different portion of this segment, or possibly in another segment such as the distal convoluted tubule. A defect in sodium chloride transport in the distal convoluted tubule would tend to decrease the lumen negative transepithelial potential difference in this segment, and might lower intracellular sodium levels and favour sodium-calcium counter transport at the basolateral cell membrane. Either of these processes might enhance calcium reabsorption, producing an effect resembling that of thiazide or amiloride administration[19]. Alternatively, the enhanced calcium reabsorption could be occurring in a non-furosemide sensitive portion of the thick ascending limb in response to hypomagnesemia[7]. Available information in our patients does not clarify whether the hypocalciuria is part of the primary tubular abnormality, or is secondary to the hypomagnesemia.

CONCLUSION

Two types of hypocalciuric disorder have been identified. In FHH there is enhanced tubular reabsorption of both calcium and magnesium, probably in the thick ascending limb of Henle's loop. This abnormality may result from an insensitivity of this nephron segment to the normal inhibitory effect of a high extracellular calcium level on calcium (and magnesium) reabsorption.

In a variant of Bartter's syndrome, and in chronic cisplatin nephropathy, hypocalciuria is associated with renal magnesium wasting. Although the magnesium reabsorptive defect in these disorders may be in the loop of Henle, and the enhanced calcium reabsorption may be secondary to hypomagnesemia, the available evidence does not exclude the different site for the defect such as the distal convoluted tubule, a site at which sodium, calcium and magnesium transport can be dissociated, for example by thiazide diuretics and amiloride.

REFERENCES

1. W.G. Robertson and D.B. Morgan, The distribution of urinary calcium excretions in normal persons and stoneformers. Clin, Chem, Acta 37: 503 (1972).
2. M.M. Popovtzer, L.I. Schainuck, S.G. Massry, and C.R. Kleeman,Divalent ion excretion in chronic kidney disease: relation to degree of renal insufficiency. Clin. Sci. 38:297 (1970).
3. R.V. Talmage, F.W. Kraintz and G.D. Buchanan, Effect of parathyroid extract and phosphate salts on renal calcium and phosphate after parathyroidectomy, Proc. Soc. Exp. Biol. Med. 88:600 (1955).
4. N.A. Edwards and A. Hodgkinson, Studies on renal function in patients with idiopathic hypercalciuria, Clin. Sci. 29:327 (1965).
5. B.A. Lamberg and B. Kuhlbach, Effect of chlorothiazide and hydrochlorothiazide on the excretion of calcium in the urine, Scandinav. J. Clin. Invest. 11:351 (1959).
6. R.A.L. Sutton, N.L.M. Wong, G.A. Quamme and J.H. Dirks, Renal tubular calcium transport: Effect of changes in filtered calcium load, Am. J. Physiol. 245:F515 (1983).
7. G.A. Quamme and J.H. Dirks, Intraluminal and contraluminal magnesium on magnesium and calcium transfer in the rat nephron. Am.J.Physiol. 238:187 (1979).
8. M. Peacock, W.G. Robertson and B.E.C. Nordin, Relation between serum and urinary calcium with particular reference to parathyroid activity, Lancet 1:384 (1969).
9. S.J. Marx, A.M. Spiegel, E.M. Brown, J.O. Koehler, D.G. Gardner, M.F. Brennan and G.D. Aurbach, Divalent cation metabolism. Familial hypocalciuric hypercalcemia versus typical primary hyperparathyroidism, Am.J. Med 65:235 (1978).
10. M.F. Attie, J.R. Gill,Jr., J.L. Stock, A.M. Spiegel, R.W. Downs, Jr., M.A. Levine and S.J. Marx, Urinary calcium excretion in familial hypocalciuric hypercalcemia. Persistence of relative hypocalciuria after induction of hypoparathyroidism, J. Clin. Invest. 72:667 (1983).
11. H. Watanabe and R.A.L. Sutton, Renal calcium handling in familial hypocalciuric hypercalcemia, Clin. Res. 29:425A (1981).
12. H. Watanabe and R.A. L. Sutton, Renal calcium handling in familial hypocalciuric hypercalcemia, Kidney Int'l., 24:353 (1983).
13. L.S. Costanzo and E.E. Windhager, Calcium and sodium transport by the distal convoluted tubule of the rat, Am. J. Physiol. 235:F492 (1978).
14. R.A.L. Sutton, Plasma magnesium concentration in primary hyperparathyroidism, Brit. Med. J. 1:529 (1970).
15. R.L. Schilsky, A. Barlock and R.F. Ozols, Persistent hypomagnesemia

following cisplatin chemotherapy for testicular cancer,Cancer Treat. Reports 66:1767 (1982).
16. R.A.L. Sutton, N.L.M. Wong, J.H. Dirks, C.M.L. Coppin and V. Mavichak, Renal magnesium wasting following cisplatin therapy, Abstr. Am. Soc. Nephrol. Annual Meeting 29A (1984).
17. G.R. Shareghi and Z.S. Agus, Magnesium transport in the cortical thick ascending limb of Henle's loop in the rabbit, J. Clin. Invest. 69: 759 (1982).
18. N.L.M. Wong, G.A. Quamme and J.H. Dirks, Effect of chlorothiazide on renal calcium and magnesium handling in the hamster, Can. J. Physiol. Pharmacol. 60:1160 (1982).
19. L.S. Costanzo, Comparison of calcium and sodium transport in early and late rat distal tubules: effect of amiloride, Am. J. Physiol. 246:F937 (1984).
20. D.A. McCredie, E. Rotenberg and A.C. Williams, Hypocalciuria in potassium-losing nephropathy: A variant of Bartter's syndrome, Aust. Paediatr. J. 10:286 (1974).
21. L. Norby, A.L. Mark, and G.J. Kaloyanides, On the pathogenesis of Bartter's syndrome: Report of studies in patients with this disorder, Clin. Nephrol. 6:404 (1976).
22. V. Delaney, J. Oliver, M. Simms, J. Costello and E. Bourke, Bartter's syndrome physiologic and pharmacologic studies, Quarterly J. Med. 50:213 (1981).
23. A. Rudin, B. Sjogren, and M. Aurell, Low urinary calcium excretion in Bartter's syndrome, N. Engl. J . Med. 310:1190 (1984).
24. R.A.L. Sutton and J.H. Dirks, Calcium and magnesium. Renal handling and disorders of metabolism, in: "The Kidney", 3rd Ed., B.M.Brenner and F.C. Rector,Jr., eds. W.B. Saunders Company, Philadelphia (1985).
25. R.A. Evans, J.N. Carter, C.R.P. George, R.S. Walls,, C.R. Newland, G.D. McDonnell, and J.R. Lawrence, The congenital 'magnesium-losing kidney'. Report of two patients, Quarterly J. Med. 197:39 (1981).
26. H.J. Gitelman, J.B. Graham, and L.G. Welt, A new familial disorder characterized by hypokalemia and hypomagnesemia, Trans. Assoc. Amer. Physcns. 79:221 (1966).
27. R.A.L. Sutton, V. Mavichak and G.E. Wilkins, Bartter's syndrome with hypocalciuria - a distal tubule defect? Clin. Res. 33:85A (1985).
28. J.R. Gill, Jr., and F.C. Bartter, Evidence for a prostaglandin independent defect in chloride reabsorption in the loop of Henle as a proximal cause of Bartter's syndrome, Am. J. Med. 65:766 (1978).
29. W.N. Suki, D. Rouse, R.C.J. Ng, and J.P. Kokko, Calcium transport in the thick ascending limb. Heterogeneity of function in the medullary and cortical segments, J. Clin. Invest. 66:1004 (1980).

following cisplatin chemotherapy for testicular cancer. Cancer Treat. Report. 68:1767 (1984).
66. [illegible]

RENAL MAGNESIUM WASTING DISORDERS

John H. Dirks and Norman L. M. Wong

Dept. of Medicine, University of British Columbia
Acute Care Unit, Health Sciences Centre Hospital
2211 Wesbrook Mall, Vancouver, B.C. V6T 1W5 Canada

INTRODUCTION

Hypomagnesemia and magnesium deficiency may be due to gastrointestinal malabsorption, poor dietary intake or renal (tubular) magnesium wasting. Magnesium wasting would result in hypomagnesemia if dietary magnesium intake is insufficient. A number of causes of magnesium wasting have been identified including congenital or primary renal magnesium wasting, hyperaldosteronism, Bartter's syndrome, diabetic ketoacidosis, acute alcohol intake and hypercalcemia with hypercalcuria. More recently attention has focussed on renal magnesium wasting due to drugs including diuretics, antibiotics, chemotherapeutic and immunosuppressive agents. One can consider renal magnesium wasting as an exaggerated physiological response due to pharmacological or toxic impairment of normal tubular magnesium reabsorption. It is clear that the ensuing hypomagnesemia is dependent on the dietary magnesium intake and the overall cellular magnesium depletion does not necessarily occur in a predictable fashion.

NORMAL RENAL TUBULAR MAGNESIUM REABSORPTION AND ITS CONTROL.

The proximal tubule reabsorbs some 15 - 30% of the filtered magnesium, which is considerably less than the rate of normal sodium and calcium reabsorption[1]. Reabsorption of magnesium in the proximal tubule is affected by changes in extracellular fluid volume, so that volume expansion reduces magnesium reabsorption and allows a greater delivery of magnesium to more distal parts of the nephron, while volume contraction has the opposite affect. The thick ascending limb of Henle's loop is the dominant site of magnesium reabsorption in the renal tubule. Some 60% or more of the filtered magnesium is reabsorbed at this site[1]. It would appear that the mechanism of magnesium reabsorption in the thick ascending limb is largely passive which is dependent on the sodium-potassium chloride cotransport system. Magnesium diffuses passively out of the lumen across the paracellular channels into the interstitial fluid. Active magnesium reabsorption at the basolateral surface may also occur. Loop diuretics such as furosemide profoundly reduce magnesium reabsorption in the thick ascending limb. Rapid flow of fluid through the loop occurs due to massive volume expansion or with osmotic diuretics which limit the positive transepithelial potential and hence magnesium transport[2].
Hypermagnesemia and hypercalcemia reduce magnesium reabsorption in the

thick ascending limb through a competitive interaction with calcium on the antiluminal side of the cell[3]. Conversely hypocalcemia and hypomagnesemia appear to increase magnesium reabsorption. Parathyroid hormone increases magnesium transport in the thick ascending limb as does alkalosis while metabolic acidosis reduces magnesium transport[4,5]. The effects of alkalosis and acidosis may extend to the distal convoluted tubule[5]. Magnesium reabsorption in the distal convoluted tubule and the collecting duct system appears to constitute only about 5% of the magnesium filtered. The precise control of distal magnesium transport remains unclear. This may, however, be the site of the magnesium leak that occurs during chronic phosphate depletion[6].

Chronic dietary deficiency of magnesium leads to a reduction in filtered load due to progressive hypomagnesemia. Hyperabsorption occurs in the thick ascending limb with concomitant reduction in renal magnesium excretion.

RENAL MAGNESIUM WASTING DUE TO DRUGS

Diuretics

The loop diuretics including furosemide, ethacrynic acid, piretanide, bumetanide and muzolimine can bring about magnesium loss particularly when the natriuretic and diuretic response are marked[7,8]. Because the loop diuretics inhibit the sodium-potassium chloride co-transport system on the luminal membrane of the thick ascending limb, a favourable gradient for passive diffusion for magnesium across the paracellular channels is progressively diminished and less magnesium is absorbed. The increase in delivery of magnesium to the distal parts of the nephron quickly saturates the distal transport system for magnesium and marked magnesuria occurs. Experiments indicate that the relative inhibition by loop diuretics is greater on magnesium than it is on calcium as compared to sodium[9]. In clinical circumstances hypomagnesemia can occur following furosemide diuresis, e.g. in the treatment of acute symptomatic hypercalcemia. Obviously marked volume expansion greatly triggers the magnesuria. Osmotic diuretics such as mannitol have interesting effects on actually increasing magnesium transport in the proximal tubule to a mild degree, but have major inhibitory effects in reducing magnesium reabsorption in the thick ascending limb of Henle's loop as the osmotic diuretic is progressively infused[10]. Microperfusion experiments have shown that as the flow rate through the loop is progressively increased, the inhibition of magnesium transport occurs presumably due to the less favourable positive potential within the thick ascending limb secondary to a reduced sodium gradient as a result of the high flow. Other factors such as the medullary washout associated with mannitol diuresis produces a marked magnesuria. Urea diuresis also results in magnesuria due to loop effects. The effects of other diuretics on magnesium transport are much less significant. Carbonic anhydrase inhibitors like actazoleamide which have their site of action in the proximal convoluted tubule have little effect on overall magnesium excretion[11]. The acute administration of thiazide diuretics have been shown in micropuncture and in clearance experiments to have little or no effect on overall magnesium excretion[12]. However, chronic thiazide administration in some patients may result in mild degrees of magnesium wasting. The natriuretic potassium sparing agents such as amiloride and triameterene result in a mild enhancement of magnesium reabsorption as they do for calcium[13].

Cisplatinum Toxicity

Much attention has focussed on the magnesium wasting associated with cisplatinum treatment for a variety of solid tumors such as testicular and ovarian carcinomas. Approximately one third of patients who have received a full course of cisplatinum are left with a residual lesion characterized by a modest reduction in glomerular filtration rate and by hypomagnesemia due to renal wasting[14]. We have embarked on both animal and human studies in an attempt to characterize the defect in magnesium excretion.

First in rats, it was observed that three intraperitoneal injections of cisplatinum (2.5 mg/kg of body weight) induced chronic hypomagnesemia which was present from the second week and persisted through eight weeks of experiment. Plasma magnesium concentrations were 0.72 ± 0.2 mM as compared to 0.83 ± 0.82 in control rats at the end of eight weeks. The renal excretion of magnesium in the cisplatinum treated rats was inappropriately high relative to the lower plasma magnesium concentration indicating the presence of renal magnesium wasting induced by cisplatinum. Generally the degree of hypomagnesemia is very dependent on the magnesium dietary intake and could be overcome by augmenting magnesium intake. Fractional magnesium excretion is higher in the treated rats than in control. The rats with renal magnesium wasting due to cisplatinum also frequently had phosphaturia and hypercalciuria. Micropuncture studies indicated the proximal tubular absorption of magnesium was normal. Morphological studies indicated histopathology predominantly in the straight portion of the proximal tubule. After initial focal necrosis in the other strip of the medulla there was dilatation of the proximal tubules. The tubules were lined by variably flattened epithelial cells devoid of brush borders and many cells of enlarged hypochromatic nucleii. These changes are most extensive at four weeks and diminished thereafter but were still present focally at 24 weeks. By electron microscopy patchy S3 segment cell necrosis was noted at one day post injection of cisplatinum and the latter S3 segments of the proximal tubules were lined with flattened cells. Tubular basement membranes were focally thickened. The morphological studies indicated that the S3 segment was the prime target of cisplatinum induced morphologic changes that regeneration was slow and incomplete suggesting a reason for the persistence of cisplatinum nephrotoxicity. However, a distal defect may still be present as furosemide diuresis resulted in exaggerated response in cisplatinum treated rats. Furthermore, superficial proximal and distal tubule micropuncture showed similar amounts of magnesium remaining in the cisplatinum as compared to controlled rats after furosemide diuresis. Thus, it remains possible that there is a distal defect in addition to the defect identified in the S3 segments.

Secondly, in six chronically hypomagnesemic patients (mean serum magnesium 1.3 mg/dl) baseline urinary magnesium excretion was not different in patients and controls. The patients also had hypocalciuria, a small increase in mean serum calcium level, mild hypokalemia and hypophosphatemia with evidence of renal potassium and phosphate wasting. Renal magnesium handling was examined during oral magnesium deprivation and the patients were observed to reduce the magnesium losses as a result of a further fall in their plasma magnesium levels. When tested with furosemide magnesuria was blunted during the first hour but was followed by an exaggerated response in the second and third hours and the overall cumulative magnesium excretion after furosemide was greater than in control patients. The response to furosemide suggests an initial defect furosemide secretion in the proximal tubule together with a defect in the distal nephron.

Additional studies were performed to determine the protective effect of sodium thiosulphate in cisplatinum treated rats. Interstrand cross-links in DNA account for a major part of cisplatinum cytotoxicity to rapidly proliferating cells. The concurrent administration of sodium thiosulphate probably reduce the toxicity by reversing the cisplatinum induced cross-linking. It was observed that thiosulphate blocks not only the mild reduction in GFR but the magnesium wasting associated with cisplatinum.

Cisplatinum also affected renal phosphate handling. Bone and muscle phosphate is significantly decreased in cisplatinum treated rats although the renal phosphate excretion was the same in cisplatinum and control animals. Sodium dependent phosphate uptake in the brush border membrane vesicles isolated from the renal cortex of cisplatinum treated rats was significantly depressed when compared to nontreated rats. There was a two fold increase in the Km in the cisplatinum rats with no effect on Vmax. Thus, cisplatinum may induce changes in the sodium dependent phosphate transporter in the luminal membrane of the proximal tubule.

Antibiotics

Periodic reports of magnesium wasting and hypomagnesemia have been associated with aminoglycoside antibiotics particularly gentamicin[15] and capreomicin. Amphotericin is sometimes associated with magnesium wasting even though proof is not conclusive. Gentamicin is well known for its effect in causing acute tubular necrosis when the doses are too high. Lower doses can cause selective tubular damage and in rats and we have been able to demonstrate magnesium wasting. This again seems to be identified with pathology in the S3 segment of the proximal tubule. Interestingly, the magnesium wasting due to gentamicin as well as its effect in reducing GFR can be blocked by pretreatment with thiosulphate.

Cyclosporine

A number of reports have suggested that cyclosporine can cause magnesium wasting. This is interesting in light of the fact that cyclosporine is nephrotoxic and can cause acute tubular necrosis. Cyclosporine is secreted by the straight portion of the proximal tubule. Thompson et al[16] reported hypomagnesemia in patients with bone marrow transplants who were treated with cyclosporine. More data is required to identify the frequency of magnesium wasting and the site within the nephron where this occurs.

CONCLUSION

Renal magnesium wasting, associated with a wide spectrum of clinical conditions, is being identified with increasing frequency. The development of hypomagnesemia is dependent not only upon the degree of toxicity but on the degree of dietary compensation of magnesium. A number of magnesium wasting conditions are exaggerations of normal physiological controls as occurs with excessive volume expansion, severe metabolic acidosis, marked hypercalcemia as in hyperparathyroidism. Drugs are a major cause of magnesium wasting of which diuretic therapy with furosemide would appear to be the chief cause. Cisplatinum toxicity has received much attention and would appear to be the major cause of magnesium wasting next to diuretic treatment and nutritional magnesium deprivation. Cisplatinum in rats causes a lesion in the S3 segment of the proximal tubule and magnesium wasting can be prevented with sodium thiosulfate. In man there may be an additional defect in the distal part of the nephron as hypocalciuria rather than hypercalciuria results. Aminoglycoside antibiotics such as gentamicin are a sporadic cause of magnesium wasting and cyclosporine may become a more frequent cause as more observations are made.

REFERENCES

1. G. A. Quamme and J. H. Dirks, Magnesium: Cellular and Renal Exchanges, in "The Kidney, Physiology and Pathophysiology." D. W. Seldin and G. Giebisch, ed., Raven Press, New York (1985).

2. G. R. Shareghi and Z. S. Agus, Magnesium transport in the cortical thick ascending limb of Henle's loop of the rabbit. J. Clin. Invest. 69:759 (1982).
3. G. A. Quamme, Effect of hypercalcemia on renal tubular handling of calcium and magnesium. Can. J. Physiol. Pharmacol. 60:1275 (1982).
4. G. R. Shareghi and Z. S. Agus, PTH - Sensitive magnesium transport in the rabbit cortical thick ascending limb of Henle. (CTAL) Kidney Int. 16:S37A (1979).
5. N. L. M. Wong, G. A. Quamme, and J. H. Dirks, Influence of Bicarbonate on the Renal Handling of Magnesium, in: "Urolithiasis, Clinical and Basic Research." L. H. Smith, W. G. Robertson, and B. Finleyson, ed., Plenum Press, New York (1984).
6. N. L. M. Wong, G. A. Quamme, R. A. L. Sutton, and T. O'Callaghan, Renal and tubular transport in phosphate depletion: A micropuncture study. Can. J. Physiol. Pharmacol. 58:1063 (1980).
7. J. Seeham and A. White, Diuretic associated hypomagneseuria. Br. Med. J. 285:1157 (1982).
8. P. O. Wester and T. Dyckner, Diuretic treatment and magnesium losses. Acta Medica. Scand. 647 (Suppl.):145 (1981).
9. G. A. Quamme, Effect of furosemide on calcium and magnesium transport in the rat nephron. Am. J. Physiol. 241:340 (1981).
10. N. L. M. Wong, G. A. Quamme, R. A. L. Sutton, and J. H. Dirks, Effect of mannitol on water and electrolyte transport in the dog kidney. J. Lab. and Clin. Med. 94:683 (1979).
11. E. S. Barker, J. R. Elkinton, and J. K. Clark, Studies in the renal excretion of magnesium in man. J. of Clin. Invest. 38:1733 (1959).
12. N. L. M. Wong, G. A. Quamme, and J. H. Dirks, Effect of chlorothiazide on renal calcium and magnesium handling in the hamster. Can. J. Physiol. Pharmacol. 60:1160 (1980).
13. J. Devane and M. P. Ryan, Evidence for a magnesium-sparing action by amiloride during renal clearance studies in rats. Br. J. of Pharmacol. 78:891 (1983).
14. R. L. Schilsky, A. Barlock, and R. F. Ozols, Persistent hypomagnesemia following cisplatin chemotherapy for testicular cancer. Cancer Treat. Reports 66:1767 (1982).
15. C. G. D. Brook, Hypomagnesemia, hypocalcemia with hypokalemia caused by treatment with high dose gentamycin. Arch. Dis. Child. 53:817 (1978).
16. C. B. Thompson, C. H. June, K. M. Sullivan, and E. D. Thomas, Association between cyclosporin neurotoxicity and hypomagnesemia. Lancet. 2(8412):1116 (1984).

ROLE OF PTH, VITAMIN D AND BONE METABOLISM IN LOW PROTEIN-INDUCED HYPERCALCIURIA

Emma Fernández-Repollet and Manuel Martínez-Maldonado

Departments of Pharmacology and Medicine, University of Puerto Rico School of Medicine and Veterans Administration Hospital, San Juan, Puerto Rico

INTRODUCTION

Protein calorie malnutrition is a world wide problem affecting 30-40% of hospitalized patients in industrial societies and hundred of thousands in developing countries.[1,2]

Few experiments, however, have dealt with the effects of low protein diet on the renal handling of calcium and phosphate. Dynamic studies performed by Le Roith and Pimstone[7] showed a large increase in the fractional calcium excretion of protein-deficient rats. Their techniques, however, did not distinguish between faecal and urinary calcium excretion.

The purpose of this study was to investigate the effects of low protein intake on renal calcium handling and whether protein deprivation alters PTH and Vitamin D secretion and/or bone metabolism.

MATERIALS AND METHODS

Experimental Animals

Male Wistar-Furth rats weighing 100 ± 10 g were used in all experiments.

Balance Studies

Balance studies were performed prior to and after 2 weeks on a normal (23%) or a low (6%) protein diet. Both diets were isocaloric and similar in electrolyte content. All rats were pair-fed during the study.

Balance studies consisted of daily measurements of body weight, food and water intake. Twenty-four hour urine samples were collected for the determination of volume, calcium, phosphorus, urea and cyclic AMP concentration. Blood samples for measurement of electrolyte concentration and plasma 1,25 $(OH)_2$ Vitamin D_3 concentration were obtained by heart puncture under light ether anesthesia. Plasma 1,25 $(OH)_2$ Vitamin D_3 concentration was measured in the laboratory of Dr. Anthony Norman in Riverside, California.

Bone Analysis

At the end of the experimental period, the animals were sacrificed and the left and right tibiae were removed for chemical and morphological analysis.

The left tibiae from each of 6 normal and 6 low protein rats were defatted with ether, then dried to constant weight at 105° C. The ash content was determined following ashing of the whole bone at 600° C for 24 hrs. Bone ash was hydrolyzed in 6N HCl and calcium, magnesium and phosphorus contents were determined colorimetrically.

The right tibiae from 3 normal and 3 low protein rats were dissected and dehydrated in ethanol. The samples were cut longitudinally to expose the endosteal surface. The specimens were then extracted with hot 1,2 ethane diamine for 48 hrs. and washed in absolute ethanol. This procedure removes most of the organic material from the bone, exposing the pattern of the mineralized collagen fibers. The sections were then air dried, and examined at 25 KV on an ETEC Autoscan scanning electron microscope.

RESULTS

As shown in Table 1, rats on a low protein diet showed a significant reduction in weight gain (%) compared to rats on a normal protein intake. Urine volume (ml/24 hr/100 g BW) was also significantly decreased in the protein-deprived rats. A significant increase in urinary calcium excretion (uEq/24 hr/100g BW) and a significant decrease in urinary phosphorus and urea excretion (mMoles/24hr/100gBW) were observed in rats on a low protein intake.

Table 1. Effects of low protein diet on body weight, urine flow rate, and the urinary excretion of calcium, phosphorus and urea.

	WEIGHT GAIN	URINE VOLUME	U V Ca	U V Pi	U V urea
NP (n=12)	88 ± 5	22 ± 3	27 ± 9	78 ± 20	1808 ± 330
LP (n=12)	49 ± 3*	16 ± 2*	85 ± 41*	13 ± 12*	88 ± 50*

*p<0.05

Determination of plasma electrolyte concentration (Table 2) revealed lower plasma calcium, urea and protein concentration in the low protein group. No significant differences in plasma phosphorus and magnesium concentration were observed between rats fed a normal or a low protein diet.

In an attempt to examine the role of PTH on the hypercalciuria induced by low protein intake, urinary cyclic AMP excretion was assessed in growing rats on a normal and low protein intake. No significiant changes in urinary cyclic AMP were observed in rats ingesting a normal protein diet 2 weeks after basal measurement (Basal: 26 $\pm$ 2 vs. 2 WKS: 24 $\pm$ 3 mMoles/ml/100gBW). In contrast, rats on a low protein intake showed a 49% reduction in urinary cyclic AMP excretion 2 weeks after basal measurements (Basal: 30 $\pm$ 1 vs. 2 WKS: 12 $\pm$ 1; $P< 0.05$).

Plasma 1,25 $(OH)_2$ Vitamin D_3 concentration was also measured in normal and low protein rats. The results indicated that 1,25 $(OH)_2$ Vitamin D_3 concentration averaged 34 $\pm$ 10 pg/ml in rats fed a low protein diet, a value significantly lower than that of 80 $\pm$12 pg/ml measured in the normal protein group ($P<0.05$).

In view of the detrimental effect on rat growth and the hypercalciuria induced by protein deprivation, we decided to evaluate the effects of low protein intake on bone length, weight and chemical composition.

As shown in Table 3, rats on a low protein diet had shorter and lighter bones than the pair-fed normal group. The difference in bone length between the 2 groups is further illustrated in Figure 1. Protein deprivation, however, did not affect ash content or calcium, phosphorus and magnesium content. Thus, protein deprivation impaired bone growth but did not alter bone quality or mineral content.

Table 2. Effects of low protein diet on plasma electrolyte concentration.

	P_{Ca}	P_{Pi}	P_{Mg}	P_{urea}	$P_{Prot.}$
	mg%				
NP (n=18)	9.2 $\pm$ 0.2	7.5 $\pm$ 0.2	2.1 $\pm$ 0.1	13.2 $\pm$ 0.4	7.1 $\pm$ 0.1
LP (n=18)	8.5 $\pm$ 0.3*	7.0 $\pm$ 0.3	2.2 $\pm$ 0.1	3.9 $\pm$ 0.2*	6.7 $\pm$ 0.1*

* $P<0.05$

Table 3. Effects of low protein diet on bone (tibiae) length, weight and chemical composition

	Length (cm)	FFDW (g)	Ash content (% ffdw)	Ca (% Ash Content)	P	Mg
NP (n=6)	3.4$\pm$0.1	0.168$\pm$0.001	64.7$\pm$0.3	36$\pm$1	18$\pm$1	0.6$\pm$0.1
LP (n=6)	2.9$\pm$0.1*	0.138$\pm$0.003*	66.0$\pm$0.5	36$\pm$1	17$\pm$1	0.6$\pm$0.1

* $P<0.05$

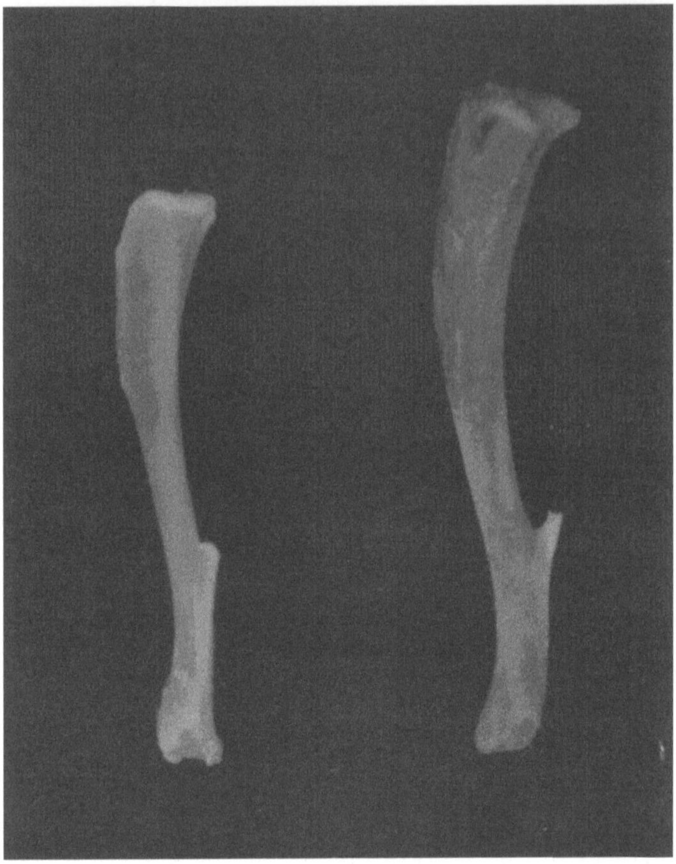

Figure 1. Photograph of tibiae bones from low protein (left) and normal (right) protein rats.

A

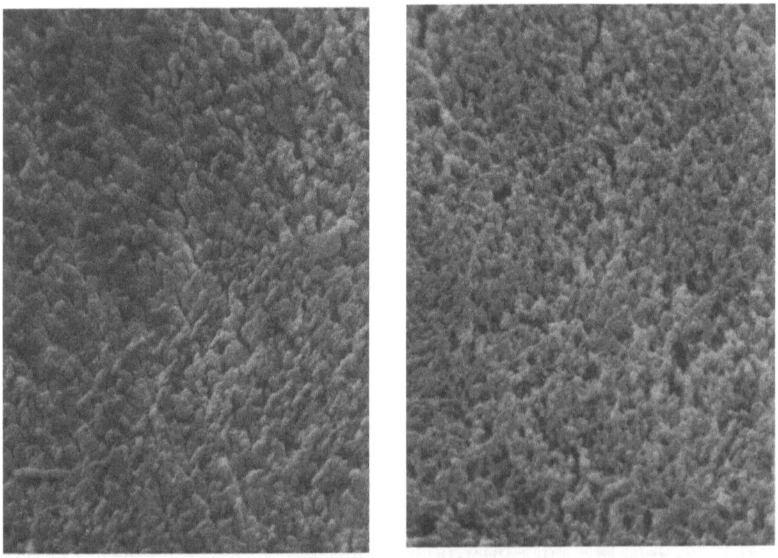

B

Figure 2. Scanning electron micrographs of the endosteal surface of tibiae bones of normal (A) and low (B) protein rats.

The morphological appearance by scanning electron microscopy of the endosteal surface of tibiae bone from animals fed a normal protein (A) or a low protein (B) diet is illustrated in Figure 2. Since the preparation has been made anorganic, what is actually observed is the pattern of calcium-phosphate crystals laid upon the collagen fibers during mineralization. It is evident that the endosteal surface of bones obtained from low protein rats lacks the highly ordered structure characteristic of collagen, which is known to act as a nucleation site for the formation and growth of calcium-phosphate crystals. These morphological data suggest that low protein diet impairs collagen synthesis.

DISCUSSION

In agreement with the findings of others[7,8] low protein intake diminished the skeletal development of growing rats. Our results also confirmed that this abnormality is a result of a decrease in the amount of bone formed rather than an altered mineral content. In addition, morphological studies revealed an abnormal pattern of calcium-phosphate crystal formation on the endosteal surface of tibiae bone, suggesting an impairment in bone collagen synthesis. These data are also in agreement with the observation of Le Roith and Primstone[7], indicating that the incorporation of ^{14}C proline into hydroxyproline was reduced in protein-deprived rats. The reduction of collagen synthesis may not only be the result of a direct effect of protein deprivation, but may also be due to the reduction of PTH secretion associated with this condition.[9] Whatever the mechanism, a decrease in collagen synthesis may contribute to the impairment in bone growth associated with protein depletion.

Despite a reduction in urine flow rate, rats on a low protein diet showed a dramatic hypercalciuria and a significantly lower plasma calcium and protein concentration. These disturbances in calcium metabolism were associated with a decrease in PTH secretion and a diminished 1,25 $(OH)_2$ Vitamin D_3 plasma concentration. The increase in urinary calcium excretion and the decrease in urinary phosphorus excretion appear to be the consequence of a reduction in PTH secretion. A direct effect of protein deprivation on PTH secretion is further supported by the fact that the hormone activity remained suppressed despite a lower plasma calcium concentration. A reduction in PTH secretion may also lead to a decreased production of 1,25 $(OH)_2$ Vitamin D_3 concentration. Low protein-induced hypocalcemia may also result from the continuous elevation in urinary calcium excretion as well as the decrease in plasma protein concentration associated with protein deprivation.

As a result of the decreased intestinal absorption of calcium, which should result from the reduction in plasma 1,25 $(OH)_2$ Vitamin D_3 concentration, the exchangeable calcium pool is expected to be reduced. Such a reduction has been shown to occur in protein deprived rats. A reduction in the exchangeable calcium pool would obviously retard bone growth.

Despite the fact that the exchangeable calcium pool is probably reduced, our data indicate that a higher fraction of this pool continues to be excreted by the kidney in protein-depleted rats. Further studies are in progress to elucidate the exact mechanism(s) by which dietary protein intake interferes with the renal handling of calcium.

ACKNOWLEDGEMENT

Supported in part by Grant # 1-R01 AM-30201 of the National Institutes of Health and the Veterans Administration Merit Review.

REFERENCES

1. B. R. Bistvian, G.L. Blackburn, J. Vitale, D. Cochran, and J. Naylor, Prevalence of malnutrition in general medical patients, JAMA 235:1567-1570 (1976).

2. G.L. Hill, I. Pickford, G.A. Young, C.J. Schorah, R.L. Blackett, L. Burkinshaw, J.V. Warren, and D.B. Morgan, Malnutrition in surgical patients. An unrecognized problem, Lancet 2:689-692 (1977).

3. P. Viart, Hemodynamic findings during treatment of protein calorie malnutrition, Am J Clin Nutr 31:911-926 (1978).

4. S. Klahr and G.A.O. Alleyne, Effect of chronic protein calorie malnutrition on the kidney, Kid Int 3:129-141 (1973).

5. S. Klahr and K. Tripathy, Evaluation of renal function in malnutrition, Arch Int Med 118:322-325 (1966).

6. I. Ichikawa, M.L. Purkerson, S. Klahr, J.L. Troy, M. Martínez-Maldonado and B.M. Brenner, Mechanism of reduced glomerular filtration rate in chronic malnutrition, J Clin Invest 65:982-988 (1980).

7. D. Le Roith and B.L. Pimstone. Bone metabolism and compositon in the protein-deprived rat, Clin Sci 44:305-319 (1973).

8. P.L. Glick and D.J. Rowe, Effects of chronic protein deficiency on skeletal development of young rats, Calcif Tissue Int 33:223-231 (1981).

9. E. Slatopolsky and M. P. Whyte, Renal regulation of extrarenal function: Bone, in: The Kidney: Physiology and Pathophysiology. D. Seldin and G. Giehisch eds., Raven Press, New York, 1985.

INTESTINAL ABSORPTION OF PHOSPHATE AND CALCIUM

MECHANISM AND REGULATION OF INTESTINAL PHOSPHATE TRANSPORT

David B. N. Lee

Medical and Research Services, Veterans Administration
Sepulveda Medical Center, and Department of Medicine, UCLA
School of Medicine, Los Angeles, California 90073

I will like to express my indebtedness to Dr. Marlin M. Walling (January 17, 1941 - November 30, 1979), a true scientist and scholar. He inspired and generated most of the data included in this presentation.

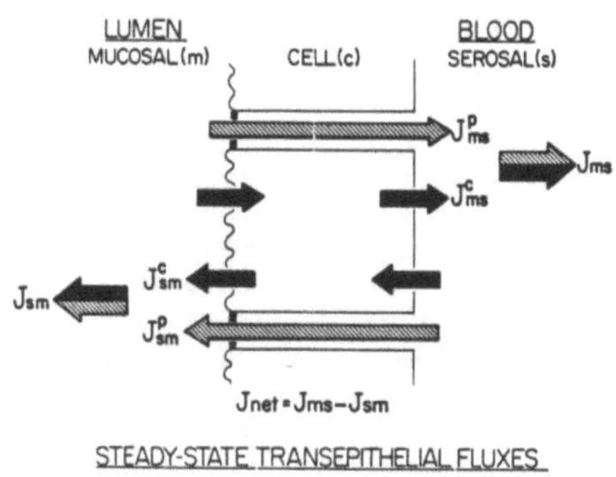

FIGURE 1.

The movement (flux) of Ca and P across the intestinal epithelium can be resolved into two unidirectional fluxes (figure 1): an absorptive, mucosal-to-serosal flux (Jms) and a secretory, serosal-to-mucosal flux (Jsm). Each unidirectional flux in turn consists of two components: thus Jms consists of a transcellular flux (Jc,ms) and a paracellular flux (Jp,ms), and the same is true for Jsm in the opposite direction. The net flux (Jnet) is the difference between Jms and Jsm, or:

$$Jnet = Jms - Jsm = (Jc,ms+Jp,ms) - (Jc,sm+Jp,sm) \quad [1].$$

A positive value for Jnet indicates net absorption, while a negative value would indicate net secretion. The assumption is that the paracellular flux represents passive, diffusional process, and is driven by transepithelial chemical and physical (electrical and hydrostatic) gradients, while the transcellular flux represents active, energy-dependent transport process, and is assumed to be unaffected by the transepithelial electro-chemical gradients. A piece of intestinal epithelium can be mounted and studied in an Ussing system which has the capability of eliminating the transepithelial physical and chemical gradients (1,2). The nulling of electro-chemical driving forces across the epithelium would equalize the oppositely directed fluxes across the paracellular pathway, i.e., Jp,ms = Jp,sm, simplifying equation [1] to:

$$Jnet = Jc,ms - Jc,sm \quad [2].$$

Thus, the demonstration of a net flux, under the constraints of the Ussing system, would strongly suggest the presence of a transcellular absorptive or secretory process.

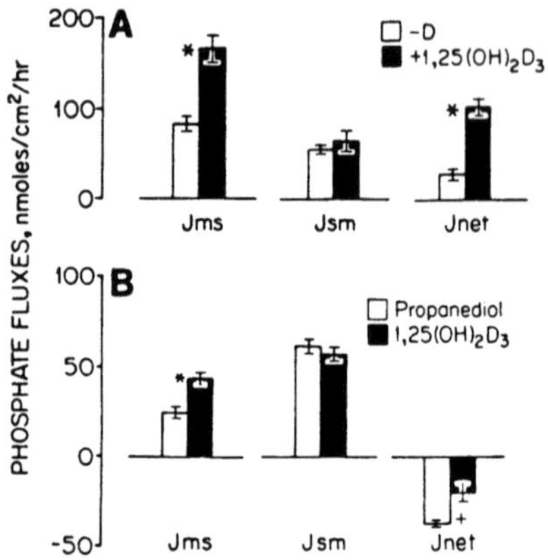

FIGURE 2. Modified from (3).

We have measured transepithelial Ca and P fluxes in young (150-180g) and older (250-280g) rats using the Ussing system. Figure 2A summarizes the steady-state transepithelial P fluxes in jejunum from, young, vitamin D-deficient (-D) rats. These rats were either treated with vehicle (-D, open colums) or 1,25-dihydroxyvitamin D_3 ($1,25(OH)_2D_3$, dark columns). $1,25(OH)_2D_3$ (1,25D)-treatment caused stimulation of Jms without affecting the Jsm, giving rise to a clear increase in net absorptive flux (Jnet). Note that Jms exceeded Jsm, giving rise to a positive net flux even in the -D rats. The presence of active P absorption in rats severely depleted of vitamin D raised the possibility of non-vitamin D-dependent factor(s) in the regulation of active P absorption. In figure 2B the transepithelial P fluxes were measured in normal, older (250-280g) rats treated with solvent (propanediol, open columns) or exogenous 1,25D ($+1,25(OH)_2D_3$, dark columns). In these normal rats, Jms was clearly lower than Jsm resulting in a negative Jnet, i.e., a net P secretory flux. Exogenous 1,25D administration led to the anticipated Jms-only effect, reducing but did not eliminate the net secretion of P. These data demonstrate on the one hand, active P absorption in jejunum from young rats, severely deprived of vitamin D, and on the other hand, active P secretion in jejunum from normal, vitamin D-replete adult rats. Taken together, these observations suggest that physiologic mechanism(s) other than that dependent on vitamin D may regulate jejunal P transport. Thus, factors such as the demand to satisfy the stress of growth (as in younger, -D rats), or the lack of demand for P when growth plateaus (as in older, non-vitamin D-deficient, adult rats), may modulate net jejunal P absorption, independent of vitamin D nutritional status.

An interesting example of vitamin D-independent regulation of P absorption is the ability of local (peri-enterocyte) factors to modify or reverse the effect of 1,25D. The enterocytes, unlike the renal tubular cells, "see" frequent and often marked fluctuations in luminal concentrations of solutes such as P, caused by quantitative and qualitative variations in the feeding habits of the individual animal. Thus it would appear advantageous for a solute, e.g., P, to exert some local regulation on its own absorption so that should the environment of the enterocytes become flooded with P in face of a pre-programed state of active P absorption, appropriate response may be elicited to avoid P overloading to both the enterocytes and the animal. Figure 3 summarizes the effect of raising extracellular P concentraion ([P]) on trans-jejunal

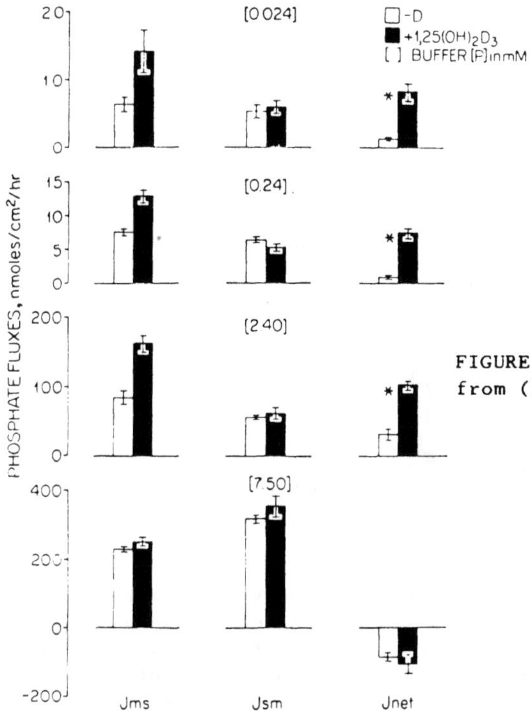

FIGURE 3. Reproduced from (3).

P fluxes in both the -D (light columns) and 1,25D-repleted (dark columns) rats. At buffer (extracellular) [P] of 0.02, 0.24 and 2.4 mM, Jms always exceeded Jsm in young, -D or 1,25D-repleted rats, so that net absorption of P was observed in all cases studied. As already noted, the effect of 1,25D was a specific stimulation of Jms leaving Jsm unaffected. Two major differences were observed in the pattern of P fluxes at buffer [P] of 7.5 mM. The first was that Jms was no longer different between -D and 1,25D-repleted rats and the second was that Jsm now exceeded Jms so that in both treatment groups there was a net secretory flux - findings similar to those observed in normal adult rats (figure 2B). Thus, at high extracellular [P] the stimulatory effect of 1,25D on active P absorption was no longer apparent, and both -D and 1,25D-stimulated jejunum behaved as P secretory epithelium. The secretory flux was not a vitamin D-dependent process, since it was activated, and to equal extent, in both -D and 1,25D-repleted rats.

To conclude this series of studies, we examined the effect of dietary P deprivation on jejunal P transport (figure 4). Rather surprisingly, in -D rats the active P absorption was not different between rats eating P-replete (+P) diet and rats eating P-deficient (-P) diet. On the other hand, in vitamin D_3 ($+D_3$) and 1,25D-replete ($+1,25(OH)_2D_3$) rats, -P diet had a clear stimulatory effect on active P absorption. This observation would suggest that the induction of P absorptive response to P-deprivation requires the presence of vitamin D or its metabolites, and provides interesting contrast to P-deprivation-induced renal adaptative response which appeared intact in -D rats (4).

We propose another explanation for the absence of changes in jejunal P transport to P-deprivation in -D rats (figure 5). Intestinal P absorption is considered to consist of a D-independent (hatched bar) and a D-dependent (solid bar) components. Adaptative response to

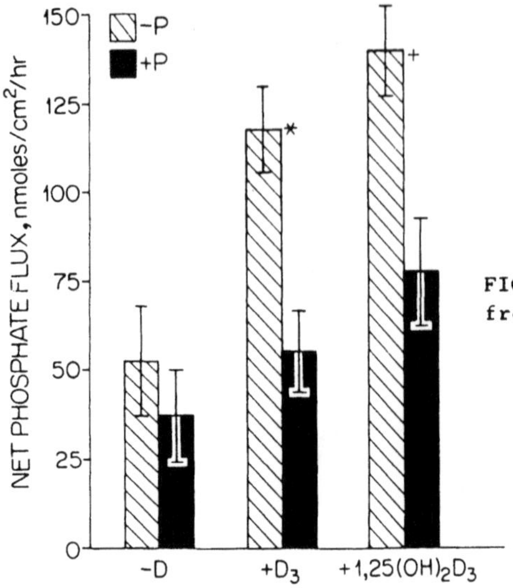

FIGURE 4. Reproduced from (3).

P-deprivation is postulated to be mediated through a D-independent mechanism - the same mechanism which is already maximally stimulated in -D rats eating an otherwise normal (+P) diet. Additional stress of dietary P-deprivation is therefore not translated into further augmentation in active P absorption. In D-replete rats fed a normal P diet, the D-sensitive active P absorption provides adequate P supply to satisfy growth demand thus leaving the postulated D-independent mechanism dormant. In such animal, the stress of dietary P-deprivation could be expected to mobilize the additional, D-independent, P absorptive mechanism, and thus could cause further increase in active P absorption.

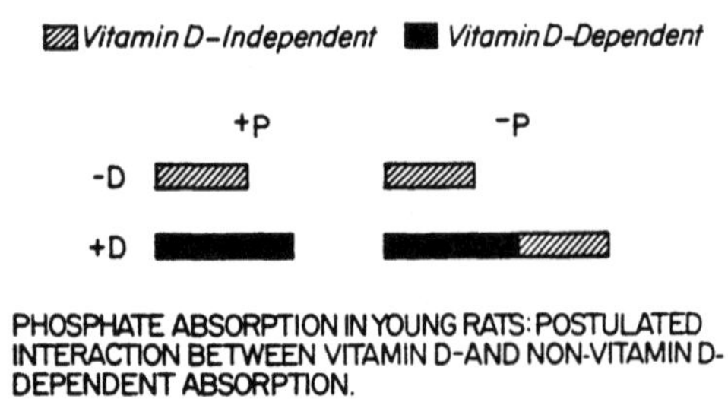

FIGURE 5.

Figure 6 provides a hypothetical model for the purpose of summarising our observations. Intra-enterocytic P content is maintained by P influx. The best known mechanism is the Na-dependent P influx across the brush border membrane (BBM) (5-9), but influx across basolateral membrane (BLM) by mechanism similar to, or different from, that described in the kidney (10) needs to be considered as well. We suggest that the enterocyte is capable of altering the relative permeabilities of these two membranes for P efflux. A state of greater BBM P permeability would result in net secretion while a greater BLM P permeability would result in net absorption. The signals for such shuttling may be tied to cellular P homeostasis and P balance of the organism. During growth or P-deprivation, states in which P demand exceeds supply the permeability is in favour of BLM resulting in net P absorption (left column). This state persists even if Na-P influx is low, e.g., in -D states (left column, top cartoon). 1,25D-repletion, as we have observed, caused further increases in net absorption. This is most likely the result of an

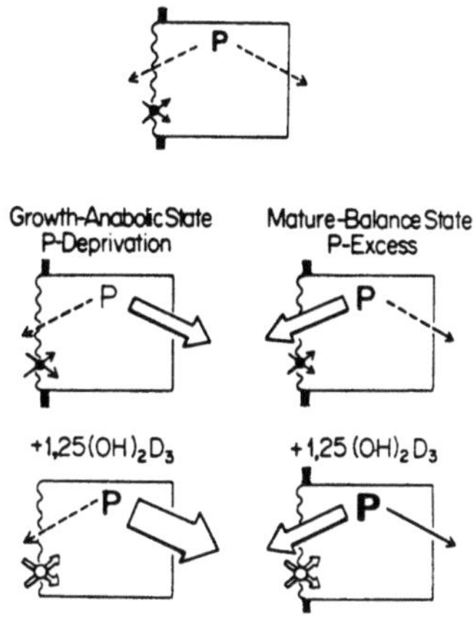

FIGURE 6.

increase in P influx across the BBM, with consequent increase in intracellular P activities and further increases in BLM efflux (left column, bottom cartoon). It is also possible that 1,25D has additional enhancing effect on BLM permeability to P.

In contrast, in grown adult rats or and in other situations in which there is an excess of P, such as high extracellular [P] - states in which supply outstrips demand and threaten P overload, the permeability is reversed to favour the BBM efflux (right column). This leads to net secretion as observed in normal adult rats or in ECF P loading. 1,25D-treatment does not further augment the secretory flux, but it may still increase BBM influx (with possible increase also in BLM permeability) of P, leading to an increase in absorptive flux, although the net transport is still in the secretory direction under such conditions. In a sense, the regulation of P absorption can be considered as incidental to the need for cellular and body P homeostasis. This cellular P homeostasis probably preceeds, evolutionary-wise, the vitamin D-regulatory mechanism, and not only can function in the absence of vitamin D but is able to override the action of this hormone on P absorption.

REFERENCES

1. Walling MW: Intestinal Ca and phosphate transport: differential responses to vitamin D_3 metabolites. Am J Physiol 233(6):E488-E494, 1977.

2. Lee DBN, MW Walling, V Silis, U Gafter and JW Coburn: Calcium and inorganic phosphate transport in rat colon. Dissociated response to 1,25-dihydroxyvitamin D_3. J Clin Invest 65:1326-1331, 1980.

3. Lee DBN, MW Walling, N Brautbar: Intestinal phosphate absorption: Influence of vitamin D and non-vitamin D factors. Am J Physiol 250:G369-G373, 1986.

4. Brautbar N, MW Walling and JW Coburn: Interactions between vitamin D deficiency and phosphorus depletion in the rat. J Clin Invest 63:335-341, 1979.

5. Murer H and B Hildmann: Transcellular transport of calcium and inorganic phosphate in the small intestinal epithelium. Am J Physiol 240:G409-G416, 1981.

6. Danisi G, J-Ph Bonjour and RW Straub: Regulation of Na-dependent phosphate influx across the mucosal border of duodenum by 1,25-dihydroxycholecalciferol. Pflugers Arch 388:227-232, 1980.

7. Hildmann B, C Storelli, G Danisi and H Murer: Regulation of Na^+-Pi cotransport by 1,25-dihydroxyvitamin D_3 in rabbit duodenal brush-border membrane. Am J Physiol 242(5):G533-G539, 1982.

8. Peterlik M and RH Wasserman: Effect of vitamin D on transepithelial phosphate transport in chick intestine. Am J Physiol 234(3):E379-E388, 1978.

9. Lee DBN, MW Walling and DB Corry: Phosphate transport across rat jejunum: influence of sodium, pH and 1,25-dihydroxyvitamin D_3. Am J Physiol (in press).

10. Schwab SJ, S Klahr and MR Hammerman: Na+ gradient-dependent Pi uptake in basolateral membrane vesicles from dog kidney. Am J Physiol 246:F663-F669, 1984.

INTESTINAL ABSORPTION OF PHOSPHATE IN MAN: EFFECT OF DRUGS

Carlo Gennari, Paolo Nardi, and Luciana Fusi

Institute of Medical Semiotics
University of Siena
Siena, Italy

In man, over the normal dietary intake, inorganic phosphate (Pi) absorbed from the gut is a linear function of Pi intake[1]. Pi absorption measured in man by the intubation-perfusion technique shows that at low Pi concentrations, Pi absorbed rises steeply with increasing Pi perfused; at increasing concentrations Pi absorption falls but never saturates[1]. According to these data it has been suggested that Pi absorption consists of two processes: the first, a saturable,carrier mediated, active transport and the second a linear concentration-dependent diffusion process.

In animal experiments it has been demonstrated that Pi is rapidly transported across all segments of the small intestine; in some species, Pi absorption was greater in the jejunum than in the duodenum[2]. In man, Pi absorption can be easily estimated following an oral dose of ^{32}P; the radiophosphorus levels in serial blood samples over the four hour period after the dose are interpreted by the kinetic two-compartment model of Marshall and Nordin[3] to obtain a measure of Pi absorption expressed as ^{32}P fx, the circulating fraction of the absorbed dose[4]. In patients, who underwent a partial gastrectomy with exclusion of the duodenum for a duodenal ulcer, the intestinal absorption of Pi was normal or low-normal, while the intestinal absorption of radiocalcium was impaired in all cases. The reinsertion of the duodenum corrected the malabsorption of calcium with no modifications of the intestinal transport of Pi. This study demonstrated that in man the active transport of calcium is greater in the duodenum than in the jejunum or ileum; on the contrary the greatest transport of Pi is confined to the jejunum or ileum but not to the duodenum.

In small intestinal epithelium, the villus with its associated crypts represent the unit of proliferative and functional activity[5]. Within the functional unit the proliferative activity varies at different levels. At the base of the crypts there is a small pool of slowly-cycling cells acting as a functional stem for the rest of the crypt. Many factors could probably affect the rate and pattern of proliferative activity in the small bowel, including dietary, luminal and systemic factors. Starvation has an atrophic effect, producing mucosal hypoplasia and a reduced rate

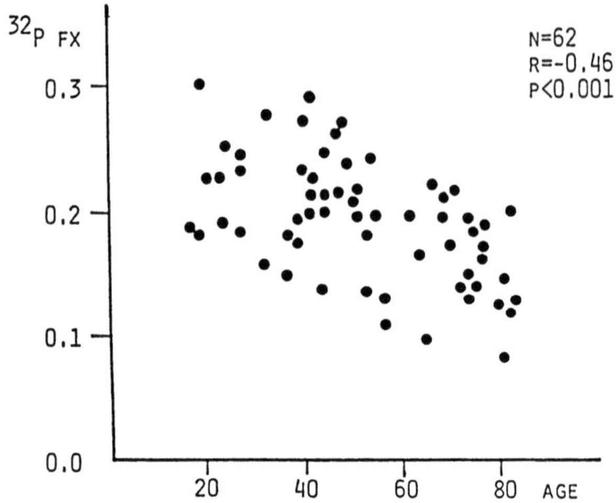

Fig. 1. Relationship between intestinal Pi absorption, expressed as the circulating fraction of the absorbed dose of ^{32}P, and age in 62 normal subjects.

of cell renewal[6] and re-feeding produces a prompt hyperproliferative response. It has been observed that age also could be accompanied by a progressive hypoplasia of the mucosa[7]. If we assume that Pi transport, like other transport processes, depends primarily on the integrity of the intestinal cell, the hypoplasia of the small bowel mucosa might reflect a decrease in the absorptive surface area for Pi. In fact the collected data from normal subjects of both sexes demonstrated that Pi absorption decreases with increasing age (Fig. 1). Malabsorption of Pi may also occur in a variety of chronic jejunal disorders, especially in adult coeliac disease. In these cases the reduction of the mucosal surface area is responsible, at least in part, of the calcium and Pi malabsorption (Fig. 2).

In animal studies it has been shown that vitamin D enhances Pi absorption after a time lag of 16 hours. The time involved for the conversion of administered vitamin D into metabolically active forms by the liver and kidney can account for the largest part of this lag period[1]. Furthermore, it has been observed that in the jejunum 25-hydroxycholecalciferol ($25(OH)D_3$) and 1.25-dihydroxycholecalciferol ($1.25(OH)_2D_3$) stimulated Pi transport and that nephrectomy prevented the response to $25(OH)D_3$ but not to $1.25(OH)_2D_3$, thus implicating $1.25(OH)_2D_3$ as the metabolically active form of vitamin D in this system[8]. In man, vitamin D certainly enhances Pi absorption in patients with vitamin D deficiency states[1]. In normal subjects $25(OH)D_3$ and $1.25(OH)_2D_3$, given at physiological doses for few days, cause a remarkable increase in the intestinal absorption of Pi[9]. At present it appears that the major hormonal regulation of intestinal Pi absorption is mediated by $1.25(OH)_2D_3$[10].

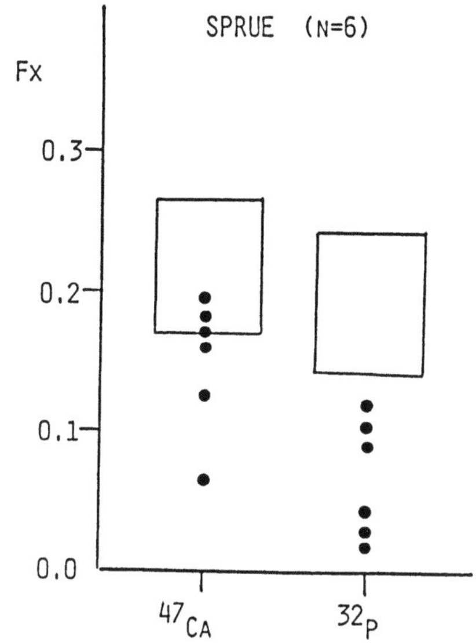

Fig. 2. Intestinal absorption of radiocalcium (^{47}Ca) and radiophosphorus (^{32}P) in six cases of coeliac disease. The rectangles represent the normal ranges (mean±standard deviation), obtained for ^{47}Ca fx in 100 and for ^{32}P fx in 62 normal adult subjects.

The mechanism by which $1.25(OH)_2D_3$ enhances intestinal Pi absorption is still unclear. This metabolite may induce increases in active Pi absorption by acting on protein synthesis as well as on diffusional mucosal permeability. However, in studies on rachitic rats it has been demonstrated that vitamin D accelerates mucosal cell turnover and increases villus height[11]. More recently it has been observed that duodenal mucosa in patients with renal failure exhibited structural abnormalities which were normalized after $1.25(OH)_2D_3$ therapy[12]. Thus, it appears that $1.25(OH)_2D_3$ stimulation of intestinal Pi absorption may be the consequence of not only its action on the Pi transport mechanisms, but also of its effect on the rate of mucosal proliferation, thus producing an increase in the effective absorptive area.

Many other hormones, besides the hormonal form of vitamin D, could directly or indirectly influence the intestinal transport processes of Pi. In patients with primary hyperparathyroidism the absorbed fraction of Pi was significantly increased with respect to normal subjects (Fig. 3). The hyper-absorption of Pi rapidly diminished after the surgical removal of parathyroid adenoma. On the other hand, in cases of primary hypoparathyroidism (surgical or idiopathic) the intestinal absorption of Pi was often impaired (Fig. 3). The stimulation of the 1α-hydroxylation of vitamin D by parathyroid hormone has been firmly established. It seems probable that the increased secretion of $1.25(OH)_2D_3$ in response to increased circulating parathyroid hormone is an essential component in the hyper-absorption of Pi observed in primary hyperparathyroidism. In three

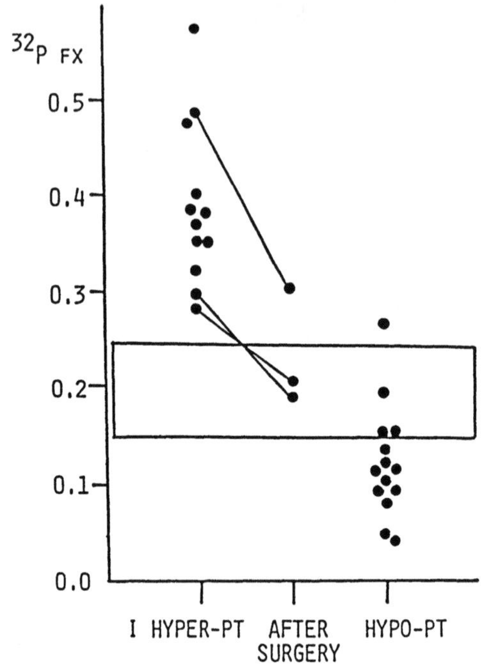

Fig. 3. Intestinal absorption of Pi in patients with primary hyperparathyroidism before and after surgery and in patients with idiopathic or surgical hypoparathyroidism.

cases of hypophosphatemia the intravenous injection of the fragment 1-34 of human synthetic parathyroid hormone determined a progressive increase in circulating $1.25(OH)_2D_3$ levels, reaching a peak after 12 hours from the injection; plasma Pi was significantly elevated above basal levels only after 24 hours (Fig. 4). These data suggest that parathyroid hormone promotes Pi absorption via stimulation of synthesis of $1.25(OH)_2D_3$ by the kidney, rather than by a direct action on intestinal mucosa.

Some years ago, in an acromegalic patient with hyperphosphatemia we have observed that the treatment with the dopaminergic agonist, bromocriptine, produced a simultaneous and parallel decrease in circulating Pi and growth hormone levels[13]. More recently, in 6 patients with acromegaly we have documented a condition of Pi hyper-absorption; in some cases a brief period of treatment with bromocriptine produced a clear impairment of intestinal Pi absorption (Fig. 5). These data seem to indicate that growth hormone promotes Pi absorption. It is still unknown if the hormone exerts this effect directly on the intestinal mucosa or indirectly by causing an increase in the concentration of circulating vitamin D metabolites.

In 1972, the intestinal absorption of Pi was measured in 15 postmenopausal women with osteoporosis before and after six months' treatment with an oral estrogen-gestagen combination; comparison with a control group indicated a significant improvement in intestinal Pi absorption after treatment[14]. In a more recent study we have observed that one year

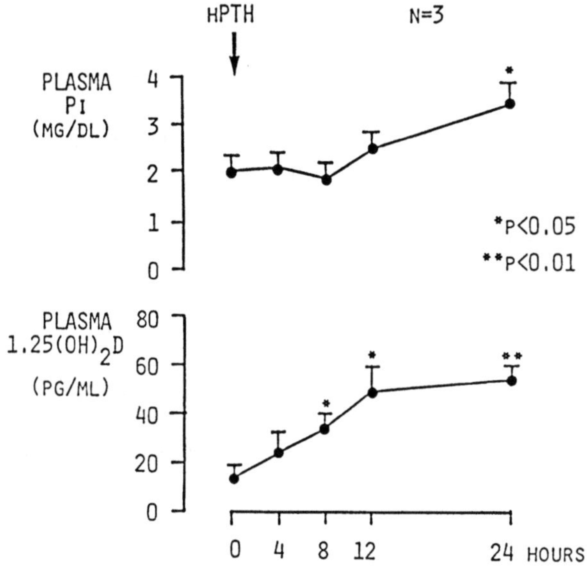

Fig. 4. Effect of an intravenous injection of the fragment 1-34 of human synthetic parathyroid hormone (200 I.U.) on plasma Pi and $1.25(OH)_2D_3$ levels in 3 hypophosphatemic patients.

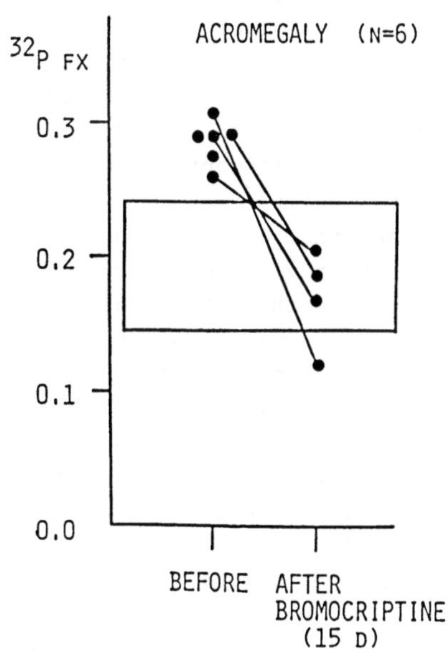

Fig. 5. Intestinal absorption of ^{32}P in patients with acromegaly before and after 15 days of treatment with bromocriptine (7.5 mg/day).

of treatment with conjugated estrogens produced a significant increase of circulating 1.25(OH)$_2$D$_3$ in postmenopausal women. In two cases the progressive increase in plasma levels of 1.25(OH)$_2$D$_3$ during the period of treatment paralleled a concomitant increase of the intestinal Pi absorption (Fig. 6). Therefore, it appears that the effect of estrogens on intestinal Pi absorption is mediated by 1.25(OH)$_2$D$_3$. However, we cannot exclude a direct effect of estrogens on intestinal cell. In fact, it has been recently demonstrated in rats that ovariectomy dramatically reduces the number of receptors for 1.25(OH)$_2$D$_3$ in enterocytes[15].

Glucocorticoids (GCs) have been shown to precociously affect intestinal absorption of calcium and Pi in animals and man, but the mechanism underlying these effects remains unclear. In patients treated chronically with prednisone the evaluation of intestinal Pi absorption gives the impression that the impairment of this function was correlated with the duration of treatment (Fig. 7). It has been suggested that GCs might directly impair the conversion of vitamin D into its active metabolites. The intestinal absorption of radiocalcium and radiophosphate has been evaluated in patients with rheumatic or allergic diseases before and after a 15-day treatment period with equiactive dose of three different GCs, betamethasone, prednisone and deflazacort, a new oxazoline derivative of prednisolone[16]. The comparison of the effects of the three different GCs demonstrated that deflazacort was ineffective in modifying the intestinal absorption of calcium and Pi, prednisone significantly reduced the intestinal transport of calcium without affecting the intestinal transport of Pi, betamethasone induced a significant decrease in both intestinal transport processes. In all cases serum concentrations of 1.25(OH)$_2$D$_3$ were not significantly altered by the three different treatments[16]. Thus, from these studies it appears likely that GC induced suppression of Pi absorption is probably the result of a direct effect on intestinal cellular function, rather than a consequence of an altered production of biologically active vitamin D metabolites. In this regard, recent studies on the effects of GCs on rat intestinal cellular structure have demonstrated that GCs significantly modify villus height and crypt depth of the intestinal cells[17].

There is evidence in man that other hormones may influence the intestinal transport of Pi, i.e. thyroxine[1,18] and calcitonin[19]. As it concerns the effects of drugs on the intestinal absorption of Pi the studies with ^{32}P oral absorption in human subjects are scarce. In animal experiments it has been observed that metabolic inhibitors, like iodoacetate, arsenite, 2,4-dinitrophenol and oubain reduced Pi influx and tissue transport in intestinal loops[1,20,21]. With human small intestinal mucosa in vitro it has been shown that sodium cyanide reduced Pi uptake[22]. Thus Pi transport depends on the integrity of the energy-producing mechanism of the intestinal cell.

The use of non-absorbable aluminium compounds as antacids impairs Pi absorption because of the binding of Pi to this compound in the gastrointestinal tract[1]. In patients with hyperphosphatemia of chronic renal failure it is possible to demonstrate that a brief treatment with aluminium hydroxide could raise the faecal Pi above the dietary Pi intake[1]. In these cases the intestinal ^{32}P absorption will result markedly reduced.

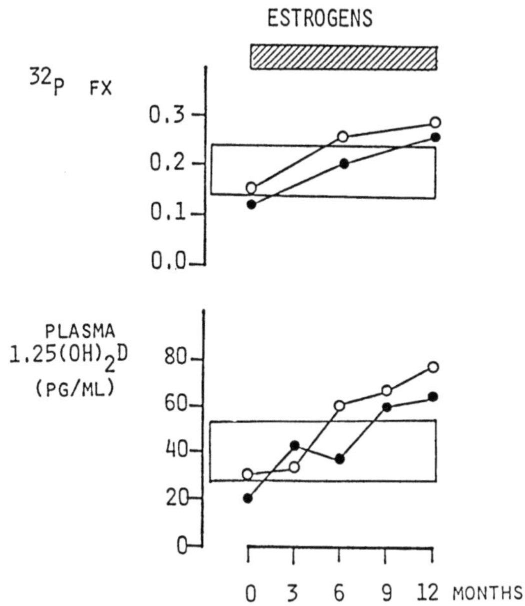

Fig. 6. Effects of one year of treatment with conjugated estrogens on intestinal absorption of Pi and on the circulating level of 1.25$(OH)_2D_3$ in two post-menopausal women.

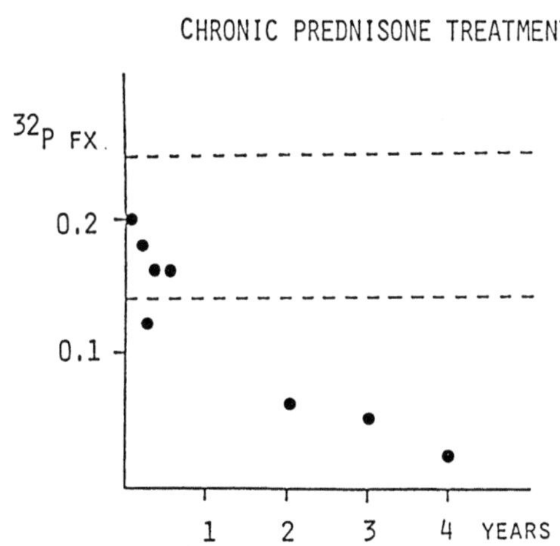

Fig. 7. Intestinal absorption of Pi in 8 patients chronically treated for different periods with prednisone (mean daily dose 15 mg).

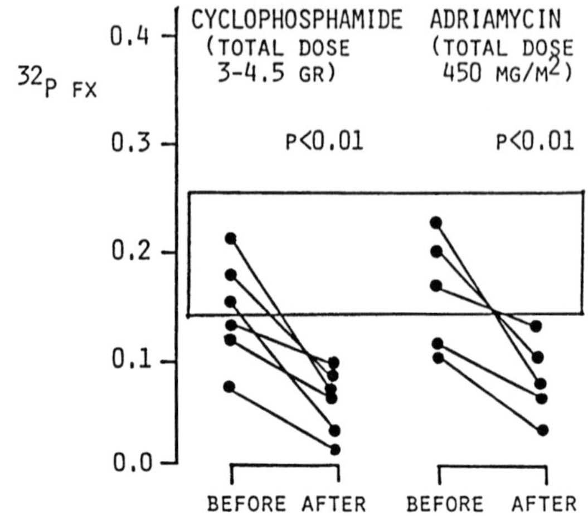

Fig. 8. Effect of cytotoxic drugs on intestinal absorption of Pi in patients with leukaemias and cancers.

In patients with leukaemias and cancers we noted malabsorption of Pi after treatment with cyclophosphamide or with adriamycin (Fig. 8). The effect of cytotoxic drugs on intestinal structure and function are well known: it has been demonstrated that some chemotherapeutic agents, like cyclophosphamide and adriamycin, induce inhibitory effects on nucleic acid synthesis and proliferative activity of the small bowel mucosa, producing a reduced rate of cell renewal and causing a considerable damage of enterocytes[23,24]. This effect is probably direct, and responsible for the serious alterations of nutritional status observed in patients treated with cytotoxic drugs. Nevertheless, is still unclear if the impairment in Pi absorption is due to a general effect on cell function or to a specific effect on Pi absorption mechanism.

In conclusion, we have observed that the intestinal absorption of Pi in man may be directly or indirectly influenced by certain hormones and drugs. Abnormalities of intestinal absorption of Pi occur when these substances provoke a structural derangement of the intestinal mucosa or a trouble of the metabolism of vitamin D. These observations do not exclude that there could be other unidentified mechanisms by which hormones and drugs could influence the Pi absorption in man.

REFERENCES

1. R. Wilkinson, Absorption of calcium, phosphorus and magnesium, in: "Calcium, Phosphate and Magnesium Metabolism", B.E.C. Nordin, ed., p. 36-112, Churchill Livingstone, Edinburgh (1972).
2. S. Kowarski, and D. Schachter, Effects of vitamin D on phosphate transport and incorporation into mucosal costituents of rat intestinal mucosa, J. Biol. Chem. 244:211-217 (1969).

3. D.H. Marshall, and B.E.C. Nordin, Kinetic analysis of plasma radioactivity after oral ingestion of radiocalcium, Nature 222:797 (1969).
4. A. Caniggia, and C. Gennari, Absorption du phosphate radioactif chez l'homme et sa regulation, in: "Phosphate et Metabolism Phosphocalcique", D.J. Hioco, ed., p. 209-235, Sandoz Editions, Paris (1971).
5. R.M. Clarke, Progress in measuring epithelial turnover in the villus of the small intestine, Digestion 8:161-175 (1973).
6. G.P. Ryan, S.J. Dudrick, and E.M. Copeland, Effects of various diets on colonic growth in rats, Gastroenterology 77:658-663 (1979).
7. P.M. Warren, M.A. Pepperman, and R.D. Montgomery, Age changes in small-intestinal mucosa, Lancet ii:849-850 (1978).
8. T.C. Chen, L. Castillo, M. Korycka-Dahl, and H.F. DeLuca, Role of vitamin D metabolites in phosphate transport of rat intestine, J. Nutr. 104:1056-1060 (1974).
9. A. Caniggia, and C. Gennari, Intestinal absorption of radiophosphate after physiological doses of $25(OH)_2D_3$ in normals, liver cirrhosis and chronic renal failure patients, in: "Vitamin D: Biochemical, Chemical and Clinical Aspects Related to Calcium Metabolism", A.W. Norman, K. Schaefer, J.W. Coburn, H.F. DeLuca, D. Fraser, H.G. Grigoleit, D.V. Herrath eds., p. 755-757, Walter de Gruyter, Berlin (1977).
10. M.W. Walling, Intestinal inorganic phosphate transport, in: "Homeostasis of Phosphate and Other Minerals", S.G. Massry, E. Ritz, A. Rapado, eds., p. 131-147, Plenum Press, New York (1978).
11. S.J. Birge, and D.H. Alpers, Stimulation of intestinal mucosal proliferation by vitamin D, Gastroenterology 64:977-982 (1973).
12. D.A. Goldstein, R.E. Horowitz, S. Petit, B. Haldimann, and S.G. Massry, The duodenal mucosa in patients with renal failure: response to $1.25(OH)_2D_3$, Kidney Int. 19:324-331 (1981).
13. C. Gennari, M. Galli, A. Vattimo, F. Lore', R. Nuti, M. Picchi, and M. Bernini, Effect of bromocriptine on phosphatemia in acromegaly, Boll. Soc. It. Biol. Sperim. 53:1173-1177 (1977).
14. A. Caniggia, and C. Gennari, Sites and modes of action of an estrogen-gestagen combination on calcium and phosphate metabolism in post-menopausal osteoporosis, Clin. Orthop. 85:184-192 (1972).
15. S.D.H. Chan, D.K.H. Chiu, and D. Atkins, Oophorectomy leads to a selective decrease in 1.25-dihydroxycholecalciferol receptors in rat jejunal villous cells, Clin. Sci. 66:745-748 (1984).
16. C. Gennari, M. Bernini, P. Nardi, L. Fusi, G. Francini, R. Nami, M. Montagnani, B. Imbimbo, and L.V. Avioli, Glucocorticoids: radiocalcium and radiophosphate absorption in man, in: "Osteoporosis: A Multidisciplinary Problem", A.St.J. Dixon, R.G.G. Russell, T.B.C. Stamp, eds., p. 75-80, Academic Press Inc., London (1983).
17. R.M. Batt, G. Wells, and T.J. Peters, The effects of prednisolone on the rat enterocyte at a subcellular level, Clin. Sci. Mol. Med. 55:435-443 (1978).
18. C. Gennari, M. Bernini, P. Nardi, L. Fusi, R. Nami, and M. Montagnani, Hormones and the intestinal absorption of calcium and phosphate in man, Hormone Res. 20:13-21 (1984).
19. A. Caniggia, C. Gennari, M. Bencini, V. Palazzuoli, G. Borrello, and F. Lenzi, Action of thyrocalcitonin on intestinal absorption of radiophosphate in man, J. Nuclear Biol. Med. 12:83-85 (1968).

20. R.H. Wasserman, and A.N. Taylor, Intestinal absorption of phosphate in the chick: effect of vitamin D_3 and other parameters, J. Nutr. 103:586-599 (1973).
21. A.N. Taylor, In vitro phosphate transport in chick ileum. Effect of cholecalciferol, calcium sodium and metabolic inhibitors, J. Nutr. 104:489-494 (1974).
22. E.M. Short, H.J. Binder, and L.E. Rosenberg, Familial hypophosphataemic rickets: defective transport of inorganic phosphates by intestinal mucosa, Science 179:700-702 (1973).
23. N.T. Shaw, M.H. Spector, and A.J. Ledman, Effect of cancer, radiotherapy and cytotoxic drugs on intestinal structure and function, Cancer Treat. Reviews 6:141-151 (1979).
24. S.S. Donaldson, and R.S. Lenon, Alterations of nutritional status. Impact of chemotherapy and radiation therapy, Cancer 43:2036-2052 (1979).

MECHANISM OF PHOSPHATE TRANSPORT ADAPTATION IN RAT INTESTINAL AND RENAL BRUSH BORDER MEMBRANES

G. Danisi, J. Caverzasio, J-Ph. Bonjour, H. Murer and R.W. Straub

Department of Pharmacology and Division of Pathophysiology
University of Geneva, Switzerland

Studies on the regulation of phosphate (P_i) transport in gut and kidney have revealed an adaptation of P_i transport under dietary restriction of P_i. In response to low P_i diet, P_i absorption is enhanced in kidney and small intestine (1,2). In kidney, this stimulation results, at least in part, from the increase in the capacity of the sodium-P_i cotransport process located in the brush border membrane of the proximal tubule. The K_m of the transport process is not modified (3). It is not known, to the best of our knowledge, whether the adaptation of P_i absorption in small intestine is expressed at the brush border membrane level and how these changes occur, if any. Furthermore, since sodium is cotransported with phosphate in the luminal membrane of renal and intestinal tissues, it could be that an alteration of the dependency of the phosphate carrier on sodium played some role in the stimulation of P_i transport.

The adaptation of renal and intestinal P_i transport has been mainly studied after rather long times of P_i deprivation as week(s). Recent experiments in brush border membrane vesicles (BBMV) from rat kidney proximal tubules, show, however, that the stimulation of sodium-dependent P_i transport in response to dietary P_i deprivation can be rapid: a highly significant increase is observed 2 hours only after a low P_i meal (4). It is not known whether the small intestine can adapt the P_i transport rate in such a short time.

The goals of this study are manifold. To test whether the adaptation of P_i transport in small intestine occurs at the brush border membrane level and how it is expressed. To investigate whether the dependency of the P_i carrier on sodium is changed in response to low P_i diet in intestinal and renal BBMV. To test whether rapid P_i transport adaptation occurs in small intestine.

METHODS

Normal male Wistar rats of 200-220 g body weight were used.They were fed a semi-synthetic diet containing 0.8 % phosphorus and 1.1 % calcium ad libitum for a week after arrival. Afterwards, all the animals received 20 g of a similar diet but half of the rats had 0.2 (LPD) instead of 0.8 % (HPD) phosphorus for another 8 days. 10 1.U. of vitamin D_3 dissolved in 0.1 ml vegetable oil was given daily on the ration. The animals had free access to distilled water. BBMV were prepared from the kidney cortex and jejunum of LPD and HPD rats by a Mg-precipitation method and uptake of P_i was measured by a rapid filtration technique (4).

RESULTS

The time course of P_i uptake into jejunal BBMV from LPD and HPD rats is shown for a representative experiment in Fig.1. There is a 3 fold increase of the P_i uptake in LPD compared to HPD animals at early time points in the presence of an inwardly directed sodium gradient. This difference decreased when longer incubation times were allowed. There was no difference in the uptakes of P_i in both groups when sodium was replaced by potassium. Further, equilibrium values in both conditions for both groups of rats were similar.

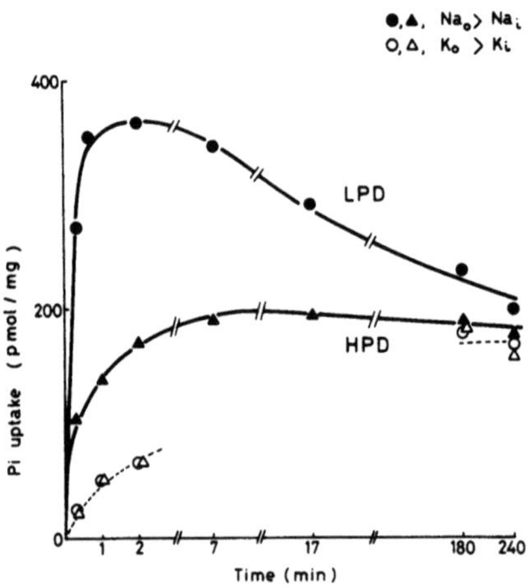

Fig.1. Time course of P_i uptake in jejunal BBMV from LPD and HPD rats. External Na and P_i concentrations were 100 mM and 0.1 mM, respectively. $pH_o = pH_i = 7.4$.

Initial rates of P_i uptake were studied as a function of external P_i concentrations. Na-dependent P_i transport was saturable in LPD and HPD jejunal BBMV (Na 100 mM, pH 7.4). The values of V_{max} and K_m obtained from Lineweaver-Burk plots for two independent experiments are shown in Table I. V_{max} is 2-3 fold higher in LPD than in HPD rats and this increase is similar to what has been observed in kidney. In contrast, however, K_m in jejunum appears to be reduced by low P_i diet in the rat.

Table I. Kinetic parameters of Na-dependent P_i transport in jejunal BBMV from LPD and HPD rats.

	V_{max} (pmol/mg.10s)	K_m (mM)
HPD	71 / 114	0.18 / 0.12
LPD	168 / 245	0.08 / 0.05

In order to investigate if the stimulation of P_i transport in renal and intestinal BBMV was due to a modification of the behaviour of the P_i carrier with sodium, BBMV were isolated from the kidney cortex and jejunum of the same LPD and HPD animals and the change of P_i uptake with varying sodium concentrations was measured at 0.1 mM P_i and $pH_o = pH_i = 7.4$. The results are shown in Fig.2. The data show that the sodium concentration for half maximal P_i transport, $K^{Na}_{0.5}$, appears unchanged by dietary P_i in either renal or intestinal BBMV. LPD, however, increased the maximal P_i transport from 26 ± 5 to 88 ± 6 in jejunal and from 1800 ± 205 to 3025 ± 380 pmol/mg.10s in kidney BBMV. The fact that $K^{Na}_{0.5}$ appears to be the same in LPD and HPD rats could suggest that the affinity of sodium for the P_i transporter is not affected by P_i deprivation in either renal and intestinal BBMV.

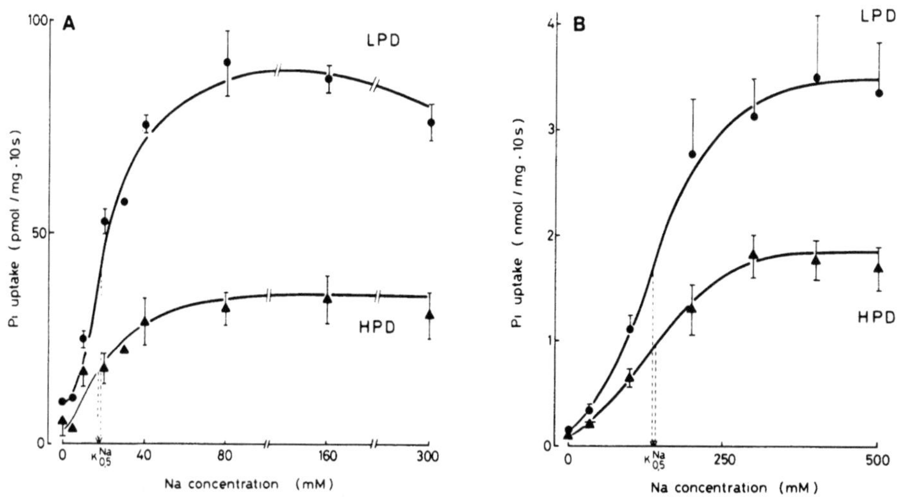

Fig.2. Change of P_i uptake with external Na concentration in jejunal (A) and kidney cortex (B) BBMV from LPD and HPD rats. $pH_o = pHi = 7.4$.

In the experiments below, rapid adaptation of P_i transport in jejunal BBMB was investigated. Rats were trained to consume their meal, either low or high in P_i, in one hour, and after 3 further hours, BBMV were prepared for the measurement of P_i uptake. These preliminary results are shown in Fig.3.

It is evident that BBMV from LPD rats have a higher transport of P_i than vesicles from HPD rats. This effect is specific for P_i, since Na-dependent D-glucose uptake, measured in the same animals, was similar in both groups. These results suggest that rat jejunum can adapt quite fast to a low P_i meal, as has been observed in kidney cortex (4). The factors which trigger this adaptation as well as the mechanism remain to be determined.

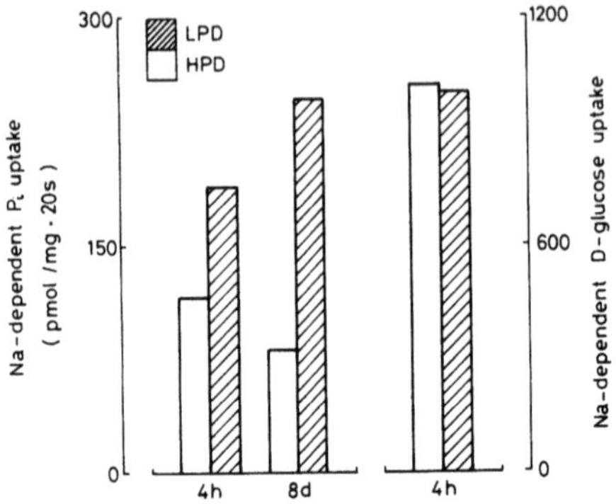

Fig.3. Rapid P_i transport adaptation in jejunal BBMV from rats which consumed a LPD or HPD meal 4 hs before sacrifice. For comparison results from rats of the same lot but adapted for 8 days are presented. Results from a single representative experiment are shown.

CONCLUSIONS

In conclusion, P_i transport adaptation in rat small intestine is expressed at the brush border membrane level. The adaptation results from an increase in the capacity of the Na-P_i cotransport system and, there appears from a reduction of K_m of the P_i transport process. In both renal and jejunal BBMV there is a clear stimulation of the maximum activation by Na of the P_i transport system by LPD. $K_{0.5}^{Na}$ appears to be unchanged. Preliminary experiments further suggest that small intestine, at least in the rat, can adapt very rapidly to a low P_i meal.

ACKNOWLEDGEMENTS

Supported by grants 3.806.0.82 and 3.437.0.83 from the Swiss National Science Foundation.

REFERENCES

1. Cramer, C.F. and J. McMillan. Amer.J.Physiol. 239 : G261-G265, 1980
2. Bonjour, J-Ph. and J. Caverzasio. Rev.Physiol.Biochem.Pharmacol. 100 : 161-214, 1984
3. Stoll, R., R. Kinne and H. Murer. Biochem.J. 180 : 465-470,1979
4. Caverzasio, J. and J-Ph. Bonjour. Pflügers Arch. 404 : 227-232,1985

ANALYSIS OF CALCIUM TRANSPORT IN RAT INTESTINE

F. Bronner, D. Pansu[1)] and W.D. Stein[2)]

Department of Oral Biology
University of Connecticut Health Center
Farmington, CT 06032

Transmural calcium absorption is the resultant of two processes: a saturable, regulatable process that traverses the cell and a non-saturable, non-regulated process that is largely paracellular. This article will summarize findings and analyses in support of this generalization and provide details of the action of vitamin D, the primary regulator of active calcium absorption.

Figure 1 shows the time course of calcium movement out of a duodenal loop for varying concentrations of calcium. For low concentrations, absorption is completed quite rapidly, whereas absorption is far from complete at the higher concentrations. If similar experiments are done with loops prepared from the ileum, absorption of all calcium concentrations instilled is as slow as when 150-200 mM calcium solutions are instilled in the duodenum. When intestinal loops are prepared from vitamin D-deficient animals, whether these are from the duodenum or ileum, absorption occurs at the slow rates characteristic of high calcium concentrations in the duodenal loops from vitamin D-replete animals. In other words the rate of absorption is enhanced manyfold as a result of the metabolic action of vitamin D.

Let us now turn to an analysis of these rates. As a first approximation let us consider the rates of absorption for duodenal loops from vitamin D-replete animals as a saturating function characterized by Michaelis-Menten kinetics. Figure 2 shows the logarithm of the calcium concentration in the loop at a given time as a fraction of the initial calcium concentration, plotted against time. This plot has the characteristics of being a straight line at low concentrations. The experimental points at low concentrations are indeed described fairly well by a straight line. At higher concentrations the function becomes curvilinear. At 25 mM calcium, theory and experiment coincide well and one can calculate a V_m of 22 umol/h/g duodenum and a K_m of 3.9 mM. These values agree well with what has been published on the basis of perfusion experiments[1].

[1)] Permanent Address: INSERM U-45 and EPHE, Hopital E. Herriot, 69374 Lyon Cedex 08, FRANCE

[2)] Permanent Address: Department of Biological Chemistry, Institute of Life Sciences, Hebrew University, Jerusalem 91904, ISRAEL

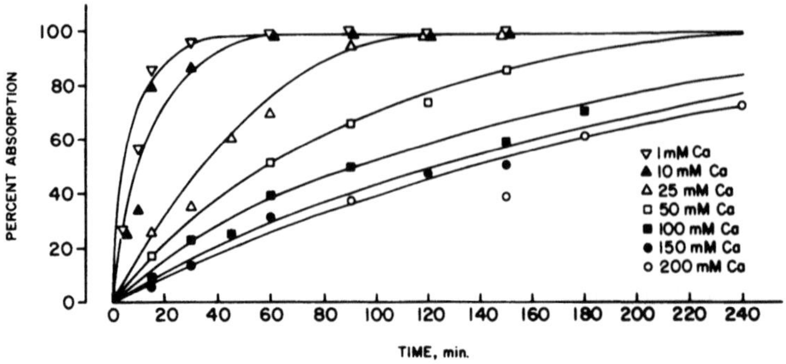

Fig. 1. Time course of calcium efflux from in situ duodenal loops. Each experimental point is the mean of 2-6 loops from male Sprague-Dawley rats, 120-150 g BW, that had been placed on a semi-synthetic diet[5] that contained 1.5% Ca and 1.5% P. Two days before the experiment the animals were placed on a low-calcium diet (0.05% Ca, 0.2% P). Calcium was instilled in the lumen at the various concentrations shown and efflux was estimated from the calcium lost in the indicated time-period.

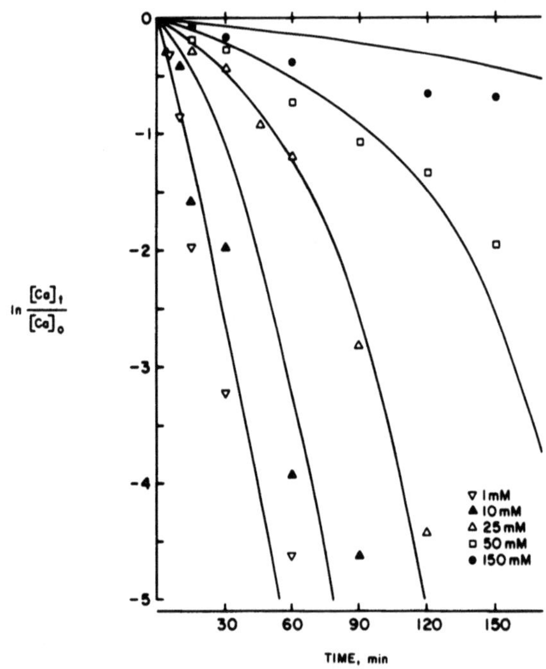

Fig. 2. A plot of ln $[Ca]_t/[Ca]_o$ against time, using the data of Fig. 1. The lines were calculated with the aid of the following equation:

$$\ln \frac{[Ca]_t}{[Ca]_o} + \frac{[Ca]_t - [Ca]_o}{K_m} = - \frac{V_m t}{K_m}$$

where $K_m = 3.85$ mM
$V_m = 22.25$ µmol/h/loop

Note, however, that at high calcium concentrations the experimentally determined rate of calcium movement is much greater than theory predicts. This means that there exists another type of movement, in addition to that described by the saturating function. This movement is the non-saturating flux referred to above.

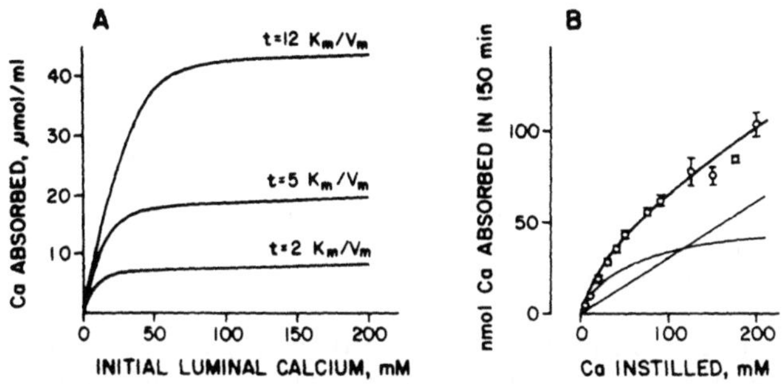

Fig. 3. Comparison of Theoretical (A) and Experimental (B) Curves of Calcium Absorption. A. The theoretical curve, calculated from the equation in the legend of Fig. 2, shows the amount of calcium absorbed as a function of initial luminal calcium concentration, at three time periods (t = 2, 5 and 12 x K_m/V_m; K_m = 3.85 mM). B. Calcium absorption in 40-day old male rats, measured by an *in situ* loop procedure 150 min. after instillation of calcium at the concentrations indicated in the graph. The slope of the straight line was obtained by fitting the data at the four highest concentrations; the hyperbolic curve was obtained with a best-fit procedure by subtracting the linear component from each of the experimental points, the line through the experimental points being the algebraic sum of the linear and saturable components. For details, see[4].

Another way to appreciate this is shown in Figure 3. Figure 3A shows the theoretical function plotted for various times expressed in K_m/V_m units, with K_m = 3.85 mM. Figure 3B depicts a typical experimental finding for an *in situ* duodenal loop. The theoretical curves (Fig. 3A) were constructed on the assumption of a pure saturable movement and at no time resemble what is found experimentally (Fig. 3B). As shown, the experimental findings can be resolved into the sum of a saturable and non-saturable process[2-5], as first proposed by Wasserman and Taylor[6]. Analysis of all experimental results has shown that the non-saturable movement, depicted here by the straight line has an average value in the rat of 16%/hour[7].

Active calcium transport is a function of age, calcium and vitamin D intake and of physiological status. For example V_m is high in animals on a low calcium diet[3], low in similar animals on a high calcium diet[4], it is high in young and low in older individuals[2], it is elevated in pregnancy[8] and diminished in glucocorticoid excess[9]. Active calcium transport occurs primarily in the proximal intestine, especially the proximal duodenum, and is absent in the ileum[5].

Transcellular calcium movement involves these steps: Entry across the brush border, movement inside the cell, binding to various intracellular organelles, and extrusion via a pump driven by the Ca^{2+}/Mg^{2+}-ATPase. Much effort in recent years has been spent on examining the entry step, often with the aid of brush-border membrane vesicles (BBMV). Calcium uptake by BBMV has been shown[10] to be time- and concentration-dependent, but to require no exogenous source of energy. Moreover, the effect of vitamin D on BBMV uptake is quite modest, only about 30%. It is likely that a membrane-bound, integral calcium-binding protein that is vitamin D-dependent[11,12] is involved in mediating this entry step. As yet it is not known whether Ca entry is mediated by a carrier or a channel. Published experiments[10,13] can be interpreted in favor of one or the other of these possibilities. However, whether entry is by carrier or channel, it can be shown that neither is rate-limiting[7].

Calcium extrusion at the basolateral pole is effected by the Ca^{2+}/Mg^{2+}-ATPase[14]. It is energy-dependent, the addition of exogenous ATP increasing calcium uptake by basolateral membrane vesicles (BBMV) by a factor of at least four[14,15]. Since ATP cannot penetrate the cell wall, stimulation by ATP added to the reaction mixture means that only inside-out vesicles can be stimulated to take up calcium. This is equivalent to in vivo calcium extrusion. Roughly half the vesicles are inside out[14].

TFP, an inhibitor of the Ca^{2+}/Mg^{2+}-ATPase, can inhibit BLMV calcium uptake in a dose-dependent manner[16]. TFP has virtually no effect on Ca uptake by BBMV[16]. This means that it is possible to inhibit the extrusion process independently of an effect on calcium entry via the brush-border membranes.

Does vitamin D then act on the extrusion process? Data from Schiffl and Binswager[17] show that Ca^{2+}/Mg^{2+} ATPase activity varies directly and proportionately with active transport, i.e. the S/M ratio in an everted sac preparation. Ghijsen et al.[14] have shown that this is also true for the BLMV preparation. However, pump activity is stimulated only two-three fold by vitamin D. Moreover, we have calculated[7] that the level of Ca pumping even in the absence of vitamin D is adequate to handle the amount that would be absorbed if vitamin D were present. Therefore Ca extrusion via the Ca^{2+}/Mg^{2+}-ATPase-driven pump does not seem a limiting rate.

We believe the limiting factor in transcellular calcium transport is the concentration of the cytosolic CaBP (Mr ~ 9000 in the rat). Figure 4A shows there exists a close linear correlation between the V_m for calcium transport, as measured by in situ intestinal loops[3,4], and CaBP content. Figure 4B shows a similar relationship exists between intestinal calcium transport, as evaluated in everted intestinal sacs, and CaBP content[16]. Figure 4A shows a correlation between steady-state calcium transport and steady-state CaBP levels; in Figure 4B, transport and CaBP were induced by treating vitamin D-deficient or vitamin D-replete animals with varying doses of 1,25-dihydroxyvitamin D_3.

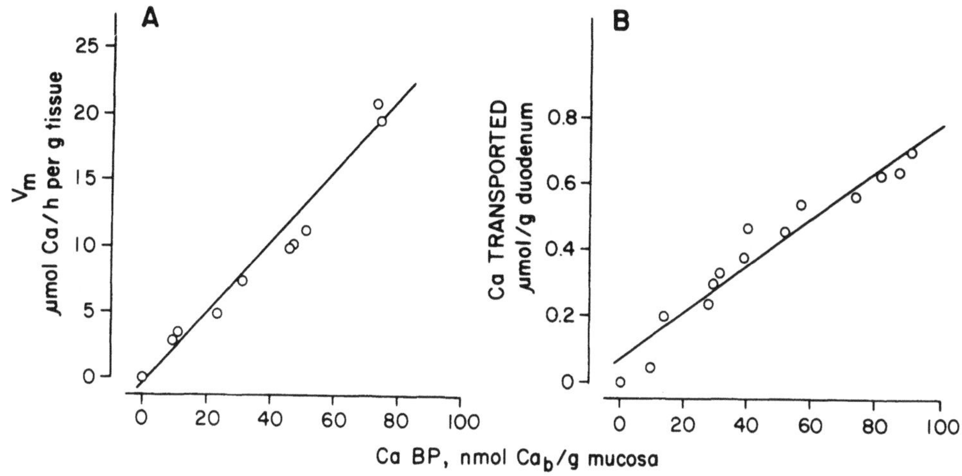

Fig. 4. The Relationship Between Intestinal Calcium Transport and CaBP Content. A. V_m, calculated from in situ duodenal, jejunal and ileal loop experiments[4], shown as a function of CaBP content. The equation describing the relationship is:
V_m = -0.59 + 0.26 CaBP, units as shown
n = 10, r = 0.98

B. Calcium transport, as evaluated from everted duodenal sac experiments shown as a function of CaBP content[16]. The equation describing the relationship is:
Ca transport = 0.070 + 0.00714 CaBP, units as shown
n = 14, r = 0.97

The data of Fig. 4B have been obtained from male Sprague-Dawley or Wistar rats, 70-150 g BW, maintained on a semi-synthetic, high-calcium diet (1.5% calcium, 1.5% P) with or without vitamin D.

CaBP in our view acts like a ferry, a concept first proposed and analyzed by Kretsinger et al.[18]. To understand how CaBP acts to promote transcellular calcium flux, it is necessary to recall that throughout the cell, free calcium and CaBP are in equilibrium, with free calcium partitioning between the bound and free states according to the equilibrium content of CaBP. As a result the very low free Ca^{2+} concentration of 0.2 uM[19] is greatly amplified and the total calcium gradient across the cell approaches the concentration of CaBP in the cell. One can calculate[7] that in the total absence of CaBP, free calcium would diffuse through the cell at a rate of only 1 percent of what it does in the presence of CaBP. The transport amplification provided by CaBP is however somewhat less than 100, because CaBP diffuses more slowly than free Ca^{2+} and because of the relationship between the K_d for CaBP and the free Ca^{2+} concentration. Other soluble calcium-binding proteins will also contribute to this facilitation. For calmodulin, which is not vitamin D-dependent, the facilitation is only about 10% of that due to CaBP[7]. Therefore intracellular calcium movement is the most likely candidate for the rate-limiting step. In turn this accounts for the close linear relationship between calcium transport and the intestinal content of CaBP shown in Figure 4.

There is one other aspect of vitamin D regulation that needs to be mentioned. Previous work[20] has shown that the calcium content of the intestinal cell varies with vitamin D status. Thus duodenal cells from vitamin D-replete animals take up twice as much calcium as cells from vitamin D-deficient rats. Moreover, ileal cells take up about the same amount of calcium, whether harvested from vitamin D-deficient or vitamin D-replete animals. Also, calcium uptake by ileal cells is about the same as by duodenal cells from vitamin D-deficient rats. Weiser and colleagues[21] have found that mitochondria and other cell organelles take up more calcium when harvested from animals treated with vitamin D than from untreated vitamin D-deficient animals. In our view, the vitamin D-dependent intracellular calcium binding found at immobile sites may be the reason for the increased calcium uptake found in cells from vitamin D-treated animals. A likely role for calcium fixation is to allow buffering to occur. It would allow intestinal cells to defend themselves against sudden calcium surges as occur when animals that have been on a low-calcium diet and therefore have a high transport capacity are suddently confronted with a high level of dietary calcium. For example, animals fed a 0.2% Ca diet have nearly twice as great a capacity to transport calcium as animals fed on a 0.44% Ca diet[3]. When patients are given exogenous vitamin D in large doses, their fractional calcium absorption increases promptly. A parallel increase in calcium buffering capacity clearly represents a useful defense against flooding.

Table 1 summarizes how vitamin D acts on transcellular calcium transport. Because intracellular calcium movement appears to be the rate-limiting step, the effect of vitamin D is most marked on intracellular diffusion, with the effect on calcium entry, calcium binding at immobile sites and calcium extrusion less dramatic.

Table 1. Effects of Vitamin D on Transcellular Calcium Transport in Rat Intestine

STEP	MECHANISM OR STRUCTURE	EFFECT OF VITAMIN D	MECHANISM
Entry Across Brush Border	Down Chemical Gradient via Channel or Carrier	Enhances 20-30%	Possibly via Integral Ca-Binding Protein
Binding to Fixed Cellular Sites (Buffering)	Golgi Apparatus RER Mitochondria	Enhances by 100%	Unknown
Intracellular Movement	Diffusion	Facilitates Diffusion In Direct Proportion (50-100 fold)	Biosynthesis of Soluble CaBP ($M_r \sim 9000$) which acts as Ca ferry
Extrusion	Pumping against a gradient Ca/Mg-ATPase	Increases Ca/Mg-ATPase 200-300%	Unknown

The difference between the rate of calcium diffusion in water and the experimentally determined rate of transmural flux can be used not only to postulate the need for a specific mechanism to speed up intracellular diffusion, but also to evaluate the probable path of the non-saturable movement. From the kind of experiments depicted in Fig. 3B, the mean rate of non-saturable transport has been estimated at 0.16/h[7]. This value is the same throughout the length of the intestine and thus prevails also in the ileum, where saturable transport is virtually non-existent and where CaBP is practically absent[5,22]. Yet a rate of 0.16/h is 160 times greater than the amount of calcium that can diffuse through an ileal cell devoid of CaBP[7]. Consequently the fraction of calcium moved transcellularly by the non-saturable pathway is necessarily negligible and most of that calcium must move paracellularly. This conclusion is in accord with earlier studies by Nellans and Kimberg[23] who, on the basis of electrical measurements, have estimated that at least 90% of the calcium absorbed in the ileum travels by a paracellular path. In our view this is equally true for the non-saturable component of calcium absorption in the proximal intestine.

This conclusion does not mean, however, that calcium moves between cells in an unhindered fashion. The experimentally determined rate of paracellular diffusion (0.16/h) is much slower than can be calculated if calcium diffused through the paracellular spaces at the rate at which it diffuses through water. It seems reasonable to assume that structures in the paracellular region, especially along the tight junctions, significantly restrict the non-saturable flow of calcium

ACKNOWLEDGMENT

Support of these studies by the University of Connecticut Research Foundation, NATO grant 594/83 and ATP INSERM No. 79106 is gratefully acknowledged.

REFERENCES

1. H.P. Schedl, D.L. Miller and N. Iskandarani, Small intestinal site, calcium transport kinetics, and calcium homeostasis in the hamster, in: "Calcium and Phosphate Transport Across Biomembranes", F. Bronner and M. Peterlik, eds., Academic Press, New York (1981).
2. A.E. Zornitzer, and F. Bronner, In situ studies of calcium absorption in rats, Am. J. Physiol., 220: 1261-1265 (1971).
3. D. Pansu, C. Bellaton and F. Bronner, The effect of calcium intake on the saturable and non-saturable components of duodenal calcium transport, Am. J. Physiol., 240: G32-G37 (1981).
4. D. Pansu, C. Bellaton and F. Bronner, Developmental changes in the mechanisms of duodenal calcium transport in the rat, Am. J. Physiol, 244: G20-G26 (1983).
5. D. Pansu, C. Bellaton, C. Roche and F. Bronner, Duodenal and ileal calcium absorption in the rat and effects of vitamin D, Am. J. Physiol., 244: G695-G700 (1983).
6. R.H. Wasserman, and A.N. Taylor, Some aspects of the intestinal absorption of calcium, with special reference to vitamin D, in: "Mineral Metabolism: An Advanced Treatise", C.L. Comar and F. Bronner, eds., Academic Press, New York and London (1969).
7. F. Bronner, D. Pansu, and W.D. Stein, An analysis of intestinal calcium transport across the intestine. "Editorial Review", Am. J. Physiol. (submitted).
8. J. Wrobel and G. Nagel, Diurnal variations of active calcium transport in the intestine of the pregnant and lactating rat, Biomedicine, 33: 143-145 (1980).

9. J.J. Feher, and R.J. Wasserman, Intestinal calcium-binding protein and calcium absorption in cortisol-treated chicks: Effects of vitamin D_3 and 1,25-dihydroxyvitamin D_3, Endocrinology, 104: 547-551 (1979).
10. A. Miller III, and F. Bronner, Calcium uptake in isolated brush-border vesicles from rat small intestine, Biochem. J., 196: 391-401 (1981).
11. A. Miller III, T.-H. Ueng, and F. Bronner, Isolation of a vitamin D-dependent, calcium-binding protein from brush borders of rat duodenal mucosa, FEBS Letters, 103: 319-322 (1979).
12. S. Kowarski and D. Schachter, Intestinal membrane calcium-binding protein: Vitamin D-dependent membrane component of the intestinal calcium transport mechanism, J. Biol. Chem., 255: 10834-10840 (1980).
13. A. Miller III, S.T. Li, and F. Bronner, Characterization of calcium binding to brush border membranes from rat duodenum, Biochem. J., 208: 773-782 (1982).
14. W.E.J.M. Ghijsen, M.D. DeJong, and C.H. van Os, ATP-dependent calcium transport and its correlation with Ca^{2+}-ATPase activity in basolateral plasma membranes of rat duodenum, Biochim. Biophys. Acta, 689: 327-336 (1982).
15. C. Roche, D. Pansu, and F. Bronner, The effect of trifluoperazine on transepithelial calcium transport in rat duodenum, in: "Epithelial Calcium and Phosphate Transport: Molecular and Cellular Aspects", F. Bronner and M. Peterlik, eds., Alan Liss, New York (1984).
16. C. Roche, D. Bellaton, D. Pansu, and F. Bronner, Simultaneous induction of CaBP and active calcium transport in rat duodenum by 1,25-dihydroxyvitamin D_3 in: "Epithelial Calcium and Phosphate Transport: Molecular and Cellular Aspects", F. Bronner and M. Peterlik, eds., Alan R. Liss, New York (1984).
17. H. Schiffl, and U. Binswager, Calcium ATPase and intestnal calcium transport in uremic rats, Am. J. Physiol., 238: G424-G428 (1980).
18. R.H. Kretsinger, J.E. Mann and J.G. Simmons, Model of the facilitated diffusion of calcium by the intestinal calcium binding protein, in: "Vitamin D. Chemical, Biochemical and Clinical Endocrinology of Calcium Metabolism", A.W. Norman, K. Schaefer, D.V. Herrath, and H.-G. Gringoleit, eds, DeGruyter, Berlin and New York (1982).
19. D.D. Bikle, D.M. Shoback, and S.J. Munson, 1,25-dihydroxyvitamin D increases the intracellular free calcium concentration of duodenal epithelial cells, "Abstracts", Sixth Workshop on Vitamin D, Merano, Italy (1985).
20. F. Bronner, D. Pansu, A. Bosshard, J.H. Lipton, Calcium uptake by isolated rat intestinal cells, J. Cell Physiol., 116: 322-328, (1983).
21. M.M. Weiser, J.H. Bloor, A. Dasmahapatra, R.A. Freedman and J.A. MacLaughlin, Vitamin D-dependent rat intestinal Ca^{2+} transport Ca^{2+} uptake by Golgi membranes and early nuclear events, in: "Calcium and Phosphate Transport Across Biomembranes", F. Bronner and M. Peterlik, eds., Academic Press, New York (1981).
22. M. Thomasset, C.-O. Parkes and P. Cuisinier-Gleizes, Rat calcium-binding proteins: distribution, development and vitamin D-dependence, Am. J. Physiol, 243: E483-E488 (1982).
23. H.N. Nellans, and D.V. Kimberg, Cellular and paracellular calcium transport in rat ileum: effects of dietary calcium, Am. J. Physiol., 236: E726-E737 (1978).

CALCIUM ABSORPTION IN THE RAT COLON MEASURED IN VIVO

Patrick Ammann, René Rizzoli and Herbert Fleisch

Department of Pathophysiology, University of Bern
Murtenstrasse 35,
CH-3010 Bern, Switzerland

ABSTRACT

Calcium absorption in the large intestine was studied under physiological conditions in the rat. When ^{45}Ca was administered directly into the caecum, the isotope was significantly absorbed. This absorption was modulated by dietary Ca and Pi as well as by the $1,25(OH)_2D_3$ status. On the other hand, calcium coming from the food, either as ^{45}Ca or as ^{40}Ca, was not absorbed to a significant amount. When the absorption of ^{45}Ca from the caecal content was assessed in duodenal loops, less ^{45}Ca was absorbed when the isotope was administered orally than when it was added to the caecal content. Thus, the fact that oral Ca is not absorbed in the colon despite the efficient Ca transport system of this part of the intestine may be due to the transformation in the digestive tract of Ca from an absorbable into a non-absorbable form.

INTRODUCTION

Studies in vitro have shown that the mucosa of the large intestine is able to transport calcium (2,5,9), a process which is influenced by dietary manipulations (2), vitamin D (2,5,7) and hormonal changes (3,8). In certain experimental circumstances calcium has also been shown to be absorbed in vivo (4,10,11), although the conditions used did not reflect those occurring normally. However, it is still not known whether calcium can be absorbed in the large intestine in vivo in normal conditions.

The aim of the present work was to investigate whether calcium is absorbed in the colon in vivo under physiological conditions.

METHODS

Animals

All animals used were male Wistar rats of our own breeding with a body weight of about 165 g.

Absorption of ^{45}Ca injected into the caecum

A catheter was implanted into the caecum, tied to the abdominal wall and exteriorated behind the neck. Three days after this operation 0.2 µCi $^{45}CaCl_2$ and 0.5 µCi ^{3}H-polyethyleneglycol (PEG), the latter administered as a non-absorble marker (followed by 0.5 ml of 2.5 mM $CaCl_2$), were injected in the catheter. The faeces were then collected over the next 24 hours. ^{45}Ca and ^{3}H-PEG were determined both in the administered solution and in the faeces, allowing the calculation of the percentage of ^{45}Ca absorbed along the large intestine.

Net absorption of Ca or of ^{45}Ca in the large intestine

The animals were fed a diet containing ^{3}H-PEG during three days. On the fourth day the rats were sacrificed and both a specimen of caecal content and of the faeces were collected and measured for Ca and ^{3}H-PEG. This allowed the determination of the percentage of Ca arriving in the caecum which was absorbed along the large intestine.

In some experiments the animals were fed ^{45}Ca together with ^{3}H-PEG and the absorption of ^{45}Ca in the colon was measured.

Absorbability of caecal Ca

The absorbability of Ca in the caecum was measured in the duodenum using an in situ duodenal loop model (1,15). For this, specimens of the caecal content were labelled with ^{45}Ca and ^{3}H-PEG by administering these isotopes either orally or by adding them in vitro to the caecal content. 0.5 ml of these caecal contents were injected into duodenal loops of 9 cm length. ^{45}Ca absorption was measured during a 15 minute incubation by determining the difference between the injected and recovered ^{45}Ca corrected for that of ^{3}H-PEG. This procedure allowed the detection of any difference in absorbability between Ca administered orally or added in situ.

RESULTS

$^{45}CaCl_2$ injected in the caecum was significantly absorbed in the large intestine. Thus, 26 $\pm$ 2.5 % disappeared when the animals were fed the normal diet of 1.2 g% Ca and 1.0 g% Pi. This absorption was influenced by the dietary conditions since it increased to 64.0 $\pm$ 4.2 % when the animals were fed 0.2 g% Ca, and also increased when dietary Pi was lowered. The absorption was vitamin D-dependent since the administration of 10 mg P/kg of 1-hydroxyethylidene-1,1-bisphosphonate (HEBP), a compound known to decrease the production of endogenous $1,25(OH)_2D_3$ (14), led to a decrease. This decrease in the absorption of ^{45}Ca could be fully corrected by administering 26 pmol/day of $1,25(OH)_2D_3$, an amount which compensates the decrease induced by HEBP (1,12,13). The changes due to $1,25(OH)_2D_3$ were shown to be independent of the amount of Ca arriving in the colon.

However, while ^{45}Ca injected in the caecum was significantly absorbed, no absorption could be detected for non-radioactive Ca along the large intestine. Altering dietary Ca or the $1,25(OH)_2D_3$ status had no influence either. In contrast, a significant although not very large absorption of about 10 % was seen for Pi.

The discrepancy between the results for the absorption of ^{45}Ca administered in situ and of Ca given orally might have been due to a secretory flux of Ca into the large intestine which would compensate the absorptive flux. However, this proved not to be the case since no significant secretory flux could be detected as assessed by injecting ^{45}Ca into the blood and measuring its appearence in the colon. Furthermore, when ^{45}Ca was given orally, no absorption of the isotope in the large intestine could be detected either.

The other possible explanation for the absence of significant absorption of Ca or ^{45}Ca given orally, despite a large absorption of ^{45}Ca directly injected into the caecum, could be that oral Ca is transformed in the digestive tract into a non-absorbable form. This proved to be the case. Indeed, when the absorbability of ^{45}Ca added to caecal contents in vitro were tested in a duodenal loop in situ, an absorption of 35.6 $\pm$ 4.6 % was found. However, the solution of Ca or ^{45}Ca administered orally was not different from zero.

DISCUSSION

This study represents the first report on calcium absorption in the large intestine in vivo performed under physiological conditions. It shows that ^{45}Ca is well absorbed when injected into the caecum and that this absorption is modulated by dietary calcium and phosphate and by the 1,25(OH)$_2$D$_3$ status of the animal. On the other hand, no significant absorption of ^{40}Ca or ^{45}Ca given orally was observed, irrespective of the diet or the 1,25(OH)$_2$D$_3$ status. This difference was not due to a secretory flux of Ca into the colon, but due to the fact that orally administered calcium is no more absorbable when it reaches the large intestine. These results thus show that the colon posesses the capacity to absorb Ca but that this mechanism does not normally operate because the Ca arriving into the large intestine is no longer in an absorbable form.

This mechanism has both pathophysiological and therapeutic implications. Thus, the possibility exists that in certain diseases this transformation of the absorbability of calcium might be disturbed leading to an increased absorption. In this respect, it is interesting that in man, after the resection of a part of the small intestine, the absorption of radio-calcium has been reported to be higher when the large intestine was conserved (6). From the therapeutic aspect, an increased absorption might be obtained if the transformation of calcium into the poorely absorbable form could be prevented in some way, or if Ca could be supplied in such a form that it escapes the transformation.

This work has been supported by the Swiss National Science Foundation and by the Ausbildungs- und Förderungsfonds der Arbeitsgemeinschaft für Osteosynthese (AO), Chur, Switzerland.

REFERENCES

1. Bonjour, J.-P., H. Fleisch, and U. Trechsel. Calcium absorption in diphosphonate-treated rats: effect of parathyroid function, dietary calcium and phosphorus. J. Physiol. 264:125-139, 1977.
2. Favus, M.J., S.C. Kathpalia, F.L. Coe, and A.E. Mond. Effects of diet calcium and 1,25-dihydroxyvitamin D$_3$ on colon calcium active transport. Am. J. Physiol. 238:G75-G78, 1980.

3. Favus, M.J., M. Berelowitz, and F.L. Coe. Effects of somatostatin on intestinal calcium transport in the rat. Am. J. Physiol. 241: G215-G221, 1981.
4. Grinstead, W.C., C.Y.C. Pak, and G.J. Krejs. Effect of 1,25-dihydroxyvitamin D_3 on calcium absorption in the colon of healthy humans. Am. J. Physiol. 247: G189-G192, 1984.
5. Harrison, H.C., and H.E. Harrison. Calcium transport by rat colon in vitro. Am. J. Physiol. 217:121-125, 1969.
6. Hylander, E., K. Ladefoged, and S. Jarnum. The importance of the colon in calcium absorption following small-intestine resection. Scand. J. Gastroenterol. 15: 55-60, 1980.
7. Lee, D.B.N., W. Walling, U. Gafter, V. Silis, and J.W. Coburn. Calcium and inorganic phosphate transport in rat colon. Dissociated response to 1,25-dihydroxyvitamin D_3. J. Clin. Invest. 65:1326-1331, 1980.
8. Mainoya, J.R. Effects of bovine growth hormone, human placental lactogen and ovine prolactin on intestinal fluid and ion transport in the rat. Endocrinology 96: 1165-1170, 1975.
9. Nellans, H.N., and R.S. Goldsmith. Transepithelial calcium transport by rat cecum: high-efficiency absorptive site. Am. J. Physiol. 240:G424-G431, 1981
10. Petith, M.M., and H.P. Schedl. Intestinal adaptation to dietary calcium restriction: in vivo cecal and colonic calcium transport in the rat. Gastroenterology 71:1039-1042, 1976.
11. Petith, M.M., H.D. Wilson, and H.P. Schedl. Vitamin D dependence of in vivo calcium transport and mucosal calcium binding protein in rat large intestine. Gastroenterology 76(1):99-104, 1979.
12. Rizzoli, R., H. Fleisch, and J.-P. Bonjour. Effect of thyroparathyroidectomy on calcium metabolism in rats: role of 1,25-dihydroxyvitamin D_3. Am. J. Physiol. 233: E160-E164, 1977.
13. Rizzoli, R., H. Fleisch, and J.-P. Bonjour. Role of 1,25-dihydroxyvitamin D_3 on intestinal phosphate absorption in rats with a normal vitamin D supply. J. Clin. Invest. 60:639-647, 1977.
14. Trechsel, U., C.M. Taylor, J.A. Eisman, J.-P. Bonjour, and H. Fleisch. Plasma levels of vitamin D metabolites in diphosphonate-treated rats. Clin. Sci. 61: 471-476, 1981.
15. Wasserman, R.H. Vitamin D and the absorption of calcium and strontium in vivo. In The Transfer of Calcium and Strontium Across Biological Membranes. R.H. Wasserman, editor. Academic Press, New York. 1963, pp. 211-228.

IMPAIRED INTESTINAL CALCIUM TRANSPORT IN THE SPONTANEOUSLY HYPERTENSIVE RAT

P.A. Lucas*, T. Drueke*, B. Lacour*, R.C. Brown**, and D.A. McCarron***

*INSERM U90, Hopital Necker, Paris, France
**Dept. of Medical Biochemistry
Univ. of Wales College of Medicine, Cardiff, U.K.
***Oregon Health Sciences University
Portland, Oregon

INTRODUCTION

The spontaneously hypertensive rat (SHR) exhibits a variety of abnormalities in its handling of calcium (1-3) and phosphorus (4,5) and calcium supplementation has been shown to reduce its arterial blood pressure (3,6). The relationship between disturbed calcium and phosphorus metabolism and hypertension, however, remains unclear. Several investigators have studied intestinal calcium transport in the SHR and have reported increased (7,8), unchanged (9) or decreased (10) absorption compared with its genetic control, the Wistar-Kyoto rat (WKY). In these studies, Ca^{2+} absorption has been determined by means of intestinal perfusion (10) or in situ uptake (8) in vivo, everted gut sacs in vitro (7,9,10), or by balance studies (8). The variety of experimental systems employed may in part explain the diversity in these findings. We have recently addressed this issue utilising the modified Ussing apparatus (11) which permits continuous monitoring of tissue viability and the study of active intestinal calcium transport under electrically controlled conditions.

METHODS

Six SHRs and six WKYs were raised until the age of 12-14 weeks on a 1 % Ca^{2+}, 0.46 % phosphorus, 0.45 % Na^+ diet (% dry weight) ; 20 % dietary protein as casein ; vitamin D, 2,200 U/kg food. Transmural intestinal calcium fluxes were then determined using the modified Ussing apparatus. Duodenal-jejunal segments 15 cm in length, immediately distal to the pylorus, were rinsed in situ with ice cold Krebs-Ringer bicarbonate solution. The segments were then excised, opened along the mesenteric border and mounted unstripped in the modified Ussing chambers. Tissue apertures were 0.5 cm^2.

Tissues from SHRs and WKYs were studied in random order, the sequence of which was unknown to the operator.

The mounted intestinal segment was bathed in a phosphate-free Krebs-Ringer bicarbonate solution maintained at pH 7.4 and 37°C (Na^+ 145, K^+ 4.7, Ca^{2+} 1.25, Mg^{2+} 1.2, Cl^- 127.2, HCO_3^- 25, SO_4^{2-} 1.2, D- glucose 11 mmol/l). Each hemichamber compartment contained 10 ml electrolyte solution and was aerated with 95 % O_2 / 5 % CO_2. The solutions bathing the mucosal and serosal sides were connected via agar bridges to calomel electrodes for the direct measurement of the potential difference (PD) and to Ag-AgCl electrodes for the passage of direct current through the tissue, using an automatic voltage clamp (Dual Voltage Clamp : 616C-3 Bioengineering, The University of Iowa). The short-circuit current (Isc) was recorded in the absence of PD after appropriate corrections for the resistance of the fluid between the two PD-sensing bridges. Conductance (G) was calculated from Isc and PD. Mucosa-to-serosa (Jms) and serosa-to-mucosa (Jsm) fluxes were paired employing the 30 % conductance matching criterion of Walling and Kimberg (12).

After the steady state had been reached (60 min), 30 µCi $^{45}Ca^{2+}$ was added to either the serosal or the mucosal side of the paired chambers. One ml samples were taken from the "cold" side at 20 minute intervals for one hour. Each aliquot removed was replaced with the same solution without $^{45}Ca^{2+}$. The aliquots removed were counted in 5 ml scintillation fluid (Pico-Fluor TM 30, United Technologies, Packard Instrument Company, Inc.) with a LKB 1212 RackBeta Wallace Liquid Scintillation Counter. Calcium fluxes were calculated as described by Schultz and Zalusky (13) and expressed as nmol. cm^{-2}. h^{-1}. Steady state values were reached after 20 minutes and are reported as the mean of the fluxes in the two subsequent 20 minute periods. Net flux was calculated on paired tissues from the same animal ($J_{net} = J_{ms} - J_{sm}$).

When the initial study of six WKYs and six SHRs had been completed, eight additional rats (three WKYs, five SHRs) were studied to verify the Jms data.

STATISTICS

The 2-tailed Mann-Whitney U test was used throughout.

RESULTS

Net intestinal calcium absorption (Jnet) was significantly lower in the SHR compared with the WKY, $p < 0.01$, resulting from a significantly lower mucosa to serosal transport (Jm-s), $p < 0.03$ (Figure 1). Additional Jm-s values from a further 5 SHRs and 3 WKYs provided verification of the reduction in Jm-s in the SHR : 64.4 ± 6.7 nmol/cm2hr vs 36.2 ± 3.7 nmol/cm2.hr, $p < 0.002$ (Figure 2). Electrical conductance did not differ between the WKY and the SHR : 28.4 ± 7.8 mho/cm2 vs 29.6 ± 7.3 mho/cm2. Short circuit current was lower in the SHR than in the WKY : 128.6 ± 9.1 uA/cm2, $p < 0.02$.

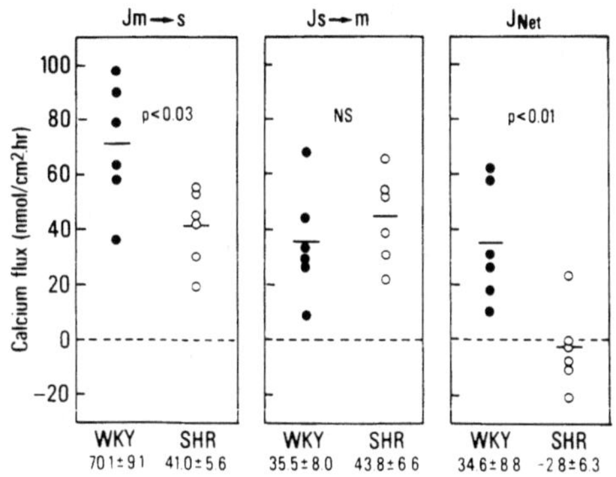

Figure 1 : Unidirectional (Jm-s, Js-m) and net (Jnet) transmural intestinal calcium fluxes in six WKY (●) and in six SHR (o) ; mean $\pm$ SEM. (Figure reprinted from (15) with permission of the Journal of Clinical Investigation)

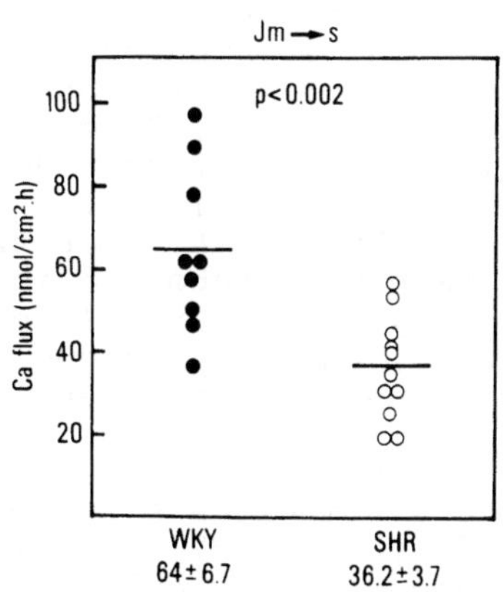

Figure 2 : Mucosa-to-serosa calcium fluxes in nine WKY (●) and in eleven SHR (o) ; mean $\pm$ SEM.

DISCUSSION

Our data supply evidence for an impairment of active calcium absorption in the duodenum-jejunum of the SHR. Our results are consistent with those of Schedl et al (10), who used both everted duodenal-jejunal sacs in vitro and duodenal perfusion in vivo to document a reduction in maximal calcium transport in SHRs of 12-14 weeks of age. Other workers have studied animals at different ages (8,9) using different experimental techniques and dietary regimes (7,8,9) and these factors may account for the fact that increased calcium absorption has been observed.

By means of the Ussing chamber we have been able to demonstrate an impairment in the SHR of the active component of calcium transport, the component which is normally vitamin D responsive. The question of the vitamin D status of the SHR is of obvious importance. Stern et al (9) and Schedl et al (10) found no difference in 1,25(OH)2D3 levels between the SHR and the WKY. The latter investigators did however find higher 25(OH) D3 in the SHR, suggesting the vitamin D axis to be somewhat abnormal.

In summary, we have shown active duodenal-jejunal calcium transport to be impaired in the SHR. This abnormality may reflect a generalised disorder of cellular calcium handling and at the same time may help to explain the reduced serum ionised calcium (2,3) in the SHR. It is probable that the SHR's reduction in blood pressure in response to increase dietary calcium is related to increased absorption by the "passive" route (14).

REFERENCES

1. J.P. Noon, P.J. Rice, and R.J. Baldessarini, Calcium leakage as a cause of the high resting tension in vascular smooth muscle from the spontaneously hypertensive rat. Proc.Natl. Acad. Sci. USA 75 : 1605 (1978).
2. G.L. Wright, and G.O. Rankin, Concentrations of ionic and total calcium in plasma of four models of hypertension. Am. J. Physiol. 243 : H365 (1982).
3. D.A. McCarron, N.N. Yung, B.A. Ugoretz, and S. Krutzik, Disturbances of calcium metabolism in the spontaneously hypertensive rat. Hypertension 3 (suppl. I) : I-162 (1981).
4. C.H. Hsu, P.S. Chen, and R.M. Caldwell, Renal phosphate excretion in spontaneously hypertensive and normotensive Wistar Kyoto rats. Kidney Int. 25 : 789 (1984).
5. R.J. Bindels, J.A. Geertsen, C.H. van Os, and J.F. Slegers, Disturbed phosphate homeostasis in sponteneously hypertensive rats. Kidney Int. 27 : 703 (1985).
6. S. Ayachi, Increased dietary calcium lowers blood pressure in the spontaneously hypertensive rat. Metabolism 28 : 1234 (1979).
7. M.A. Toraason, and G.L. Wright, Transport of calcium by duodenum of spontaneously hypertensive rat. Am. J. Physiol. 241 : G344 (1981).
8. K. Lau, J. Zikos, J. Spirnak, and B. Eby, Evidence for an intestinal mechanism in hypercalciuria of the spontaneously hypertensive rat. Am. J. Physiol. 247 : E625 (1984).

9. N. Stern, D.B.N. Lee, V. Silis, F.W. Beck, L. Deftos, S.C. Manolagas, and J.R. Sowers, Effects of high calcium intake on blood pressure and calcium metabolism in young SHR. Hypertension 6 : 639 (1984).
10. H.P. Schedl, D.L. Miller, J.M. Pape, R.L. Horst, and H.D. Wilson, Calcium and sodium transport and vitamin D metabolism in the spontaneously hypertensive rat. J. Clin. Invest. 73 : 980 (1984).
11. M.W. Walling, and D.V. Kimberg, Active secretion of calcium by adult rat ileum and jejunum in vitro. Am. J. Physiol. 225 : 415 (1973).
12. M.W. Walling, and D.V. Kimberg, Effects of 1,25-dihydroxy-vitamin D3 and Solanum Glaucophyllum on intestinal calcium and phosphate transport and on plasma Ca, Mg, and phosphorus levels in the rat. Endocrinol. 97 : 1567 (1978).
13. S.G. Schulz, and R. Zalusky, Ion transport in isolated rabbit ileum, short circuit current and Na+ fluxes. J. Gen. Physiol. 47 : 567 (1964).
14. P. Ireland and J.S. Fordtran, Effect of dietary calcium and age on jejunal calcium absorption in humans studied by intestinal perfusion. J. Clin. Invest. 52 : 2672 (1973).
15. D.A. McCarron, P.A. Lucas, B. Lacour, R.J. Shneidman, and T. Drüeke, Blood pressure development of the spontaneously hypertensive rat following concurrent manipulations of dietary Ca2+ and Na+ : relation to intestinal Ca2+ fluxes. J. Clin. Invest. (in press).

ACKNOWLEDGEMENT

We acknowledge the excellent technical assistance of Lydie Berthet.

BONE METABOLISM

CELLULAR MECHANISMS OF BONE REMODELING EVALUATED AT THE INTERMEDIARY LEVEL OF ORGANIZATION OF BONE

Pierre J. Meunier

INSERM Unit 234, Faculty Alexis Carrel

69008 Lyon, France

INTRODUCTION

It is now well-known that the remodeling system which operates in adult lamellar bone is a quantal phenomenon due to the coordinated activity of two continually renewing cell populations that are opposite in action : osteoclasts and osteoblasts.They are tightly coupled in space and time, as shown by H.Frost[1] who proposed the concept of basic multicellular unit (BMU).It is less known that the original morphological observations of H.Frost which have directly demonstrated the cellular coupling between osteoclastic resorption and osteoblastic formation were made at the intermediary organization of bone level, i.e. on the cortical basic structure units (BSU), or osteons. H.Frost made the observation that the cement line limiting the outer circumference of the osteon was always scalloped, thus representing always the limits of a preceding resorption phase. This was the first observation made at the BSU level which considerably enlightened the cellular mechanisms of bone remodeling. Despite H.Frost [2] described next year the possibility to measure the mean wall thickness (MWT) of cortical osteons and to derive from this parameter - and from the osteoblastic appositional rate - the activation frequency and the lifespan of the active osteoblastic cells, this intermediary level of organization of bone was temporarly forgotten in many laboratories. During about 15 years most studies of bone histomorphometry were concentrated on the percentage of bone surfaces occupied by formation or resorption sites, and most studies at the individual cellular level having used tissue cultures or electronmicroscopy. These studies provided informations which unfortunately cannot be extrapolated to the tissue- and organ-levels, without taking into account the intermediary level of organization of bone, i.e. the basic stucture units of cortical and trabecular bone. They represent the end-products of the osteoblastic activity and are bound on one side by the bone-marrow interface or the haversian canal, and on the other side by the scalloped cement line which represents the limits of the resorption phase (Figure 1). The average distance between the cement line and the surface of the completed packet is the mean wall thickness and is measured under polarized light. This parameter provides a direct indication of the amount of bone formed by the BMU. These basic structure units cannot be studied by any other tool than bone histomorphometry and have been compared by H. Frost[3] to the nephrons of the kidney.

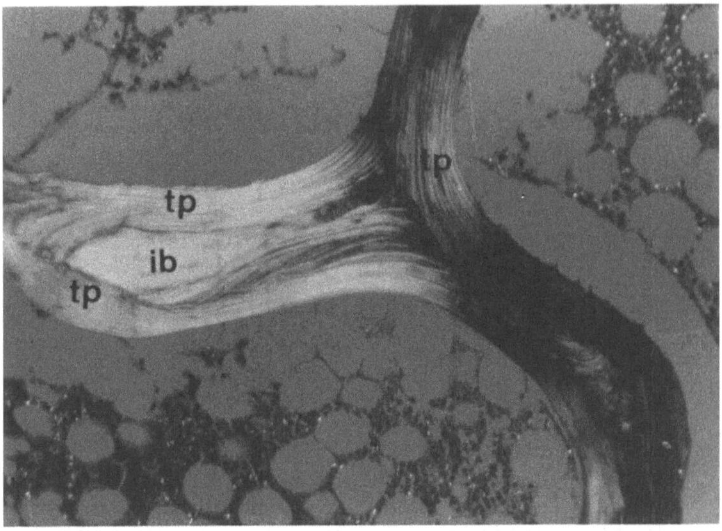

Fig. 1. Trabecular packets (tp) or basic structure units of trabecular bone. Interstitial bone (ib) occupies the center of the trabecula. Undecalcified section of human iliac bone .Stain: Solochrome cyanin R. Polarized light. 140 x.

COUPLING AND BONE CELLS

The cells involved in the remodeling process of adult lamellar bone comprise osteoclasts and osteoblasts.

Osteoclasts. In response to systemic or local stimuli,osteoclasts, which originate from immature mononuclear bone marrow cells, differentiate and remove a certain volume of bone. They are multinucleated and motile giant cells with a diameter from 20 to 100 micrometers. The mononuclear and proosteoclastic cells can be identified by acid phosphatase staining. The osteoclasts drill pits on bone surface and the surface of the cell in contact with bone tissue shows a ruffled border characterized by numerous membranous folds. Containing acid phosphatases, osteoclasts destroy bone by removing the mineral and the matrix. The sites of present or past osteoclastic erosion are identifiable by a scalloped interruption of bone tissue on the endhaversian, subcortical or trabecular endosteal envelopes.

Osteoblasts. Derived from undifferentiated mesenchymatous cells, these are mononucleated cells with a diameter of 20 to 30 micrometers. They are aligned in a monolayer on bone surfaces, normally in close contact with a thin layer of osteoid. The osteoblasts are responsible for the deposition of this osteoid and its subsequent mineralization. Few osteoblasts are progressively embedded in bone matrix and become osteocytes, each being buried inside a periosteocytic lacuna.

Coupling. Osteoclasts and osteoblasts do not differentiate and function independently in the human adult, and there is very likely a chemical communication system[4,5] between both types of cells, with still unidentified factors coupling resorption and osteoblast differenciation, and, reciprocally, osteoblastic factors capable to stimulate the resorptive activity of osteoclasts at the cell-level. According to H.Frost[6], the term "coupling" only refers to time and designates a stereotyped sequence: resorption - formation at the BMU level. It does not refer to the amount of bone resorbed or formed, or to the tissue balance at the BSU-level[7]. In contrast, "uncoupling" refers to the complete failure of bone formation to follow locally the resorption process[8], such as seen in severe disuse osteoporosis or multiple myeloma. The term "uncoupling" has also been used by others in a sense of imbalance between bone resorption and formation[9], i.e. referring to a disequilibrium in bone tissue balance at the BSU-level. These two different meanings for the same term may be confusing for the understanding of articles discussing the histopathogenetic mechanisms of bone loss.

AGE-RELATED CHANGES IN BONE REMODELING AT THE BSU-LEVEL

The MWT measurement provides a direct indication of the amount of bone formed during the osteoblastic phase of the BMU. In our laboratory the MWT of both trabecular and cortical bone packets were measured on sections stained with Solochrome cyanin R under polarized light (Figure 1). For trabecular bone we used either the measurement at 4 equidistant points on each packet of the distance between the cement line and the trabecular bone surface[10], or the orthogonal projection technique described by J.Kragstrup[11]. For cortical bone packets- or osteons- we used the method described by H.Frost[2] and P.Broulik[12]. We found that the MWT of trabecular packets, measured either with P.Lips' method[10] or J.Kragstrup's one[13], decreased significantly with age in both sexes. This age-related reduction of the MWT has also been found in cortical osteons[13] of control females, and a positive relationship was found in the same control subjects between the MWT of cortical and trabecular packets [13].

This age-related thinning of cortical and trabecular BSUs suggests that, with increasing age, the osteoblastic population becomes progressively less able to reconstitute previously resorbed bone, and that there is a net decrease of bone formation at the BMU-level. Thus, the decreased MWT may be due (i) either to a reduction of the local osteoblastic population resulting from a decreased lifespan of the cells or from a decreased recruitment of the cells, (ii)or from a decreased osteoblastic appositional rate, i.e. a decreased daily synthesis of bone matrix by an osteoblastic population having a normal lifespan(or formation period or $sima_f$ in the Frost's terminology). By using tetracycline double labelling which permits the evaluation of the osteoblastic appositional rate, it is possible to explore further the mechanisms of this defective bone formation. This osteoblastic appositional rate (OAR) does not change with age in both sexes, as we [14] and F. Melsen's group[15] have shown. Consequently, as the MWT of the trabecular packets decreases with age and the OAR does not change, it seems likely that a reduction in the lifespan (or MWT/OAR) of the active osteoblastic population at the BMU-level was the cause of the reduced thickness with age of the trabecular packets. In other words there is in aged controls an early dedifferentiation of the osteoblastic cells and this appears as a fundamental characteristic of bone ageing.

This age-related osteoblastic dysfunction is associated with a reduction in the depth of the resorption cavities. This has been shown in our Unit by P.Courpron[16] who measured the total thickness of the trabeculae, including the thickness of two packets and of the interstitial bone (Figure 1). From the mean trabecular thickness and the MWT mean values at 20, 50 and 80 years of age, P.Courpron obtained indirectly the interstitial bone thickness at the corresponding ages, and found an unexpected age-related increase of the interstitial bone thickness. E.F. Eriksen recently confirmed from direct measurements of the trabecular resorption depth a decrease with age of the depth of resorption cavities in both sexes[17].

Thus, at the BMU-level, the quantity of bone resorbed is always greater than the quantity of bone formed, but it also decreases with age, resorption decreasing less than formation.

BONE FORMATION IN OSTEOPOROTIC STATES

.<u>In post-menopausal osteoporosis</u> we have measured the MWT and the OAR in a group of patients having normal resorption and formation surfaces as well as a normal OAR [18]. The MWT of these osteoporotic patients showed no correlation with age, but the mean was reduced by 20 per cent and significantly different from the control group($p < 0.01$). Sigma$_f$ which was calculated without correction by using the formula MWT/OAR and was expressed in days, was 23 per cent shorter in osteoporotics than in controls($p < 0.01$). Analysis of the relationship of sigma$_f$ with age did not show a significant correlation in osteoporotics.

Table 1. Mean values of bone remodeling parameters in patients with post-menopausal osteoporosis (PMOP) and corticosteroid-induced osteoporosis (CSOP) compared to age- and sex-matched controls. (°: number of cases)

	CONTROLS	PMOP	CSOP
MWT	48.2 ± 8.2 (9°)	38.7 ± 3.1 (9)	-
(µm)	48.0 ± 6.2 (20)	-	32.7 ± 4.3 (20)
OAR	.70 ± .07 (9)	.73 ± .06 (9)	-
(µm/day)	.68 ± .05 (7)	-	.62 + .48 (12)
Sigma$_f$	68.9 ± 12.4 (9)	52.8 ± 4.9 (9)	-
(days)	71.3 ± 8.2 (7)	-	55.9 ± 14.4 (12)

MWT:mean wall thickness/OAR:osteoblastic appositional rate

.<u>In corticosteroid-induced osteoporosis</u>, the mean osteoblastic appositional rate, measured in 50 patients doubly labelled with tetracycline prior to the transiliac bone biopsy (.34± .20 µm/day in 40 females and .40± .19 µm/day in 10 males)[19], was found significantly decreased in both sexes as compared to controls (.72 ± .12 µm/day). This indicates a decrease of osteoblastic activity at the individual cell level. In addition, we have measured MWT of trabecular packets in a subgroup of 12 patients selected as having no significant reduction in OAR, in order to separately analyze the effects of corticosteroids on the formation period ($sigma_f$) and on the daily activity of the osteoblasts [20]. In this subgroup (Table 1), the average MWT in the CS-treated patients was 32 per cent lower than the mean control values, and the difference was significant ($p < 0.001$). The formation period was found significantly shorter in the CS-treated osteoporotics than in the controls ($p < 0.01$).

These results obtained in post-menopausal osteoporosis and in corticosteroid-induced osteoporosis suggest that a major contribution to the negative skeletal balance existing in both conditions is a decrease in bone formation caused by a reduction in the lifespan of the osteoblastic population at the BMU-level, i.e. a failure in the osteblastic recruitment. There is in addition a reduction in the osteoblastic appositional rate in some patients with post-menopausal osteoporosis[21] and in most patients with corticosteroid-induced osteoporosis[19].

BONE FORMATION IN PRIMARY HYPERPARATHYROIDISM

Parathyroid hormone excess does not induce a significant reduction in trabecular bone volume in males and females older than 50 years of age with primary Hyperparathyroidism (HPT 1), in spite of an increased turnover of bone tissue demonstrated by the extension of resorption and formation surfaces [22]. In order to determine the effects at the BMU-level of a parathyroid hormone excess on osteoblastic bone formation, MWT of trabecular packets was measured on iliac bone biopsies from 26 patients with HPT 1 (12 males and 14 females). Mean value of MWT in HPT 1 patients (45.2 ± 5.8 µm) was found to be significantly lower ($p < 0.001$) than in 35 controls (50.1 ± 8.5 µm)[23]. This difference is due to a significant reduction of MWT in HPT 1 females before 50 years of age(46.0 ± 2.3 µm) when compared to controls (56.2 ± 7.1 µm), whereas there is no significant difference of average MWT in males of any age or in femeales after 50. This reduction of MWT could explain the rarefaction of trabecular bone volume noticed in females younger than 50 years[22]. In addition, the osteoblastic formation period ($sigma_f$) was determined to be shortened in females before 50 and lengthened in older females and in males at any age. This could explain the non rarefaction of spongy bone in most patients with HPT 1 and possibly the anabolic effect of 1-34 hPTH fragment on trabecular osteoporosis in post-menopausal females with crush fractures [24].

FLUORIDE INCREASES BONE FORMATION AT THE BMU-LEVEL IN OSTEOPOROSIS

When the trabecular bone mass has become lower than the fracture threshold, the ideal therapeutis approach for trabecular osteoporosis (OP) responsible for vertebral compression fractures (type I OP), is the use of an agent which not merely will slow down the rate of bone loss, but rather which will substantially increase bone mass so that the fracture rate will decrease because the trabecular bone mass will become higher than the fracture threshold. If we remember that in post-menopausal OP with crush fractures, the mean trabecular bone volume is 30 to 50 per cent lower than in age-matched controls [21], we may realize how large should be the gain of trabecular bone induced by a treatment for reducing the vertebral fracture rate. The reduced MWT and formation period of trabecular bone packets in post-menopausal OP [18], and the reduced OAR in 23 out of 109 patients with vertebral fractures due to primary OP[21], both suggest that a major mechanism underlying the negative skeletal balance in a number of patients with OP is an osteoblast dysfunction. Fluoride is one of the few agents known for being capable of increasing the osteoblast population and of inducing a positive bone balance at the BMU-level.

We performed a histomorphometric analysis of trabecular bone on pre- and post-treatment transiliac biopsies after 2 years of combined therapy using NaF (50 mg/day), calcium (1 gm /day) and vitamin D2 (8000 iu/day)[25,26]. The most striking change was a marked increase in the osteoid surfaces (20.7 ± 9.9 % to 50.7 ± 20.6 %; p 0.001; n= 37), and in surfaces ered by osteoblasts 26. In contrast, osteoclastic resorption surfaces did not increase significantly and the result of this increase of bone formation was a marked augmentation of trabecular bone volume (9.9 ± 3.8 % to 15.9 ± 8.6 % ; p 0.001) reaching plus 60 per cent from initial values. In 11 patients who were classified as good responders on the basis of an increase in trabecular volume higher than 33 per cent, measurements of the mean wall thickness of trabecular packets and of the mean thickness of a pair of lamellae were performed. The newly formed bone was lamellar and the augmentation of TBV was caused by a significant increase in the MWT of trabecular packets (35.1 ± 8.6 µm to 40.8 ± 9.4 µm; $p < 0.01$), associated to an increase of the mean thickness of matrix lamellae (6.50 ± 0.45 µm to 6.94 ± 0.42 µm; $p < 0.001$). The mean calcification rate did not change significantly and this suggests an increase in the formation period of trabecular bone packets, in other words an increase in the lifespan of the active osteoblastic cells under the influence of sodium fluoride.

<u>In conclusion</u>, the histomorphometric analysis of bone structure units -including measurements of the mean wall thickness of cortical or trabecular packets, of the bone erosion depth and of the osteoblastic appositional rate after tetracycline double labelling - demonstrates that it is fundamental to take into account the intermediary level of organization of bone for elucidating the cellular mechanisms of bone remodeling.H.Frost[3] and ZFG Jaworski[7] were right when they recently reminded us that any direct inferences and applications from the molecular level to the whole skeleton, without considering the BMU- and BSU-levels, could be misleading. Consequently, morphometric studies at the intermediary level of organization of bone should be developped in further studies in animals- at the exclusion of the rat - and in humans.

REFERENCES

1. H.M. Frost,"Bone Remodeling Dynamics, "Charles C.Thomas, Springfield,Ill.(1963).
2. H.M. Frost, "Mathematical Elements of Bone Remodeling" Charles C. Thomas,Springfield,Ill.(1964).
3. H.M. Frost, The skeletal intermediary organization: a review,<u>Metab. Bone Dis. rel. Res.</u>,4 : 281 (1983).
4. J.R. Farley and D.J. Baylink, A putative coupling factor from human bone: effects on bone cells in vitro,<u>Calcif. Tissue Int.</u>,33:292 (1981).
5. M.R. Urist,R.J.De Lange and G.A.M.Finerman,Bone cell differentiation and growth factors,<u>Science</u>,220:680 (1983).
6. H.M. Frost,Dynamics of bone remodeling,<u>in</u> "Bone Biodynamics",H.M.Frost ed.,Little,Brown and <u>Co</u>,Boston(I964).

7. Z.F.G. Jaworski,Coupling of bone formation to bone resoption: a broader view,Calcif. Tissue Int.,36:53I (I984).
8. H.M. Frost,A.R.Villanueva and L. Ilnicki, Tetracycline-based studies of bone remodeling in primary and secondary hyperparathyroid states in man, in " Parathyroid Hormone and Thyrocalcitonin ",R.V. Talmage and R.F.Belanger, eds,Excerpta Medica,Amsterdam (I968).
9. J.L. Ivey and D.J. Baylink,Postmenopausal osteoporosis: proposed roles of defective coupling and estrogen deficiency,Metab. Bone Dis.rel.Res,3:3(1981).
10. P.Lips,P.Courpron and P.J.Meunier,Mean wall thickness of trabecular bone packets in the human iliac crest:changes with age,Calcif. Tissue Res.,26:13(1978).
11. J.Kragstrup,H.J.G. Gundersen,F.Melsen and L.Mosekilde, Estimation of the three-dimensional wall thickness of completed remodeling sites in iliac trabecular bone, Metab. Bone Dis. rel. Res.,4:113 (I982).
12. P.Broulik,J.Kragstrup,L.Mosekilde and F.Melsen,Osteon cross-sectional size in the iliac cest,Acta Pathol. Microbiol. Immunol. Scand.(A), 90:339 (1982).
13. L.G. Sainte Marie,R.Marcus and P.J. Meunier,Correlations between mean wall thicknesses of trabecular and cortical bone structure units in human iliac crest in"4th International Workshop on Bone Histomorphometry", Aarhus,Denmark,10-13.6.I984 (abstract).
14. P.J.Meunier,P.Courpron,C.Edouard,C.Alexandre,C.Bressot, P.Lips and B.Boyce,Bone histomorphometry in osteoporotic states,in "Osteoporosis II",U.S.Barzel, ed.,Grunne & Stratton,New York (1979).
15. F.Melsen and L.Mosekilde,Tetracycline double-labeling of iliac trabecular bone in 41 normal adults,Calcif. Tissue Res.,26:99(1978).
16. P.Courpron,Bone tissue mechanisms underlying osteoporoses, Orthop. Clin.North Am.,12:513 (1981).
17. E.F.Eriksen,L.Mosekilde and F.Melsen,Trabecular bone resorption depth decreases with age,Bone, 6:141 (1985).
18. A.J.Darby and P.J.Meunier,Mean walla thickness and formation periods of trabecular bone packets in idiopathic osteoporosis,Calcif. Tissue Int., 33:199 (1981).
19. C.Bressot,P.J.Meunier,M.C.Chapuy,E.Lejeune,C.Edouard and A.J.Darby,Histomorphometric profile,pathophysiology and reversibility of corticosteroid-induced osteoporosis, Metab. Bone Dis.and rel. Res.,1:303 (1979).
20. D.W.Dempster,M.A.Arlot and P.J.Meunier,Mean wall thickness and formation periods of trabecular bone packets in corticosteroid-induced osteoporosis,Calcif.Tissue Int.,35: 410 (1983).
21. P.J.Meunier,S.Sellami,D.Briancon and C.Edouard,Histological heterogeneity of apparently idiopathic osteoporosis, in "Osteoporosis,Recent Advances in Pathogenesis and Treatment",H.F.DeLuca,H.M.Frost,W.S.Jee,C.C.Johnston and A.M.Parfitt eds. University Park Press,Baltimore (1981).
22. P.J.Meunier and C.Bressot,Endocrine influences on bone cells and bone remodeling evaluated by clinical histomorphometry,in "Endocrinology of Calcium Metabolism", J.A.Parsons ed.,Raven Press, New York (1982).
23. S.A.Charhon,C.M.Edouard,M.E.Arlot and P.J.Meunier,Effects of parathyroid hormone on remodeling of iliac trabecular bone packets in patients with primary hyperparathyroidism,Clin.Orthop.and rel.Res.,162:255(1982).

24. J.Reeve,P.J.Meunier,J.A.Parsons,M.Bernat,O.L.M.Bijvoet, P.Courpron,C.Edouard,L.Klenerman,R.M.Neer,J.C.Renier, D.Slovik,F.J.F.E.Vismans and J.T.Potts,Anabolic effect of human parathyroid hormone fragment on trabecular bone in involutional osteoporosis: a multicentre trial,Br. Med.J.,280:1340 (1980).
25. D.Briancon and P.J.Meunier,Treatment of osteoporosis with fluoride,calcium and vitamin D.,Orth.Clinics North America,12:629 (1981).
26. P.J.Meunier,K.Galus,D.Briancon,J.Reeve,R.Podbesek,C.Edouard,M.Arlot,S.Charhon,P.Delmas,B.Benevise,A.Valentin, M.C.Chapuy,Treatment of primary osteoporosis with drugs that increase bone formation:sodium fluoride,hPTH 1-34, ADFR concept, in "Osteoporosis",C.Christiansen et al., eds,Glostrup Hospital,Copenhagen (1984).

THE LOCAL REGULATION OF BONE REMODELING

William A. Peck, Leonard Rifas, Su-Li Cheng,
and Victor Shen

Department of Medicine
Washington University at Jewish Hospital
St. Louis, Missouri

Mounting experimental evidence indicates that skeletal remodeling is under local as well as systemic control. Local environmental factors which may well be significant include mechanical (loading and fluid pressure) and electrical forces, and locally elaborated humoral factors such as prostaglandins, regulatory proteins, and components of the organic matrix. The local ionic milieu and direct cell-cell interactions may also contribute. The effects of systemically elaborated hormones such as parathyroid hormone (PTH) and $1\alpha,25\text{-}(OH)_2\ D_3$ under physiological conditions may only be permissive, at least under physiological conditions, establishing the proper environment and ensuring an adequate supply of cellular participants for remodeling. High levels of these agents sharply enhance the likelihood of resorption and remodeling. By contrast, local forces seem to be important determinants of the location and frequency of remodeling activity, and may well perpetuate the remodeling cascade so that it proceeds inevitably through an entire coupled sequence of resorption and formation.

Currently held concepts indicate that a team of cells and their precursors are responsible for remodeling a quantum of skeletal tissue at each discrete remodeling locus, whether it be on a trabecular or endosteal surface or along a Haversian vascular channel.[1] These multicellular units provide the functional framework for understanding the regulatory processes that determine the initiation of remodeling (activation), its perpetuation, and finally, its termination.[2] It should be stressed that considerations of local regulation or control cannot perforce be the sole basis for understanding skeletal physiology or disease. As Frost has pointed out, their comprehension requires understanding of organ and tissue levels of organization.[3]

The multicellular unit of remodeling subserves a cascade of events which includes the resorption of bone and the subsequent repair of the resorption cavity. One can divide this cascade into at least six stages: a) an inciting event; b) an initial response, probably involving preparation of the bone surface for osteoclast attack; c) osteoclast recruitment; d) resorption; e) osteoblast recruitment; and f) repair. There is reason to believe that each step inevitably succeeds the preceding one. That is, all steps will occur automatically beginning with the inciting event. Although systemic and local environmental factors may determine the rate, quantity and quality of those steps, as well as the statistical likelihood of an initial response, local factors are the perpetuators of the remodeling sequence, ensuring its automaticity. Whether aborted remodeling sequences occur is not known, though the complete osteoclastic perforation of individual spicules of trabecular bone, probably commonplace, would essentially preclude a subsequent repair phase.

A local cellular supervisory system may influence all remodeling steps. The leading candidate for that role is a macrophage-like cell, posited in the region of a remodeling locus.[4,5] Macrophages are highly qualified for such a calling; they are extraordinarily versatile. Macrophages can migrate through tissues and through the circulation, they can be phagocytic and they elaborate a strikingly broad array of biologically active substances: prostanoids, enzymes, nutrients, peptide growth factors, and others.

The inciting event may be defined as a local change in steady state which triggers a remodeling cascade. In theory, it is possible to visualize two requirements for effective incitement: <u>first</u>, an environmental change and <u>second</u>, a mechanism for targeting (geographically restricting the inciting event or the response to it) so that not all surfaces in a given milieu will begin to resorb simultaneously. Environmental changes that could predispose a given region to remodeling are alterations in mechanical loading, fluid forces, and chemical forces (ionic composition and concentrations, including hydrogen ion, partial pressure of dissolved gasses, hormones). Targeting could occur at the locus of maximum change, or alternately, could arise when local susceptibility to targetting is at its peak, due to enhanced responsivity of bone-lining cells or the availability and responsivity of local primary transducer or effector cells. The proximity of remodeling loci to blood channels and hematopoietic marrow is consistent with the second possibility. Candidates to serve as the local primary transducer cells might be the bone-lining cells themselves, osteocytes, or vicinal macrophages, or perhaps a combination of these cellular elements. It is the lining cell, however, which has received greatest attention as the local transducer of remodeling.

The classical view that the osteoclast or its immediate precursor is the initial target cell for resorption stimuli has been challenged and now seems most unlikely. Highlighted in the provocative theoretical considerations of Rodan and Martin,[6] an impressive body of evidence has now accumulated which points to a more indirect mechanism of activation, involving an intermediate cell type, most likely the osteoblast-derived bone-lining cell. Accordingly, resorption stimuli are envisaged as altering the structure and function of this cell which in turn readies the bone surface for osteoclast attraction and activation. Among the specific effects of resorption stimuli on the lining cell are: a) acute contraction, exposing acellular bone surfaces which contain substances that are chemotactic for osteoclasts[7-9]; and b) the elaboration of collagenase which denudes the bone surface of its most superficial stratum of organic matrix, thus permitting osteoclasts to degrade bone mineral.[10-12] In addition, these cells may elaborate a second extracellular messenger(s) for resorption. This messenger may serve as an autocrine modifier of the lining cells themselves and may also aid in eliciting an osteoclast response. A leading candidate for that messenger role is a prostaglandin of the E series (PG). Bone cells are known to produce PG,[13] and PG in turn can promote shape change,[14] collagenase release,[15] and osteoclast-mediated bone resorption.[16] Moreover, there is evidence that PGs mediate the resorption-stimulating effects of platelet-derived growth factor,[17] epidermal growth factor,[18] and transforming growth factor alpha.[19] PGs also stimulate lymphocytes to produce the bone-resorbing lymphokine osteoclast activating factor (OAF).[20] An alternative messenger might be an interleukin 1 (IL1)-like substance (a resorption stimulator which is produced by macrophages and perhaps by periosteal tissue as well).[21,22]

Once stimulated by a product of the lining cell or by the newly exposed matrix, osteoclast precursors migrate to the now prepared bone surface. Although the actual skeletal chemoattractant is still elusive, osteocalcin (a peptide component of bone matrix) and its amino acid components have been shown to be chemotactically active for cells of the monocyte/macrophage series.[8,9]

Whereas the mature osteoclast and the monocyte/macrophage appear to share a common lineage, at least in part, the nature of the osteoclast precursor remains a matter of controversy.[12] Differentiated macrophages can fuse to form inflammatory giant cells (polykaryons), but there are sufficient biological differences between osteoclasts and those cells to preclude their common immediate ancestry. Rather, preosteoclasts may arise from primitive monocytic progenitors by a separate (but closely related) branch, or represent an early offshoot of the monocyte series.[23] Of interest is the fact that one of the physiological roles of $1\alpha,25\text{-}(OH)_2$ vitamin D_3, which also promotes calcium accumulation and bone resorption, is to enhance the differentiation of osteoclasts and of macrophages.[24,25] This effect may represent the heretofore unexplained link between vitamin D action and bone formation, quite apart from the ability of vitamin D to provide calcium for mineralization.

The resorption-formation interval is key, for it represents the conceptual time compartment when "coupling" or resorption-formation linkage occurs. Since bona fide stimulators of bone formation unique to the systemic circulation have yet to be identified (for example, somatomedins are known to be elaborated locally by many tissues[26]), locally elaborated "coupling factors" must be emphasized. There is no dearth of candidates. Osteoblast growth-stimulators have been isolated from devitalized bone tissue itself,[27,28] from the incubation media bathing cultured bones and bone cells,[29,30] and from isolated cultured macrophages.[31] Furthermore, separate factors have been identified which can promote growth (mitogenic activity) and differentiation (specialization) of osteoblasts.[32] The participation of macrophages in coupling has received experimental support. Macrophage-like cells inhabit resorption bays[4,5] and are rich sources of a bone cell mitogen which appears to differ from IL1, though that difference may only be evident under highly selective experimental circumstances.[31]

The nature of the osteoblast precursor which is responsive to one or more of these mitogenic factors remains unclear. That osteoblasts extend ultimately from mitotically active cells of the marrow stroma seems a most reasonable possibility.[33] One is impressed, however, that the workaday generation of new osteoblasts occurs at the circumferential borders of resorption bays from adjacent surface lining cells, which migrate (or proliferate) laterally to fill the resorption defect. Hence, one does not need to invoke the migratory ingress of a distant stromal preosteoblast population to explain the formation phase. Derivation of active osteoblasts from nearby surface-lining cells (or, as has been hypothesized) from deeper osteocytes, excavated during resorption, would obviate the necessity of a long distance "coupling factor." Merely a change in the bone surface would suffice to initiate osteoblast precursor proliferation. Autocrine and paracrine factors would do the rest.

This hypothetical schema pivots on the intimate participation of local factors--factors which are elaborated or stored (or both) and which act locally. Although some factors have been identified, doubtless other candidates for regulatory significance will be uncovered. It may well be that the nature of regulation differs among different remodeling areas (e.g., trabecular surface, endosteal surface) and at different times. Furthermore, the governance of geometric extents, of rates, and of quantitative coupling has yet to be elucidated. Finally, as mentioned previously, the mechanism whereby individual remodeling foci relate to one another in the maintenance of regional or organ-level integrity is a most important unanswered question.

REFERENCES

1. H.M. Frost, "Mathematical Element of Bone Remodeling," Charles C. Thomas, Springfield, Illinois (1964).
2. H.M. Frost, Bone dynamics in metabolic bone disease, J. Bone Joint Surg. 48A:1192 (1966).
3. H.M. Frost, The skeletal intermediary organization; a synthesis, in "Bone and Mineral Research/3," pp. 49-108, W.A. Peck, ed., Elsevier, Amsterdam (1985).
4. P. Tran Van, A. Vignery, and R. Baron, Cellular kinetics of the bone remodeling sequence in the rat, Anat. Rec. 202:445 (1982).
5. P. Tran Van, A. Vignery, and R. Baron, An electron microscopic study of the bone remodeling sequence in the rat, Cell Tissue Res. 225:283 (1982).
6. G.A. Rodan and T.J. Martin, Role of osteoclasts in hormonal control of bone resorption - a hypothesis, Calc. Tiss. Int. 33:349 (1981).
7. S.S. Miller, A.M. Wolf, and C.D. Arnaud, Bone cells in culture: morphologic transformation by hormones, Science 192:1340 (1976).
8. J.D. Malone, S.L. Teitelbaum, G.L. Griffin, R.M. Senior, and A.J. Kahn, Recruitment of osteoclast precursors by purified bone matrix constituents, J. Cell Biol. 92:227 (1982).
9. G.R. Mundy and J.W. Poser, Chemotactic activity of the γ-carboxyglutamic acid containing protein in bone, Calc. Tiss. Int. 35:164 (1983).
10. S. Sakamoto and M. Sakamoto, Isolation and characterization of collagenase synthesized by mouse bone cells in culture, Biomed. Res. 5:39-46 (1984).
11. M. Sakamoto and S. Sakamoto, Immunocytochemical localization of collagenase in isolated mouse bone cells, Biomed. Res. 5:29-38 (1984).
12. R.L. Jilka and J.W. Hamilton, Evidence for two pathways for stimulation of collagenolysis in bone, Calc. Tiss. Int. 37:300 (1985).
13. L.G. Raisz, J.Y. Vanderhoek, H.A. Simmons, B.E. Kream, and K.C. Nicolaou, Prostaglandin synthesis by fetal rat bone in vitro: evidence for a role of prostacyclin, Prostaglandins 17: 905 (1979).
14. Aubin, J.E., Alders, E., and Heersche, J.N.M. A primary role for microfilaments but not microtubules in hormone-induced cytoplasmic retraction. Exp. Cell Res. 143: 439 (1983).
15. S. Sakamoto, M. Sakamoto, P. Goldhaber, and M.J. Glimcher, Collagenase activity and morphological and chemical bone resorption induced by prostaglandin E_2 in tissue culture, Proc. Soc. Exp. Biol. Med. 161:99 (1979).
16. D.C. Klein and L.G. Raisz, Prostaglandins: stimulation of bone resorption in tissue culture, Endocrinology 86:1436 (1970).
17. A.H. Tashjian, Jr., E.L. Hohmann, H.N. Antoniades, and L. Levine, Platelet-derived growth factor stimulates bone resorption via a prostaglandin-mediated mechanism, Endocrinology 111:118 (1982).
18. A.H. Tashjian, Jr. and L. Levine, Epidermal growth factor stimulates prostaglandin production and bone resorption in cultured mouse calvaria, Biochem. Biophys. Res. Commun. 85:966.
19. K.J. Ibbotson, J. Harrod, M. Gowen, S. D'Souza, M. Winkler, G. Carpenter, R. Derynck, and G.R. Mundy, Effects of human transforming growth factor (TGF)α on bone resorption and formation in vitro, abstract 64 in "Program and Abstracts, Seventh Annual Scientific Meeting of the American Society for Bone and Mineral Research," Kelseyville, California, Society for Bone and Mineral Research.
20. T. Yoneda and G.R. Mundy, Monocytes regulate osteoclast-activating factor production by releasing prostaglandins, J. Exp. Med. 150:338 (1979).
21. M. Gowen, D.D. Wood, E.J. Ihrie, M.K.B. McGuire, and R.G.G. Russell, An interleukin 1 like factor stimulates bone resorption in vitro, Nature 306:378 (1983).

22. J.K. Heath, M.C. Meikle, S.J. Atkinson, and J.J. Reynolds, A factor synthesized by rabbit periosteal fibroblasts stimulates bone resorption and collagenase production by connective tissue cells in vitro, Biochim. Biophys. Acta 800:301 (1984).
23. G.R. Mundy, Monocyte-macrophage system and bone resorption, Lab. Invest. 49:119 (1983).
24. K.J. Ibbotson, G.D. Roodman, L.M. McManus, and G.R. Mundy, Identification and characterization of osteoclast-like cells and their progenitors in cultures of feline marrow mononuclear cells, J. Cell Biol. 99:471 (1984).
25. M.J. Pharoah and J.N.M. Heersche, 1,25-Dihydroxyvitamin D_3 causes an increase in the number of osteoclastlike cells in cat bone marrow cultures, Calc. Tiss. Int. 37:276 (1985).
26. A.J. D'Ercole, G.T. Applewhite, and L.E. Underwood, Evidence that somatomedin is synthesized by multiple tissue in the fetus. Dev. Biol. 75:315 (1980).
27. J.R. Farley and D.J. Baylink, Purification of a skeletal growth factor from human bone, Biochemistry 21:3502 (1982).
28. J.R. Farley, T. Masuda, J.E. Wergedal, and D.J. Baylink, Human skeletal growth factor: characterization of the mitogenic effect on bone cells in vitro, Biochemistry 21:3508 (1982).
29. E. Canalis, W.A. Peck, and L.G. Raisz, Stimulation of DNA collagen synthesis by autologous growth factor in cultured fetal rat calvaria, Science 210:1021 (1980).
30. R.T. Turner, J.E. Puzas, M.D. Forte, G.E. Lester, T.K. Gray, G.A. Howard, and D.J. Baylink, In vitro synthesis of 1α,25-dihydroxycholecalciferol and 24,25-dihydroxycholecalciferol by isolated calvarial cells, Proc. Natl. Acad. Sci. USA 77:5720 (1980).
31. L. Rifas, V. Shen, K. Mitchell, and W.A. Peck, Macrophage-derived growth factor for osteoblast-like cells and chondrocytes, Proc. Natl. Acad. Sci. USA 81:4558 (1984).
32. V. Shen, L. Rifas, G. Kohler, and W.A. Peck, Fetal rat chondrocytes sequentially elaborate separate growth- and differentiation-promoting peptides during their development in Vitro, Endocrinology 116:920, 1985.
33. M. Owen, Lineage of osteogenic cells and their relationship to the stromal system, pp. 1-26, in "Bone and Mineral Research/3," W.A. Peck, ed., Elsevier, Amsterdam (1985).

IMMUNE CELLS AND BONE RESORPTION

M. Gowen, B.R. MacDonald, D.E. Hughes, H. Skjodt and R.G.G. Russell

Department of Human Metabolism & Clinical Biochemistry
University of Sheffield Medical School
Beech Hill Road
Sheffield
S10 2RX, UK

INTRODUCTION

The concept of interactions between cells of the immune system and bone in the control of bone turnover is far from new. The relationship between haematopoietic precursor cells and the osteoclast has long been recognised and studies on animals with congenital bone diseases such as osteopetrosis have identified defective components of the immune system as causing or contributing to the disease. For example, the op/op (osteopetrotic) rat exhibits thymic atrophy and the disease can be cured by either administration of normal bone marrow cells or by thymus transplant[1]. Several mutant rodents have been described in which the osteopetrosis can be cured by injection of bone marrow, spleen or thymocytes[1]. These studies suggest that a functioning immune system is necessary for the normal activity of the bone resorbing cells.

In pathological conditions which feature chronic inflammation, such as rheumatoid arthritis and periodontal disease, increased bone resorption and indeed, bone loss occurs. The possibility that the infiltrating immune cells release products which act locally to stimulate bone resorption was proposed, although early work concentrated on mediators of inflammation rather than specific catabolic factors for connective tissues. However, the work of Horton et al[2] changed this emphasis by demonstrating the production of bone resorbing proteins by peripheral blood cells *in vitro*.

We review below some of the *in vitro* observations made more recently which illustrate the potential importance of the immune system to normal bone cell function.

THE ORIGIN OF THE OSTEOCLAST

The earliest link between cells of the immune system and the control of bone resorption was the suggestion that the osteoclast precursor, rather than being of mesenchymal origin, was a haematopoietic cell[3]. However, it was not until much later that studies, including parabiosis experiments between osteopetrotic and normal mice[4], and the use of chick-quail chimerae[5], demonstrated that the osteoclast precursor is a circulating cell. Further investigations led to the belief that the osteoclast precursor, present in the bone marrow, is of the mononuclear phagocyte (MNP) lineage. Use of intrinsic markers (the giant lysosomes found in the monocytes, granulocytes and osteoclasts of beige mutant mice) showed that mi/mi (osteopetrotic) recipients of beige bone marrow cells[6] or spleen cells[7] were cured and their osteoclasts contained the beige giant lysosomal granules. However, in neither study were the injected cells fractionated in order to identify precisely the cell type(s) which formed osteoclasts. The more recent work of Burger et al[8], employing coculture of non-adherent bone marrow cells and bone explants from mouse embryos before development of osteoclasts, strongly suggests that osteoclasts arise from haematopoietic stem cells of the MNP series and, furthermore, that differentiated adherent monocyte/macrophages cannot develop further into osteoclasts. Factors derived from immune cells which encourage the proliferation and differentiation of such precursors are discussed below.

Despite the large body of evidence (not reviewed here) supporting the MNP lineage as the osteoclast origin, recent studies have raised objections to this hypothesis. Horton et al[9] have failed to demonstrate the presence of a large number of granulocyte-monocyte antigens on osteoclasts present in osteoclastomas, although three monocyte determinants were present. Furthermore, generation of monoclonal antibodies to osteoclastoma osteoclasts did not cross-react with macrophages[10]. However, Oursler et al[11] have also isolated monoclonal antibodies to chick osteoclasts, a number of which do recognise cells of the MNP system. Baron et al[12] have demonstrated the presence of the same antigen on the ruffled border plasma membrane of the osteoclast and the lysosomal membrane of the monocyte. Although such conflicting results must raise doubts concerning the origin of the osteoclast they are not in themselves conclusive. Cells at different stages of differentiation, or mature cells from the same precursor having differentiated along different pathways (such as the macrophage and the osteoclast) would not necessarily be expected to express common antigens. The exact identity of the osteoclast precursor is thus still in doubt.

Several methods of studying osteoclast function _in vitro_ have recently been developed[13,14,15,16] but they are of little use in the study of the origin of the osteoclast, since they employ the harvesting and purification of preformed osteoclasts from bone. Studies of osteoclasts would be greatly facilitated if they could be generated and cultured _in vitro_. Testa et al first reported the appearance of multinucleated cells in feline long-term bone marrow cultures[17]. This approach has been extended to the use of cat[18], baboon[19] and human[20] bone marrow cultures. Cells forming after two to three weeks in these cultures resemble osteoclasts morphologically and biochemically, although their functional activity has yet to be proven. The formation of these cells can be modulated by the osteotropic hormones 1,25-dihydroxyvitamin D_3, parathyroid hormone and calcitonin, and the system provides a valuable model for the investigation of the mode of action of various bone-active agents, and the study of metabolic bone disease[20].

REGULATION OF BONE RESORPTION BY PRODUCTS OF IMMUNE CELLS

1. Immune cell-derived bone resorbing factors

Horton et al[2] initiated the search for the identity of bone resorbing factors elaborated by immune cells and demonstrated bone resorbing activity in the culture medium of stimulated peripheral blood leukocytes. This was followed by numerous attempts at purification and characterisation of the molecule(s) involved[21-25]. It was accepted until quite recently that a single molecule, called "osteoclast activating factor" (OAF), was responsible for the bone resorbing activity in these supernatants. This was probably the most important reason for the confusion and controversy generated by contradictory reports about the properties of "OAF", for example the molecular weight[21,22], indomethacin sensitivity[23,24], and ability to stimulate cAMP accumulation in bone cells[22,25]. Although the lymphocyte was generally regarded to be the source of "OAF", the particular lymphocyte subset has also been disputed[26,27]. However, the work of Horowitz et al[27], using specific antisera to different mouse lymphocyte populations, has shown convincingly that T lymphocytes, in the presence of monocytes and an antigenic stimulus, produce bone resorbing activity, whereas B cells, or T cells alone, do not. Purification and identification of this factor has not yet been accomplished, so its relationship to the cytokines described below is not determined.

It is now clear that monocytes themselves also produce a bone resorbing factor(s). The first demonstrations of this second type of "OAF"[28,29] were followed by a report of its close similarity with the immunomodulator, interleukin 1 (IL-1)[30]. This protein has certain physicochemical properties, such as molecular weight and heat sensitivity, in common with those ascribed to "OAF"[21,28,30].

Recent advances in lymphokine research have made it possible to test pure recombinant lymphocyte and monocyte products for bone resorbing activity. Recombinant murine IL-1-α[31] is a potent stimulator of bone resorption, its activity detectable at 10^{-12}M and half-maximal at 10^{-11}M[32]. Inhibitors of prostaglandin synthesis had no effect on IL-1-stimulated bone resorption in this system. In contrast calcitonin, a specific osteoclast inhibitor, markedly depressed the enhanced bone resorption caused by IL-1.

Exposure of bone (mouse calvaria) to IL-1 for as little as one hour resulted in a marked stimulation of bone resorption detectable 48 hours later. This suggests triggering of early events, such as DNA or protein synthesis, by IL-1 which leads to formation and activation of bone resorbing cells. Indeed, histological examination of calvariae cultured for 24 hours with IL-1 revealed slightly increased osteoclast activity, whereas those cultured for 48 hours had large numbers of active osteoclasts and a greatly reduced amount of mineralised and non-mineralised bone matrix. By 72 hours bone matrix was almost completely absent and the resulting spaces were filled with proliferating fibroblast-like cells, either from the marrow or the bone lining cells or the periosteum. This finding may be related to the previously described bone cell proliferative activity of IL-1[33].

It has been suggested that IL-1 acts in bone via the osteoblast, since no direct activation of mature osteoclasts by IL-1 could be demonstrated[34]. While this remains a possibility, it is equally likely that its site of action is the osteoclast progenitor cell, or an intermediate immune cell regulator within the bone marrow. The synthesis of lymphokines, such as the bone resorbing factor described by Horowitz et al[27], is usually dependent on the presence of monocytes, and this is due, at least in part, to their production of IL-1. An obvious mediator of IL-1 action in bone is interleukin 2 (IL-2), since immune regulation by IL-1 appears to be via stimulation of IL-2 production by T helper cells. However, IL-2 had no effect on bone resorption when added alone or in combination with IL-1[32].

It seems very likely that treatment of bone with IL-1 stimulates the proliferation and/or fusion of osteoclast precursors, whether directly or indirectly. This is supported by our histological findings and is, after all, the most obvious way of producing a sustained resorptive response. Activation of pre-existing mature osteoclasts is unlikely to result in a marked stimulation of bone resorption. Evidence that IL-1 does indeed stimulate the formation of osteoclasts has been obtained using the long-term marrow culture described above. Treatment of marrow cells with concentrations of IL-1 which stimulate bone resorption in organ culture led to the formation of enhanced numbers of multinucleated osteoclast-like cells (MacDonald BR, Takahashi N and Gowen M, unpublished results).

A recent report describing the purification of "OAF", in which the predominant bone resorbing factor in stimulated leukocyte culture medium was obtained in a homogeneous form and subjected to N-terminal amino acid sequencing, has revealed its identity with IL-1[35]. However, although this appears to be the most potent and/or abundant bone resorbing protein in such culture supernatants, it is certainly not the only one. Other blood cell-derived factors have been demonstrated to stimulate bone resorption *in vitro*, including platelet derived growth factor[36] and transforming growth factor beta[37]. In addition, two more recombinant immune cell products, tumour necrosis factor α (from monocytes) and tumour necrosis factor β (from lymphocytes; previously known as 'lymphotoxin') have been shown recently to have bone resorbing activity[38]. Thus there appears to be a family of bone resorbing factors produced by lymphocytes and monocytes in vitro (see Figure 1). Their production, possible interactions, and relevance to the control of bone turnover in vivo is as yet unclear.

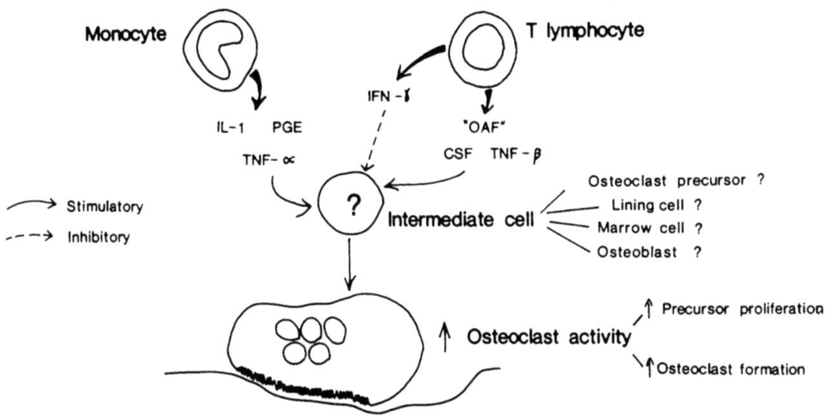

Fig. 1. Production of factors by monocytes and lymphocytes which stimulate and inhibit bone resorption.

The monokines and lymphokines shown in this figure have been demonstrated to have stimulatory (⟶) or inhibitory (- - -→) effects on bone resorption in vitro. The involvement and identity of an intermediate cell is at present speculative. IL-1 - interleukin 1; TNF-α-tumour necrosis factorα; PGE - prostaglandin E; IFN-γ-interferon gamma; TNFβ-tumour necrosis factor β; CSF-colony stimulating factor; "OAF" -"osteoclast activating factor".

2. Colony stimulating factors

The suggestion that the osteoclast may be derived from the monocyte-macrophage series raises the possibility that the generation of osteoclast precursors may be regulated by haematopoietic colony stimulating factors (CSFs). The existence of soluble factors capable of regulating the formation of blood cells from their precursor stem cells has been recognised for many years[39,40]. They are glycoproteins[41,42] which cause the formation of colonies in bone marrow cells cultured in semisolid medium. Each colony results from the division of a single committed stem cell of that particular lineage. Committed stem cells cultured in the absence of CSFs do not divide.

Experiments utilising the long term bone marrow culture model of osteoclast formation have demonstrated that recombinant human CSF-GM and highly purified CSF-1, both of which stimulate monocyte/macrophage precursors, increase the number of osteoclast-like cells found in these cultures[43]. This effect was seen two weeks after the removal of CSF from the cultures and was potentiated by subsequent addition of 1,25-dihyroxyvitamin D_3. Autoradiographic studies indicated that CSF stimulated formation of osteoclast-like cells by increased proliferation of the precursor cell, and that 1,25-dihydroxyvitamin D_3 caused fusion of the increased numbers of precursors. These studies suggest that colony stimulating factors and their interaction with circulating osteotropic hormones may be important in the control of osteoclast formation and bone resorption (see Figure 2).

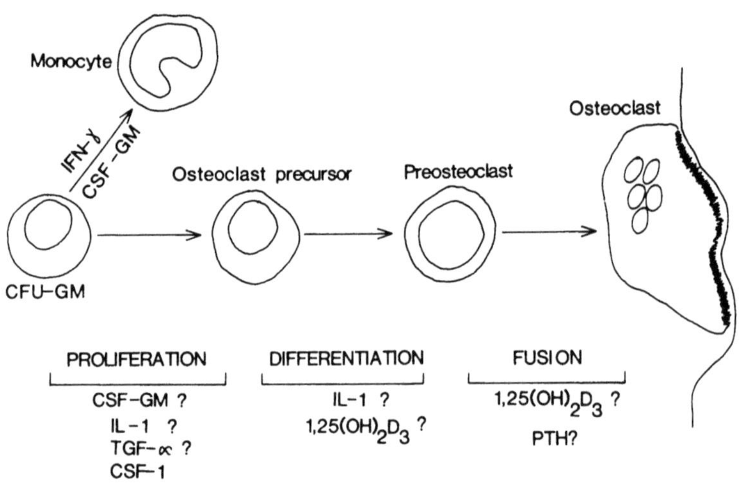

Fig. 2. Osteoclast formation pathway and possible points of action of bone-resorbing factors.

CSF-GM - granulocyte/macrophage colony stimulating factor;
$1,25(OH)_2D_3$ - 1,25-dihydroxyvitamin D_3
PTH - parathyroid hormone
CFU-GM - granulocyte/macrophage colony forming unit

Evidence that CSFs may play a role in the regulation of bone resorption in vivo was reported by Wiktor-Jedrzejczak et al[44] who showed that a particular osteopetrotic mouse mutant (op/op) had normal osteoclast precursors, but that these were not able to differentiate, presumably because of a defect in the bone marrow microenvironment. When CSF was administered to the animals these cells resumed their normal differentiation and functioning osteoclasts appeared. In addition, Burger et al[8] have demonstrated that CSF-stimulated non-adherent bone marrow cells will form osteoclasts in vitro.

3. Interferon gamma

Although the existence of immune cell-derived stimulators of bone resorption is now well-established, the identification of inhibitors has occurred only recently. We have demonstrated that pure recombinant interferon gamma[45] (IFN), a lymphocyte product, inhibits IL-1-stimulated bone resorption at concentrations of 100-300 U/ml[46]. A very interesting although unexpected observation was that this inhibition seems to be specific to bone resorption induced by local cytokines since the actions of IL-1, TNF α and TNF β were markedly inhibited whereas bone resorption stimulated by PTH and 1,25-dihydroxyvitamin D_3 was unaffected using the same interferon concentrations and length of treatment. PTH-induced bone resorption could be partially inhibited, but this required higher concentrations of IFN and a doubling of the incubation period. This effect on PTH is similar to that of interferon α reported by Jilka et al[47]. Further investigation into the nature of inhibition of IL-1-stimulated bone resorption by IFN has revealed that, in order for it to exercise its inhibitory effect, it must be present during the early phase of the culture. If IFN is added after eight hours of treatment with IL-1 no inhibitory effect is observed.

These data suggest that IL-1 induces some early event(s), such as the expression of specific genes in the osteoclast precursor, which is inhibitable by IFN. One recently described activity of IFN is its anti-proliferative effect on mesenchymal cells. Another reported effect is the stimulation of maturation of monocyte-macrophages[48]. If osteoclasts and monocytes do have common precursors IFN may inhibit osteoclast formation by directing precursor cells along another differentiation pathway, thus reducing the number of available osteoclast precursors and so inhibiting bone resorption (see Figure 2).

The identification of an immune cell-derived inhibitor of bone resorption is of great importance if a physiological role is postulated for the locally produced factors described above. Clearly the control of bone turnover is dependent upon the ability to both initiate and terminate a phase of bone resorption. The intracellular mechanisms of action of these cytokines, and the control of their production and interactions are still areas of great interest deserving future study.

COMMUNICATION BETWEEN DIFFERENT CELL TYPES AND SYSTEMIC HORMONES

Systemic hormones such as calcitonin, 1,25-dihydroxyvitamin D_3 and parathyroid hormone (PTH) have major effects on bone resorption in vivo.

Although these effects may in part be due to actions on tissues other than bone, such as kidney and intestine, their direct effects in organ culture studies suggest that they also act on bone cells. 1,25-dihydroxyvitamin D_3 and PTH have been shown to affect the function of osteoblasts in a number of studies, yet their ability to directly stimulate osteoclast activity is controversial. This is largely as a result of the lack of good evidence demonstrating the presence of receptors for these hormones on osteoclasts. However, as has been discussed previously, an important mechanism of stimulation of bone resorption is the stimulation of osteoclast proliferation and differentiation. Hormone receptors could therefore be present on precursor cells, which are difficult to identify, and be lost on their maturation to the large multinucleated osteoclast. This hypothesis is supported by a number of reports of interactions between osteotropic hormones and monocytes, the putative osteoclast precursors. For example, Minkin et al have shown effects of PTH and calcitonin on adenylate cyclase activity of immature macrophages[49], and many studies have demonstrated the ability of 1,25-dihydroxyvitamin D_3 to cause differentiation of precursor mononuclear cells into macrophages[50,51] and multinucleated osteoclast-like cells[18,19] (see Figure 2).

However, the most popular current hypothesis explaining hormone action on bone resorption requires the involvement of an intermediate cell type, namely the osteoblast[52]. An alternative view of this theory is that immune cells in the bone marrow respond to the hormones and provide the secondary signals to the osteoclast. Again there is a good deal of evidence <u>in vitro</u> in support of this suggestion.

Lymphocytes have been shown to possess receptors for PTH[53,54], calcitonin[55] and 1,25-dihydroxyvitamin D_3[56], and these hormones have effects on lymphocyte proliferation and lymphokine production. Thus a systemic hormone such as 1,25-dihydroxyvitamin D_3 could exert its action on the osteoclast by stimulating the production of a bone resorbing factor by a marrow lymphocyte or monocyte. 1,25-dihydroxyvitamin D_3 has been shown to stimulate the production of IL-1 by a human monocyte cell line, U937, in conjunction with a lymphocyte factor, as yet unidentified[51]. Unpublished data by Vignery et al[57] indicate that 1,25-dihydroxyvitamin D_3 can stimulate the production of a bone resorbing factor from activated spleen cells.

Another means of communication between cell types not involving the release of soluble factors is direct cell-cell contact. In a number of immune reactions this is achieved via HLA antigen expression on the cell surface. MHC class II or HLA-D antigens are found constitutively on B lymphocytes, some monocytes and certain T lymphocyte subsets and are generally associated with helper and suppressor T cell recognition of antigen presented by macrophages and certain other cell types. Thus stimulation of T helper cell proliferation requires presentation of antigen in a MHC class II-restricted manner.

A number of cell types has recently been shown to be capable of HLA-D antigen expression, namely endothelial cells[58] and synovial cells[59]. In studies using human osteoblast-like trabecular bone cells we have demonstrated the constitutive expression of HLA-D antigens on 20-25% of the cell population. Recombinant interferon gamma increases this expression and also induces expression in human articular chondrocytes (Skjodt H, unpublished data). It is not yet known whether the HLA-DR-positive cells represent a specific subpopulation of cells, or if all cells are capable of HLA-DR antigen expression given a sufficient stimulus. It is possible that cells within the cultures are capable of the production of interferon-like factors which can act in an autocrine or paracrine manner. The relevance of these findings to bone function is as yet unknown. However, it is interesting to speculate that they may reflect a potential means of communication between bone cells and immune cells, that osteoblast-like cells may be involved in the elaboration of lymphokines in bone, and that the immune system may be involved in the control of bone formation as well as bone resorption.

In summary, an extremely complex set of interactions may exist between systemic osteotropic hormones, immune cells and bone cells, involving the production of a numerous stimulatory and inhibitory factors. Interactions between different cascades of events could also involve feedback in a positive or negative fashion on previous steps. A role of the immune system in the control of bone resorption is unproven, but the body of evidence supporting such a role is mounting. However, the complexities of the immune system and the difficulties encountered in studying bone turnover in vitro suggest that an understanding of the relevance of any or all of these observations may lie some way in the future.

REFERENCES

1. G. Milhaud and M.L. Labat, Thymus and osteopetrosis, Clin. Orthop. Rel. Res. 135:260-271 (1978).
2. J.E. Horton, L.G. Raisz and H.A. Simmons, Bone resorbing activity in supernatant fluid from cultures of human peripheral blood leukocytes. Science 177:793-795 (1972).
3. D.A. Fishman and E.D. Hay, Origin of osteoclasts from mononuclear leukocytes in regenerating newt limbs, Anat. Record. 143:329-334 (1962).
4. D.G. Walker, Congenital osteopetrosis in mice cured by parabiotic union with normal siblings, Endocrinol. 91:916-920 (1972).
5. A.J. Khan and D.J. Simmons, Investigation of cell lineage in bone using chimera of chick and quail embyronic tissue, Nature 258:323-327 (1975).
6. P. Ash, J.F. Loutit and K.M.S. Townsend, Osteoclasts derived from haemopoietic stem cells, Nature 283:669-670 (1980).
7. S.C. Marks and D.G. Walker, The haematogenous origin of osteoclasts: experimental evidence from osteopetrotic (microphthalmic) mice treated with spleen cells from beige mouse donors, Am. J. Anat. 161:1-10 (1981).

8. E. Burger, J.W.M. Van der Meer, J.S. Van Der Geuil, J.C. Gribnau, C.W. Thesingh and R. Van Furth, In vitro formation of osteoclasts from long term culture of bone marrow mononuclear phagocytes, J. Exp. Med. 156:1604-1614 (1982).
9. M.A. Horton, E.F. Rimmer, D. Lewis, J.A.S. Pringle, K. Fuller and T.J. Chambers, Cell surface characterisation of the human osteoclast: phenotypic relationship to other bone marrow-derived cell types, J. Pathol. 144:282-294 (1984).
10. M.A. Horton, D. Louis, K. McNulty, J.A.S. Pringle and T.J. Chambers, Characterisation of monoclonal antibodies specific for human osteoclasts, Calcif. Tissue Int.(in press).
11. M.J. Oursler, L.V. Bell, B. Clevinger and P. Osdoby, Identification of osteoclast-specific monoclonal antibodies, J. Cell Biol. 100:1592-1600 (1985).
12. R. Baron, L. Neff, D. Ouvard and P. Courtoy, Acidification and bone resorption : immunocytochemical localisation of a lysosomal membrane protein at the ruffled border of osteoclasts, Calcif. Tissue Int. 36(4) (1984).
13. A.Z. Zambonin-Zallone, A. Teti and M.V. Primavera, Isolated osteoclast in primary culture : first observations on structure and survival in tissue culture media, Anat. Embryol. 165:405-413 (1982).
14. P. Osdoby, M.C. Martini and A.I. Caplan, Isolated osteoclasts and their presumed progenitor cells, the monocyte, in culture, J. Exp. Zool., 224:331-344 (1982).
15. T.J. Helfey and P.H. Stern, Isolation of osteoclasts from fetal rat long bones, Calcif. Tissue Int. 34:480-487 (1982).
16. T.J. Chambers and A. Moore, The sensitivity of isolated osteoclasts to morphological transformation by calcitonin, J. Clin. Enocrinol. Metab. 57:819-824 (1983).
17. N.G. Testa, T.D. Allen, L.G. Lathja, D. Onions and O. Jarrett, Generation of osteoclasts in vitro. J. Cell Science 47:127-137 (1981).
18. K.J. Ibbotson, G.D. Roodman, L.M. McManus and G.R. Mundy, Identification and characterisation of osteoclast-like cells and their progenitors in cultures of feline marrow mononuclear cells, J. Cell Biol.99:471-480 (1984).
19. G.D. Roodman, K.J. Ibbotson, B.R. MacDonald, T.J. Kuehl and G.R. Mundy, $1,25(OH)_2$ vitamin D_3 causes formation of multinucleate cells with several osteoclast characteristics in cultures of primate marrow, Proc. Natl. Acad. Sci USA (in press).
20. B.R. MacDonald, N. Takahashi, L. McManus, G.R. Mundy and G.D. Roodman, Formation of multinucleated cells with osteoclastic characteristics in human long-term marrow cultures is controlled by calcium regulating hormones, Manuscript submitted.
21. G.R. Mundy and L.G. Raisz, Big and little forms of osteoclast activating factor, J. Clin. Invest. 60:122-128 (1977).
22. R.A. Luben, M C-Y Chen, D.M. Rosen and M.A. Mohler, Effects of osteoclast activating factor from human lymphocytes on cyclic AMP concentrations in isolated mouse bone and bone cells, Calcif. Tissue Int.28:32-32 (1979).

23. R.S. Bockman and M. Repo, Lymphokine-mediated bone resorption requires endogenous prostaglandin synthesis, J. Exp. Med. 154:529-534 (1981).
24. T. Yoneda and G.R. Mundy, Prostaglandins are necessary for OAF production by activated peripheral blood leukocytes, J. Exp. Med. 149:279-283 (1979).
25. R.G. Josse, T.M. Murray, G.R. Mundy, D. Jez and J.N.M. Heersche, Observations on the mechanism of bone resorption induced by multiple myeloma culture fluids and partially-purified OAF, J. Clin. Invest. 67:1472-1481 (1981).
26. P. Chen, C. Trummel, J. Horton, T.J. Barker and T.J. Oppenheim, Production of osteoclast-activating factor by human peripheral blood rosetting and non-rosetting lymphocytes, Eur. J. Immunol. 6:723-736 (1976).
27. M. Horowitz, A. Vignery, R.K. Gershon and R. Baron, Thymus-derived lymphocytes and their interactions with macrophages are required for the production of osteoclast-activating factor in the mouse, Proc. Natl. Acad. Sci. USA 81:2181-2185 (1984).
28. M. Gowen, M.C. Meikle and J.J. Reynolds, Stimulation of bone resorption in vitro by a non-prostanoid factor released by human monocytes in culture, Biochim. Biophys. Acta.762:471-474 (1983).
29. B. Rutherford and C.L. Trummel, Monocyte-mediated bone resorption involves release of non-dialysable substances in addition to prostaglandin. J. Reticuloendothelial Soc. 33:175-184 (1983).
30. M. Gowen, D.D. Wood, E.J. Ihrie, M.K.B. McGuire and R.G.G. Russell, An interleukin-1-like factor stimulates bone resorption in vitro, Nature 306:378-380 (1983).
31. P.T. Lomedico, U. Gubler, C.P. Hellmann, M. Pukovich, J.G. Giri, Y-C.E. Pan, K. Collier, R. Semionow, A.O. Chiva and S.B. Mizel, Cloning and expression of murine interleukin 1 cDNA in Escherichia coli, Nature 312:458-462 (1984).
32. M. Gowen and G.R. Mundy, Actions of recombinant interleukin 1, interleukin 2 and interferon gamma on bone resorption in vitro, Manuscript submitted.
33. M. Gowen, D.D. Wood and R.G.G. Russell, Stimulation of the proliferation of human bone cells in vitro by human monocyte products with interleukin 1 activity, J. Clin. Invest. 75:1223-1228 (1985).
34. B.M. Thomson and T.J. Chambers, Osteoblastic cells are induced by interleukin 1 (IL-1) to stimulate osteoclastic bone resorption, Calcif. Tissue Int. (in press).
35. P.P. Stashenko and F.E. Dewhirst, Purification of osteoclast-activating factor, Calcif. Tissue Int.(in press).
36. A.H. Tashjian, E.L. Hohman, H.N. Antoniades and L. Levine, Platelet-derived growth factor stimulates bone resorption via a prostaglandin-mediated mechanism, Endocrinol. 111:118-123 (1982).
37. A.H. Tashjian, E.F. Voelkel, M. Lazzaro, F.S. Singer, A.B. Roberts, R. Derynck, M.E. Winkler and L. Levine, and human transforming growth factors stimulate prostaglandin production and bone resorption in cultured mouse calvaria, Proc. Natl. Acad. Sci. USA 82:4535-4538 (1985).

38. Bertolini, G.E., Nedwin, T.S., Bringman and G.R. Mundy, Recombinant human cytokines with tumour necrosis activity stimulate osteoclastic bone resorption in vitro, Manuscript submitted.
39. D.H. Pluznik and L. Sachs, The induction of clones of normal mast cells by a substance from conditioned medium, Exp. Cell. Res. 43:553-563 (1966).
40. T.R. Bradley and D. Metcalf, The growth of mouse bone marrow cells in vitro, Aust. J. Exp. Biol. Med. Sci. 44:287-300 (1966).
41. E.R. Stanley, Colony stimulating factors. In: The Lymphokines : biochemistry and biological activity (W.E. Stewart and J.W. Hodder eds). Humana Press, New Jersey, pp 102-132 (1981).
42. A.W. Burgess and D. Metcalf, The nature of granulocyte macrophage colony-stimulating factors, Blood 56:947-958 (1980).
43. B.R. MacDonald, G.R. Mundy, S. Clark, E.A. Wong, T.J. Kuehl, E.R. Stanley and G.D. Roodman, Human recombinant CSF-GM and highly purified CSF-1 stimulate the formation of multinucleated cells with osteoclast characteristics in long-term bone marrow cultures. Manuscript submitted.
44. W. Wiktor-Jedrzejzcak, A. Ahmed, C. Szczylik and R.R. Skelly, Hematological characterisation of congenital osteopetrosis in the op/op mouse, J. Exp. Med. 156:1516-1527 (1982).
45. P.W. Gray and D.V. Goeddel, Cloning and expression of murine immune interferon cDNA, Proc. Natl. Acad. Sci. USA 80:5842-5846 (1983).
46. M.Gowen, G.E. Nedwin and G.R. Mundy, Preferential inhibition of cytokine-stimulated bone resorption by recombinant interferon gamma, Manuscript submitted.
47. R.L. Jilka and J.W. Hamilton, Inhibition of parathormone-stimulated bone resorption by type I interferon, Biochem. Biophys. Res. Commun. 120:553-558 (1984).
48. M. Takei, K. Takeda and K. Konno, The role of interferon in induction of differentiation of human myeloid leukemia cell lines ML-1 and HL-60, Biochem. Biophys. Res. Commun. 124:100-105 (1984).
49. C. Minkin, L. Blackman, J. Newbrey, S. Potress, R. Posek and M. Walling, Effects of parathyroid hormone and calcitonin on adenylate cyclase in murine mononuclear phagocytes, Biochem. Biophys. Res. Commun. 76:875-881 (1977).
50. H. Tanaka, E. Abe, C. Miyaura, Y. Shira and T. Suda, $1,25(OH)_2D_3$ induces differentiation of human promyelocytic leukemia cells into monocyte macrophages but not into granulocytes, Biochem. Biophys. Res. Commun. 117:86-93 (1983).
51. E.P. Amento, A.K. Bhalla, J.T. Kurnick, R.L. Kradin, T.L. Clemens, S.A. Holick, M.F. Holick and S.M. Krane, 1,25-dihydroxyvitamin D_3 induces maturation of the human monocyte cell line U937, and in association with a factor from human T lymphocytes augments production of the monokine, mononuclear cell factor, J. Clin. Invest.73:731-739 (1984).
52. G.A. Rodan and T.J. Martin, Role of osteoblasts in hormonal control of bone resorption - a hypothesis, Calcif. Tiss. Int. 33:349-352 (1981).

53. H.M. Perry, J.C. Chappel, E. Bellorin-Font, K.J. Martin and S.L. Teitelbaum, Parathyroid hormone receptors on circulating human mononuclear leukocytes, J. Biol. Chem. 259:5531-5535 (1984).
54. I. Yamamoto, J.T. Potts and G.V. Segre, Circulating bovine lymphocytes contain receptors for parathyroid hormone, J. Clin. Invest. 71:404-408 (1983).
55. S.J. Marx, G.D. Aurbach, J.R. Gavin, D.W. Buell, Calcitonin receptors on cultured human lymphocytes, J. Biol. Chem. 149:6812-6816 (1974),
56. W.F.C. Rigby, T. Stacy and M.W. Faviger, Inhibition of T lymphocyte mitogenesis by 1,25 dihydroxyvitamin D_3, J. Clin. Invest.74: 1451-1455 (1984).
57. R. Baron, A. Vignery and M. Horowitz, Lymphocytes, macrophages and the regulation of bone remodeling. In: Bone and Mineral Research, Annual 2. (Peck WA ed) Elsevier Science Publishers pp 175-246 (1983).
58. G. Groenewegen, W.A. Buurman and C.J. Van der Linden, Lymphokine dependence of in vivo expression of MHC class II antigens by endothelium, Nature 326:361-363 (1985).
59. E.P. Amento, A.K. Bhan, K.G. McCullagh and S.M. Krane, Influences of gamma interferon on synovial fibroblast-like cells, J. Clin. Invest, 76: 837-848 (1985).

BISPHOSPHONATES CAN REDUCE OSTEOCLASTIC BONE RESORPTION BY TWO DIFFERENT MECHANISMS

C.W.G.M. Löwik, P.M. Boonekamp, G. van de Pluym, L. van de Wee-Pals, H. Bloys van Treslong-de Groot and O.L.M. Bijvoet
Clinical Investigation Unit of the department of Clinical Endocrinology, University Hospital Leiden, the Netherlands

INTRODUCTION

Bisphosphonates (P-C-P's) inhibit osteoclastic bone resorption (1). Their inhibitory potency has been clinically exploited with considerable benefit. The P-C-P's have simplified the management of Paget's disease (2,3) and of hypercalcemia of malignancy (4,5), both originating from excessive bone resorption. Three P-C-P's have been investigated to a large extent, EHDP ((1-hydroxyethylidene)-1,1-bisphosphonate), Cl_2MDP ((dichloromethylidene)-bisphosphonate) and APD ((3-amino-1-hydroxypropylidene)-1,1-bisphosphonate), yet their mechanism of action has remained uncertain. We have always found it disturbing that their relative potencies in-vitro and in-vivo differ (6,7)- In in-vitro cultures of explanted calvaria or radii Cl_2MDP and EHDP exhibited the greatest potency (6) while the amino-bisphosphonate, APD, was the more potent clinically (3,4,8).

Bone resorption is the ultimate result of a sequence of events that includes activation, chemotaxis, attachment and differentiation of osteoclast-precursors and osteoclasts to bone (9,10). If one assumes that P-C-P's modify the action of the resorbing osteoclast itself (11), it becomes difficult to account for the different relative potencies in-vivo and in-vitro. We now report that the aminobisphosphonate APD, inhibits the resorption sequence at an earlier phase than the non-aminobisphosphonates, Cl_2MDP and EHDP. This action was elucidated by using explants of embryonic long bones devoid of mature osteoclasts and of their precursors and in which initiation of osteoclastic resorption can be manipulated by the experimentator through exposure of the explant to a source of osteoclast-precursors.

The results reveal a dual mechanism of action of bisphosphonates each depending on previous P-C-P binding to bone. One mechanism concerns inhibition of the mature resorbing osteoclast itself and involves cytotoxicity. In this respect all P-C-P's have comparable potencies suggesting that the action may be secondary to ingestion by the osteoclasts of P-C-P that is bound to the mineralized matrix. The second mechanism prevents accession of osteoclast-precursors to bone and their subsequent functioning as mature osteoclasts. P-C-P's, dependent on their structure, differ much in potency and relative potencies in-vitro reflect relative potencies in-vivo. This suggests that the second action involves specific binding to sites that are actively involved in directing the accession of osteoclasts to bone-matrix or in the induction of their differentiation.

MATERIALS AND METHODS

Pregnant Swiss mice were injected with 30 μCi ^{45}Ca on day 16 and killed one day later. Explantation of embryonic radii and culture were carried out according to Raisz (7). On day 17 the periosteum of the metacarpals that contain precursors (13) are removed according to the method of Burger et al. (10). Seventeen-day-old fetal livers, which serves as a source of osteoclast-precursors (8,10), were removed and diced. Bone or liver can be precultured in the presence of additives before coculturing for 6 days in semi-solid medium (α-MEM) + 20% cock plasma + 10% chicken embryonic extract. In some cases the cocultured fetal liver and 'stripped' metacarpal are separated by interposition of millipore filters with different pore sizes. Media are changed every 2-3 days. Resorption is measured as ^{45}Ca released with correction for non-cell mediated release measured in killed bones. Histology was performed using standard methods.

RESULTS AND DISCUSSION

Difference between the two bone culture systems

Not all bones of 17-day-old mice embryo's are at the same stage of development. The calvarium or radius, when explanted are found to contain calcified matrix and amongst other cells, actively resorbing osteoclasts. When the periosteum, which is a source of osteoclast-precursors (13), is removed from the radius there is no effect on subsequent resorption (see table 1). Metatarsal or metacarpal bones are less developed. Their matrix, though already mineralized, does not yet contain mature osteoclasts. During culture, however, osteoclasts develop from precursors present in the periosteum (10, 13). The explants fail to develop osteoclasts when the periosteum is removed. Resorption of these 'stripped' bones can be initiated by adding an outside source of haemopoietic cells that may serve as a source of precursors and which during coculture develop into mature resorbing osteoclasts.

Table 1. Difference between the radius and coculture system.

culture	Net % ^{45}Ca release ± S.E.M. (n=5)
Control radius	56.9 ± 4.8
'stripped' radius	61.4 ± 8.5
'stripped' metacarpal	8.3 ± 1.7
'stripped' metacarpal + fetal liver	69.3 ± 5.1

Resorption was measured as release over the whole 6-day culture period. Note that removal of periosteum of the radius has no effect.

Chemotaxis, attachment and differentiation of osteoclast-precursors

We here demonstrate for the first time actual migration of osteoclast-precursors from the haemopoietic tissue to the mineralized matrix through interposed filters with different pore sizes. The design of the experiment is shown in figure 1. Figure 2 shows the dependence of resorption on pore size. When the haemopoietic tissue and bone explant are separated by a filter of 0.2 μm pore size no resorption occurs and no osteoclasts develop. In the histological section reproduced in figure 3 one can actually see the upward migration of precursors through an 8 micron pore size filter and their transformation into multinucleated tartrate-resistant-acid-phosphatase-positive cells, when they reach the calcified matrix. Tartrate-resistant-acid-phosphatase-positive cells are never found in the filter itself. Figure 3 therefore distinguishes between 2 different phenomena, one is the migration towards the mineralized matrix and the other is the differentiation into mature osteoclasts on contact with the matrix.

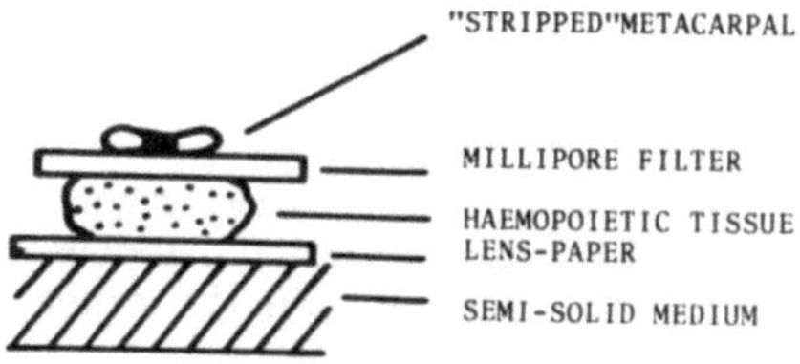

fig. 1: Schematic presentation of the design of the experiment.

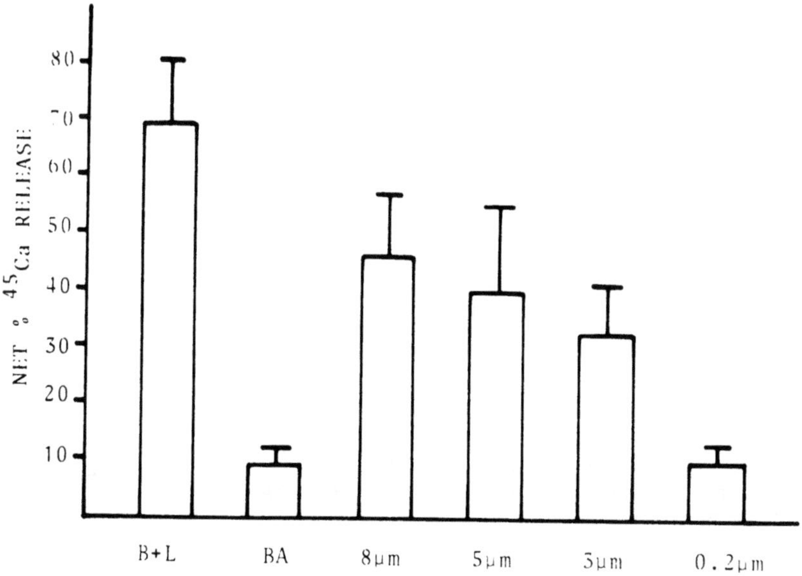

fig. 2: In this experiment the 'stripped' metacarpals and fetal livers are separated by millipore filters with different pore sizes. The effect of interposition of these filters on bone resorption is shown.

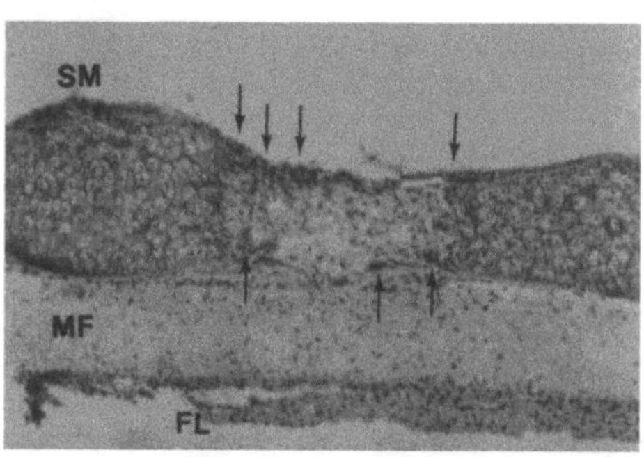

fig. 3: Histological section of a coculture of a 'stripped' metacarpal (SM) with a fetal liver (FL) separated by a millipore filter (MF) with a pore size of 8 µm. Arrows indicate multinucleated tartrate resistant acid phosphatase positive osteoclasts.

Effects of P-C-P's on bone resorption in radius explants

Additions of APD, EHDP or Cl_2MDP to the culture medium results in a dose-dependent inhibition of osteoclastic resorption of the mineralized matrix of cultured fetal radii (fig.4). APD and EHDP are equipotent whereas Cl_2MDP is about 10 times more potent than the other two P-C-P's.

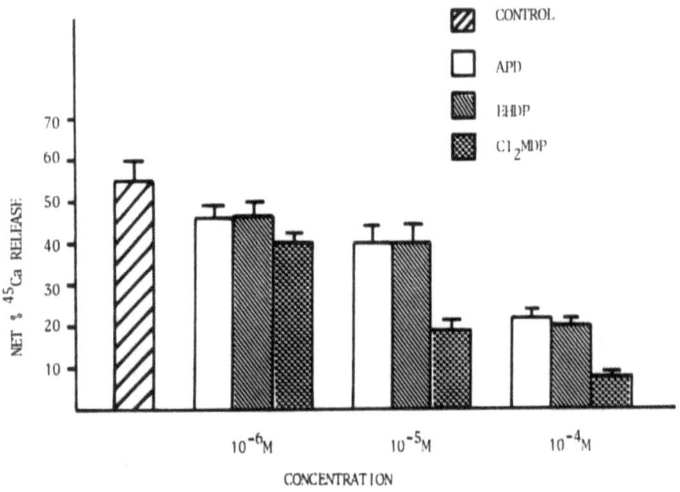

fig.4: Effect of various concentrations of P-C-P's on the resorption of radii. P-C-P's were continuously present. Data are expressed as % ^{45}Ca release ± S.E.M. (n=5).

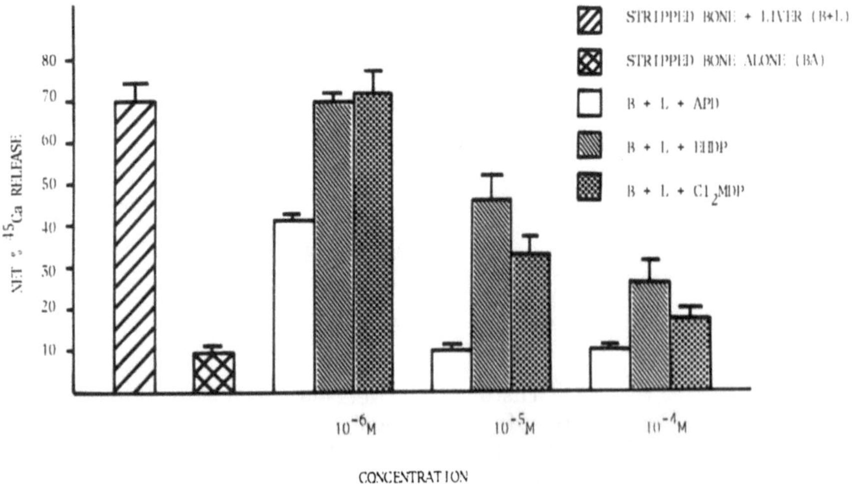

fig. 5: Effect of various concentrations of P-C-P's on the resorption in co-cultures of 'stripped' metacarpalia with fetal liver.

Effect of P-C-P's on bone resorption in cocultures of metacarpals with an extraneous source of precursors

Additions of APD, EHDP and Cl_2MDP to the medium of cocultures of metacarpals with embryonic liver also cause dose-dependent inhibition of resorption (fig.5), but in this system APD is 10 times more potent than Cl_2MDP or EHDP and is around 50-100 times more potent than it was in the radius explant.

Effect of delayed addition of APD and Cl_2MDP on the resorption measured in both bone culture systems

The addition of P-C-P's to metacarpal long bones can be delayed until after mature osteoclasts have developed. If the high potency of the amino-P-C-P is due to a specific action on osteoclast accession, the potency should be lost when addition is delayed until after the proces of accession is completed.

Table 2: The effect of delayed addition of APD and Cl_2MDP on the resorption of cocultured metacarpalia and liver, or in radii.

	% ^{45}Ca release $\pm$ S.E.M. (n=5) coculture-system	radius-system
control	76.4 $\pm$ 3.8	63.5 $\pm$ 6.4
Cl_2MDP (10^{-5}M)	38.6 $\pm$ 4.1	36.7 $\pm$ 4.9
APD (10^{-5}M)	46.5 $\pm$ 3.3	37.6 $\pm$ 3.1

Cocultures and radii were cultured for 3 days in semi-solid medium. The medium was then replaced by P-C-P containing medium and cultures were prolonged for another 5 days. Resorption was measured as release over the whole culture period.

Table 2 shows that the potency of Cl_2MDP remains unchanged in whatever system its effect is studied. Efficacy in the radius system establishes that inhibition of the mature functioning osteoclast is part of its action.
When APD was applied in a later phase the increased potency is indeed lost. We conclude that the increased potency of APD is due to a specific effect on osteoclast accession.

Separate treatment of metacarpal bone or of liver previous to coculture

In order to discriminate between a bone mediated or a direct effect of P-C-P's on the precursors, either the metacarpal or the liver were pretreated separately with P-C-P and then cocultured in the absence of P-C-P (fig. 6).

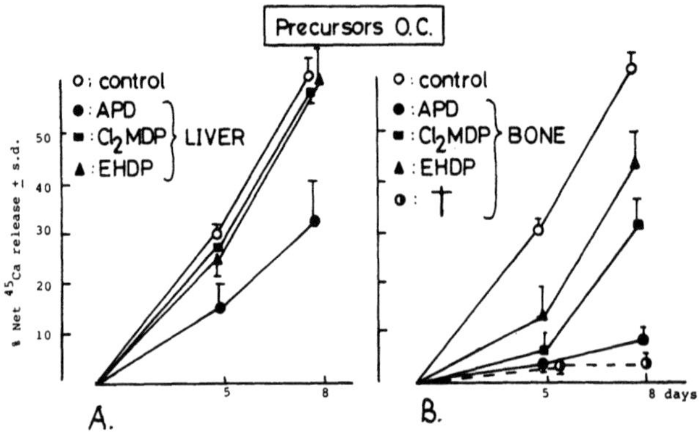

fig. 6: Effect of pretreatment of liver or metacarpals for 24 hours with 10^{-4}M P-C-P's as indicated. Data are expressed as % net. ^{45}Ca release ± S.E.M. (n=6).

As we can see from figure 6 pretreatment of osteoclast precursors at a dose of 10^{-4}M is without effect except for APD. Binding studies with ^{14}C-APD have shown that with 10^{-4}M APD there is a carry-over of 2% from the liver to the bone. Addition of a comparable amount to a coculture, that is 10^{-6}M APD, is sufficient to inhibit resorption to a same extent (see figure 5). When precursors are pretreated with 10^{-5}M APD there is no carry-over and subsequently no effect on bone resorption. (results not shown). On the contrary, pretreatment of metacarpal bone with P-C-P's inhibits resorption to the same extent as does continuous addition during coculture.

It is interesting to note that osteoclasts do appear in the peripheral cartilage areas of metacarpals pretreated with APD, that mineralize after explantation. Since these areas are not labeled with ^{45}Ca, such resorption is not measured. Apparently resorption of mineralized matrix only occurs when it has not been exposed to APD (results not shown). This allows us to conclude that the effect of P-C-P's involves binding to the calcified matrix.

Failure to form differentiated osteoclasts in metacarpals during P-C-P treatment

Migration of osteoclast precursors from the liver to the bone through a millipore filter with a pore size of 8 µm and subsequent differentiation into mature multinucleated tartrate-resistant-acid-phosphatase-positive osteoclasts, was also studied in the presence of an amino-bisphosphonate (fig.7).

Although normal celmigration was observed (compare with fig. 3), no differentiated multinucleated tartrate-resistant-acid-phosphatase-positive osteoclasts developed on the contact with the bone matrix. This suggests that attachment of osteoclast-precursors and final differentiation cannot be induced by mineralized matrix when this is occupied with low dose amino-bisphosphonate. The specificity of this effect suggests that it might replace from the matrix a natural substance that is essential for attachment and induction of differentiation into functional osteoclasts.

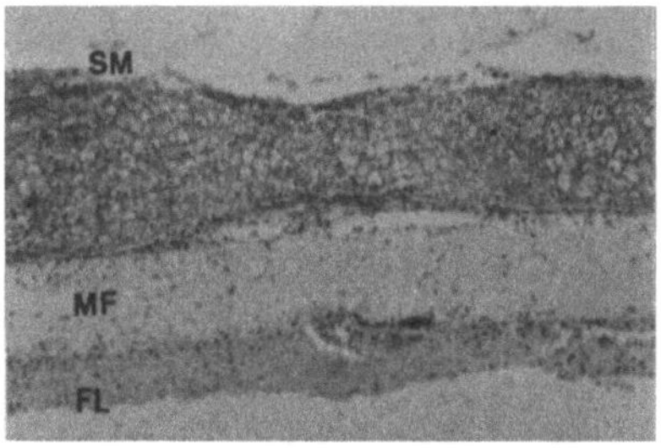

fig. 7: Histological section of a coculture of a 'stripped' metacarpalia (SM) with a fetal liver (FL) separated by a millipore filter (MF) with a pore size of 8 µm in the presence of an amino-bisphosphonate. Note that in contrast to fig. 3 no marrow cavity is formed and no multinucleated tartrate-resistant acid-phosphatase-positive osteoclasts are formed, although there is normal celmigration from the liver towards bone.

In conclusion P-C-P's exhibit two modes of action, both dependent on their affinity for mineralized matrix. One, an aspecific action, is a direct inhibitory effect on mature resorbing osteoclasts and may involve cytotoxicity. A second action of much greater specificity involves inhibition of accession of osteoclast-precursors to bone matrix or of the induction of their differentiation.

References

1. H. Fleisch. In: Bone and Mineral Research Annual 1, p. 319. Ed. W.A. Peck, Excerpta Medica, Amsterdam-Oxford-Princeton (1982).
2. W.B. Frijlink, O.L.M. Bijvoet, J. te Velde and G. Heynen. Lancet i: 799-803 (1979).
3. P.J. Meunier, M.C. Chapuy, C. Alexandre, C. Bressot, C. Edouard, E. Vignon, L. Mathieu and U. Trechsel. Lancet ii:489-492 (1979).
4. M.C. Chapuy, P.J. Meunier, C.M. Alexandre and E.P. Vignon. Eur. J. Clin. Invest. 65:1243-1247 (1980).
5. H.P. Sleeboom, O.L.M. Bijvoet, A.T. van Oosterom, J.H. Gleed and J.C.H. O'Riordan. Lancet i:239-243 (1983).
6. T.P. Spiro and G.R. Mundy, Calcif. Tiss. Int. 31:66 (1980).
7. M. Shinoda, G. Adamek, R. Felix, H. Fleisch, R. Schenk and P. Hagan. Calcif. Tiss. Int. 35:87 (1983).
8. H.I.J. Harinck, O.L.M. Bijvoet, H.J. Blanksma and P.J. Dahlinghaus-Nienhuys (in preparation).
9. T.J. Chambers. J. Clin. Pathol. 38:241 (1985).
10. E.H. Burger, J.W.M. van der Meer, J.S. van de Gevel, J.C. Gribnau, C.W. Thesingh and R. van Furth. J. Exp. Med. 156:1604 (1982).
11. P.H. Reitsma, S.L. Teitelbaum, O.L.M. Bijvoet and A.J. Kahn. J. Clin. Invest. 70:927 (1982).
12. L.G. Raisz. J. Clin. Invest. 44:103 (1965).
13. C.W. Thesingh and E.H. Burger. Devel. Biol. 95:429 (1983).

COMPARISON OF ALUMINUM RELATED WITH VITAMIN D RELATED OSTEOMALACIA BY TETRACYCLINE BASED BONE HISTOMORPHOMETRY

A.M. Parfitt, D. Rao, J. Stanciu and A.R. Villanueva

Bone and Mineral Research Laboratory
Henry Ford Hospital
Detroit, Michigan

Previous reports of aluminum related osteomalacia (ARO) described distinctive clinical, radiographic, biochemical and therapeutic features (1-4), but few distinctive histologic features except for the presence of aluminum (5-8). According to these reports, indices of bone resorption are reduced (5-8), but this has been based either on qualitative appraisal alone (8) or on measurement without correction for the availability of mineralized bone accessible to resorption (5-7). Furthermore, the comparison has invariably been with renal osteitis fibrosa or with normal subjects, not with other forms of osteomalacia. More recently, the term low turnover osteomalacia has been introduced (9,10), but although bone turnover in ARO is lower than in renal osteitis fibrosa, it has not been demonstrated to be lower than in other forms of osteomalacia. We have developed rigorous kinetic criteria for the diagnosis of classical osteomalacia due to vitamin D depletion (11-13); we now apply these criteria to renal bone disease and for the first time directly compare the two forms of osteomalacia.

PATIENTS AND METHODS

We studied 40 patients with osteomalacia due to vitamin D depletion (D-) due either to nutritional deficiency or intestinal malabsorption (9 males and 31 females, mean age (±SD) 57.0±13.3 y.) and 27 patients with osteomalacia on maintenance hemodialysis (HD), of whom 7 had undergone subtotal parathyroidectomy (12 males and 15 females, mean age 56.6±14.6 y.). All patients were suspected of having metabolic bone disease on the basis of their symptoms and/or radiographic and biochemical findings. Reference values (13) were obtained from 40 white females without metabolic bone disease (18 pre- and 22 post-menopausal, mean age 47.8 y.). All subjects had a transilial bone biopsy after *in vivo* double tetracycline labeling (13,14), with examination of undecalcified sections using the Zeiss MOP-III digitizing system (13). In all patients, relative osteoid volume (ROV) was >10%, mean osteoid seam width (MOSW; uncorrected for section obliquity) was >15 μm and mineralization lag time (MLT; MOSW divided by effective apposition rate (EAR)) was >100 days (11-13). None of the D- and all of the HD patients had extensive aluminum deposition at the osteoid:bone interface (15); the latter will subsequently be referred to as Al+. The patients were classified as stage II (some double labels) or stage III (no double labels); on this classification stage I corresponds to hyperparathyroid bone disease without a significant mineraliz-

Table 1. Frequency of different labeling patterns in two types of osteomalacia.

		Vitamin D depletion	Aluminum accumulation
Stage II	(DL+)	11 (27.5%)	5 (18.5%)
Stage III	(DL-)	29 (72.5%)	22 (81.5%)
	(SL)	20 (50.0%)	4 (14.8%)
	(UL)	9 (22.5%)	18 (66.7%)

Proportions of stage II and stage III do not differ significantly (χ^2 = 0.7, NS), but in stage III, proportion with no label (UL) is significantly higher in Al+ (χ^2 = 12.9, $p < 0.0005$).

ation defect (11-13). Stage III patients were further subdivided according to the presence (SL) or absence (UL) of a single label. In the patients with D- there was a significant hyperbolic regression of MOSW on relative osteoid surface (ROS), such that

$$\text{MOSW} (\mu m) = 9.0/(1-0.0075 \text{ ROS} (\%))$$

This relationship was used to compute predicted MOSW in the Al+ patients with stage III osteomalacia.

RESULTS

There was no significant difference in the proportions of patients in stages II or III between the two types of osteomalacia, but in stage III the proportion with no detectable tetracycline fixation in bone was significantly higher in the Al+ than in the D- patients (Table 1). When present, a single label in the absence of double label was invariably the second label. In those with stage II osteomalacia there was no significant difference in any histomorphometric index between the Al+ and D- patients (Table 2). In particular, bone resorption indices were similarly increased, and bone formation rate similarly depressed. In those with stage III osteomalacia (either SL only or UL) ROV, ROS and osteoclast surface/unit of mineralized bone (non-osteoid) surface (OCMB) were all significantly lower in the Al+ than in the D- patients (Table 2). Observed MOSW was not different in the two groups, but was significantly greater than predicted from ROS in the Al+ patients (30.1 v. 22.8; $p < 0.05$). Bone formation rate by definition was zero in both groups, and with stages II and III combined was still not significantly lower in Al+ (2.2) than in D- (3.4).

DISCUSSION

When bone is examined while the ECF contains tetracycline, all bone surfaces accessible to the circulation are labeled (16). Retention of the second tetracycline label in the absence of mineralization may reflect increased affinity for some constituent of the cement line or delayed escape of tetracycline from bone after disappearance of tetracycline from the ECF due to increased thickness of the osteoid seam (16). Our data suggest that aluminum at the osteoid:bone interface either blocks the initial reversible uptake of tetracycline or reduces binding to the cement line, and are consistent with the view that aluminum impairs osteoid mineralization, at least in the presence of incompletely

Table 2. Comparison of bone histomorphometry in two types of osteomalacia.

	Normal (40)	Stage II (DL+)		Stage III (DL-)	
		D- (11)	AL+ (5)	D- (29)	Al+ (22)
ROV (% TB)	2.09±1.1	27.6±8.9	27.1±13.6	41.4±17.1	31.1±17.3 *
ROS (% TS)	15.8±7.6	85.3±8.5	81.9±9.8	90.1±8.4	76.2±18.6 ***
MOSW (μm)	11.3±2.5	25.2±9.6	25.1±7.9	32.5±12.2	30.1±13.4
EAR (μm/d)	0.40±0.2	0.05±0.03	0.04±0.03	0	0
BFR ($\mu m^3/\mu m^2/y$)	17.1±8.8	12.3±8.4	11.9±8.4	0	0
OCS (% TS)	0.54±0.5	1.67±1.33	1.50±1.58	1.48±1.50	1.56±2.52
OCMB (% NOS)	0.68±0.7	14.3±12.7	9.38±4.59	21.2±14.3	10.3±16.7 **

D- = vitamin D depletion; Al+ = Aluminum accumulation during hemodialysis. Data expressed as mean ±SD. EAR = Effective apposition rate (corrected for fractional labeling of osteoid, uncorrected for section obliquity). In stage II, all indices in both groups deviate significantly from the reference values but there are no significant differences between the groups. In stage III, significant differences between D- and Al+ as indicated. * - $p<0.05$ **$p<0.02$ ***$p<0.001$

corrected uremia (17-19). This is supported also by the increase in MOSW relative to ROS, which implies an additional focal element to the osteomalacia (20).

During the evolution of D- osteomalacia, ROS increases earlier than MOSW because of the stimulation of remodeling activation and bone turnover by secondary hyperparathyroidism (11-13). Consequently, both the lower ROS and ROV and the lower OCMB are consistent with lower PTH levels in ARO than in renal osteitis fibrosa even in the absence of parathyroid surgery (21). In D-, ROS is greater in stage III than in stage II, whereas in Al+, ROS is lower in stage III than in stage II. This suggests that in Al+ the complete cessation of mineralization that determines the transition from stage II to stage III occurs earlier in the course of the disease than in D- and that further progression of PTH-dependent abnormalities is arrested.

Bone formation rate is significantly lower in both forms of osteomalacia than in normal female subjects, but is not lower in Al+ than in D-. Since low bone turnover is characteristic of all forms of osteomalacia, there is no justification for restricting this qualification to ARO, especially since this usage has given rise to the erroneous notion that there exists a condition of high turnover osteomalacia. This is found only very rarely in vitamin D depletion during the transition from stage I to stage II, provided that the evaluation of bone turnover is based on measurement of bone formation rate by double tetracycline labeling. A similar transition must occur during the evolution of ARO, since almost all patients with chronic renal failure have at least some degree of osteitis fibrosa before dialysis treatment is begun (22). However, we have not yet found bone formation rate to be increased in any patient with ARO, so that the transition must occur before the onset of symptoms.

CONCLUSIONS

In its early stages, aluminum related osteomalacia does not differ histologically from the osteomalacia of vitamin D depletion, but when more advanced the histologic expression of hyperparathyroidism is less severe and the defect in bone mineralization is more severe and occurs earlier in the course of the disease. Bone turnover is reduced in all forms of osteomalacia and is no lower in the aluminum related than in the vitamin D related forms. There is no such condition as high turnover osteomalacia except occasionally and transiently during the evolution of vitamin D depletion.

REFERENCES

1. Alvarez-Ude F, Feest TG, Ward MK, Pierides AM, Ellis HA, Peart KM, Simpson W, Weightman D, Kerr DNS. Hemodialysis bone disease: Correlation between clinical, histologic, and other findings. Kidney International 1978;14:68-73.

2. Hodsman AB, Sherrard DJ, Wong EGC, Brickman AS, Lee DBN, Alfrey AC, Singer FR, Norman AW, Coburn JW. Vitamin-D-resistant osteomalacia in hemodialysis patients lacking secondary hyperparathyroidism. Ann Intern Med 1981;94:629-637.

3. Ott SM, Maloney NA, Coburn JW, Alfrey AC, Sherrard DJ. The prevalence of bone aluminum deposition in renal osteodystrophy and its relation to the response to calcitriol therapy. N Engl J Med 1982;307:709-713.

4. Wills MR, Savory J. Aluminium poisoning:dialysis encephalopathy, osteomalacia, and anaemia. Lancet 1983;ii:29-34.

5. Cournot-Witmer G, Zingraff J, Plachot JJ, Escaig F, Lefèevre R, Boumati P, Bourdeau A, Garabedian M, Galle P, Bourdon R, Drüke, Balsan S. Aluminum localization in bone from hemodialyzed patients: Relationship to matrix mineralization. Kidney International 1981;20:375-385.

6. Hodsman AB, Sherrard DJ, Alfrey AC, Ott S, Brickman AS, Miller NL, Maloney NA, Coburn JW. Bone aluminum and histomorphometric features of renal osteodystrophy. J Clin Endocrinol Metab 1982;54:539-546.

7. Boyce BF, Fell GS, Elder HY, Junor BJ, Elliot HL, Beastall G, Fogelman I, Boyle IT. Hypercalcaemic osteomalacia due to aluminium toxicity. Lancet 1982;ii:1009-1013.

8. Ellis HA. Aluminum induced osteomalacia in patients with chronic renal failure and in animals. Nieren-und Hochdruckkrankheiten 1983;12:S198-206.

9. Faugere MC, Abreo K, Smith A, Malluche HH. Bone aluminum accumulation: Etiology, prevalence and its influence on vitamin D therapy in patients with mild to advanced renal failure. In: Cohn DV, Potts JT Jr, Fujita T, eds. Endocrine control of bone and calcium metabolism, Vol. 8B. Elsevier Science Publishers B.V., 1984:151-152.

10. Malluche HH, Smith AJ, Abreo K, Faugere M-C. The use of deferoxamine in the management of aluminum accumulation in bone in patients with renal failure. N Engl J Med 1984;311:140-144.

11. Rao DS, Villanueva A, Mathews M, Pumo B, Frame B, Kleerekoper M, Parfitt AM. Histologic evolution of vitamin D depletion in patients with intestinal malabsorption or dietary deficiency. In: Frame B, Potts JT Jr, eds. Clinical Disorders of Bone and Mineral Metabolism. Amsterdam, Excerpta Medica, 1983:224-226.

12. Parfitt AM. The cellular mechanisms of osteoid accumulation in metabolic bone disease. In: Mineral Metabolism Research in Italy, Vol. 4. Milano, Wichtig Editore, 1984:3-9.

13. Parfitt AM, PØdenphant J, Villanueva AR, Frame B. Metabolic bone disease with and without osteomalacia after intestinal bypass surgery: A bone histomorphometric study. Bone 1985;6:211-220.

14. Rao DS. Practical approach to bone biopsy. In: Bone Histomorphometry. Techniques and Interpretations. Recker R, ed. CRC Press, Boca Raton, FL, 1983:3-11.

15. Maloney NA, Ott SM, Alfrey AC, Miller NL, Coburn JW, Sherrard DJ. Histological quantitation of aluminum in iliac bone from patients with renal failure. J Lab Clin Med 1982;99:206-216.

16. Parfitt AM. The physiologic and clinical significance of bone histomorphometric data. In: Recker R, ed. Bone Histomorphometry. Techniques and Interpretations. Boca Raton, CRC Press, 1983:143-223.

17. Hodsman AB, Anderson C, Leung FY. Accelerated accumulation of aluminum by osteoid matrix in vitamin D deficiency. Mineral Electrolyte Metab 1984; 10:309-315.

18. Quarles DL, Dennis VW, Gitelman HJ, Harrelson JM, Drezner MK. Aluminum deposition at the osteoid-bone interface: An epiphenomenon of the osteomalacic state in vitamin D deficient dogs. J Clin Invest 1985;75:1441-1447.

19. PØdenphant J, Salem N, Sypitkowski C, Frame B, Parfitt AM. Reversal of aluminum related dialysis osteomalacia after transplantation. Proc Fourth International Workshop on Bone Histomorphometry. BONE (in press).

20. Boyce BF, Smith L, Fogelman I, Johnston E, Ralston S, Boyle IT. Focal osteomalacia due to low-dose diphosphonate therapy in Paget's disease. Lancet 1984;1:821-824.

21. Andress D, Felsenfeld AJ, Voigts A, Llach F. Parathyroid hormone response to hypocalcemia in hemodialysis patients with osteomalacia. Kidney International 1983;24:364-370.

22. Malluche HH, Ritz E, Lange HP, Kutschera J, Hodgson M, Seiffert U, Schoeppe W. Bone histology in incipient and advanced renal failure. Kidney International 1976;9:355-362.

PARATHYROID HORMONE AND 1.25 VITAMIN D_3 EXERT OPPOSITE EFFECTS ON BONE

Hartmut H. Malluche, Clifford Matthews,
Marie-Claude Faugere, Paolo Fanti, and
Robert M. Friedler

Division of Nephrology, Bone & Mineral Metabolism
University of Kentucky Medical Center
800 Rose Street
Lexington, Kentucky 40536-0084

Histologic studies of bone in patients with hyperparathyroid bone disease reveal that the hallmark of the pathologic changes is an increase in bone forming and resorbing cells (1,2). In contrast, administration of vitamin D to patients with secondary hyperparathyroidism results in a decrease in number of bone forming and resorbing cells (3). Based on these data we advanced the hypothesis that vitamin D and parathyroid hormone may have opposite effects on bone i.e., parathyroid hormone induces an increase and 1.25 Vit D_3 a decrease in number of bone cells. This hypothesis was tested in an animal model in which experimental beagle dogs were rendered deficient in 1.25 Vit D_3 and parathyroid hormone by 5/6 nephrectomy and thyroparathyroidectomy. Thyroid hormone was replaced and then both parathyroid hormone and 1.25 Vit D_3 were exogenously substituted to produce all combinations between the two hormones, i.e., 1.25D+/PTH+ (n=5), 1.25D-/PTH- (n=7), 1.25D+/PTH- (n=9), 1.25D-/PTH+ (n=5). Six normal control dogs were sham operated and injected with vehicle. Bone biopsies were done after eight months of experimental administration of these hormones. Histomorphometry of bone for static and dynamic parameters of bone cells revealed that deficiency in parathyroid hormone results in a decrease in the number of bone forming and resorbing cells. This decrease in cell number was observed independent of the presence or absence of deficiency in 1.25 Vit D_3. In addition, administration of parathyroid hormone increased the number of bone cells independent of the status in 1.25 Vit D_3. In contrast, deficiency in 1.25 Vit D_3 resulted in a decreased cellular activity without an alteration in cell number and administration of 1.25 Vit D_3 increased the cellular activity without an increase in numbers of bone cells.

These data confirm that 1.25 Vit D_3 represents a trophic hormone for bone that promotes cellular activity, i.e., differentiation whereas parathyroid hormone affects number of bone cells most likely through an effect on cellular mitosis rate. In addition, our data show that parathyroid hormone and 1.25 Vit D_3 exert effects on bone which are different but increases in one or the other hormone may compensate for a deficiency in either.

REFERENCES

1. H.H. Malluche, E. Ritz, H.P. Lange, J. Kutschera, M. Hodgson, U. Seiffert and W. Schoeppe, Bone Histology in Incipient and Advanced Renal Failure.
Kidney Int., 9:355-362 (1976).

2. H.H. Malluche, D.A. Goldstein and S.G. Massry, The Value of 1.25 Dihydroxy Vitamin D_3 in the Management of Uremic Osteodystrophy. In: Proc. II Int. Symp: "Bone Structure Function and Disease", Adelaide, Australia, p. 8, (1982).

3. D.J. Sherrard, D.J. Baylink, J.W. Wergedal and N.A. Maloney, Quantitative Histological Studies on the Pathogenesis of Uremic Bone Disease. J. Clin. Endocrin. Metab., 39:119-135 (1976).

SERUM BONE GLA-PROTEIN COMPARED TO BONE HISTOMORPHOMETRY IN HEMODIALYZED PATIENTS

Sam A. Charhon, Pierre D. Delmas, Luc Malaval, Pascale M. Chavassieux, Marie Claire Chapuy and Pierre J. Meunier

INSERM U 234, Faculté A. Carrel, Lyon, France

INTRODUCTION

Bone gla-protein (BGP), also called osteocalcin[1,2], is the most abundant non-collagenous bone protein. BGP is synthesized by the osteoblasts, is concentrated in the bone matrix and also found circulating in the blood. The measurement of BGP in serum by radioimmunoassay[3] has proved to be a useful tool for assessing bone turnover in a variety of metabolic or endocrine bone diseases[4-6] and more specifically bone formation in osteoporosis[7].

Renal osteodystrophy (ROD) is a heterogenous bone disorder which includes various degrees of bone resorption, bone matrix formation and bone mineralization[8,9]. The purpose of the present study is to compare the serum BGP values with the histomorphometric parameters of iliac bone biopsies in patients undergoing chronic hemodialysis in order to determine the clinical value of the serum BGP measurement in renal osteodystrophy.

PATIENTS AND METHODS

Patients : Forty two patients underwent a transiliac bone biopsy. There were 25 males and 17 females, aged 16-73 years (mean ± SD : 42.8 ± 15.0 yrs). They were on chronic maintenance hemodialysis for 63.8 ± 36.1 months (range : 12-149 mo.) for various kidney diseases. Five patients had undergone a subtotal parathyroidectomy.

Methods

Biochemistry

- Serum BGP levels were determined using a sensitive radioimmunoassay according to the methods published previously[10]. This assay uses a rabbit antiserum (R 102 M) raised against bovine BGP at a final dilution of 1:40 000. Antibody bound and free 125I-labeled BGP are separated by the double antibody technique using a complex of sheep antirabbit IgG and polyethylene glycol (PR, CEA-ORIS, Gif sur Yvette, France). The sensitivity of the assay is 0.05 ng of bovine BGP per tube. Duplicate values of two appropriate dilutions of serum samples were assayed for each patient. We have previously shown that standard curves obtained with bovine BGP and with dilutions of human plasma samples from patients with advanced chronic renal failure are parallel[11]. Mean serum BGP obtained in 170 normal adults (107 women and 63 men), aged 56 ± 19 years, was 6.2 ± 2.2 ng/ml.

- Total serum alkaline phosphatase (AP) levels were measured by a colorimetric method. The mean value was 2.7 ± 1.0 Bodansky Units in 41 normal adults.

- Serum immunoreactive parathyroid hormone (iPTH) determinations were performed by radioimmunoassay using a rabbit antibovine PTH antiserum recognizing the carboxyl terminal end of the molecule[12]. The mean value was 3.1 ± 0.7 mU/ml in 62 normal adults.

- Serum aluminum (Al) levels were measured by flameless atomic absorption spectrophotometry (Perkin Elmer 5000 spectrophotometer). The normal mean value was 34.1 ± 7.0 µg/l[13].

Bone histomorphometry

Prior to bone biopsy, a tetracycline double labeling was performed with demethylchlortetracycline (600 mg per day) or tetracycline chlorhydrate (500 mg per day) administered following the schedule : 2 days on - 12 days off and then 4 days on. The biopsy was taken 3 to 7 days after the end of the second tetracycline labeling. The bone specimens were processed without decalcification and embedded in methylmetacrylate. Measurements of the static and dynamic parameters were carried out according to the methods published previously[8,14].

The static parameters included : the trabecular bone volume (TBV, percentage of the total spongy bone space occupied by trabecular bone), the resorption surfaces (RS, percentage of trabecular surfaces scalloped by Howship's lacunae), the number of osteoclasts (Ocl.) per mm^2 of spongy bone section, the osteoid volume (OV, percentage of trabecular bone occupied by osteoid tissue), the osteoid surfaces (OS, percentage of trabecular surfaces covered with osteoid), the thickness index of the osteoid seams (TIOS = (OV/OS) x 100), the osteoblastic surfaces (Obl. S, percentage of trabecular surfaces lined with osteoblasts).

The dynamic parameters included : the double labeled surfaces (DLS, percentage of trabecular surfaces showing a double tetracycline label), the calcification rate (CR, which is the rate of mineral deposition in the osteoid expressed in micrometers per day) and the bone formation rate at the tissue level (BFR_{tissue} = CR x DLS) expressed in $\mu m^3/\mu m^2$/day.

The presence of aluminum deposits in bone was revealed by histochemical method (Aluminon staining) according to the method of Maloney et al[15].

Statistical methods : The methods employed included Student's t test and linear correlation.

The results are expressed as the mean ± 1 SD.

RESULTS

Serum BGP levels were significantly increased (64.0 ± 74.8 ng/ml vs 6.2 ± 2.2 ng/ml in controls, $p < 0.001$) and correlated with serum AP and iPTH levels ($r = 0.53$ and $r = 0.55$ respectively, $p < 0.001$). As shown in Table 1, serum BGP levels were highly significantly correlated with the osteoblastic surfaces ($r = 0.63$), the calcification rate ($r = 0.62$), the double labeled surfaces ($r = 0.77$) and the bone formation rate at the tissue level ($r = 0.81$). A significant correlation was also found between serum BGP and the resorption parameters ($r = 0.57$-0.60). Serum AP and iPTH levels were correlated with the bone formation rate but with lower coefficients of correlation.

Based on the individual values of bone formation rate, bone lesions were classified as high turnover renal osteodystrophy (HT-ROD ; BFR_{tissue} greater than the normal mean value + 2 SD, n = 14) and low turnover renal osteodystrophy (LT-ROD, n = 28). The resorption parameters and the osteoblastic surfaces were significantly greater in patients with HT-ROD than in patients with LT-ROD. Only 2 patients with HT-ROD had Al deposits in bone whereas 25/28 patients with LT-ROD showed a positive histochemical staining for Al. Serum BGP values were significantly higher ($p < 0.001$) in HT-ROD compared with LT-ROD (Table 2) and the discrimination between the 2 groups was better with serum BGP measurement than that obtained with serum AP levels. Only 3 patients with HT-ROD had serum BGP values within the same range as for patients with LT-ROD.

Table 1 : Coefficients of linear correlation between serum biochemical parameters and histomorphometric parameters of iliac bone biopsy.

		S-BGP	S-AP	iPTH
Static parameters				
TBV	(%)	0.14	0.22	0.28
RS	(%)	$0.60^{\dagger}$	0.36^{*}	$0.67^{\dagger}$
Ocl.	($/mm^2$)	$0.57^{\dagger}$	$0.59^{\dagger}$	$0.73^{\dagger}$
OV	(%)	0.06	0.41^{**}	0.26
OS	(%)	0.07	0.36^{*}	0.31^{*}
TIOS		0.06	0.40^{**}	0.26
Obl. S	(%)	$0.63^{\dagger}$	$0.59^{\dagger}$	$0.78^{\dagger}$
Dynamic parameters				
CR	(µm/day)	$0.62^{\dagger}$	0.25	$0.59^{\dagger}$
DLS	(%)	$0.77^{\dagger}$	$0.64^{\dagger}$	$0.71^{\dagger}$
BFR_{tissue}	($\mu m^3/\mu m^2$/day)	$0.81^{\dagger}$	$0.62^{\dagger}$	$0.74^{\dagger}$

Significant at $* p < 0.05$; $** p < 0.01$; $\dagger p < 0.001$

Table 2 : Serum biochemical parameters in hemodialyzed patients with high turnover renal osteodystrophy (HT-ROD) and patients with low turnover renal osteodystrophy (LT-ROD).

	HT-ROD (n=14)	LT-ROD (n=28)
Serum BGP, ng/ml (N = 6.2 ± 2.2)	138.5 ± 90.8	26.8 ± 14.8 †
Serum AP, Bod. U. (N = 2.7 ± 1.0)	15.4 ± 10.0	7.1 ± 5.9*
Serum iPTH, mU/ml (N = 3.1 ± 0.7)	36.3 ± 16.8	11.3 ± 7.4 †
Serum Al, µg/l (N = 34.1 ± 7.0)	100.2 ± 57.5	130.4 ± 139.9

Significantly different from HT-ROD at * $p < 0.01$; † $p < 0.001$

Patients with LT-ROD were divided into 2 groups : one group with osteomalacia (LT-OM) as defined by an increased thickness index of the osteoid seams and one group without osteomalacia (non OM-LT). As shown in Table 3, serum BGP levels were similar in the 11 patients with osteomalacia and in the 17 patients without osteomalacia. However, serum iPTH and Al levels were significantly higher in patients with osteomalacia than in patients without osteomalacia.

Table 3 : Serum biochemical parameters in low turnover renal osteodystrophy with osteomalacia (OM-LT) or without osteomalacia (non OM-LT).

	OM-LT (n=11)	non OM-LT (n=17)
Serum BGP, ng/ml (N = 6.2 ± 2.2)	26.5 ± 7.4	27.0 ± 18.3
Serum AP, Bod. U. (N = 2.7 ± 1.0)	9.5 ± 5.8	5.5 ± 5.5
Serum iPTH, mU/ml (N = 3.1 ± 0.7)	15.5 ± 8.9	7.7 ± 3.3**
Serum Al, µg/l (N = 34.1 ± 7.0)	202.2 ± 190.6	78.0 ± 45.1*

Significantly different from OM-LT at * $p < 0.05$; ** $p < 0.01$

DISCUSSION

As previously found in patients with chronic renal failure[6,11,16,17], patients undergoing chronic hemodialysis had increased values of serum BGP. However, they showed a wide range of serum BGP values which could not be explained by a difference in the degree of renal failure since all patients had been on chronic maintenance hemodialysis for at least one year. In fact, as reported by Malluche et al[6], this wide range reflects a difference in bone remodeling levels. In the present study, serum BGP levels were significantly correlated with the histomorphometric parameters of bone remodeling, and specially those reflecting bone formation such as the extent of the double labeled surfaces and the calcification rate. The highest coefficient of correlation was found between serum BGP and the bone formation rate at the tissue level, a parameter which reflects both the number and the activity of the individual osteoblasts. Serum BGP was also significantly correlated with the resorption parameters and this is consistent with the well-known coupling between bone resorption and bone formation. Serum BGP measurements allowed a better discrimination between patients with high bone turnover and patients with low bone turnover than serum AP. Indeed, serum BGP is a more specific marker of bone metabolism than serum AP but it must be noted that liver contribution to serum AP levels was not assessed in our study. In patients with low turnover renal osteodystrophy, however, serum BGP did not discriminate between patients with or without osteomalacia.

In conclusion, these data show that serum BGP is a valuable marker for evaluating bone remodeling in hemodialyzed patients and suggest that the serum BGP measurements could be helpful in assessing the effects of treatments that alter bone remodeling in hemodialyzed patients.

This work was supported in part by the Commissariat à l'Energie Atomique (ORIS, contrat n° 098520).

REFERENCES

1. P.F. Hauschka, J.B. Lian and P.M. Gallop, Direct identification of the calcium binding amino acid-carboxyglutamate in mineralized tissue, Proc. Natl. Acad. Sci. USA 72 : 3925 (1975)
2. P.A. Price, A.S. Otsuka, J.W. Poser, J. Kristaponis and N. Ramin, Characterization of a gamma-carboxyglutamic acid-containing protein from bone, Proc. Natl. Acad. Sci. USA 73 : 1147 (1976)
3. P.A. Price and S.K. Nishimoto, Radioimmunoassay for the vitamin K-dependent protein of bone and its discovery in plasma, Proc. Natl. Acad. Sci. USA 77 : 2234 (1980)
4. P.A. Price, J.G. Parthemore and L.J. Deftos, New biochemical marker for bone metabolism. Measurement by radioimmunoassay of bone gla-protein in the plasma of normal subjects and patients with bone disease, J. Clin. Invest, 66 : 878 (1980)
5. P. Delmas, H.W. Wahner, K.G. Mann and B.L. Riggs, Assessment of bone turnover in postmenopausal osteoporosis by measurement of serum bone gla-protein, J. Lab. Clin. Med. 102 : 470 (1983)

6. H.H. Malluche, M.C. Faugere, P. Fanti and P.A. Price, Plasma levels of bone gla-protein reflect bone formation in patients on chronic maintenance hemodialysis, Kidney Int. 26 : 869 (1984)
7. J.P. Brown, P.D. Delmas, L. Malaval, C. Edouard, M.C. Chapuy and P.J. Meunier, Serum bone gla-protein : a specific marker for bone formation in post menopausal osteoporosis, Lancet, i : 1091 (1984)
8. S.A. Charhon, Y.F. Berland, M.J. Olmer, E. Delawari, J. Traeger and P.J. Meunier, Effects of parathyroidectomy on bone formation and mineralization in hemodialyzed patients, Kidney Int., 27 : 426 (1985)
9. S.A. Charhon, P.M. Chavassieux, M.C. Chapuy, G.Y. Boivin and P.J. Meunier, Low rate of bone formation with or without histological appearance of osteomalacia in patients with aluminum intoxication, J. Lab. Clin. Med. (1985, in press)
10. P.D. Delmas, D. Stenner, H.W. Wahner, K.G. Mann and B.L. Riggs, Increase in serum bone gamma-carboxyglutamic acid protein with aging in women, J. Clin. Invest. 71 : 1316 (1983)
11. P.D. Delmas, D.M. Wilson, K.G. Mann and B.L. Riggs, Effect of renal function of plasma levels of bone gla-protein, J. Clin. Endocrinol. Metab. 57 : 1028 (1983)
12. H.H. Conaway and C.S. Anast, Double antibody radioimmunoassay for parathyroid hormone, J. Lab. Clin. Med. 83 : 129 (1974)
13. Y. Pegon, Dosage de l'aluminium dans les liquides biologiques par absorption atomique sans flamme, Anal. Chim. Acta. 101 : 385 (1978)
14. P.J. Meunier, Bone biopsy in diagnosis of metabolic bone disease, in : "Hormonal control of calcium metabolism" D.V. Cohn, R.V. Talmage and J. Les Matthews, eds, Excerpta Medica, Amsterdam (1980)
15. N.A. Maloney, S.M. Ott, A.C. Alfrey, N.L. Milliner, J.W. Coburn and D.J. Sherrard, Histological quantitation of aluminum in iliac bone from patients with renal failure, J. Lab. Clin. Med. 99 : 206 (1982)
16. A.K. Cheung, S.C. Manolagas, B.D. Catherwood, C.A. Mosely Jr, J.A. Mitas III, R.C. Blantz and L.J. Deftos, Determinants of serum 1,25(OH)$_2$D levels in renal disease, Kidney Int. 24 : 104 (1983)
17. S. Epstein, H. Traberg, R. Raja and J. Poser, Serum and dialysate osteocalcin levels in hemodialysis and peritoneal dialysis patients and after renal transplantation, J. Clin. Endocrinol. Metab. 60 : 1253 (1985)

EFFECTS OF $1,25(OH)_2D_3$ ADMINISTRATION ON SERUM BONE GLA PROTEIN IN PREDIALYSIS CHRONIC RENAL FAILURE

Giorgio Coen, Sandro Mazzaferro, Giuseppe Donato, Carlo Massimetti, Paola Ballanti*, Franco Bondatti, Carlo Della Rocca*, Antonella Smacchi, Flavia Mantazzoli and Giulio A. Cinotti

Division of Nephrology and Department of Human Biopathology* "La Sapienza" University, Rome, Italy

INTRODUCTION

The bone GLA protein (BGP) or osteocalcin, is a vitamin K-dependent protein produced by the osteoblasts (1), which accumulates in the non-collagenous protein fraction of bone, but it is also measurable by radioimmunoassay in the serum (2). In vitro (1) and experimental in vivo studies (3) have shown that production of this protein is modulated by $1,25(OH)_2D_3$. However a basal production of BGP can be observed even in the vitamin D depleted animal (3). In predialysis chronic renal failure serum BGP levels are higher than normal (4) in spite of the known $1,25(OH)_2D_3$ deficit (5).

The aim of this study was to explore the effects of short- and long-term $1,25(OH)_2D_3$ administration on serum BGP levels in predialysis chronic renal failure. The results suggest that osteoblasts respond to $1,25(OH)_2D_3$ by increasing the production of BGP. In the long-term treatment serum BGP levels decrease probably as a consequence of the reduction in number of the osteoblastic cells.

MATERIALS AND METHODS

The study was performed on 39 patients with slowly evolving predialysis chronic renal failure from the out-patient renal clinic.

12 patients (age 48.5 ± 15.3 years, serum creatinine 4.44 ± 1.07 mg/dl;M,SD) were treated with $1,25(OH)_2D_3$ p.o., 0.5 ug twice daily at 8 A.M. and 8 P.M. for two weeks. Blood for calcium, phosphate, alkaline phosphatase (AP), BGP, iPTH and $1,25(OH)_2D_3$ was drawn basally and at the end of treatment.

In a second group of 12 patients comparable for age and serum creatinine (age 55.3 ± 14.2 years, serum creatinine 4.31 ± 1.45 mg/dl), $1,25(OH)_2D_3$ 0.25 ug daily, was administered for about one year. A lower dose of the metabolite was chosen for the long-term administration to avoid hypercalcemia and renal damage (6).

A third group of 12 patients (age 39.2 ± 21.7 years, serum creatinine

3.99±1.65 mg/dl) which did not receive the vitamin D metabolite, served as control to the second group. In both second and third group the same blood parameters as for the short-term experiment were explored basally and every four months for one year.

In addition 15 patients (group 2 + 3 patients) following long-term treatment with $1,25(OH)_2D_3$, 0.25 ug daily for an average period of 16 months, were subjected to transiliac bone biopsy for histomorphometric evaluation. Their serum creatinine at the time the biopsy was performed, was 7.17±3.4 mg/dl. The results of the histomorphometric evaluation were compared to those obtained in 23 patients with predialysis chronic renal failure with comparable rates of decline of renal function, never treated with vitamin D metabolites (age 42.3±17.7 years, serum creatinine 6.11±2.38 mg/dl).

All patients were on a diet moderately restricted in protein (0.8 g/kg b.w.) and phosphorus (12 mg/kg b.w.). Part of the patients were treated with antihypertensive medication (frusemide, alphamethyldopa, clonidine). None of them received phosphate binders, corticosteroids, anticonvulsive and anticoagulant medication. The normality of the prothrombin time was considered indicative of a normal vit. K supply.

Serum BGP was measured with a radioimmunoassay employing bovine standard and antiserum against bovine BGP, according with Price and Nishimoto (2). Normal values are 3.89±1.45 ng/ml. Blood samples for BGP determination were stored at -30°C and measurements for each patient were carried out in the same assay in order to avoid interassay variations. Serum iPTH was measured with a radioimmunoassay based on an antiserum against the C-terminal portion of the molecule. Normal values are 0.34±0.19 ng/ml. Serum $1,25(OH)_2D_3$ was measured by a receptorial method, with cytosol receptor obtained from chicken intestinal mucosa as described by Eisman et al. (7). Details concerning purification procedure and the assay have already been reported (8,9). Normal values are 34.8±10.2 pg/ml.

Calcium was determined with atomic absorption spectrophotometry. Serum creatinine, phosphate and alkaline phosphatase were measured with standard laboratory techniques.

The procedure for the transiliac bone biopsy, for the fixation and inclusion of the sample and the technique employed for histomorphometric evaluation have previously been reported (9). The parameters evaluated for this study were: Active Resorption Surface (ARS), the percentage of the total bone surface undergoing resorption with visible osteoclasts; Resorption Surface (RS), the percentage of the total surface consisting of Howship's lacunae; Active Osteoblastic Surface (AOS), the percentage of the trabecular surface covered by osteoid lined by active osteoblasts.

RESULTS

Serum calcium, phosphate and $1,25(OH)_2D_3$ levels basally and following short- and long-term administration of $1,25(OH)_2D_3$ are reported in Table 1. Short- and long-term treatments induced comparable changes of these parameters. Administration of $1,25(OH)_2D_3$ for two weeks induced a significant decrease in serum iPTH levels ($p<0.002$), an increase in serum BGP ($p<0.005$) and a small but significant increment ($p<0.01$) also in the serum AP (Fig.1). Long-term treatment induced a significant fall in serum iPTH, AP and BGP levels ($p<0.05$, 0.002 and 0.001 respectively, at the end of treatment) already noticeable 4 months following the start of $1,25(OH)_2D_3$ administration, while in the control group no change or an increase of the parameters was found (Fig. 2-4).

The long-term treatment with $1,25(OH)_2D_3$, at the dose of 0.25 ug daily, was effective in keeping within normal limits the parameter ARS and in reducing toward normal levels the AOS (Table 2).

Table 1. Calcium, phosphate and $1,25(OH)_2D_3$ serum levels in the short- and long-term observation

	Basal	2 weeks	Basal	6-12 months
Ca (mg/dl)	9.05±0.45	9.49±1.05[a]	9.03±0.58	9.56±0.75[b]
P (mg/dl)	3.46±0.62	4.19±0.81[a]	3.84±0.49	4.42±0.81[a]
$1,25(OH)_2D_3$(pg/ml)	16.36±11.06	30.54±21.37[c]	16.58±7.81	32.32±8.94[c]

a=$p<0.05$; b=$p<0.01$; c=$p<0.005$. Values are the mean ± SD.

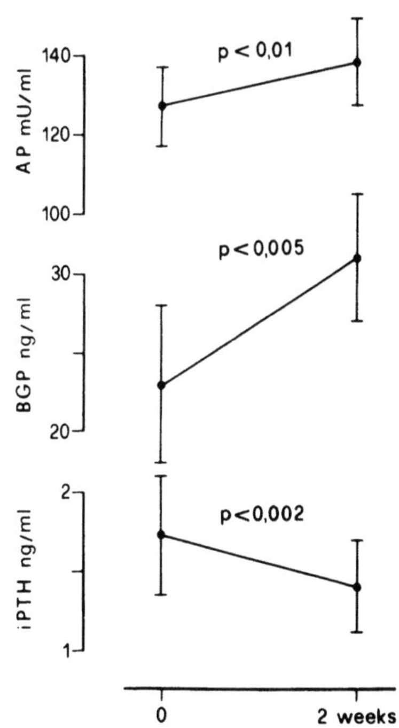

Figure 1. Serum levels of AP, BGP and iPTH basally and following 2 weeks treatment with $1,25(OH)_2D_3$.

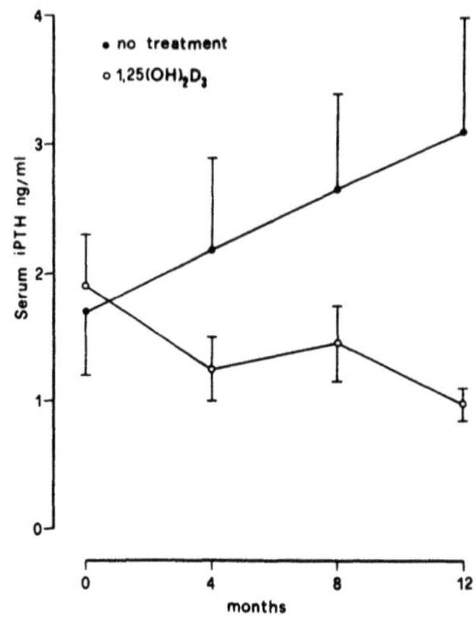

Figure 2. Timecourse of serum iPTH levels during long-term $1,25(OH)_2D_3$ administration and in the control group.

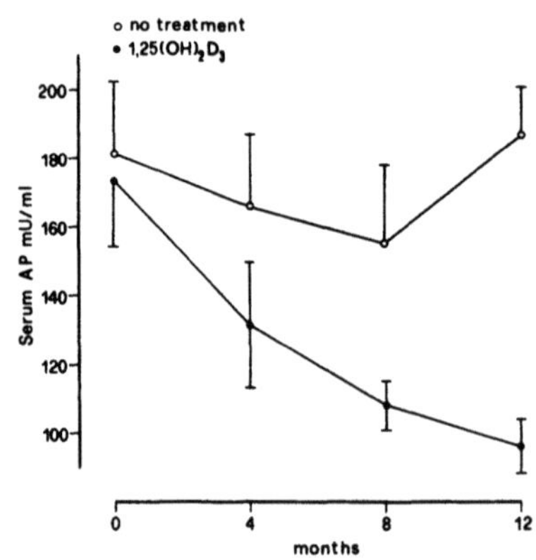

Figure 3. Timecourse of serum AP levels during long-term $1,25(OH)_2D_3$ administration and in the control group

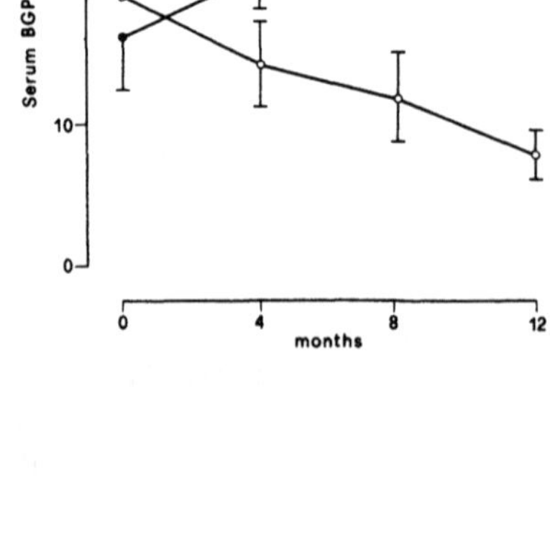

Figure 4. Timecourse of serum BGP levels during long-term $1',25(OH)_2D_3$ administration and in the control group.

Table 2. Histomorphometric parameters of treated ($1,25(OH)_2D_3$) and untreated patients with predialysis CRF

	$1,25(OH)_2D_3$ (15)	Controls (23)	p	Normal values (57)
RS	2.30±2.37	5.63±4.76	<0.01	1.77±1.5
ARS	0.24±0.27	1.11±1.22	<0.01	0.19±0.30
AOS	1.36±2.00	3.36±3.39	<0.05	0.27±0.64

Number of subjects in parenthesis.

DISCUSSION

Our data indicate that short-term $1,25(OH)_2D_3$ administration is able to provoke a fall in serum iPTH along with a sizeable significant increase in BGP levels and a slight increment in serum AP. On the contrary long-term administration of the vitamin D metabolite induces a concensual fall in serum iPTH, AP and BGP. The difference in the response in BGP levels does not seem to be due to the use of different doses of the metabolite since the levels of serum $1,25(OH)_2D_3$, calcium and phosphate following the short and long-term treatment were of the same order of magnitude.

The results of the short-term treatment suggest that also in man, at least in this experimental setting, $1,25(OH)_2D_3$ can stimulate the osteoblasts to an enhanced synthesis of BGP, as already observed in vitro and in animal experiments (1,3). The increment in serum BGP after the short-term $1,25(OH)_2D_3$ administration occurred in conjunction with a fall in iPTH serum levels. This finding suggests that parathyroid hormone does not play any direct stimulatory effect on BGP production by the osteoblasts. Accordingly parathyroid hormone was found to inhibit rather than stimulate BGP production by the osteoblasts in vitro (1).

This conclusion may appear in contradiction with the known elevated levels of osteocalcin along with secondary hyperparathyroidism, found in chronic renal failure (4,9). We have already shown that the serum BGP levels in predialysis chronic renal failure are more closely correlated to the dimension of the active osteoblastic population and to serum iPTH levels than to the renal damage expressed by serum creatinine levels (10). Consequently the elevated levels of serum BGP in chronic renal failure mainly reflect the extent of the osteoblastic pool, which is dependent to a large degree on the duration and severity of secondary hyperparathyroidism.

Therefore long-term $1,25(OH)_2D_3$ administration progressively lowers serum BGP levels through suppression of secondary hyperparathyroidism and reduction in number of the active osteoblasts. This conclusion is in agreement with our histomorphometric data. In facts they clearly show that the long-term administered dose of $1,25(OH)_2D_3$ was able to depress the number of osteoblasts and osteoclasts.

According with a recent report (11), also the decrease in number of the osteoclasts may play a role in the drop in serum BGP levels following the long-term treatment. Weinstein and Gundberg (11) have found several immunoreactive fractions of BGP in the sera of patients with secondary hyperparathyroidism of chronic renal failure, a fact which may indicate in the liberation of fragments of BGP from the demolitive side of bone turnover a cause of the increased levels of the immunoassayable protein in chronic renal insufficiency. We have already reported (10) a strict correlation between BGP serum levels and the parameter ARS in predialysis chronic renal failure, outweighing in significance the correlation between BGP and AOS, a finding in agreement with Weinstein and Gundberg's hypothesis.

In conclusion an increment in BGP synthesis may be the result either of stimulation of osteoblastic synthesis of the protein or of an increased number of active osteoblasts. $1,25(OH)_2D_3$ is a direct stimulator of BGP synthesis while parathyroid hormone seems to act mainly through an increase in the osteoblastic pool.

ACKNOWLEDGEMENTS

This work was supported by funds of Ministero della Pubblica Istruzione.

REFERENCES

1. J.N.Beresford, J.A.Gallagher, J.W. Poser and R.G.G. Russell, Production of osteocalcin by human bone cells in vitro. Effects of $1,25(OH)_2D_3$, $24,25(OH)_2D_3$, parathroid hormone and glucosteroids. Metab. Bone Dis. & Rel.Res. 5: 229 (1984).
2. P.A.Price and S.K.Nishimoto, Radioimmunoassay for the vitamin K-dependent protein of bone and its discovery in plasma. Proc. Natl. Acad. Sci., USA 77: 2234 (1980).
3. P.A.Price, M.K.Williamson and S.A.Baukol, The vitamin K-dependent bone protein and the vitamin D-dependent biochemical response of bone to dietary calcium deficiency. In: Vitamin D, Chemical, Biochemical and Clinical Endocrinology of Calcium Metabolism. A.W.Norman et al. eds, W. de Gruyter, Berlin-New York, p.351 (1982).
4. A.K.Cheung, S.C.Manolagas, B.D.Chaterwood, C.A.Mosely Jr., J.A.Mitas II, R.C.Blantz and L.J.Deftos, Determinants of serum $1,25(OH)_2D_3$ levels in renal disease. Kidney Int. 24: 104 (1983).
5. E.B.Mawer, J.Backhouse, C.M.Taylor, G.A.Lumb and S.W.Stanbury, Failure of formation of 1,25dihydroxycholecalciferol in chronic renal insufficiency. Lancet 1: 626 (1973).
6. G.Coen, F.Messa, M.Donnetti, C.Massimetti, S.Mazzaferro, G.Donato, G. Giuliano, D.Finistauri and G.A.Cinotti, 1-alpha-OHD_3 and low phosphate diet in predialysis chronic renal failure: effects on secondary hyperparathyroidism and on renal function. Acta Vitam.Enzym. 6: 129 (1984).
7. J.A.Eisman, A.J.Hamstra, B.E.Kream and H.F.DeLuca, 1,25-dihydroxyvitamin D in biological fluids: a simplified and sensitive assay. Science 193: 1021 (1976).
8. A.Smacchi, A.R.Bianchi, G.Donato, D.Palmiotti, B.Bacchielli, L.Chini and G.Coen, A simplified assay of 1,25dihydroxycholecalciferol in serum,

Mineral Metab.Res. in Italy 4: 179 (1983).
9. G.Coen, S.Mazzaferro, E.Bonucci, F.Taggi, P.Ballanti, A.R.Bianchi, G.Donato, C.Massimetti, A.Smacchi and G.A.Cinotti, Bone GLA protein in predialysis chronic renal failure: effects of $1,25(OH)_2D_3$ administration in a long-term follow-up, Kidney Int. (in press).
10. G.Coen, S.Mazzaferro, E.Bonucci, P.Ballanti, C.Massimetti, G.Donato, F.Bondatti, A.Smacchi, G.A.Cinotti and F.Taggi, Serum bone GLA protein (BGP) and secondary hyperparathyroidism in predialysis chronic renal failure . In: Vitamin D, Chemical, Biochemical and Clinical Update. A.W.Norman et al. eds, W. de Gruyter, Berlin-New York, p.1147 (1985).
11. R.S.Weinstein and C.M.Gundberg, Multiple immunoreactive forms of osteocalcin in the sera of dialysis patients.(Abs) VIIth Annual Scientific Meeting, American Society for Bone and Mineral Research, June 15-18 (1985).

SKELETAL MASS IN LONG TERM HEMODIALYSIS PATIENTS

Joseph Letteri, Purshotam Nagwani and Stanton Cohn

Division of Renal Disease, Nassau County Medical Center
SUNY at Stony Brook, East Meadow, N.Y.
Brookhaven National Laboratories, Upton, N.Y.

INTRODUCTION

Histomorphometric analyses of bone, bony radiographic analysis, bone densitometry and total body mineral content in hemodialyzed patients indicate that skeletal mass is quite variable in patients receiving hemodialysis for one to three years.[1,2,3] No long term studies of skeletal mass are available in patients receiving hemodialysis treatments for five to ten years. This study estimates skeletal mass by measurements of total body calcium by neutron activation analysis.

NEUTRON ACTIVATION ANALYSIS

Neutron activation analysis is a nondestructive technique which has been used widely for quantitative determination of the large number of elements in the past 20 years. The technique is based on the measurement of prompt grays from an excited compound nucleus. The activation occurs when neutrons collide with and are absorbed into the nuclei of atoms present in the tissues. If the new isotopes thus produced are radioactive, their presence can be detected by a whole body counter. Therefore, to determine absolute levels of such elements as calcium, sodium chloride, and phosphorus by neutron activation, a uniform neutron flux density throughout the body is required to activate all of the individual elements in the body. In addition, a sensitive whole body counter with an invariant counting response to radionuclide distribution and body size is required to measure the absolute levels of the y emissions of the induced radionuclides. The activations in practice are performed by radiating the human subject with partially moderated 14 MeV neutrons from a generator or neutrons from a portable α N neutron source and bilateral exposure in a radiation cell. One can measure absolute levels of calcium, sodium, and chloride along with relative measurements of nitrogen and phosphorus in patients exposed to these neutrons. The neutron sources (238 Pu-Be) provide a fairly uniform broad beam of neutrons with an average energy of 5 MeV from the following reaction: $^{9}Be + ^{4}He = ^{12}C + N$.

At the termination of radiation exposure, the patient is removed from the radiation cell and quickly transported to the whole body counter. A 15-min count is taken on each subject beginning 3 min post radiation. The data are processed and analyzed with an on-line computer. The levels of Ca, Cl, and Al (from phosphorus) induced following exposure to the moderated Pu-Be neutrons, are determined by whole body counting. We estimate the precision of reproducibility of whole body neutron activation has the potential of a 1% error for calcium.

NORMALIZATION OF TOTAL BODY STORES OF ELECTROLYTES

In order to calculate the deficit or excess in total body content of certain elements in individual patients from the absolute measurement of total body elements, it is necessary to normalize the data for each patient with a correction based on sex, age, weight and body habitus. The variations, for example, in the total body calcium are large because of the variation in the size of the individual, the amount of skeletal mass, and the extent of bone mineralization. Therefore, in addition to accurate methods of measuring whole body composition, a standard of reference is needed against which the measurements can be compared. Calculation of relative total body calcium or excess may be estimated by equation.

$$\text{TB Ca} = \alpha \times H\sqrt{K}$$
$$\alpha \text{ males} = 54.5$$
$$\alpha \text{ females} = 56.6$$

which is used to preduct normal body calcium based on weight, height, sex, size and age of the patient. This equation is calculated from a large series of normal patients who had had neutron activation analysis. The measured total body Ca expressed in terms of the predicted normal total body Ca is referred to as the calcium ratio. The relative deficit for an individual patient is equal to 1- Ca ratio. In normal subjects, the calcium ratio is 1.0 + 0.1; (+2S.D.). In the above equation, the total body K measured by whole body counting is an index of lean body mass, i.e. the body weight minus the variable fat content of the body. Since there are variations among patients in the adipose tissue of the body which contains no Ca, Na, or K, expressing body composition or compartment sizes relative to weight or height is subject to a greater variance than if the variables were expressed in terms of lean body mass.

CLINICAL STUDIES IN LONG TERM DIALYSIS PATIENTS

Eight patients receiving hemodialysis with a 1 sq meter dialyzer 12-18 hours weekly for as long as 10 years were studied. Dialysate calcium concentration was 6.50 mg/dl. All subjects received phosphate binding gels, therapeutic quantities of either 25(OH)D3 or 1,25(OH)2D3 and their diets were supplemented with 1 gram of elemental calcium. Measurements of total body calcium were performed at intervals of 3-12 months in most patients and compared to age adjusted normal subjects. Table I lists the clinical characteristics of the patients. Seven males and one female, varying in age from 23-58 years, were studied for 60-148 months with a minimum of 3 measurements to as many as 11 measurements. Table II lists the blood biochemical and skeletal radiographic severity of renal osteodystrophy of these patients. All patients demonstrated mild metabolic acidosis, modest hyperphosphatemia, and low normal total serum calcium. Alkaline phosphatase levels were elevated in most patients. Modest to markedly elevated serum immunoreactive C terminal PTH values were noted. The severity of the radiographic index for renal osteodystrophy was only modestly increased in some patients. Two patients required subtotal parathyroidectomy for control of renal osteodystrophy.

TABLE I

CLINICAL CHARACTERISTICS OF THE PATIENTS

PT NAME	SEX	AGE AT ONSET OF STUDY	WEIGHT (lb)	HEIGHT (cms)	ETIOLOGY OF C.R.F.	FIRST HD / NO HRS OF DIALYSIS PER WEEK	FIRST STUDY	DURATION OF STUDY (MONTHS)	NO. OF OBSERVATIONS
S.F.	M	50	178	178	UNKNOWN	5-1972 / 15	JAN 1972	60	5
R.A.	M	23	136	178	MEMBRANO PROLIF. G.N.	9-1968 / 15	MAR 1970	82	9
M.J.	M	38	157	171	MALIG. HYPER-TENSION	12-1973 / 12-15	MAR 1974	96	3
S.I.	M	37	151	166	CHRONIC G.N.	6-1972 / 12-15	AUG 1972	104	9
V.N.	M	58	154	170	CHRONIC PYELO-NEPHRITIS	9-1968 / 15-18	APR 1970	104	11
D.J.	M	35	139	160	CHRONIC G.N.	2-1973 / 12-15	OCT 1972	130	8
B.J.	M	49	174	164	MEMBRA-NOUS NEPHROPATHY	12-1969 / 12-18	APR 1970	148	8
G.A.	F	32	182	168	POLYCYSTIC KIDNEY DISEASE	7-1968 / 12-18	MAR 1970	71	11

TABLE II

BLOOD BIOEHEMISTRY AND SKELETAL CHARACTERISTICS OF THE PATIENTS

PT. NAME	SERUM Na MEAN RANGE	SERUM K MEAN RANGE	SERUM CL MEAN RANGE	SERUM HCO_3 MEAN RANGE	SERUM Ca MEAN RANGE	SERUM PHOS MEAN RANGE	SERUM A.P. MEAN RANGE	PTH	SEVERITY OF R.O.D. ONSET OF STUDY 0-4	SEVERITY OF R.O.D. END OF STUDY 0-4
S.F.	140 137-143	5.44 5.2-5.9	104 101-105	17 12-20	8.2 7.2-9.4	7.4 6.4-8.6	222 220-224	282	0	0
R.A.	138 132-145	5.0 4.2-6.3	96 83-103	22 20-28	9.6 8.1-10.4	6.1 3.8-8.0	115 75-320	411	2+	0
M.J.	142 137-148	4.8 3.6-6.1	102 97-111	19 14-26	8.9 7.5-10.1	6.5 3.8-8.3	89 52-131	1104	0	0
S.I.	140 127-147	5.2 3.5-6.3	98 96-101	19 13-24	10.0 9-10.6	5.8 3.9-7.8	68 63-163	305	0	0
V.N.	139 135-145	5.3 4.1-6.7	98 93-105	21 16-32	9.3 7.4-12.5	4.4 2.8-7.2	76 34-138	287	1+	1+
D.J.	136 134-139	5.5 4.9-6.4	101 93-109	15 10-24	9.6 8.7-10.7	5.0 1.8-12.0	194 81-368	1294	0	3+
B.J.	137 134-140	6.0 5.7-6.2	98 94-102	19 16-21	9.6 9.5-9.9	5.8 5-7.4	34 27-45	1486	0	0
G.A.	137 127-142	5.8 3.7-7.6	97 90-103	N.A.	8.7 7-10.8	8.0 6.7-10.8	96 51-290	N.A.	0	2+

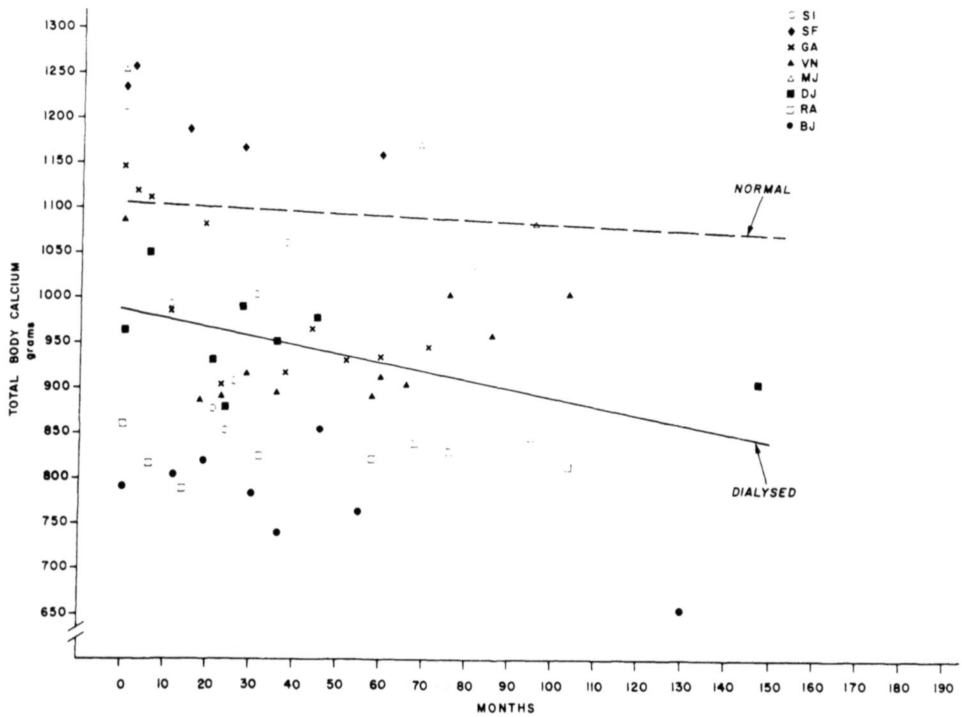

Figure 1 Total body calcium in dialyzed patients and normal subjects. The dashed line represents the regression equation for normal subjects. The solid line represents the regression equation for dialyzed patients.

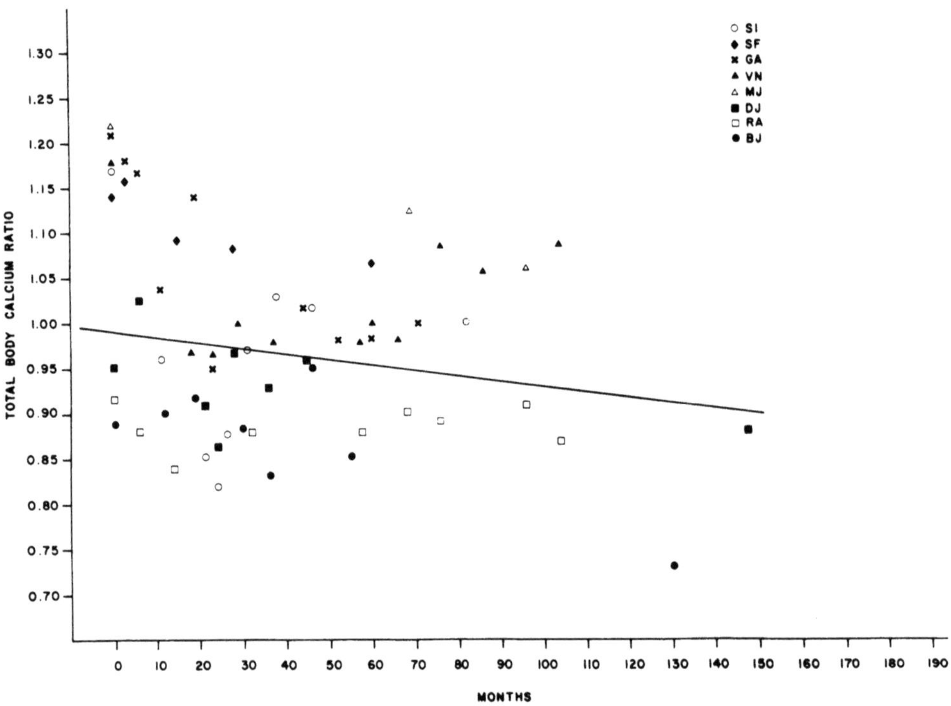

Figure 2 Total body calcium ratio in dialyzed patients. Solid line depicts the regression equation for dialyzed patients.

RESULTS

As depicted in Figure 1, total body calcium decreases progressively with time. In normal males, age 40-60, the annual loss of total body calcium was 6.0 grams. In the eight dialyzed patients, the calculated annual loss was 12 grams. Figure 2 depicts the same measurements with a different frame of reference i.e. the observed whole body calcium in relation to the predicted whole body calcium normalized for age, sex and body habitus. These data indicate that no major whole body calcium loss does occur with long term hemodialysis treatments.

DISCUSSION

Using modern strategies to protect the skeleton from the factors that aggravate bone loss and impair bone formation, these studies clearly demonstrated that skeletal mass as estimated from total body calcium is well maintained in long term dialysis patients receiving supplemental calcium in their diet, analogues of Vitamin D, phosphate binding gels and a dialysate calcium concentration that limits net flux of calcium from the patient. Indeed, the loss of calcium on an annual basis is only 6 g/year greater than normal and clearly of limited biologic importance.

It is unwise to conclude from these observations that certain parts of the skeleton may not be affected in the dialysis patient. Thus, a major redistribution of bone mass may occur in dialyzed patients i.e. between spongy and cortical bone or between axial and appendicular skeleton. We have previously noted that in some dialysis patients, changes in forearm bone mineral content did not always correlate with changes in total body calcium suggesting redistribution of minerals between spongy and cortical bone and between axial and appendicular skeleton.[3]

CONCLUSIONS

1. Total body calcium is not significantly different from normal in patients receiving dialysis for 5-10 years.
2. Whole body calcium decreased modestly in long term dialysis patients and is of limited biologic importance.
3. Redistribution of calcium between different parts of skeleton e.g. spongy and cortical bone may occur and may be a factor in development of osteodystrophy.

REFERENCES

1. S.H. Cohn and C.S. Dombrowski: Measurement of total body calcium, sodium chloride, nitrogen and phosphorus in man by in-vitro neutron activation analysis. J. Nucl Med 12:499-505 (1971)
2. S.H. Cohn and C.S. Dombrowski: Absolute measurement of whole body potassium by gamma ray spectrometry. J Nucl Med 11:239-246 (1970)
3. J.M. Letteri and S.H. Cohn: Total body neutron activation analysis in the study of mineral homeostasis in chronic renal disease, in "Calcium Metabolism in Renal Failure and Nephrolithiasis", S. David, ed., John Wiley & Sons, New York (1977)

PARATHYROID HORMONE

PARATHYROID HORMONE ANTAGONISTS EFFECTIVE *IN VIVO*

Lynn H. Caporale and Michael Rosenblatt

Merck Sharp & Dohme Research Laboratories
West Point, Pennsylvania

INTRODUCTION

There has long been great impetus for preparing peptide hormone antagonists effective *in vivo*. Pure hormone antagonists can be used to determine the mechanism of action of peptide hormones and to elucidate both the role of a hormone in normal physiology and its contribution to pathophysiology. For parathyroid hormone in particular, the impetus for generating antagonists effective *in vivo* includes a wide variety of therapeutic and diagnostic clinical uses in hyperparathyroid conditions (Table 1) and diagnostic utility in disorders of calcium metabolism.

In addition, if parathyroid hormone is found to play a critical or permissive role in the development of osteoporosis, it is possible that parathyroid hormone antagonists may serve a useful role in this common medical disorder.

Table 1. Parathyroid Hormone Antagonists: Potential Utility

A. Acute treatment of hypercalcemic crisis ("parathyroid storm" due to parathyroid adenoma, hyperplasia, or carcinoma); pre-operative optimization of blood calcium levels

B. Chronic treatment of hyperparathyroidism
   1. High-risk surgical candidates
   2. Patients who have undergone unsuccessful surgery
   3. Parathyroid carcinoma (inoperable)
   4. ? General medical management of hyperparathyroidism

C. ? Adjunct treatment
   1. Renal failure + secondary hyperparathyroidism
   2. Post-renal transplantation
   3. ? Osteoporosis

THE GENERAL PROBLEM OF PEPTIDE HORMONE ANTAGONIST DESIGN

The design of peptide hormone antagonists contrasts with the design of enzyme inhibitors in several important ways. Firstly, the crystal structure of a substrate in the active site of certain enzymes is known, as is the chemical mechanism by which catalysis occurs.[1] Structural information regarding peptide hormone--receptor activation has not been obtained at the same level of detail for any hormonal system. The task is further complicated by the existence of multiple forms of receptor for a given hormone within either a single tissue or in different tissues in the body, some of which we may want to affect differentially, as is illustrated by the endogenous opioid peptides[2] and vasopressin[3] systems. These factors, in part, have made the design of effective hormone antagonists a more difficult task than that of developing inhibitors of proteolytic enzymes; the latter has now proven highly successful for several clinically important enzymes, such as angiotensin-converting enzyme[4,5] and renin.[6,7]

Another difficult problem in obtaining effective antagonists _in vivo_ has been the design of specialized bioassays for evaluation of analogs. It is extremely important to "cast a wide net" of assays to build a high confidence level concerning inhibitory properties, and to carefully design bioassays which facilitate the large jump from _in vitro_ to complex _in vivo_ systems, where multiple factors such as metabolism and clearance play a large role in antagonist efficacy.[8] Many of these issues are understood poorly, if at all.

THE DESIGN AND BIOLOGICAL EVALUATION OF PARATHYROID HORMONE ANTAGONISTS

Having discussed some of the general issues involved in the preparation of peptide hormone antagonists, a closer examination is provided of the design and synthesis of parathyroid hormone antagonists effective _in vivo_.

Structure--Activity Studies

Over the past decade, we have chemically synthesized nearly 100 fragments and analogs of pre-proparathyroid hormone[9,10] using modifications of the solid-phase method of peptide synthesis developed by Merrifield,[11] for which he received the Nobel Prize in Chemistry in 1984. Taken together, these fragments comprise, in overlapping fashion, the entire 115-amino acid sequence of pre-proparathyroid hormone. Since, in most cases, the peptides that were synthesized were 30-40 amino acids in length (and thus close to the limits of the synthetic methodology available at the time), care was taken to assure complete amino acid incorporation during synthesis by special monitoring techniques,[12] and equal attention was given to issues of purification of the synthetic product.[13] At least two independent purification modalities were employed, followed by a battery of analytical methods.[14] Using this approach, the synthetic peptides which were prepared could be demonstrated to accurately represent the desired sequence. It is only by using authentic and highly purified peptides that conclusions drawn from biological experiments can be valid.

In addition, the development of a radioreceptor assay by Segre et al.[15] proved instrumental in the effort to design antagonists. The advent of radioreceptor binding assays has made it possible to evaluate hormone--receptor interactions at very high concentrations of weak affinity hormone fragments or analogs. These concentrations are so high as to be clearly remote from the realm of normal physiology. Nevertheless, studying the effect of structural manipulation on binding affinity provides unique and invaluable insights into the first step of hormonal action, namely interaction of hormone with receptor. Ultimately, this approach has led to design of hormone antagonists of the requisite receptor avidity for in vivo usage.

The structure--activity studies directed at designing parathyroid hormone antagonists began with an effort to identify the minimum sequence of the hormone molecule necessary for biological activity. Firstly, Potts et al.[16] demonstrated, using a synthetic peptide, that all of the biological activity of parathyroid hormone (in terms of conventional parameters of mineral ion flux and cyclic AMP generation) reside within the amino terminal one-third of the molecule, the sequence 1--34 (Figure 1). What followed were a series of studies by Tregear et al.[17] delineating precisely the minimum number of contiguous amino acids required for bioactivity. Stepwise deletions from the

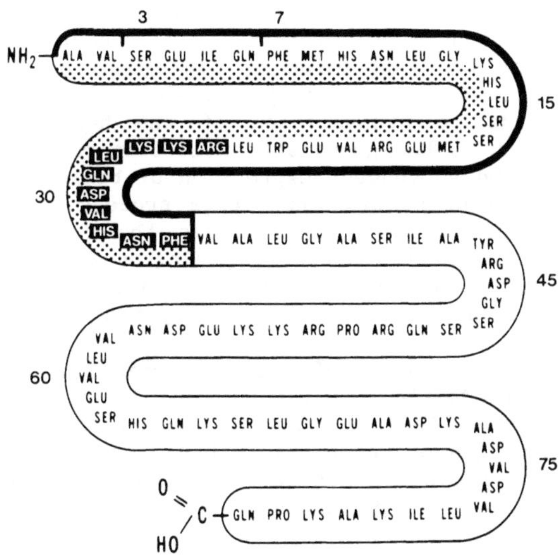

Figure 1. The 84-amino acid sequence of bovine PTH-(1--84). The fully biologically active region 1--34 is indicated by stipling. The 1--34 sequence is largely separable into 1) a region that can bind to PTH receptors without causing activation of adenylate cyclase and 2) a small amino-terminal segment that is important for activation of second messenger system(s) once binding to the receptor has occurred. The precise extent of amino-terminal truncation required for loss of receptor-activation has not been determined. Also indicated on this figure are the amino termini of the 3--34 and 7--34 analogs described later in the text. Shaded residues depict the 25--34 region, which satisfies the structural requirements for receptor occupancy and represents a principal binding domain of the hormone molecule.

carboxyl-terminal position 34 toward the amino terminus produced a progressive decline in potency until position 25 was reached. The remaining sequence, namely the segment 1--25, did not bind to parathyroid hormone receptors at a detectable level and lacked bioactivity completely in the available assays.

There were striking differences between carboxyl-terminal and amino-terminal deletions. At the amino terminus bioactivity was lost in short order. Instead of requiring removal of 10 amino acids, the simple removal of the residues at positions 1 and 2 produced an apparent complete loss in bioactivity in the same in vitro assays. More importantly from the viewpoint of design directions for antagonists, the remaining fragment comprising the sequence 3--34 still binds, although weakly, to parathyroid hormone receptors.[15,18] This single finding provided the conceptual wedge for dividing the hormone molecule into "binding" and "receptor-activation" domains, and spurred attempts to generate more potent parathyroid hormone antagonists.

This concept fits the model that the peptide hormone--receptor interaction is allosteric in nature[19]. In the first stage, hormone binds to receptor. Binding to a target-tissue receptor is required, but not alone sufficient for expression of biological action.[3,18] Binding is followed by a second event, namely a conformational shift within the receptor which leads to transmission of the hormonal signal to an effector system (such as the guanylate nucleotide regulatory subunit and adenylate cyclase) and subsequent expression of biological activity. It is this fundamental concept of a two-stage process leading to an allosteric change in the receptor which forms the foundation of hormone antagonist design, an ideal antagonist being a molecule which retains strong binding capability, but is unable to produce the conformational shift leading to activation of the effector system.

Thus, the separation of a "binding" domain from a "receptor-activation" domain for the hormone by deletions of critical regions of the hormone leads to identification of structural features necessary for activation of receptors. This information, in turn, leads to the synthesis of hormonal fragments which bind to receptors without activating effector and second messenger systems, i.e. antagonists.

A Potent In Vitro Parathyroid Hormone Antagonist

Because the 3--34 analog, which had a dramatic loss of potency, bound only weakly, a series of structure--activity studies were undertaken to enhance avidity of the 3--34 analog for the parathyroid hormone receptor. The details of these studies will not be discussed here, although several years of effort were required. In these studies the effects on biological activity of combining several structural alterations in the same analog approximated the result of adding the effects of individual modifications. What emerged for parathyroid hormone was the most potent in vitro antagonist for any peptide hormone at the time[20]. The analog [Nle-8,Nle-18,Tyr-34]bPTH-(3--34)amide was able to compete effectively for occupancy of parathyroid hormone receptors on a one-to-one molar basis with fully active parathyroid hormone. This analog combined structural features which enhance affinity for the receptor, such as the placement of a tyrosine amide at the carboxyl terminus[21], with modifications which provide stability (in the form of resistance to oxidation), such as substitution for methionine with norleucine (a non-naturally occurring amino acid)[22]. The inhibitory potency of this analog was approximately 200-fold greater than the parent unsubstituted peptide, bPTH-(3--34), and it appeared to completely inhibit parathyroid hormone-stimulated adenylate cyclase activity in available renal[20,23,24] and bone-derived in vitro assays.[25-28] This analog gained widespread acceptance and utility in many laboratories studying parathyroid hormone action and disorders of calcium metabolism.[28-32] Given the high potency of this antagonist and its efficacy across multiple in vitro assay systems, we were encouraged to evaluate its biological properties in vivo. We undertook large scale synthesis of this compound to obtain sufficient material for study in dogs.

The Difficult Problem of Predicting In Vivo Action from In Vitro Assays

For parathyroid hormone, our most effective competitive in vitro antagonist displayed weak, but definite, PTH-like agonist properties in vivo.[33] When administered intravenously in dogs together with parathyroid hormone (at a 200:1 molar ratio of analog to native hormone), a large increase, rather than the desired decrease, in serum calcium levels was observed. Even at a 20:1 molar ratio of analog to parathyroid hormone, elevations in blood calcium over those expected for parathyroid hormone administration alone were observed. In fact, each of the major parameters of parathyroid hormone action were displayed by the 3--34 analog alone (when sufficiently high doses were administered): elevation of blood calcium and promotion of urinary phosphate and cyclic AMP excretion. The results obtained in multiple assay systems indicated that while the analog bound as well as the agonist PTH 1--34, it lacked antagonist activity in some systems[34,35] and it had a potency as an agonist of approximately 1/300th that of parathyroid hormone in others,[36,37] precluding its use as an antagonist in vivo.

These findings highlight a major and generic issue in the effort to design peptide hormone antagonists effective in vivo, namely the problem of retained weak agonist properties or partial agonism in in vitro antagonists revealed when the transition is made to in vivo bioassays.[3,38,39]

What are the possible reasons for the observed discrepancies between the in vitro and in vivo results for the parathyroid hormone analog? It was possible that the adenylate cyclase systems extensively studied in vitro were not directly related to parathyroid hormone mechanism of action in vivo, i.e. cyclic AMP was not the principal second messenger for the physiologically relevant actions of the hormone, although there is no definitive evidence for another second messenger for PTH at this time.

It is also important to keep in mind that when one moves to in vivo systems, the small size of many peptide analogs leads to rapid enzymatic degradation and clearance. Differences in peptide availability at the receptor due to differences in distribution and degradation tremendously complicate the move from in vitro to in vivo assays, but this could not be the explanation for the discrepancy observed in this case; the peptide was clearly available, yet it appeared to behave differently at the receptors in vivo. It is likely that differences exist between the environment of the receptors in vitro and in vivo. It is also possible that receptors or the effector system might somehow be damaged in the course of preparation of the in vitro assay system in such a way that a weak agonist would be unable to elicit a hormonal response, and thus appear to be an antagonist.

In fact it is not surprising that a compound which may cause the proposed receptor conformational change at a low rate may reveal this agonism under appropriate sensitive assay conditions, in this case in vivo, yet antagonize the action of an agonist more efficient at causing the conformational change (i.e. one with a higher ratio of potency to binding affinity), under less optimal conditions.

Thus the next important goal has been to redesign in vitro assays to improve their ability to predict in vivo behavior. Good methods to predict peptide distribution and stability are not yet available, but significant progress has occurred in elucidating the biochemical steps by which agonists activate second messenger systems. We altered the cyclase assay in light of current models of hormone activation of cyclase[40,41]. These changes have enabled us to demonstrate in vitro the agonism of 3--34.

The cyclase assay conditions were altered with consideration for the important role of Mg^{++} in the activation of adenylate cyclase in PTH target tissues[42] and other systems.[40,41] The enzyme adenylate cyclase is now thought to be activated by binding to the alpha subunit of a "nucleotide regulatory" (N) protein bound GTP. This free alpha subunit can be dissociated from the $N_{\alpha\beta\gamma}$ complex in the presence of Mg^{++} and GTP. This dissociation can occur in the absence of receptor at high Mg^{++} concentrations. Because high Mg^{++} levels increase dissociation of the N complex and activation of cyclase in the absence of hormone, cyclase stimulation is more hormone-dependent at low Mg^{++} concentrations.

Figure 2 illustrates the effects of the Mg^{++} concentration on fold-stimulation above basal for the full agonist [Nle-8,Nle-18,Tyr-34]-bPTH-(1--34)amide. Although the actual quantity of cAMP produced is greater at high Mg^{++} concentrations (not shown), the fold-stimulation above basal is greatly enhanced at the lower Mg^{++} concentrations, as expected, because basal cyclase stimulation is more Mg^{++}-dependent than hormone-stimulated cyclase activity. Therefore, the use of low [Mg^{++}] should lower hormone-independent basal cyclase activity and

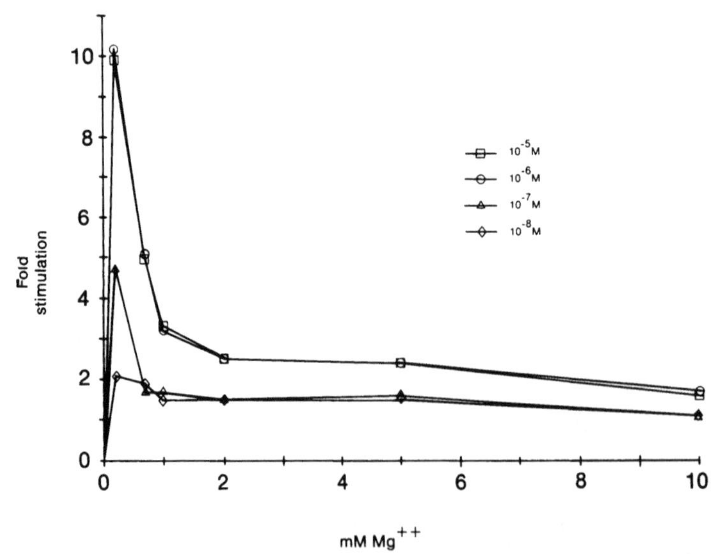

Figure 2. Effect of magnesium concentration on fold stimulation above basal for [Nle-8,Nle-18,-Tyr-34]bPTH-(1--34)amide at concentrations of --□-- 10^{-5} M; --○-- 10^{-6} M; --△-- 10^{-7} M; --·◇·-- 10^{-8} M.

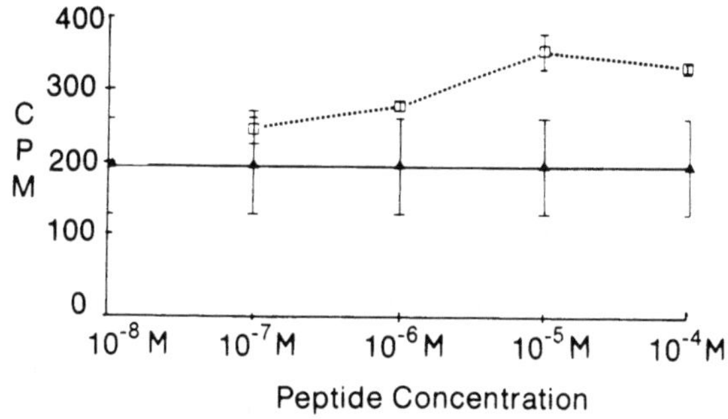

Figure 3. *In vitro* agonism of [Nle-8,Nle-18,Tyr-34]-bPTH-(3--34)amide. Cpm ^{32}P incororated into cAMP in the presence of increasing concentrations of the 3--34 analog in the presence of 0.2 mM Mg^{++}.

thus permit detection of small amounts of cyclase stimulation by very weak agonists. As a first test of the redesigned assay, it was important to demonstrate that it is now able to detect agonism of the 3--34 analog is clearly revealed at low [Mg^{++}], which may in fact better correspond to the receptor environment *in vivo* (Figure 3).

Thus, assays designed for increased hormone-dependency, rather than increased cAMP production, should provide a reliable *in vitro* measure of agonism and more accurately reflect relevant *in vivo* properties of PTH analogs.

A Parathyroid Hormone Antagonist Effective In Vivo

What then was necessary to allow us to proceed to develop an *in vivo* antagonist? Both new chemical design approaches and the development of new *in vivo* bioassays were required.

Because of the dramatic effect of amino terminal truncation on potency, we continued to focus our attention on alterations in this region. Substitutions were made at position 3 and assayed *in vitro*.[43] Together with Dr. F. Richard Bringhurst, we evaluated a series of 3--34 analogs containing subtle structural modifications at the amino terminus (unpublished data). All of these compounds contained residual agonist-like activity.

Nussbaum et al.[44] in our laboratory pursued the most promising remaining design pathway: further truncation of the 3--34 analog at the amino terminus. We suspected that truncation would be accompanied by a major liability, namely loss of affinity for the receptor.[45] A series of amino-terminally truncated analogs were generated from a single synthesis so that they could be validly compared in a parathyroid hormone radioreceptor assay. Several critical findings emerged. Stepwise truncation from the amino terminus toward the carboxyl terminus did lead to a progressive decline in receptor affinity. However, fragment analogs such as 7--34, 10--34, 15--34, etc. still bound to the parathyroid hormone receptor when added in sufficiently high concentration. In fact, a striking observation was made: for parathyroid hormone, a region (the segment 25--34) of only 10 amino acids (out of the complete native hormone sequence of 84 amino acids) could occupy receptors and completely inhibit binding of a fully active radioligand.

We also found that norleucine-for-methionine substitution was poorly tolerated in vivo, producing a 15-fold decline in bioactivity.[9,10,34] Hence, we returned to incorporating the native methionine at positions 8 and 18. From the receptor binding studies, it was known that the affinity of the 7--34 analog for parathyroid hormone receptors was at least 10-fold weaker than parathyroid hormone itself or the previously studied 3--34 analog. Nevertheless, in the hope that remaining agonist-like properties had been eliminated, we undertook large scale synthesis of [Tyr-34]bPTH-(7--34)amide in order to evaluate its biological properties in vivo.

Knowing that the inhibitory potency would be relatively weak and assuming the clearance would be rapid, the amount of material needed for in vivo studies became a major issue. It compelled us to carefully design in vivo assays which reflected major parameters of parathyroid hormone bioactivity and in which the effects of a true antagonist could be ascertained rapidly and definitively, utilizing minimum amounts of antagonist.

A variation of the Munson assay[46] was designed by Horiuchi et al.[47] This new assay reflects the renal effects of parathyroid hormone in vivo using the parameters of increased urinary phosphate and cyclic AMP excretion in response to hormone. In addition, the assay was designed so as to afford the analog the greatest opportunity to reveal antagonist properties. Hence, intravenous infusion of the 7--34 analog was begun one hour prior to and then simultaneously with the administration of parathyroid hormone in rats. The 7--34 analog inhibited the parathyroid hormone-mediated phosphaturic and cyclic AMP responses in a dose-dependent manner. When administered at a molar dose ratio of antagonist to parathyroid hormone of 200:1, complete inhibition of the parathyroid hormone response was observed (Figure 4).[47] In addition, when analog was administered alone at these doses, no parathyroid hormone-like agonist activity was observed. These studies demonstrated for the first time that parathyroid hormone action could be inhibited completely in vivo.

Having demonstrated effective antagonism for renal responses to parathyroid hormone, we next attempted to inhibit hormone action on bone in vivo. This appeared to be a more difficult task. For the renal parameters of parathyroid hormone action, the elapsed time between interaction of the hormone with receptor and the expression of biological activity (urinary excretion of phosphate and cyclic AMP) is only seconds to minutes. For bone, there is a relatively long lag period (several hours) between the time that parathyroid hormone interacts with receptors and calcium is released from bone, resulting in increased blood calcium levels in vivo. We were concerned that inhibitory properties, even for a highly potent antagonist, could be overlooked in a bone-based in vivo assay due to a comparatively short half-life in the circulation. It seemed theoretically possible to interrupt hormone--receptor interaction completely (using the antagonist) for several minutes, yet this transient antagonism would go undetected several hours later because of the long lag time from hormone--receptor interaction to the integrated target organ response.

Therefore, a new bone-based in vivo assay was designed collaboratively with Dr. Samuel H. Doppelt and Dr. Robert M. Neer. The assay was both highly sensitive and rapid. Rats were placed on a low calcium diet for one week, during which period they developed secondary hyperparathyroidism: maintenance of normal blood calcium levels was dependent on parathyroid hormone action on bone. When animals were thyroparathyroidectomized, calcium levels declined rapidly, producing

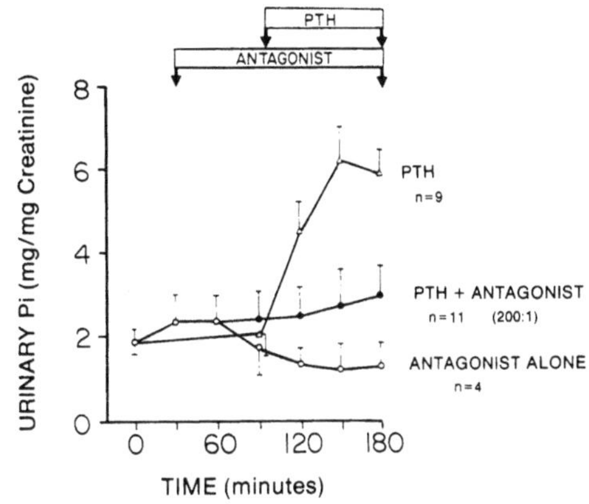

Figure 4. Evaluation of the effects of [Tyr-34]-bPTH-(7--34)NH_2 on urinary phosphate excretion in rats. The data shown are compiled from a total of 24 animals. Nine control rats received native bPTH-(1--84) alone intravenously at a rate of 0.27 nmole/hour. Eleven rats received [Tyr-34]-bPTH-(7--34)NH_2 and the native peptide through separate intravenous cannulas at rates of 54 nmole and 0.27 nmole per hour, respectively (a molar ratio of 200 to 1). Zeros in the abscissa represent values prior to peptide infusion. Infusion of the analog was begun 1 hour before the beginning of the infusion of native parathyroid hormone. Four animals received antagonist alone.

tetany and death within 6-12 hours. However, one hour after thyroparathyroidectomy, animals could be "rescued" by administration of a bolus of parathyroid hormone intravenously, which restored calcium levels to normal or higher than basal. The "rescue" effect was shown to reflect parathyroid hormone action on mobilizing calcium stores from bone (using radiolabeling of bone). The design of the assay for evaluation of the antagonist was to administer the 7--34 analog 15 seconds prior to, simultaneous with, and 15 minutes after the single bolus of parathyroid hormone. In preliminary experiments, when [Tyr-34]bPTH-(7--34)amide was administered in three intravenous doses (each containing a 30-fold molar excess of antagonist to parathyroid hormone), complete inhibition of the parathyroid hormone-mediated calcemic response was observed[48].

CLINICAL GOALS AND FUTURE PLANS

These combined in vivo studies[47,48] demonstrated that each of the major parameters of parathyroid hormone action in vivo, in terms of the serum calcium, urine phosphate, and urine cyclic AMP response, could be inhibited completely with the 7--34 antagonist. However, future animal studies are required to determine the potential clinical utility of parathyroid hormone antagonists in disorders of calcium metabolism. We have planned such studies for models of hyperparathyroidism and the hypercalcemia of malignancy.

Although the 7--34 antagonist will be evaluated for its ability to inhibit parathyroid hormone action under conditions which more closely mimic clinical hyperparathyroidism, it is likely that this antagonist lacks the requisite potency and metabolic stability to be effective in such a system. Therefore, we have also begun the design of "second generation" antagonists which may prove more potent and have a longer half-life in the circulation, as we move closer to use in models of pathophysiology. Chronic diminishment of parathyroid hormone-dependent hypercalcemia in animals by an antagonist will provide important data to support clinical evaluation of parathyroid hormone antagonists.

Another major therapeutic category for which parathyroid hormone antagonists may prove useful is the hypercalcemia of malignancy.[49] It appears as though in a substantial number of cancer patients hypercalcemia is produced by tumor-secreted factors, rather than by direct osteolysis of bone by metastases. These factors, secreted into the blood by malignancies, act on bone to stimulate resorption. In many cases, a clinical syndrome similar to hyperparathyroidism is produced,[50] yet it is clear that in many of these cases, the factor secreted by the tumor is not parathyroid hormone by both radioimmunoassay and physicochemical criteria.[50,51] It has not been established whether all tumors produce the same bone-resorbing factor, but it is possible that different factors exist or must be present concommitantly to produce hypercalcemia. In some instances, there is evidence to support the concept that malignancies produce factors which, although not parathyroid hormone, act like parathyroid hormone by interacting with and stimulating parathyroid hormone receptors. If tumor factors cause hypercalcemia by interacting with parathyroid hormone receptors, then antagonists which occupy parathyroid hormone receptors may have clinical utility in treating patients who suffer from hypercalcemia of malignancy.

Our studies demonstrate the feasibility of generating effective antagonists of parathyroid hormone action *in vivo* and establish a principal design direction for generating such antagonists. These efforts are facilitated by development of *in vitro* bioassays which appear to discriminate between true antagonists and weak agonists with rapidity, high sensitivity, and minimal consumption of peptide.[52]

Efforts are now underway to design more potent antagonists and to evaluate their utility first in animal models and ultimately in the treatment of several disorders of calcium metabolism. However, these parathyroid hormone antagonists and the peptide antagonists of other hormonal systems must be administered intravenously, which will limit their clinical utility. In order to gain wide utility as therapeutic agents, progress is needed in understanding how to design such compounds to be effective after oral, trans-mucosal (intranasal), or continuous subcutaneous (pump) administration.[8,53,54]

ACKNOWLEDGMENTS

Among the many colleagues who have contributed to the elucidation of the structural requirements for PTH activity, and to the improvement of bioassays we would especially like to thank Samuel H. Doppelt, Steven R. Goldring, Noboru Horiuchi, Robert M. Neer, Samuel R. Nussbaum, John T. Potts, Jr., George A. Tyler and Patricia DeHaven. Sandy Camburn provided expert secretarial assistance.

REFERENCES

1. A. Fersht, "Enzyme Structure and Mechanism," W. H. Freeman & Company, New York (1985).
2. W. R. Martin, Pharmacology of opioids, Pharmacology Reviews 35:283 (1984).
3. W. H. Sawyer and M. Manning, The development of vasopressin antagonists, Federation Proc. 43:87 (1984).
4. M. A. Ondetti and D. W. Cushman, Enzymes of the renin-angiotensin system and their inhibitors, Annu. Rev. Biochem. 51:283 (1982).
5. C. S. Sweet and E. H. Blaine, Angiotensin-converting enzyme and renin inhibitors, in "Cardiovascular Pharmacology," M. Antonaccio, ed., Raven Press. New York, (1984).
6. E. Haber, R. Zusman, J. Burton, V. J. Dzau, and A. C. Barger, Is renin a factor in the etiology of essential hypertension?, Hypertension 5:V-8 (1983).
7. J. Boger, Renin inhibition, in "Annual Reports in Medicinal Chemistry," Chapter 25 (in press).
8. D. F. Veber and R. M. Freidinger, The design of metabolically stable peptide analogs, Trends Neuro. Sci. 8 (in press) (1985).
9. J. T. Potts, Jr, H. M. Kronenberg, and M.Rosenblatt, Parathyroid hormone: chemistry, biosynthesis, and mode of action, Adv. Protein Chem. 35:323 (1982).
10. M. Rosenblatt, Parathyroid hormone: intracellular transport, secretion, and receptor interaction, Peptide Protein Rev. 2:209 (1984).
11. R. B. Merrifield, Solid phase peptide synthesis. I. The synthesis of a tetrapeptide, J. Am. Chem. Soc., 85:2149 (1963).
12. A. M. Felix and M. H. Jimenez, Rapid fluorometric detection for completeness in solid phase coupling reactions, Anal. Biochem. 52:377 (1973).
13. G. A.Tyler and M. Rosenblatt, Semi-preparative high-performance liquid chromatographic purification of a 28-amino acid synthetic parathyroid hormone antagonist, J. Chromatogr. 266:313 (1983).
14. M. Rosenblatt, G. W. Tregear, G. L.Shepard, G. A. Tyler, M. Veroni, and J. T. Potts, Jr., Comparison of two solid-phase peptide syntheses of a 32-amino acid carboxyl-terminal fragment of human parathyroid hormone, hPTH-(53-84), Arch. Biochem. Biophys. 199:286 (1980).
15. G. V. Segre, M. Rosenblatt, B. L. Reiner, J. E. Mahaffey, and J. T. Potts, Jr., Characterization of parathyroid hormone receptors in canine renal cortical plasma membranes using a radioiodinated sulfur-free hormone analogue: correlation of binding with adenylate cyclase activity, J. Biol. Chem. 254:6980 (1979).
16. J. T. Potts, Jr., G. W. Tregear, H. T. Keutmann et al., Synthesis of a biologically active N-terminal tetratriacontapeptide of parathyroid hormone, Proc. Natl. Acad. Sci. USA 1971; 68:63 (1971).
17. G. W. Tregear, J. van Rietschoten, E. Greene, H. T. Keutmann, H. D. Niall, B. Reit, J. A. Parsons, and J. T. Potts, Jr., Bovine parathyroid hormone: minimum chain length of synthetic peptide required for biological activity, Endocrinology 93:1349 (1973).
18. D. Goltzman, A. Peytremann, E. Callahan, G. W. Tregear, and J. T. Potts, Jr., Analysis of the requirements for parathyroid hormone action in renal membranes with the use of inhibiting analogues, J. Biol. Chem. 250:3199 (1975).
19. R. J. Lefkowitz, M. G. Caron, and G. L. Stiles, Mechanisms of membrane-receptor regulation, New Engl. J. Med. 310:1570 (1984).
20. M. Rosenblatt, E. N. Callahan, J. E. Mahaffey, A. Pont, and J. T. Potts, Jr., Parathyroid hormone inhibitors: design, synthesis, and biologic evaluation of hormone analogues, J. Biol. Chem. 252:5847 (1977).

21. M. Rosenblatt and J. T. Potts, Jr., Design and synthesis of parathyroid hormone analogues of enhanced biological activity. Endocrine Res. Commun. 14:115 (1977).
22. M. Rosenblatt, D. Goltzman, H. T. Keutmann, G. W. Tregear, and J. T. Potts, Jr., Chemical and biological properties of synthetic, sulfur-free analogues of parathyroid hormone, J. Biol. Chem. 251:159 (1976).
23. D. A. Ausiello, M. Rosenblatt, and J-M. Dayer, Parathyroid hormone modulates protein kinase in giant cell tumors of human bone. Am. J. Physiol. 239:E144 (1980).
24. T. C. Chen, M. Rosenblatt, and J. B. Puschett, Effects of calcium on a parathyroid hormone-sensitive adenylate cyclase inhibitor, Biochem. Biophys. Res. Commun. 94:1227 (1980).
25. S. R.Goldring, J. E. Mahaffey, M. Rosenblatt, J-M. Dayer, J. T. Potts, Jr. and S. M. Krane, Parathyroid hormone inhibitors: comparison of biological activity in bone and skin-derived tissue, J. Clin. Endocrinol. Metab. 48:655 (1979).
26. S. R. Goldring, J-M. Dayer, and M. Rosenblatt, Factors regulating the response of cells cultured from human giant cell tumors of bone to parathyroid hormone, J. Clin. Endocrinol. Metab. 53:295 (1981).
27. T. Takano, M. Takigawa, E. Shirai, F. Suzuki, and M. Rosenblatt, Effects of synthetic analogs and fragments of bovine parathyroid hormone on adenosine 3',5'-monophosphate level, ornithine decarboxylase activity, and glycosaminoglycan synthesis in rabbit costal chondrocytes in culture: structure-activity relations, Endocrinology 116:2536 (1985)
28. M. P. M. Herrmann-Erlee, J. N. M. Heersche, J. W. Hekkelman et al., Effects on bone in vitro of bovine parathyroid hormone and synthetic fragments representing residues 1--34, 2--34, and 3--34, Endocrine Res. Comm. 3:21 (1976).
29. R. A. Nissenson, S. R. Abbott, A. P. Teitelbaum, O. H. Clark, and C. D. Arnaud, Endogenous biologically active human parathyroid hormone: measurement by a guanyl nucleotide-amplified renal adenylate cyclase assay, J. Clin. Endocrinol. Metab. 52:840 (1981).
30. G. J. Strewler, R. D. Williams, and R. A. Nissenson, Human renal cancer cells produce hypercalcemia and a novel protein recognized by parathyroid hormone receptors, J. Clin. Invest. 71:769 (1983).
31. A. F. Stewart, K. L. Insogna, D. Goltzman, and A. E. Broadus, Identification of adenylate cyclase-stimulating activity and cytochemical glucose-6-phosphate dehydrogenase-stimulating activity in extracts of tumors from patients with humoral hypercalcemia of malignancy, Proc. Natl. Acad. Sci. USA 80:1454 (1983).
32. R. E. Rizzoli, M. Somerman, T. M. Murray, and G. D. Aurbach, Binding of radioiodinated parathyroid hormone to cloned bone cells. Endocrinology 113:1832 (1983).
33. G. V. Segre, M. Rosenblatt, G. L. Tully, III, J. Laugharn, B. Reit, and J. T. Potts, Jr., Evaluation of an in vitro parathyroid hormone antagonist in vivo in dogs, Endocrinology 116:1024 (1985).
34. D. A. Gray, J. A. Parsons, J. T. Potts, Jr., M. Rosenblatt, and R. W. Stevenson, In vivo studies on an antagonist of parathyroid hormone [Nle-8,Nle-18,Tyr-34]bPTH-(3--34)amide, Br. J. Pharmacol. 76:259 (1982).
35. J. A. McGowan, T. C. Chen, J. Fragola, J. B. Puschett, and M. Rosenblatt, Parathyroid hormone: effects of the 3--34 fragment in vivo and in vitro, Science 219:67 (1983).
36. N. Horiuchi, M. Rosenblatt, H. T. Keutmann, J. T. Potts, Jr., and M. F. Holick, A multiresponse parathyroid hormone assay: an inhibitor has agonist properties in vivo. Am. J. Physiol. 244:E589 (1983).

37. K. J. Martin, E. Bellorin-Font, J. Freitag, M. Rosenblatt, and E. Slatopolski, The arterio-venous difference for immunoreactive parathyroid hormone and the production of adenosine 3',5'-cyclic monophosphate by isolated perfused bone: studies with analogs of parathyroid hormone, Endocrinology 109:956 (1981).
38. D. G. Johnson, C. V. Goebel, V. J. Hruby, M. D. Bregman, and D. Trivedi, Science 215:1115 (1982).
39. J. A. Vilchez-Martinez, A. V. Schally, D. H. Coy, E. J. Coy, C. M. Miller, III, and A. Arimura, An in vivo assay for anti-LH-RH and anti-FSH-RH activity of inhibitory analogues of LH-RH, Endocrinology 96:1130 (1975).
40. R. Iyengar and L. Birnbaumer, Hormone receptor modulates the regulatory component of adenylyl cyclase by reducing its requirements for Mg^{2+} and enhancing its extent of activation by guanine nucleotides, Proc. Natl. Acad. Sci. USA 79:5179 (1982).
41. A. Gilman, G proteins and dual control of adenylate cyclase, Cell 36:577 (1984).
42. S. B. Rodan and G. A. Rodan GA, The role of Mg^{2+} in hormone stimulation of rat osteosarcoma adenylate cyclase, Biochim. Biophys. Acta 673:55 (1981).
43. M. Rosenblatt and J. T. Potts, Jr., Analogues of an in vitro parathyroid hormone inhibitor: modifications at the amino terminus, Calcif. Tissue Int. 33:153 (1981).
44. S. R. Nussbaum, M. Rosenblatt, and J. T. Potts, Jr., Parathyroid hormone renal receptor interactions: demonstration of two receptor-binding domains, J. Biol. Chem. 255:10183 (1980).
45. M. Rosenblatt, G. V. Segre, G. A. Tyler, G. L. Shepard, S. R. Nussbaum, and J. T. Potts, Jr., Identification of a receptor-binding region in parathyroid hormone, Endocrinology 107:545 (1980)
46. P. L. Munson, Studies on the role of the parathyroids in calcium and phosphorus metabolism, Ann. NY Acad. Sci. 1955; 60:776 (1955).
47. N. Horiuchi, M. F. Holick, J. T. Potts, Jr., and M. Rosenblatt, A parathyroid hormone inhibitor in vivo: design and biological evaluation of a hormone analog, Science 220:1053 (1983).
48. S. Doppelt, P. Federico, G. A. Tyler, M. Rosenblatt, and R. M. Neer, Inhibition of the calcemic effects of parathyroid hormone in vivo by a hormone analogue, Calcif. Tissue Int. 35:681 (1983).
49. G. R. Mundy, K. J. Ibbotson, S. M. D'Souza, E. L. Simpson, J. W. Jacobs, and T. J. Martin, The hypercalcemia of malignancy, New Engl. J. Med. 310:1718 (1984).
50. A. F. Stewart, R. Horst, L. J. Deftos, E. C. Cadman, R. Lang, and A. E. Broadus, Biochemical evaluation of patients with cancer-associated hypercalcemia: evidence for humoral and nonhumoral groups, New Engl. J. Med. 303:1377 (1980).
51. D. Goltzman, A. F. Stewart, and A. E. Broadus, Malignancy-associated hypercalcemia: evaluation with a cytochemical bioassay for parathyroid hormone, J. Clin. Endocrinol. Metab. 53:899 (1981).
52. L. H. Caporale, P. DeHaven, G. A. Tyler, and M. Rosenblatt, Discrimination between parathyroid hormone (PTH) agonists and antagonists in vitro: modification of the renal membrane adenylate cyclase assay to reflect in vivo activity of PTH analogs. American Society for Bone and Mineral Research Seventh Annual Scientific Meeting Program, Abstract 304 (1985).
53. J. Rudinger, The design of peptide hormone analogs, in "Drug Design," Vol. 2, E. J. Ariens, ed., Academic Press, New York (1971).
54. D. F. Veber, R. Saperstein, R. F. Nutt et al., A super active cyclic hexapeptide analog of somatostatin, Life Sci. 34:1371 (1984).

SOLUBILIZATION OF A GUANINE NUCLEOTIDE-SENSITIVE PARATHYROID HORMONE-RECEPTOR COMPLEX FROM CANINE RENAL CORTEX

Robert A. Nissenson, Elaina Mann, Jane Winer, Anne Teitelbaum, and Claude D. Arnaud

Department of Medicine, Veterans Administration Medical Center and the Departments of Medicine and Physiology, University of California, San Francisco, California

INTRODUCTION

The physiologic actions of PTH on kidney and bone are initiated by hormonal binding to receptors in the plasma membrane of responsive cells. Occupation of receptors by PTH leads to activation of adenylate cyclase and increased cellular levels of cAMP. Cyclic AMP in turn activates a protein kinase(s), which results in specific substrate phosphorylation and ultimately biologic effects. Most, though not all, evidence supports such a mechanism of PTH action (1). Perhaps most compelling in this regard is the functional hypoparathyroidism seen in a subgroup of patients with pseudohypoparathyroidism (type Ia) who have a genetic deficiency in the stimulatory guanine nucleotide binding component (N_s) of adenylate cyclase (2,3).

Attention has therefore been focused on the identification and characterization of adenylate cyclase-coupled PTH receptors in bone and kidney. Workers in several laboratories have used biologically active, radiolabeled PTH peptides to label renal (4-12) and skeletal (13-15) binding sites with the high affinity (K_d generally 10^{-9}-10^{-8} M) and peptide specificity expected of adenylate cyclase-coupled receptors. Bound PTH is dissociated rapidly from canine renal cortical membranes with GTP or its analogs (9,10), suggesting hormone-induced coupling of the receptor with N_s. Further study of the functional PTH receptor has been limited by its apparent resistance to solubilization from the membrane in an active form.

In this study we report the use of a zwitterionic detergent, 3-[(3-cholamidopropyl)dimethylammonio] -2-hydroxy-1-propansulfonate (CHAPSO), to solubilize an [^{125}I] bPTH(1-34)-membrane protein complex that is dissociated specifically by guanine nucleotides. Evidence is presented that this complex constitutes the PTH receptor associated with N_s.

MATERIALS AND METHODS

Materials

Bovine (b)PTH(1-34) was obtained from Bachem Fine Chemicals (Torrance, CA) and had a potency exceeding 4000 U/mg. Nucleotides were obtained from Boehringer Mannheim Biochemicals (Indianapolis, IN). CHAPSO, 3-[(3-cholamidopropyl)dimethylammonio] -1-propanesulfonate (CHAPS), and octyl-β-D-glucopyranoside were from Calbiochem-Behring, San Diego, CA. BSA preparations (Fraction V and Fraction V fatty acid free) were obtained from Miles Scientific, Naperville, IL. Digitonin and Lubrol PX were from Sigma Chemical Co, St. Louis, MO. Triton X-100 was from Packard, Downers Grove, IL. Deoxycholate (recrystallized) was a generous gift from Dr. Thomas Boyer. Other chemicals were of reagent grade.

Preparation of [^{125}I] bPTH(1-34)

Bovine PTH(1-34) was radiolabeled using an electrolytic iodination procedure (16,17), with minor modifications of the technique previously described (7). Unreacted $^{125-}$I was removed by chromatography on an 0.8 X 15 cm column of Sepharose LH-20 equilibrated with 1 M pyridine acetate (pH 5.5)/3% ethanol/0.5% β- mercaptoethanol, and pretreated with 3 mg polylysine type II, followed by 3 mg BSA. The first peak of radioactivity consisted of ^{125}I-labeled peptide (1.0 mCi/1.5 ml) and was further purified by reverse-phase high-performance liquid chromatography using a Waters C-18 μ Bondapak column equilibrated with 0.1% trifluoroacetic acid in water. Labeled peptides were eluted with a 0-95% gradient of n-propanol in 0.1% trifluoroacetic acid; fractions (0.5 ml) were collected into glass tubes containing 50 μl of 1% BSA in 10 mM acetic acid. Two major peaks of radioactivity were eluted, the second possessing the greater biologic activity, as assessed by binding to canine renal PTH receptors. Fractions comprising this peak were pooled and rechromatographed on Sephadex LH-20 as before, except elution was carried out with 10 mM acetic acid/0.1% BSA. Fractions within the single peak of [^{125}I] bPTH(1-34) were pooled and stored at -20C in aliquots. Total yield of [^{125}I] bPTH(1-34) was approximately 0.1 mCi.

Solubilization of PTH receptors

PTH receptors were solubilized from canine renal cortical plasma membranes isolated by a modification (7) of the procedure of Fitzpatrick et al. (18). These membranes are enriched 7 to 10-fold (over crude homogenates) in the specific activity of ouabain-sensitive Na^{+}-K^{+} ATPase, and contain 1-2 pmol of high affinity PTH receptors (Kd ≅ 3 nM) per mg membrane protein (10). PTH receptors were labeled prior to solubilization by incubating membranes (250 μg/ml) for 45 min at 30C in 50 mM Tris-HCl/50 mM HEPES (pH 7.5) containing 0.1% BSA, 2.0 mM $MgCl_2$, 62 μM sucrose, 0.25 mM EDTA, and 10^5 cpm/ml of [^{125}I] bPTH(1-34). The reaction mixture was chilled at 4C for 15 min and centrifuged at 20,000 xg for 15 min at 4C. The clear supernatant was discarded, and the pellet was resuspended by vortex mixing in 2.0 ml of ice-cold 50 mM Tris-HCl/50 mM HEPES (pH 7.5)/ 0.1% BSA/2.0 mM $MgCl_2$ and recentrifuged as above. The pellet was again resuspended in the same buffer, transferred to a clean test tube, and recentrifuged. Typically, the membrane pellet at this stage contained 25-30% of the added [^{125}I] bPTH(1-34). Inclusion of excess unlabeled bPTH(1-34) (5 μg/ml) during the initial incubation reduced this membrane-associated [^{125}I] bPTH(1-34) to <1% of that added. Thus, <3% of the radioactivity bound to membranes at this stage could be attributed to trapping and/or to nonspecific binding of [^{125}I] bPTH(1-34) to the membranes.

Prelabeled membranes were solubilized as follows: ice-cold solubilization buffer routinely (unless otherwise stated) consisting of 50 mM Tris-HCl/50 mM HEPES/0.5% CHAPSO/5.0% BSA (fatty acid-free) was added to achieve a membrane protein concentration of 1.25 mg/ml. The suspension was mixed at 4C using a vortex mixer, followed by 30 minutes of mixing with a magnetic stirrer and/or vigorous shaking at 4C. The mixture was then transferred to thick-walled polycarbonate tubes and diluted 10-fold with solubilization buffer. These tubes were centrifuged at 100,000 xg_{av} for 45 min at 4C. The clear supernatant containing about 25% of the [^{125}I] bPTH(1-34) orginally bound to the membranes was carefully removed and recentrifuged at 240,000 xg_{av} for 45 min at 4C. The clear, pale yellow "supernatant" was removed (no pellet was visible at this stage) and used for further studies. This soluble fraction routinely contained 6000-8000 cpm/ml of ^{125}I-radioactivity.

Assessment of "bound" vs. free [^{125}I] bPTH(1-34)

[^{125}I] bPTH(1-34) associated with macromolecular components of solubilized renal membranes was determined by exclusion from dextran-coated charcoal, essentially as described (19). Data are presented as percent bound, calculated as

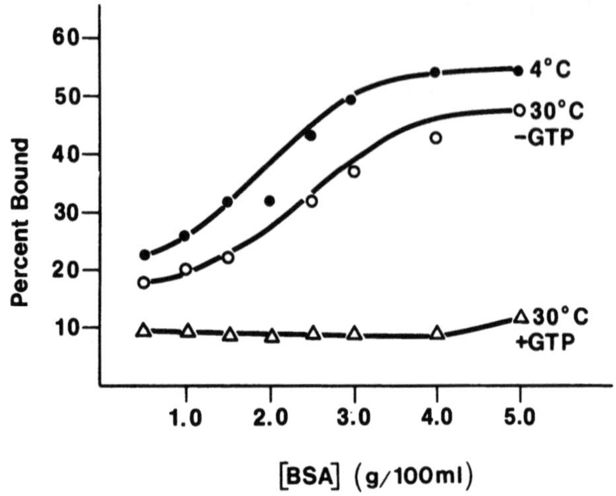

Fig. 1. Effect of BSA on solubilization of PTH-receptor complexes. Canine cortical membranes were incubated with [^{125}I] bPTH(1-34) and solubilized with 0.5% CHAPSO containing various concentrations of BSA, as described in Experimental Procedures. Aliquots of soluble extracts were assessed for binding with no incubation (maintenance at 4C), or after incubation for 10 min at 30C with or without GTP (10^{-4} M). Values are the mean of duplicate determinations that agreed ± 5% of the mean value.

the percent of ^{125}I-radioactivity in the soluble extract excluded from dextran-coated charcoal. Under standard conditions (0.5% CHAPSO/5.0% BSA), > 94% of unbound [^{125}I] bPTH(1-34) co-sedimented with dextran-coated charcoal.

RESULTS

In initial experiments, we found that [^{125}I] bPTH(1-34)-protein complexes were recovered when prelabeled membranes were solubilized with the zwitterionic detergent CHAPSO. Recovery of soluble complexes depended markedly on the presence of high concentrations of BSA during the solubilization procedure (Figure 1). In the absence of BSA, approximately 40% of the total membrane protein and a similar percentage of membrane-bound ^{125}I-radioactivity were solubilized with 0.5% CHAPSO. However, under these conditions, less than 20% of the [^{125}I] bPTH(1-34) radioactivity in the soluble fraction was excluded from dextran-coated charcoal and thus was associated with a macromolecular complex (not shown). When assessed at 4C, the bound fraction increased to a maximum of 55% with increasing concentrations of BSA during solubilization. Incubation of the

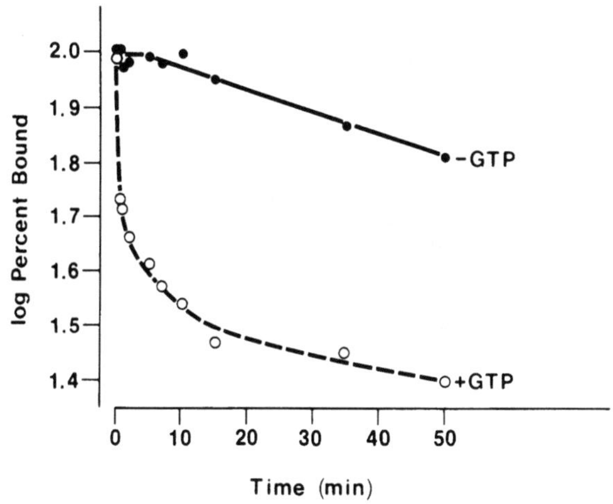

Fig. 2. Effect of GTP on dissociation of PTH from its soluble receptor. Aliquots of a solubilized extract were incubated at 30C for 1 min, then GTP (10^{-4} M) or control buffer was added and bound [^{125}I] bPTH(1-34) was assessed at various times thereafter. Results are representative of two separate kinetic experiments. Kinetic constants were calculated by a curve-stripping technique (20).

soluble fraction for 10 min at 30C resulted in a slight diminution in the bound fraction, presumably due to dissociation of [^{125}I] bPTH(1-34). Inclusion of GTP (10^{-4} M) during this incubation dramatically enhanced the dissociation of labeled hormone. The effect of GTP on the kinetics of dissociation of [^{125}I] bPTH(1-34) from the soluble complex at 30C is shown in Figure 2. In the absence of GTP, labeled PTH dissociated slowly and monoexponentially (reverse rate constant k_{-1} = 0.0092 min^{-1}, t 1/2 = 75 min). In the presence of 10^{-4} M GTP, dissociation was biexponential with a rapid component (k_{-2} = 0.30 min^{-1}, t 1/2 = 2.3 min) superimposed on a slow component (k_{-1} = 0.0066 min^{-1}, t 1/2 = 100 min). The addition of excess unlabeled bPTH(1-34)(5 μg/ml) to the extracts did not influence the kinetics of dissociation of [^{125}I] bPTH(1-34) either in the absence or presence of GTP (not shown).

The effect of GTP on the kinetics of dissociation of PTH from solubilized receptors is remarkably similar to the effect of the nucleotide in intact renal cortical plasma membranes (10), and suggests that the solubilized receptor is associated with a functional stimulatory guanine nucleotide-dependent regulatory component of adenylate cyclase (N_s). The nucleotide sensitivity and specificity of

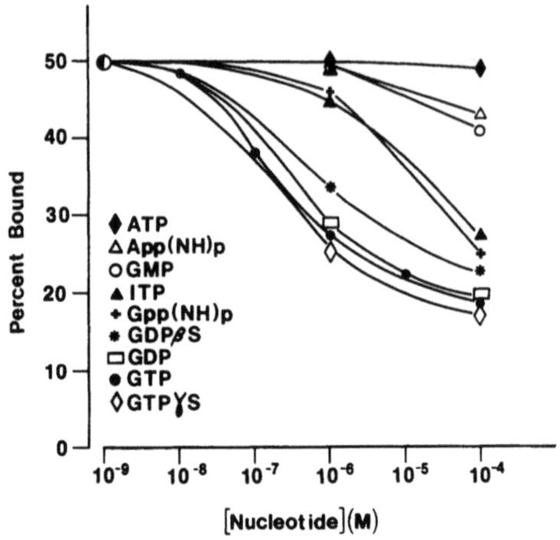

Fig. 3. Nucleotide specificity for dissociation of $[^{125}I]$ bPTH(1-34) from soluble receptors. Aliquots of a soluble extract were incubated at 30C for 10 min in the presence of various nucleotides at the concentrations indicated. Results are the mean of duplicates that agreed $\pm$ 5% of the mean value.

this effect support such a conclusion. Thus, half-maximal dissociation of $[^{125}I]$ bPTH(1-34) was evident at approximately 2 X 10^{-7} M GTP (Figure 3). The rank order of potency of a variety of nucleotides was GTP $\cong$ GTPγS $\cong$ GDP > GDPβS > ITP $\cong$ Gpp(NH)p >> GMP $\cong$ App(NH)p. ATP was inactive up to 10^{-4} M.

Previous studies have demonstrated that N_s-mediated regulation of membrane-associated β-adrenergic (21) and PTH receptors (10) by guanine nucleotides is attenuated by exposing the agonist-occupied receptors to the sulfhydryl-alkylating reagent NEM prior to nucleotide addition. Similarly, GTP-induced dissociation of $[^{125}I]$ bPTH(1-34) from the soluble complex was attenuated by the presence of 1 mM NEM during the incubation of membranes with labeled PTH (not shown). Refractoriness to GTP occurred only when PTH receptors were occupied at the time of exposure to NEM.

Gel filtration chromatography on Sepharose 6B of soluble extracts of renal membranes prelabeled with $[^{125}I]$ bPTH(1-34) revealed three major peaks of ^{125}I-radioactivity (Figure 4). One of these represented about 50% of the total ^{125}I in the extract, and was eluted in the position of free $[^{125}I]$ bPTH(1-34). Approximately two-thirds of the remaining ^{125}I eluted as a broad included peak in the position expected of a spherical, detergent-free protein of ~180,000 daltons. A third peak was eluted in the void volume and may represent an aggregated state of

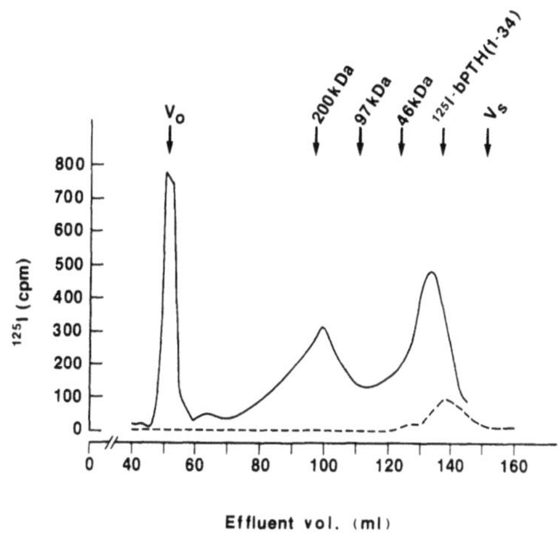

Fig. 4. Gel filtration profile of [^{125}I] bPTH(1-34) in a soluble extract. Plasma membranes were incubated with [^{125}I] bPTH(1-34) in the absence (-) or presence (- - -) of excess unlabeled bPTH(1-34) (5μg/ml). Equal volumes of solubilized extracts were fractionated on a column of Sepharose 6B, and total ^{125}I-radioactivity in column effluent fractions was assessed. Recovery of applied ^{125}I-radioactivity exceeded 80%. Column markers were dextran blue (Vo), myosin (200 kDa), phosphorylase B (97.4 kDa), ovalbumin (43 kDa), and [^{14}C] glucose (Vs). Also shown is the elution postion of [^{125}I] bPTH(1-34).

the receptor complex. Inclusion of excess unlabeled bPTH(1-34) (5μ g/ml) during the initial membrane binding reaction prevented the labeling of the 180,000 dalton and void volume components, and markedly reduced the recovery of free [^{125}I] bPTH(1-34) (Figure 4). Thus, most of the free [^{125}I] bPTH(1-34) in the soluble extract must have been due to dissociation from receptors during solubilization. Gel filtration chromatography of soluble extracts after incubation with GTP (10^{-4} M) for 10 min at 30C demonstrated a marked decrease in [^{125}I] bPTH(1-34) associated with both macromolecular components and a simultaneous increase in free [^{125}I] bPTH(1-34) (Figure 5).

A variety of detergents were screened for efficacy in the solubilization of GTP-sensitive PTH binding sites. CHAPSO was optimally effective at a concentration of 0.5%; lower concentrations failed to solubilize the receptor efficiently whereas, at higher concentrations, little or no bound [^{125}I] bPTH(1-34)was found. Virtually identical results were obtained with CHAPS. Digitonin (0.5%) was as effective as 0.5% CHAPSO in maintaining soluble GTP-sensitive

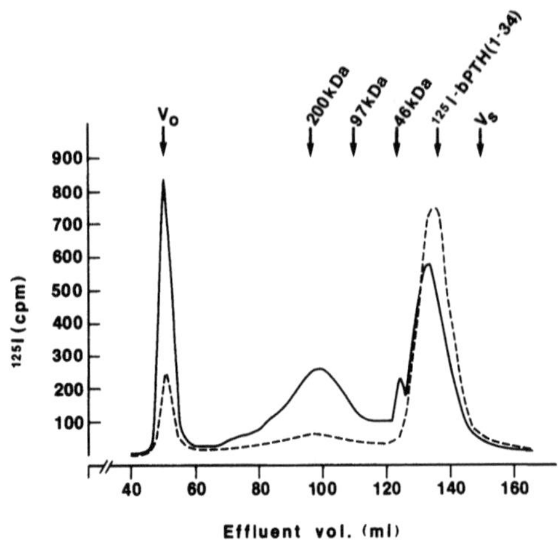

Fig. 5. Effect of GTP on the gel filtration profile of solubilized PTH receptors. Soluble extracts were incubated at 30C for 10 min with (- - -) or without (-) GTP (10^{-4} M), and fractionated by gel filtration chromatography as described in Figure 4. Column recoveries of ^{125}I-radioactivity exceeded 80% of that applied.

[^{125}I] bPTH(1-34) binding, and did so with a substantially greater yield (46% vs. 27% solubilization of membrane receptor-associated [^{125}I] bPTH(1-34)). Other detergents were unsatisfactory either because of poor maintenance of binding (Triton X-100, Lubrol PX, high concentrations of octyl β-D-glucopyranoside and deoxycholate) or poor yield (low concentrations of octyl β-D-glucopyranoside and deoxycholate).

DISCUSSION

We have previously provided evidence that PTH-dependent activation of canine renal membrane adenylate cyclase requires agonist-mediated association between PTH receptors and N_S (10). In intact membranes, the ternary PTH-receptor-N_S intermediate complex so formed represents a "high affinity" receptor state in that hormone dissociation at 30C is an exceedingly slow process (t 1/2 ≅ 200 min) that is markedly enhanced (t 1/2 ≅ 1 min) by guanine nucleotides capable of binding to N_S. Several lines of evidence presented in this study indicate that we have successfully solubilized this ternary complex.

Thus the rate constants for dissociation of [^{125}I] bPTH(1-34) from the solubilized complex resemble those obtained for the receptor in intact membranes. At 30C in the absence of guanine nucleotides, the dissociation of labeled PTH followed monoexponential kinetics (k_{-1} = 0.0092 min^{-1}, t 1/2 = 75 min); in the presence of 10^{-4} M GTP, a rapid component of dissociation (k_{-2} = 0.30 min^{-1}, t 1/2 = 2.3 min) was superimposed on a slower component (k_{-1} = 0.0066 min^{-1}, t 1/2 = 100 min). The experimental conditions (e.g., the presence of 0.5% CHAPSO, 5% BSA) might explain the quantitative deviation of these values from those in intact membranes. Nonetheless, the similarity between the soluble complex and receptors in intact membranes in dissociation kinetics and GTPsensitivity is striking. In the present study, the nucleotide specificity for dissociation of [^{125}I] bPTH(1-34) from the soluble complex was similar to that observed for the nucleotide binding site of purified rat liver N_s (22). Thus, triphosphate nucleotides of inosine and guanosine, but not of adenosine, were potent displacers with GTPγS $\cong$ GTP > Gpp(NH)p $\cong$ ITP >> App(NH)p, ATP. GDP and its stable analog GDPβS were nearly as active as GTP, whereas GMP was only weakly active. The dose-response relationship for the dissociating effect of GTP demonstrated a half-maximal effect of the nucleotide at 2 X 10^{-7} M, a value consistent with the reported affinity of GTP for purified rat liver N_s (22).

There is evidence that turkey erthrocyte membrane N_s contains a critical NEM-sensitive site that is exposed upon association of the protein with receptors occupied by β-adrenergic agonists (21). Similarly, we have shown that guanine nucleotide-induced dissociation of PTH from canine renal membrane receptors is blocked by the presence of 1 mM NEM during the binding and dissociation reactions (10). The results of the present study indicate that treatment of membranes with NEM during hormone-receptor association also locks the solubilized complex into a GTP-resistant state. This effect does not occur if the PTH receptor is unoccupied during exposure to NEM. These results are consistent with a PTH-induced conformational change in either the receptor or N_s, leading to exposure of an NEM-sensitive site, conceivably a sulfhydryl group, the alkylation of which leads to refractoriness of the ternary PTH-receptor-N_s complex to the dissociating effect of GTP.

Taken together, these data strongly imply that the solubilized complex associated with [^{125}I] bPTH(1-34) contains both receptor and N_s and that these components are associated functionally and structurally. While the criteria for "solubilization" of membrane proteins are somewhat arbitrary, our finding that receptor activity released by 0.5% CHAPSO did not sediment during a 45 min centrifugation at 240,000 xg_{av} represents a stringent confirmation of solubilization (23). In addition, bound [^{125}I] bPTH(1-34) was not removed by filtration of the extract through 0.2 μm cellulose acetate Millipore filters (data not shown). Finally, gel filtration chromatography (Sepharose 6B) of the extract yielded a major peak of bound [^{125}I] bPTH(1-34) that was eluted within the included volume of the column, at the position expected of a 180,000 dalton globular, detergent-free protein. A second peak representing one-third of the bound [^{125}I] bPTH(1-34) appeared consistently in the void volume. Pretreatment of the extract for 10 min with GTP reduced the magnitude of both peaks by approximately 70%, indicating that both peaks contain functional ternary PTH-receptor-N_s complexes. It is likely that the excluded peak represents a CHAPSO-induced aggregate containing ternary complexes and perhaps other membrane components.

Interestingly, a significant proportion of bound [^{125}I] bPTH(1-34) was not rapidly dissociated from the soluble complex by GTP. The results are consistent with a two-site model in which, after a 60 min incubation of [^{125}I] bPTH(1-34) with renal membranes at 30C, approximately 75% of the solubilized sites are responsive to GTP, whereas the remaining 25% are unaffected. Similar findings were previously reported in intact canine renal membranes (10), indicating that solubilization as such did not greatly enrich one site relative to the other. Although the nature of such a postulated GTP-insensitive site is uncertain, it is not a "non-specific" membrane binding site for PTH since, under our conditions, non-specific [^{125}I] bPTH(1-34) binding was undetectable (Figure 4). The observation that such sites co-migrated on gel filtration with GTP-sensitive sites implies that a subtle alteration may result in GTP insensitivity. In fact, as described above, alkylation with NEM mimicked this phenomenon. It may be that the NEM-sensitive site exposed by association of PTH with its receptor is also susceptible to processes intrinsic to the membrane. In this regard, it is noteworthy that Corin et al. (24) have reported that the component of high affinity [^{125}I] glucagon binding to liver membranes that was refractory to the dissociating effect of GTP was increased with increasing time of exposure of membranes to hormone, indicating a time-dependent conversion of receptors from a GTP-sensitive to a GTP-insensitive state. Recently, Stiles et al. (25) have provided evidence that N_s in rat lung membranes can be converted by an intrinsic protease to a form capable of promoting high affinity binding of β-adrenergic agonists, but defective in mediating GTP-dependent agonist dissociation and adenylate cyclase activation. It will be important to ascertain whether such transformation occurs for the PTH receptor, and if so, to assess its role in regulatory processes such as receptor desensitization.

Previous studies by our group (26) and by Coltrera et al. (27) have demonstrated a 60,000-70,000 dalton protein in canine renal membranes that is covalently and specifically labeled by photoreactive analogs of PTH. It appears that this protein constitutes the PTH binding moiety of the PTH receptor. Therefore, it is likely that, at a minimum, the solubilized PTH receptor complex consists of [^{125}I] bPTH(1-34) (4,200 daltons), the PTH binding moiety (60,000-

70,000 daltons), and N_s (90,000-100,000 daltons if present in $\alpha\beta\gamma$ form (28,29)). Given the approximation that the binding component that was eluted in the included volume of Sepharose 6B corresponds to a 180,000 dalton protein, it is reasonable to speculate that these three proteins may be the sole constituents of the complex. If so, the PTH receptor would belong to a subclass of relatively low molecular weight (50,000-70,000 dalton) proteins that couple to N_s and mediate the activation of adenylate cyclase by glucagon (30) and β-adrenergic agents (31). So far, purification of a receptor component that is sufficient to initiate agonist-mediated stimulation of adenylate cyclase has been achieved only for the β-adrenergic receptor (32).

In a previous study, Malbon and Zull (33) reported specific binding of labeled bPTH(1-84) to a Triton X-100-solubilized component of bovine renal cortical membranes. However, in the present study, agonist occupancy of the receptor prior to solubilization was requisite to obtain a soluble PTH-receptor complex. We have been unable to demonstrate high affinity binding of [^{125}I] bPTH(1-34) when added to a soluble fraction from canine renal cortical plasma membranes even when solubilization was accomplished with Triton X-100 or with 0.5% CHAPSO/5.0% BSA (R.A. Nissenson, unpublished observations). This can be explained by the inability of agonist binding to induce the interaction of soluble receptor with soluble N_s that would be required for the induction of high affinity. The affinity of [^{125}I] bPTH(1-34) for "free" receptors is apparently too low to be detected under our assay conditions. Another critical factor in our studies was the choice of detergent. Most effective in solubilizing and preserving the ternary [^{125}I] bPTH(1-34)-receptor-N_s complex were CHAPS, CHAPSO, and digitonin, all of which have relatively rigid hydrophobic regions, high critical micelle concentrations, and low aggregation numbers. CHAPS and CHAPSO were effective over a very narrow range of concentrations with maximal activity at about 0.5%. Lower concentrations failed to solubilize labeled receptors effectively, while higher concentrations induced ligand dissociation. Digitonin (0.5%) preserved PTH-receptor complexes as effectively as CHAPS or CHAPSO, and with a substantially greater yield.

Most intriguing is the dependence of solubilization of hormone-receptor complexes upon the presence of high concentrations of BSA in the solubilization buffer. This was not an artifactual effect of BSA on the dextran-coated charcoal technique used to separate bound from free [^{125}I] bPTH(1-34) since (a) in the presence of 5% BSA, estimates by gel filtration of the fraction of [^{125}I] bPTH(1-34) bound (Figures 4 and 5) agreed closely with the fraction of [^{125}I] bPTH(1-34) in the applied samples excluded from dextran-coated charcoal, (b) qualitatively identical results were obtained using polyethylene glycol precipitation in the presence of 5% BSA (data not shown), and (c) similar estimates of binding were obtained when extracts were diluted 10-fold with BSA-free (and detergent-free) buffer prior to addition of charcoal (data not shown). The mechanism whereby BSA exerts its stabilizing effect is unknown. We speculate that 5% BSA (equivalent to 0.76 mM) can act as a detergent "buffer." Thus, serum albumin is one of the few soluble proteins with binding sites for amphiphilic ligands including deoxycholate, Triton X-100, and other detergents (34). Albumin has as many as 10 amphiphile binding sites/molecule, with affinities in the millimolar range (35). Under our solubilization conditions, 8 mM (0.5%) CHAPSO and 0.76 mM BSA, it is likely that a significant fraction of CHAPSO is associated (relatively weakly) with BSA. Under these circumstances, the concentration of free CHAPSO could be maintained constant during solubilization by release of the BSA-associated detergent as free detergent associates with membrane components. If this is correct, the use of BSA as a detergent buffer might have widespread applicability for the preservation of biologic activity during solubilization of detergent-sensitive intrinsic membrane proteins.

In summary, we have developed a method for solubilizing PTH receptors in a functional ternary PTH-receptor-N_s complex that, from our previous work (10), is an obligatory intermediate in PTH-stimulation of adenylate cyclase. The structural and biochemical features of the adenylate cyclase-coupled PTH receptor are currently under investigation.

REFERENCES

1. J.F. Habener, M.Rosenblatt, and J.T. Potts, Jr, Parathyroid hormone: biochemical aspects of biosynthesis, secretion, action, and metabolism, Physiol Rev 64:985 (1984)
2. Z. Farfel, A.S. Brickman, H.R. Kaslow, V.M. Brothers, and H.R. Bourne, Defect of receptor-cyclase coupling protein in pseudohypoparathyroidism, N Engl J Med 303:237 (1980)
3. M.A. Levine, R.W. Downs, M. Singer, S.J. Marx, G.D. Aurbach, and A.M. Spiegel, Deficient activity of guanine nucleotide regulatory protein in erythrocytes from patients with pseudohypoparathyroidism, Biochem Biophys Res Commun 94:1319 (1980)
4. H.S. Sutcliffe, T.J. Martin, J.A. Eisman, and R. Pilczyk, Binding of parathyroid hormone to bovine kidney-cortex plasma membranes, Biochem J 134:913 (1973)

5. F.P. DiBella, T.P. Dousa, S.S. Miller, and C.D. Arnaud, Parathyroid hormone receptors of renal cortex: specific binding of biolgically active, ^{125}I-labeled hormone and relationship to adenylate cyclase activation, Proc Natl Acad Sci USA 71:723 (1974)
6. J.E. Zull, C.C. Malbon, and J. Chuang, Binding of tritiated bovine parathyroid hormone to plasma membranes from bovine kidney cortex, J Biol Chem 252:1071 (1977)
7. R.A. Nissenson and C.D. Arnaud, Properties of the parathyroid hormone receptor-adenylate cyclase system in chicken renal plasma membranes, J Biol Chem 254:1469 (1979)
8. G.V. Segre, M. Rosenblatt, B.L. Reiner, J.E. Mahaffey, and J.T. Potts, Jr, Characterization of parathyroid hormone receptors in canine renal cortical plasma membranes using a radioiodinated sulfur free hormone analogue, J Biol Chem 254:6980 (1979)
9. E. Bellorin-Font and K.J. Martin, Regulation of the PTH receptor-cyclase system of canine kidney: effects of calcium, magnesium, and guanine nucleotides, Am J Physiol 241:F364 (1981)
10. A.P. Teitelbaum, R.A. Nissenson, and C.D. Arnaud, Coupling of the canine renal parathyroid hormone receptor to adenylate cyclase: modulation by guanyl nucleotides and N-ethylmaleimide, Endocrinology 111:1524 (1982)
11. L.R. Forte, S.G. Langeluttig, R.E. Poelling, and M.L. Thomas, Renal parathyroid hormone receptors in the chick: down-regulation in secondary hyperparathyroid animal models, Am J Physiol 242:E154 (1982)
12. A.P. Teitelbaum and G.J. Strewler, Parathyroid hormone receptors coupled to cyclic adenosine monophosphate formation in an established renal cell line, Endocrinology 114:980 (1984)
13. N.B. Pliam, K.O. Nyiredy, and C.D. Arnaud, Parathyroid hormone receptors in avian bone cells, Proc Natl Acad Sci USA 79:2061 (1982)
14. C.M. Silve, G.T. Hradek, A.L. Jones, and C.D. Arnaud, Parathyroid hormone receptors in intact embryonic bone: characterization and cellular localization, J Cell Biol **94**:379 (1982)
15. R.E. Rizzoli, M. Somerman, T.M. Murray, and G.D. Aurbach, Binding of radioiodinated parathryoid hormone to cloned bone cells, Endocrinology 113:1832 (1983)
16. D.J. Sammon, J.S. Brand, W.F. Neuman, and L.G. Raisz, Metabolism of labeled parathyroid hormone. I. Preparation of biologically active ^{125}I-labeled parathyroid homone, Endocrinology 92:1596 (1973)
17. W.F. Neuman, M.W. Neuman, P.J. Sammon, and K. Lane, The metabolism of labeled parathyroid hormone. II. Methodological studies, Calcif Tissue Res 18:241 (1975)

18. D.F. Fitzpatrick, G.R. Davenport, L. Forte, and E.J. Landon, Characterization of plasma membrane proteins in mammaliam kidney. I. Preparation of membrane fraction and separation of the protein, J Biol Chem 244:3561 (1969)

19. C.D. Arnaud, H.S. Tsao, and T. Littledike, Radioimmunoassay of human parathyroid hormone in serum, J Clin Invest 50:21 (1971)

20. G.W. Snedecor and W.G. Cochran, "Statistical Methods," Iowa State University Press, Ames, Iowa (1967)

21. M. Korner, C. Gilon, and M. Schramm, Locking of hormone in the β-adrenergic receptor by attack on a sulfhydryl in an associated component, J Biol Chem 257:3389 (1982)

22. J.K. Northrup, M.D. Smigel, and A.G. Gilman, The guanine nucleotide activating site of the regulatory component of adenylate cyclase. Identification by ligand binding, J Biol Chem 257:11416 (1982)

23. L.M. Hjelmeland and A. Chrambach, Solubilization of functional membrane-bound receptors, in: "Receptor Biochemistry and Methodology, Vol. 1," Alan R. Liss, Inc., New York (1984)

24. R.E. Corin, P. Ferriola, and D.B. Donner, Hepatic glucagon-receptor complexes lose sensitivity to the dissociating effect of GTP, J Biol Chem 257:1626 (1982)

25. G.L. Stiles, R.H. Strasser, B.F. Kilpatrick, S.R. Taylor, and R.J. Lefkowitz, Endogenous proteinases modulate the function of the β-adrenergic receptor-adenylate cyclase system, Biochim Biophys Acta 802:390 (1984)

26. M.W. Draper, R.A. Nissenson, J. Winer, J. Ramachandran J, and C.D. Arnaud, Photoaffinity labeling of the canine renal receptor for parathyroid hormone, J Biol Chem 257:3714 (1982)

27. M.D. Coltrera, J.T. Potts Jr, and M. Rosenblatt, Identification of a renal receptor for parathyroid hormone by photoaffinity radiolabeling using a synthetic analogue, J Biol Chem 256:10555 (1981)

28. J.K. Northrup, P.C. Sternweis PC, M.D. Smigel, L.S. Schleifer, E.M. Ross, and A.G. Gilman, Purification of the regulatory component of adenylate cyclase, Proc Natl Acad Sci USA 77:6516 (1980)

29. J.D. Hildebrandt, J. Codina, R. Risinger, and L. Birnbaumer, Identification of a subunit associated with the adenyl cyclase regulatory proteins N_s and N_i, J Biol Chem 259:2039 (1984)

30. R. Iyengar and J.T. Herberg, Structural analysis of the hepatic glucagon receptor, J Biol Chem 259:5222 (1984)

31. R.G.L. Schorr, R.J. Lefkowitz, M.G. Caron, Purification of the β-adrenergic receptor, J Biol Chem 256:5820 (1981)

32. R.A. Cerione, B. Strulovici, J.L. Benovic, R.J. Lefkowitz, M.G. Caron, Pure β-adrenergic receptor: the single polypeptide confers catecholamine responsiveness to adenylate cyclase, Nature 306:562 (1983)
33. C.C. Malbon and J.E. Zull, Studies of binding of parathyroid hormone to a detergent-dispersed preparation from bovine kidney cortex plasma membranes, J Biol Chem 252:1079 (1977)
34. J.A. Reynolds, Interactions between proteins and amphiphiles, in: "Lipid-Protein Interactions," P.C. Jost and O.H. Griffith, eds., John Wiley and Sons, New York
35. S. Makino, J.A. Reynolds, and C. Tanford, The binding of deoxycholate and Triton X-100 to proteins, J Biol Chem 248:4926 (1973)

ACKNOWLEDGMENTS

The authors are grateful to Katalin Nyiredy for expert technical assistance; to Cheryl Holzaepfel for editorial suggestions and manuscript preparation; and to Dr. Donald Vessey for helpful discussions. This work was supported by the Research Service of the Veterans Administration and by the United States Public Health Service, National Institutes of Health grants AM 35323 (to R.A. Nissenson) and AM 21614 (to C.D. Arnaud).

PARATHYROID HORMONE SECRETORY RESPONSES TO PERORAL PHOSPHATE AND STIMULABILITY OF SERUM LEVELS OF CARBOXYL-TERMINAL FLANKING PEPTIDE (PDN-21) OF THE HUMAN CALCITONIN GENE BY CALCIUM IN NORMAL SUBJECTS AND OSTEOPOROTIC PATIENTS

Maximilian A. Dambacher, Joachim Ittner, Roman Muff and Jan A. Fischer*

Research Laboratory for Calcium Metabolism, Departments of Orthopedic Surgery and Medicine, University of Zurich Zurich, Switzerland

OSTEOPOROTIC PATIENTS AND NORMAL SUBJECTS

The roles of calciotropic hormones in the pathogenesis and treatment of osteoporosis are not clearly understood. Here we have assessed serum levels of immunoreactive parathyroid hormone (PTH) and urinary cyclic AMP excretion before and after acute peroral phosphate administration in osteoporotic patients and normal subjects of the same age. In some osteoporotic patients serum levels of PTH were measured after prolonged (5 days) treatment with peroral phosphate. Serum levels of immunoreactive carboxyl-terminal flanking peptide (PDN-21) (katacalcin is a synonym) of the human calcitonin gene which are closely related to calcitonin levels were also measured before and after iv calcium injections (1). The diagnosis of osteoporosis was established on the basis of at least one vertebral compression fracture. Control subjects had normal serum levels of calcium, phosphate and creatinine, and they presented no history of renal, hepatic and intestinal disease. None of the subjects were treated with glucocorticoids or were alcoholics.

ACUTE PARATHYROID HORMONE RESPONSES TO PERORAL PHOSPHATE

Parathyroid hormone secretory responses were investigated in 8 osteoporotic patients (56±5 yr; mean±SEM; 2 men and 6 women, aged 37-74 yr) and compared to 7 age-matched normal subjects (56±5 yr; 2 men and 5 women, aged 38-73 yr) at 15, 45, 120, 180 and 480 min after peroral phosphate (48 mmol, 1.5 g phosphorus, Sandoz). Serum phosphate was increased in osteoporotic and normal subjects (ΔPO_4 at 120 min 0.46±0.07 mM, $P < 0.001$, and 0.54±0.05 mM, $P < 0.001$, respectively). As a consequence serum ionized calcium was similarly decreased (ΔCa^{++} at 15 min 0.06±0.02 mM, $P < 0.02$, and 0.05±0.01 mM, $P < 0.05$, respectively; and at 120 min 0.05±0.02 mM, $P < 0.02$, and 0.06±0.01 mM, $P < 0.001$, respectively). Serum total calcium levels did not change consistently. Serum levels of immunoreactive PTH (N-RIA recognizing predominantly intact PTH on gel permeation chromatography (2)) were increased in normal subjects within 15 min from 156±25 pgeq/ml to 260±37 pgeq/ml ($P < 0.05$) and remained elevated after 120 min (215±42 pgeq/ml) ($P<0.05$).

*Correspondence: Dr. J.A. Fischer, Klinik Balgrist, Forchstrasse 340, 8008 Zurich, Switzerland

In osteoporotic patients PTH remained largely unaltered (basal: 104±26 pgeq/ml, at 15 min 127±33 pgeq/ml, at 120 min 146±31 pgeq/ml) ($P > 0.1$). Analysis of variance revealed a significant difference of PTH responses between normal and osteoporotic subjects at 15 to 480 min after phosphate administration ($P < 0.01$). Moreover, a statistically significant rise of urinary cyclic AMP (3) occurred in normal subjects after 4 h from 432±52 to 663±114 pmol/mmol creatinine ($P < 0.02$), but not in osteoporotic patients (basal: 433±65 pmol/mmol creatinine, after 4 h 493±70 pmol/mmol creatinine, $P > 0.1$).

PARATHYROID HORMONE RESPONSES TO PROLONGED PERORAL PHOSPHATE

Three osteoporotic patients, one man aged 52 yr and two women aged 37 and 56 yr, were treated with 48 mmol phosphate (1.5 g phorphorus) daily for 5 days. Serum levels of total calcium (basal: 2.31±0.03 mM, after 5 days 2.29±0.08 mM) and of PTH (C-RIA recognizing predominantly carboxyl-terminal PTH fragments (1)) (basal: 18±1 ngeq/ml, after 5 days 16±1 ngeq/ml) were unaltered.

EFFECTS OF IV CALCIUM ON SERUM LEVELS OF PDN-21

Basal serum levels of calcium and of PDN-21 were the same in normal subjects and osteoporotic patients (Table 1). 3 min after 1 min iv infusions of calcium (Sandoz) (2 mg/kg body weight) serum levels of calcium were similar in male and female, and normal and osteoporotic subjects. The PDN-21 responses to iv calcium were, however, lower both in normal and osteoporotic women than men of the same age ($P < 0.01$). Normal subjects and osteoporotic patients were grouped in less and more than 50 yr olds. Male and female osteoporotic patients less than 50 yr old had more pronounced PDN-21 responses to iv calcium than normal men and women of the same age ($P < 0.05$). The excessive release of PDN-21 was no more visible in older (> 50 yr) patients.

CONCLUSIONS

Parathyroid hormone secretion is stimulated in normal and hyperparathyroid man in parallel with falls of the serum ionized and total calcium concentrations (4-6). Basal serum PTH levels were within our normal range in osteoporotic patients. They are rarely raised in osteoporotic patients in comparison to other elderly subjects without compression fractures (7). Our results reveal that serum PTH and urinary cyclic AMP excretion were increased in normal subjects in response to a fall of serum ionized calcium concentrations evoked by the peroral administration of phosphate, whereas in osteoporotic patients of the same age, despite of a similar fall of ionized calcium, PTH and cyclic AMP remained unaltered. Moreover, serum levels of PTH remained unchanged in osteoporotic patients after treatment for 5 days with peroral phosphate which was the maximal tolerated dose. Circulating PTH levels could not be stimulated in osteoporotic patients with peroral phosphate administration and as a consequence in response to mild hypocalcemia.

The rational of the present study was an attempt to stimulate endogenous PTH secretion in osteoporotic patients. By these means we had expected to enhance new bone formation with a physiological stimulation of PTH secretion. This has appeared especially pertinent, since parenteral administration of exogenous human PTH-(1-34) is difficult to dose. Low amounts apparently raise bone formation and higher doses cause an increase in bone resorption (8, 9). Our attempt to stimulate PTH secretion with peroral phosphate at least within 5 days has failed. Rasmussen et al. (10), using the

Table 1. Effects of 1 min iv infusions of calcium (2 mg/kg body weight) on serum levels of calcium and PDN-21

	Normal subjects				Osteoporotic patients			
Group	men		women		men		women	
N	30	5	5	12	4	4	6	13
Age (yr)	31±2*	59±2	32±2	61±2	36±4	57±2	41±3	62±2
Calcium (mM)								
before	2.30±0.02	2.25±0.06	2.23±0.03	2.22±0.04	2.30±0.05	2.27±0.03	2.22±0.05	2.21±0.02
3 min after iv calcium	2.60±0.04[a]	2.55±0.04[c]	2.50±0.03[a]	2.59±0.06[c]	2.60±0.06[b]	2.62±0.06[b]	2.58±0.07[b]	2.60±0.03[a]
PDN-21 (pgeq/ml)								
before	14±1	18±9	11±1	16±3	15±1	16±2	11±1	11±1
3 min after iv calcium	96±10[a,d,e]	57±21	24±4[b,d,e]	29±4[b,e]	210±50[c,d,e]	77±20[c,d]	41±11[c,d,e]	18±2[b,d,e]

*Mean ± SEM

[a]P < 0.001, [b]P < 0.01, [c]P < 0.05 by paired t-test before and after iv calcium

[d]P < 0.01 by analysis of variance between men and women, and [e]P < 0.05 between normal and osteoporotic subjects

same dose of phosphate together with calcitonin in cyclic manner, have reported after 6 months treatment stimulation of serum levels of immunoreactive PTH associated with a rise of trabecular bone mass as revealed by quantitative histomorphometry. The latter results were also obtained by Anderson et al. (11) using peroral phosphate and the diphosphonate sodium etidronate (EHDP) in cyclic manner.

We have been unable to measure consistently basal serum levels of calcitonin in unextracted serum of normal subjects (12). A more sensitive RIA measuring a carboxyl-terminal flanking peptide (PDN-21), also encoded by the calcitonin gene and cosecreted with calcitonin, has recently been developed (1, 13-15). Much like with circulating calcitonin concentrations and secretory responses to iv calcium, the levels of PDN-21 and PDN-21 responses to iv calcium were lower in women than men (1, 14, 16-18). Deficient calcitonin responses to calcium have been thought to be involved in the development of postmenopausal osteoporotis (19). In the present study we have recognized excessive serum PDN-21 responses to iv calcium in some male and female osteoporotic patients younger than 50 yr in relation to normal subjects of the same age. In older patients (> 50 yr old) PDN-21 responses to iv calcium were similar in normal and osteoporotic subjects. Along these lines calcitonin secretion was found to be normal in patients with postmenopausal osteoporosis (20). Nonetheless, lower calcitonin production in women than men may predispose women towards osteoporosis.

Other factors, such as low estrogen production, decreased calcium intake and low physical activity are pathogenetically important (21-24).

ACKNOWLEDGEMENTS

The studies were supported by grants (3.932-0.82, 3.813-0.81, 3.957-0.84) from the Swiss National Science Foundation, the Kanton of Zurich and the Schweizerische Verein Balgrist.

REFERENCES

1. J. Ittner, M. A. Dambacher, W. Born, J.-M. Ketelslegers, M. Buysschaert, P. M. Albert, A. E. Lambert, and J. A. Fischer, Diagnostic evaluation of measurements of carboxyl-terminal flanking peptide (PDN-21) of the human calcitonin gene in human serum, J. Clin. Endocrinol. Metab. 61: in press (1985).
2. C. Nagant de Deuxchaisnes, J. A. Fischer, M. A. Dambacher, J.-P. Devogelaer, C. E. Arber, J. M. Zanelli, J. A. Parsons, N. Loveridge, L. Bitensky, and J. Chayen, Dissociation of parathyroid hormone bioactivity and immunoreactivity in pseudohypoparathyroidism type I, J. Clin. Endocrinol.Metab. 53:1105 (1981).
3. P. Olles, F. Tschopp, D. W. Dempster, P. H. Tobler, R. Muff, and J. A. Fischer, Potassium·stimulates parathyroid hormone release in the absence of extracellular calcium. Mol. Cell. Endocrinol. 32:1 (1983).
4. F. Albright, W. Bauer, D. Claflin, and J. R. Cokrill, Studies in parathyroid physiology. III. The effect of phosphate ingestion in clinical hyperparathroidism, J. Clin. Invest. 11:411 (1932).
5. E. Reiss, J. M. Canterbury, M. A. Bercovitz, and E. L. Kaplan, The role of phosphate in the secretion of parathyroid hormone in man, J. Clin. Invest. 49:2146 (1970).
6. U. Binswanger, and J. A. Fischer, Reponse of immunoreactive parathyroid hormone to intravenous phosphate infusion in primary hyperparathyroidism, Klin. Wschr. 52:30 (1974).
7. J. C. Gallagher, B. L. Riggs, C. M. Jerpbak, and C. D. Arnaud, The effect of age on serum immunoreactive parathyroid hormone in normal and

osteoporotic women, J. Lab. Clin. Med. 95:373 (1980).
8. J. Reeve, P. J. Meunier, J. A. Parsons, M. Bernat, O. L. M. Bijvoet, P. Coupron, C. Edouard, L. Klenerman, R. M. Neer, J. C. Renier, D. Slovik, F. J. F. E. Vismans, and J. T. Potts Jr., Anabolic effect of human parathyroid hormone fragment on trabecular bone in involutional osteoporosis: A multicentre trial, Brit. Med. J. ii:1340 (1980).
9. D. M. Slovik, R. M. Neer, and J. T. Potts Jr., Short-term effects of synthetic human parathyroid hormone-(1-34) administration on bone mineral metabolism in osteoporotic patients, J. Clin. Invest. 68: 1261 (1981).
10. H. Rasmussen, P. Bordier, L. Auquier, J. B. Eisinger, D. Kuntz, F. Caulin, B. Argemi, J. Gueris, and A. Julien, Effect of combined therapy with phosphate and calcitonin one bone volume in osteoporosis, Bone 2:107 (1980).
11. C. Anderson, R. D. T. Cape, R. G. Crilly, A. B. Hodsman, and B. M. J. Wolfe, Preliminary observations of a form of coherence therapy for osteoporosis, Calc. Tissue Int. 36:341 (1984).
12. P. H. Tobler, F. A. Tschopp, M. A. Dambacher, and J. A. Fischer, Salmon and human calcitonin-like peptides in man, Clin. Endocrinol. (Oxf.) 20:253 (1984).
13. R. K. Craig, L. Hall, M. R. Edbrooke, J. Allison, and I. MacIntyre, Partial nucleotide sequence of human calcitonin precursor mRNA identifies flanking cryptic peptides, Nature (Lond.) 295:345 (1982).
14. C. J. Hillyard, G. Abeyasekera, R. K. Craig, C. Myers, J. C. Stevenson, and I. MacIntyre, Katacalcin: A new plasma calcium-lowering hormone, Lancet i:846 (1983).
15. B. A. Roos, M. B. Huber, R. S. Birnbaum, D. C. Aron, A. W. Lindall, K. Lips, and S. B. Baylin, Medullary thyroid carcinomas secrete a noncalcitonin peptide corresponding to the carboxyl-terminal region of preprocalcitonin, J. Clin. Endocrinol. Metab. 56:802 (1983).
16. C. J. Hillyard, J. C. Stevenson, and I. MacIntyre, Relative deficiency of plasma-calcitonin in normal women, Lancet i:961 (1978).
17. L. J. Deftos, M. H. Weisman, G. W. Williams, D. B. Karpf, A. M. Frumar, B. J. Davidson, J. G. Parthemore, and H. L. Judd, Influence of age and sex on plasma calcitonin in human beings, N. Engl. J. Med. 302: 1351 (1980).
18. J.-J. Body, and H. Heath III, Estimates of circulating monomeric calcitonin: Physiological studies in normal and thyroidectomized man, J. Clin. Endocrinol. Metab. 57:897 (1983).
19. H. M. Taggart, J. L. Ivey, K. Sisom, C. H. Chesnut III, D. J. Baylink, M. B. Huber, and B. A. Roos, Deficient calcitonin response to calcium stimulation in postmenopausal osteoporosis?, Lancet i:475 (1982).
20. R. D. Tiegs, J.-J. Body, H. W. Wahner, J. Barta, B. L. Riggs, and H. Heath III, Calcitonin secretion in postmenopausal osteoporosis, N. Engl. J. Med. 312:1097 (1985).
21. R. Lindsay, D. M. Hart, J. M. Aitken, E. B. MacDonald, J. B. Anderson, and A. C. Clarke, Long-term prevention of postmenopausal osteoporosis by oestrogen: Evidence for an increased bone mass after delayed onset of oestrogen treatment, Lancet i:1038 (1976).
22. L. S. Richelson, H. W. Wahner, L. J. Melton III, and B. L. Riggs, Relative contributions of aging and oestrogen deficiency to postmenopausal bone loss, N. Engl. J. Med. 311:1273 (1984).
23. R. P. Heaney, J. C. Gallagher, C. C. Johnston, R. Neer, A. M. Parfitt, and G. D. Whedon, Calcium nutrition and bone health in the elderly, Am. J. Clin. Nutr. 36:986 (1982).
24. J. F. Aloia, S. H. Cohn, J. A. Ostuni, R. Cane, and K. Ellis, Prevention of involutional bone loss by exercise, Ann. Intern. Med. 89: 356 (1978).

RENAL ADENYLATE CYCLASE STIMULATING ACTION OF NORMAL AND OXIDIZED PARATHYROID HORMONE (1-34)

J.J. Helwig[(1)], N. Burgmeier[(1)], A. Gairard[(2)], M.C.M. Yang[(3)] and P.K.T. Pang[(3)]

[(1)]Centre de Recherches R. Leriche, Pav. Poincare, Hospices Civils, ULP, Strasbourg (France); [(2)]Laboratoire de Physiologie, INSERM U243-CNRS 787, Fac. Pharmacie, ULP, Strasbourg (France) and [(3)]Dept. of Pharmacology, Texas Tech University Health Sciences Center, Lubbock, TX 79430

Parathyroid hormone (PTH) has a number of actions which can be demonstrated *in vivo* and *in vitro* assays. It has been shown to be hypercalcemic, hypocalciuric, hypophosphatemic, hyperphosphaturic, renal cortical membrane adenylate cyclase stimulating, renal 1-hydroxylase stimulating, and bone calcium mobilizing (see review by Parsons, 1979). Recently, PTH has also been studied extensively for its action on contraction of various smooth muscles. Although it has been repeatedly reported that PTH is hypotensive or vasodilating (Collip & Clark, 1925; Handler & Cohn, 1952; Charbon, 1966; Charbon & Hulstaert, 1974), recent studies from various laboratories have firmly established the vasorelaxing action of PTH (Berthelot & Gairard, 1975; Schleiffer et al., 1979; Pang et al., 1980a; Pang et al., 1985). In addition, the relaxing effect of PTH has been demonstrated in the uterus (Pang et al., 1981; Shew & Pang, 1984a), vas deferens (Zhang et al., 1985), trachea (Yen et al., 1983) and the gastrointestinal tract (Yang et al., 1981). Such a great variety of PTH actions then lead to the question as to whether the receptors are the same in all these target tissues and whether these actions are mediated by the same molecular structural component of the peptide hormone. Some previous attempts to answer such questions will be summarized below. We have recently begun a series of investigations on the adenylate cyclase stimulating action of PTH in rabbit renal microvessel, and tubule preparations. These data will be discussed here in relation to the question raised above.

Oxidation of the whole PTH molecule, PTH-(1-84), resulted in the loss of biological activities in all the assays tried (Rassmussen & Craig, 1962; Tashijan et al., 1964; Potts et al., 1966; Dacke & Kenny, 1973; Rosenblatt et al., 1976). The same treatment of the active fragment PTH-(1-34) essentially abolished the renal cortical membrane adenylate cyclase stimulating action of the fragment (Rosenblatt et al., 1976). In a recent series of studies, oxidized PTH-(1-34) was shown to be largely inactive in the hypotensive (Pang et al., 1983), hyperphosphaturic (Kenny & Pang, 1981), uterine smooth muscle (Shew et al., 1984b), and tracheal smooth muscle (Yen et al., 1983) relaxation assays. However, such oxidized PTH-

(1-34) has been shown to have normal activity in Japanese quail hypercalcemic (Pang et al., 1983), chronic parathyroidectomized rat hypercalcemic and hypocalciuric (Kenny & Pang, 1982a; Kenny & Pang, 1982b) assays while the oxidized PTH-(1-84) had no effect in the same assays. These data on the oxidized PTH-(1-84) indicated that the oxidation procedure was effective and the different responses to the oxidized PTH-(1-34) may reflect the involvement of at least 2 different receptors and the molecular structural requirements of PTH for these responses are different. In another assay, bPTH-(1-34) stimulated Japanese quail renal 1-hydroxylase and inhibited the 24-hydroxylase activities. However, oxidized bPTH-(1-84) and bPTH-(1-34) both retained their full activities in these two renal enzyme assays. These data suggest that the renal vitamin D metabolizing enzymes may involve a third receptor class (Kenny & Pang, 1982c). The current hypothesis suggests that hydrogen peroxide oxidizes only the two methionines in positions 8 and 18. The resulting charge changes lead to substantial conformational changes in the peptide molecules. Because of the different sizes of PTH-(1-84) and PTH-(1-34), the extent of the conformational changes may be different so that the responses to the three receptor types can be differentiated (Kenny and Pang, 1982a). Indeed amino acid analysis of the oxidized PTH-(1-34) revealed the oxidation of the methionines only and the loss of effect in some of the assay was not related to changes in other amino acids since oxidation of the methionine free analog, $[Nle^{8,18}$ $Tyr^{34}]$bPTH-(1-34), did not inactivate the analog (Pang et al., 1983). It has also been shown recently that oxidation of bPTH-(1-34) resulted in substantial secondary and tertiary structural changes in the peptides molecule which might be related to the loss of some of the properties of PTH (Pang et al., 1984). A recent study by Galceran et al. (1984) described the loss of the phosphaturic effect of PTH-(1-34) after oxidation. All the above data support the hypothesis. However, in the Galceran et al. (1984) study, the oxidized PTH-(1-34) had no hypercalcemic activity in both normal dog and acutely parathyroidectomized rat. These results are opposite to those obtained by Kenny and Pang (1981) in chronically parathyroidectomized rats and intact Japanese quail. It is possible that different PTH target tissues are involved in the hypercalcemia produced in chronic versus acute parathyroidectomized or normal animals.

In our laboratory, we have succeeded in isolating microvessels, glomeruli and tubules from rabbit renal cortex and these pure fractions from the same kidneys all responded to bPTH-(1-34) with an increase in adenylate cyclase activity (Helwig et al., 1984a; Helwig et al., 1984b; Helwig et al., 1984c). We feel that these tissue preparations will provide an unique opportunity to study the effect of different preparations of PTH on the vascular, glomerular and tubular components of the kidney separately. *In vivo* blood flow studies have identified the renal vascular bed as one that is very sensitive to PTH (Charbon & Hulstaert, 1974; Pang et al., 1980b and Wang et al., 1984). The renal microvessels should therefore be responsive to bPTH-(1-34) and they also should not respond to oxidized bPTH-(1-34). Our data indeed showed that. However, in the tubular fraction, although the ED50 of bPTH-(1-34) for the adenylate cyclase activation is essentially the same as that in the microvessel fraction, there are substantial adenylate cyclase activation by the oxidized bPTH-(1-34). It is important to note that the same dose of oxidized bPTH-(1-34) had little or no effect in the microvessels and glomeruli preparations. Similar to the results obtained in blood pressure studies, the methionine free analog

Table 1. Renal adenylate cyclase stimulating actions of untreated and oxidized parathyroid hormones (1-34)

Conditions	Microvessels	Glomeruli pMoles cAMP/min/mg	Tubules
bPTH-(1-34)			
Control	45.3 ± 6.4(9)	107.6 ± 13.6(10)	9.5 ± 1.1(4)
10^{-6}M untreated	197.6 ± 13.1(9)	194.7 ± 21.5(10)	35.3 ± 2.4(6)
10^{-6}M oxidized	55.5 ± 8.1(9)*	118.5 ± 13.6(10)*	16.3 ± 1.3(6)*
10^{-5}M untreated	206.2 ± 21.8(4)	253.0 ± 23.3(4)	38.1 ± 3.3(4)
10^{-5}M oxidized	67.3 ± 8.5(8)*	158.8 ± 20.8(8)*	25.7 ± 1.4(4)*
Sulfur-free bPTH-(1-34) analog			
Control	33.1 ± 2.0(11)	114.1 ± 6.3(11)	24.5 ± 1.2(6)
10^{-6}M untreated	116.1 ± 13.9(7)	221.4 ± 10.5(8)	134.1 ± 21.7(3)
10^{-6}M oxidized	105.2 ± 10.9(11)	203.0 ± 17.3(8)	137.8 ± 15.6(4)
10^{-5}M untreated	132.0 ± 5.1(4)	386.9 ± 52.2(6)	165.5 ± 40.1(3)
10^{-5}M oxidized	110.1 ± 8.8(6)	296.0 ± 33.3(6)	185.4 ± 21.1(4)

Means are shown ± S.E. with the numbers of experiments in parentheses. Adenylate cyclase activities were measured in the presence of 10^{-4}M of GTP. *$P<0.05$ as compared with the corresponding untreated hormone, in the student's t-test.

[$Nle^{8,18}Tyr^{34}$]bPTH-(1-34) had full adenylate cyclase stimulating activity in the microvessel, glomeruli and tubules and oxidation did not affect such an activity. All these data are summarized in Table 1.

Our data on renal adenylate cyclase stimulation by PTH are important for two reasons. First, the failure of the oxidized bPTH-(1-34) to stimulate the adenylate cyclase in the rabbit renal microvessels but a substantial effect of the same preparation in the renal tubules suggest the possibility that there may be a receptor in the tubules different from that in the microvessels. It is possible that the tubules may contain two types of receptors; one responding to oxidized PTH and the other not. Such a hypothesis is feasible in view of the various tubular effects of PTH in relation to phosphate and calcium handling. In another study, we have reported that the adenylate cyclase systems in the microvessels and tubules showed substantial difference in relation to PTH activation. In the microvessels, exogenous GTP stimulated the cyclase by itself and it also potentiated the stimulating effect of PTH. However, in the tubules GTP by itself had little effect and it also failed to potentiate the effect of PTH (Helwig et al., 1984b). These studies together strongly suggest the presence of a receptor in the tubules different from the one in the microvessels. Second, the presence of substantial effect of oxidized bPTH-(1-34) in the tubule can be extrapolated to explain the seemingly controversial data obtained with the oxidized fragment in the various hypercalcemic assays. In Japanese quails and chronically parathyroidectomized rats, the oxidized bPTH-(1-34) had full hypercalcemic action but in the normal dog and acutely parathyroidectomized rats, the oxidized peptide had no effect. Is it possible that the normal quail and the chronically parathyroidectomized rats are highly dependent on the kidney to produce the hypercalcemia. Then, if the residual adenylate cyclase effect by the oxidized bPTH-(1-34) in the tubules is related to renal calcium balance, the hypercalcemic effect in these animals can perhaps be explained. It has been suggested that renal tubular cAMP may be related to the hypocalciuric action of PTH (Aubach and Chase, 1976). Furthermore, in the chronically parathyroidectomized rats, the oxidized bPTH-(1-34) had full hypocalciuric effect. Is it then also possible that the normal dog and acutely parathyroidectomized rats may depend more on the bone turn over for the hypercalcemic action of PTH? It has been shown that oxidized bPTH-(1-34) lost its ability to mobilize calcium in bone culture and in bone adenylate cyclase stimulation (Yee et al., 1983). This then perhaps can explain the absence of a hypercalcemic effect of the oxidized peptide fragment in these animals.

In conclusion, our studies on the renal microvessel, glomeruli and tubule adenylate cyclase activation with the various preparations of bPTH-(1-34) strongly support the presence of more than one type of PTH receptors in various target tissues.

Acknowledgments

The authors thank Miss J. Krill, Miss E. Friedman and Mrs. S. Wendling for their skilled technical assistance. This work was supported by a "Contrat de Recherches Externes" no 835005 I.N.S.E.R.M. and by the G.R.E.C.O.-C.N.R.S. no 24 and NIH grant HL32879-01.

References

Aubach, G.D. and Chase, L.R., 1976, Cyclic nucleotides and biochemical actions of PTH and calcitonin. In: Handbook of Physiology, Endocrinology, vol. VII, Parathyroid Gland (Aubach, G.D., ed.) p. 344.

Berthelot, A. and Gairard, A., 1975, Effet de la parathormone sur la pression arterielle et la contraction de l'aorte isolee. Experientia, 31: 457.

Charbon, G.A., 1966, A diuretic and hypotensive action of a parathyroid extract. Acta Physiol. Pharmac. Neerl. 14: 52.

Charbon, G.A. and Hulstaert, P.F., 1974, Augmentation of arterial hepatic and renal flow by extracted and synthetic parathyroid hormone. Endocrinol., 95: 621.

Collip, J.B. and Clark, E.P., 1925, Further studies on physiological action of parathyroid hormone. J. Biol. Chem., 64: 133.

Dacke, C.G. and Kenny, A.D., 1973, Avian bioassay method for parathyroid hormone. Endocrinol., 92: 463.

Galceran, T., Lewis-Finch, J., Martin, K.J. and Slatopolsky, E., 1984, Absence of biological effects of oxidized parathyroid hormone-(1-34) in dogs and rats, Endocrinol., 115: 2375.

Handler, P. and Cohn, D.V., 1952, Effect of parathyroid extract on renal function. Am. J. Physiol., 169: 188.

Helwig, J.J., Judes, C., Bollack, C. and Mandel, P., 1984a, Isolation and characterization of a microvascular fraction from rabbit kidney cortex. Renal Physiol., 7: 146.

Helwig, J.J., Judes, C., Gairard, A. and Bollack, C., 1984b, J. Physiol. (Paris), 79: 166.

Helwig, J.J., Schleiffer, R., Judes, C. and Gairard, A., 1984c, Distribution of parathyroid hormone-sensitive adenylate cyclase in isolated rabbit renal cortex microvessels and glomeruli. Life Sci., 35: 2649. 35: 2649.

Kenny, A.D. and Pang, P.K.T., 1981, Dissociation of the hypercalcemic and hyperphosphaturic actions of bovine parathyroid hormome (1-34). Fed. Proc., 40: 399 (Abstract).

Kenny, A.D. and Pang, P.K.T., 1982a, Structure activity relationship of parathyroid hormone: effects of oxidation. In: C. Oguro and P.K.T. Pang (eds.). Comparative Endocrinology of Calcium Regulation, Japanese Scientific Societies Press, Tokyo, p. 195.

Kenny, A.D. and Pang, P.K.T., 1982b, Effect of oxidation on the urinary calcium response in rats to bovine parathyroid hormone. Calc. Tissue Int., 34: S41.

Kenny, A.D. and Pang, P.K.T., 1982c, Response of the renal vitamin D endocrine system to oxidized parathyroid hormone (1-34). Proc. Soc. Exptl. Biol. Med., 171: 191.

Pang, P.K.T., Tenner, T.E., Jr., Yee, J.A., Yang, M.C.M. and Janssen, H.F., 1980a, Hypotensive action of parathyroid hormone preparations on rats and dogs. Proc. Natn. Acad. Sci., USA, 77: 675.

Pang, P.K.T., Janssen, H.F. and Yee, J.A., 1980b, Effects of synthetic parathyroid hormone on vascular beds of dogs. Pharmacol., 21: 213.

Pang, P.K.T., Shew, R.L. and Sawyer, W.H., 1981, Inhibition of uterine contraction of synthetic parathyroid hormone fragment. Life Sci., 28: 1317.

Pang, P.K.T., Yang, M.C.M., Keutman, H.T. and Kenny, A.D., 1983, Structure activity relationship of parathyroid hormone: separation of the hypotensive and the hypercalcemic properties. Endocrinol., 112: 284.

Pang, P.K.T., Hong, B.S., Yen, L. and Yang, M.C.M., 1984, Parathyroid hormone. A specific potent vasodilator. Contri. Nephrol., 41: 137.

Pang, P.K.T., Yang, M.C.M., Shew, R. and Tenner, T.E., Jr., 1985, The vasorelaxant action of parathyroid hormone fragments on isolated rat tail artery. Blood Vessels, 22: 57.

Parsons, J.A., 1979, Physiology of PTH, Endocrinology, vol. 2, p. 621, edited by L.J. Degroot et al., Grune and Statton, N.Y.

Potts, Jr., J.T., Aurbach, G.D. and Sherwood, L.M., 1966, Parathyroid hormone; chemical properties and structural requirements for biological and immunological activity. Recent Prog. Horm. Res., 22: 101.

Rasmussen, H. and Craig, L.C., 1962, The parathyroid polypeptides. Recent. Prog. Horm. Res., 18: 269.

Rosenblatt, M., Goltzman, D., Keutmann, H.T., Tregear, G.W. and Potts, Jr., J.T., 1976, Chemical and biological properties of synthetic, sulfur-free analogues of parathyroid hormone, J. Biol. Chem., 251: 159.

Schleiffer, R., Berthelot, A. and Gairard, A., 1979, Action of parathyroid extract on arterial blood pressure and on contraction and ^{45}Ca exchange in isolated aorta or the rat. Evr. J. Pharmacol., 58: 163.

Shew, R.L. and Pang, P.K.T., 1984a, Effects of bPTH fragments (1-34), (3-34) and (7-34) on bovine contraction. Peptides, 5: 485.

Shew, R.L., Kenny, A.D. and Pang, P.K.T., 1984b, Uterine relaxing action of parathyroid hormone: effect of oxidation and methionine substitution. Proc. Soc. Exp. Biol. Med., 175: 444.

Tashjian, A.H., Ontjes, D.A. and Munson, P.L., 1964, Alkylation and oxidation of methionine in bovine parathyroid hormone: effects on hormonal activity and antigenicity. Biochemistry, 3: 1175.

Wang, H.H., Drugge, E.D., Yen, Y.C., Blumenthal, M.R. and Pang, P.K.T., 1984, Effect of synthetic parathyroid hormone on hemodynamics and regional blood flow. European J. Pharmacol., 97: 209.

Yang, M.C.M., Kenny, A.D. and Pang, P.K.T., 1984, Effect of parathyroid hormone on rat gastrointestinal contraction in vivo. Proc. IX Int. Symp. Comp. Endocrinol., Dec. 7-11, 1981, Hong Kong, in press.

Yen, Y.C., Yang, M.C.M., Kenny, A.D. and Pang, P.K.T., 1983a, Parathyroid hormone (PTH) fragments relax the guinea pig trachea in vitro. Can. J. Physiol. Pharmacol., 61: 1324.

Yee, J.A., Shew, R.L. and Pang, P.K.T., 1983b, Effects of oxidized synthetic parathyroid hormone on bone in vitro. Calc. Tissue Int., 35: 44.

Zhang, R.H., Yang, M.C.M. and Pang, P.K.T., 1985, The relaxing effect of parathyroid hormone on rat vas deferens. Fed. Proc., 44: 1650.

HYPERCALCEMIA OF MALIGNANCY

John W. Jacobs and Ellen Simpson
Biological Chemistry
Merck Sharp & Dohme Research Laboratories
West Point, PA 19486

INTRODUCTION

Hypercalcemia is frequently associated with malignant disease (Stewart et al, 1980, Mundy et al, 1984). The primary underlying cause is an increase in osteoclastic bone resorption but the mechanisms which lead to this increase are still the focus of current research. A number of investigators have utilized different approaches to identify factors which are secreted from tumors associated with hypercalcemia and which are responsible for increased bone resorption. It is now apparent that different classes of protein factors are responsible for the hypercalcemia observed in different types of malignancies. For example, in patients with myeloma the lymphokine, osteoclast-activating factor, is most likely the humoral factor secreted from myeloma cells and causing bone resorption (Mundy et al, 1974). Hypercalcemia can also occur in patients who have solid tumors without bone metastases and bone resorption in this type of hypercalcemia is often associated with hypophosphatemia and an increase in nephrogenous cyclic AMP excretion (Stewart et al, 1980).

Existing calcium-regulating hormones, the most notable being parathryoid hormone, have been implicated in the humoral hypercalcemia of malignancy. Ectopic parathyroid hormone production by tumors has been suggested to cause hypercalcemia in certain patients. As discussed below, the techniques of molecular biology can be utilized to directly address this question by examining the production of parathyroid hormone mRNA in tumors associated with hypercalcemia (Simpson et al, 1983).

Table I below lists the types of hypercalcemia which are observed in malignancy. These include solid tumors with metastasis to bone with lung and breast carcinomas as representatives of this group.

TABLE I

TYPES OF HYPERCALCEMIA IN MALIGNANCY

1. <u>Solid Tumor With Metastasis to Bone</u>
   Mechanism: Osteolysis
   Tumors: Lung, Breast

2. <u>Solid Tumor Without Metastases to Bone</u>
   Mechanism: Humoral Mediators
   Tumors: Lung, renal, pancreas, skin

3. <u>Hematologic Malignancy</u>:
   Mechanism: Secretion of Osteoclast-Activating Factor
   Tumors: Myeloma

In the first class of hypercalcemia there is probably direct resorption of bone by hydrolytic enzymes secreted from the tumor cells. Hypercalcemia can also occur with solid tumors without evidence of bone metastases and, in these patients, the hypercalcemia is mediated by a humoral factor. A wide range of tumors have been associated with this syndrome named the "humoral hypercalcemia of malignancy" with lung and renal carcinomas being representative of the group. Finally, hematologic malignancies such as myeloma, secrete bone-resorbing lymphokines and are frequently associated with hypercalcemia.

Recently there has been an intense research effort to identify the humoral factor which causes hypercalcemia in patients with solid tumors without metastasis to bone (Class II of Table I). In this syndrome, the humoral hypercalcemia of malignancy, tumor cells secrete a polypeptide factor which activates osteoclasts resulting in increased bone resorption and entry of calcium into the extracellular fluid. As shown in Table II, the mediators of the humoral hypercalcemia of malignancy which have received the most attention to date include 1) parathyroid hormone; 2) PTH-like factors; proteins which activate adenylate cyclase and may act through the PTH receptors but are not PTH; and 3) transforming growth factors. The evidence supporting the role of each of these factors as mediators of the hypercalcemia of malignancy will be discussed below.

TABLE II

MEDIATORS OF MALIGNANT HYPERCALCEMIA

1. Parathyroid Hormone
   - Ectopic PTH secretion rarely, if ever, associated with malignant hypercalcemia

2. PTH-like Factors
   - Stimulate adenylate cyclase through PTH receptor
   - Secreted by a number of tumors associated with hypercalcemia

3. Transforming Growth Factors
   - Potent *in vitro* bone resorbers
   - Relevance *in vivo* unknown

PARATHYROID HORMONE

For many years, ectopic parathyroid hormone (PTH) secretion was thought to be responsible for the hypercalcemia of malignancy, since in many respects the clinical picture resembles that of primary hyperparathyroidism (Mundy, et al, 1984). (See Figure 1 below.) This includes stimulation of osteoclastic bone resorption, stimulation of renal cyclic AMP production and inhibition of renal phosphate reabsorption. However, subsequent evidence has revealed differences between the two syndromes. Calcium absorption from the gut is decreased in the humoral hypercalcemia of malignancy but not in primary hypeparathyroidism (See Figure 1) and 1,25-dihydroxy vitamin D concentrations are decreased in malignant hypercalcemia but not in primary hyperparathyroidism (Breslau et al, 1984).

FEATURES OF HUMORAL HYPERCALCEMIA OF MALIGNANCY

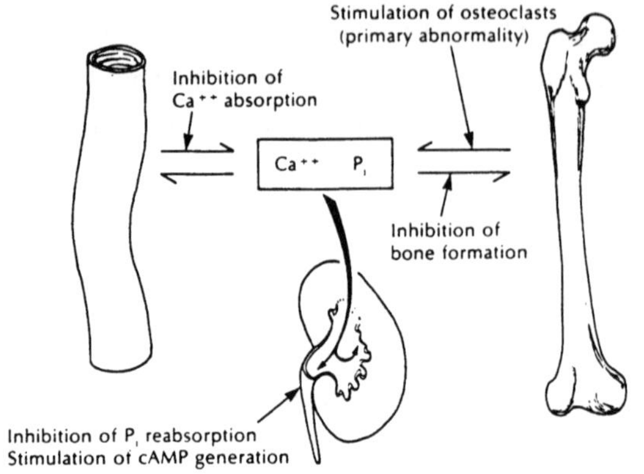

Figure 1. Calcium and phosphate homestasis in malignant hypercalcemia. (From Jacobs and Mundy, Endocrine Society Postgraduate Course, Dallas, TX, 1984)

Studies with parathyroid hormone radioimmunoassays have been controversial because with some assays concentrations of immunoreactive PTH from serum and tumor extracts of hypercalcemic patients have been increased or inappropriately normal. Other studies utilizing different assays have shown that the concentration of immunoreactive parathyroid hormone is suppressed (reviewed in Simpson et al, 1983).

To address this issue, namely, whether ectopic production of parathyroid hormone can mediate the humoral hypercalcemia of malignancy, we have used a sensitive and specific assay which employs base pair hybridization of cloned parathyroid hormone DNA with parathyroid hormone mRNA. We have examined mRNA extracted from over 20 tumors associated with the humoral hypercalcemia of malignancy and have found no parathyroid mRNA in these tissues. The ability to detect a relatively small quantity of an ectopically expressed mRNA clearly depends on the sensitivity of the methods employed. The sensitivity of the assays employed in these studies is such that we should have been able to find parathyroid mRNA in these tumors if they contained at least 1/1,000 of the concentration of PTH mRNA normally found in the parathyroid gland (Simpson et al, 1983).

The possibility exists that an abnormal or altered form of parathyroid hormone is produced by the tumors which can bind to the PTH receptor but is not detectable by PTH antisera. However, the recombinant DNA probe which was utilized in these studies was over 300 base pairs in length and should have detected PTH-like mRNAs, even if they were only distantly related.

FACTORS WHICH BIND TO THE PTH RECEPTOR

Recently, a number of tumors associated with the humoral hypercalcemia of malignancy have been found to produce factors which can be distinguished from parathyroid hormone by immunological criteria but which bind to the parathyroid hormone receptor and mimic the actions of PTH (Strewler et al, 1983; Stewart et al, 1983; Rodan et al, 1983). These tumor-derived factors have been shown to stimulate adenylate cyclase of renal membranes or osteosarcoma cells. Both of these tissues are also responsive to PTH. Furthermore, the biological effects of these tumor-derived factors can be blocked by synthetic parathyroid hormone antagonists suggesting a receptor mediated basis for the action of these factors. As does parathyroid hormone, these tumor-derived factors may have important effects on the kidneys and may be responsible for renal effects seen in this syndrome including an increase in nephrogenous cyclic AMP and renal phosphate wasting (See Figure 1). In addition, more recent evidence indicates that bone resorbing activity co-purifies with adenylate cyclase-stimulating activity and, thus, these PTH-like factors may act directly on bone (Klein et al, 1985). It should be stressed that these factors, although related to PTH in terms of their competition for similar receptors, are structurally distinct from PTH based on their molecular weights which generally range from 15,000 daltons and larger (Strewler et al, 1983; Stewart et al, 1983; Klein et al, 1985) versus 9500 daltons for PTH, and the inability of PTH-specific antisera to recognize these factors.

TRANSFORMING GROWTH FACTORS (TGFs)

Previous studies have indicated that epidermal growth factor (EGF) is a stimulator of osteoclastic bone resorption, in vitro, in organ culture assays (Raisz, 1980). Tumors are known to secrete polypeptides known as transforming growth factors, which have many of the biological properties of EGF. In selected animal tumor models of the humoral hypercalcemia of malignancy, bone resorbing activity was found to co-purify on gel filtration columns with transforming growth factor activity (Ibbotson et al, 1983). Subsequent work indicated that synthetic preparations of transforming growth factors could stimulate bone resorption in vitro (Ibbotson et al, 1985; Tashjian et al, 1985). The mechanisms whereby transforming growth factors stimulate bone resorption in vitro is controversial but recent evidence indicates that TGFs stimulate prostaglandin production in bone leading to increased resorption (Tasjian et al, 1985). It should be stressed that although TGFs increase bone resorption in vitro, there is no evidence at present that these factors mediate hypercalcemia in vivo. Many tumors which secrete transforming growth factors do not cause hypercalcemia and, thus, it is likely that TGFs are not the sole mediator of this syndrome and that additional factors, such as PTH-like factors, may be involved. The recent molecular cloning of recombinant transforming growth factors should allow the production of these growth factors for in sufficient quantities, in vivo testing.

Since epidermal growth factor is a potent resorber of bone in vitro, we have examined whether tumors associated with the humoral hypercalcemia of malignancy can produce factors which interact with EGF receptors on bone cells. To perform these studies messenger RNA was extracted from a human renal carcinoma associated with hypercalcemia and translated in Xenopus oocytes. The oocyte translation assay has a number of distinct advantages over cell-free translation assays including the production of relatively large amounts of mRNA-directed translation products and the fact that these products are secreted from the oocytes and can be harvested for use in subsequent bioassays or receptor assays. A radioreceptor assay was devised (Figure 2) to examine the production of tumor-derived products which could engage the EGF receptor.

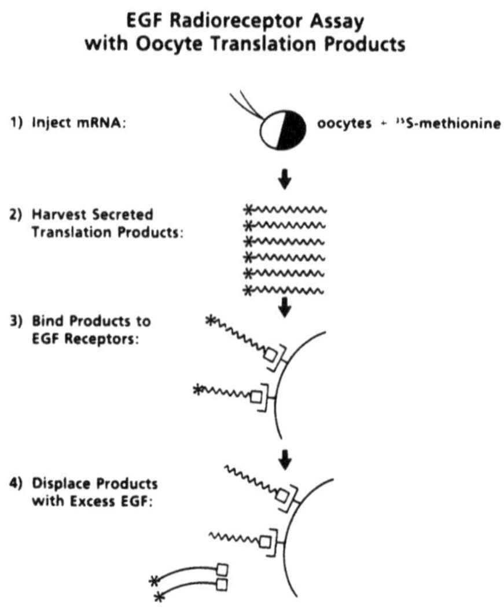

Figure 2. Diagram of Epidermal Growth Factor (EGF) radioreceptor assay.

As shown in Figure 2, the assay utilizes ^{35}[S]-methionine to label newly synthesized translation products programmed by renal carcinoma mRNA which are subsequently harvested from the oocyte medium. The translation products are then bound to cultured bone cells of the osteoblast phenotype which contain EGF receptors. The presence of an

EGF-related translation product was determined by carrying out the receptor binding assays in the absence and presence of unlabeled epidermal growth factor.

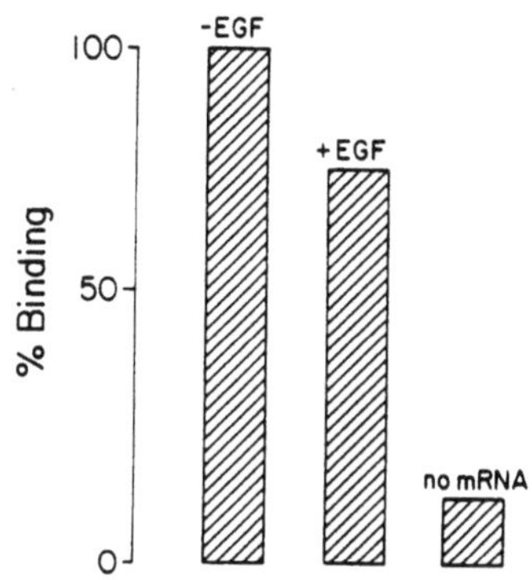

Figure 3. EGF radioreceptor assay of oocyte translation products. Renal carcinoma mRNA was microinjected into Xenopus oocytes as outlined in Figure 2 and radiolabelled translation products incubated with osteosarcoma cell membranes in the absence (-EGF) and presence (+EGF) of 15 ug unlabelled EGF. As a control, oocyte products from from water-injected eggs (no mRNA) were utilized in the assay.

As shown in Figure 3, microgram quantities of unlabelled, homogeneous EGF specifically displaced radioactivity from the membranes. In this type of experiment, the unlabelled EGF is coincubated with the radiolabelled oocyte translation products and binding is carried out at $4^{o}C$ for four hours. After binding, the osteosarcoma cell membranes are washed, harvested by centrifugation and solubilized before loading onto SDS-polyacrylamide gels.

In Figure 4, translation products displaced by unlabelled EGF are visualized after SDS-polyacrylamide gel electrophoresis and autoradiography and a major product of 17,000 daltons is observed which is specifically displaced by unlabeled EGF. (Arrow, Figure 4) A second, minor translation product of higher molecular weight is also displaced from osteosarcoma cell membranes by the EGF. From these studies we conclude that in this particular renal carcinoma associated with hypercalcemia there is at least one gene product of 17,000 daltons which can interact with EGF receptors on bone cells and is a likely candidate for a bone-resorbing factor.

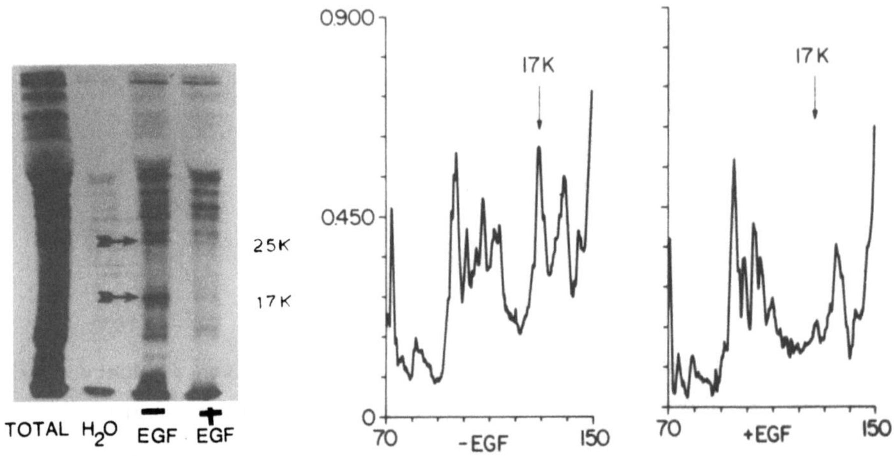

Figure 4. SDS-polyacrylamide gel electrophoresis of oocyte translation products incubated with osteosarcoma membrances in the presence and absence of epidermal growth factor. Total=total translation products from renal carcinoma mRNA injected oocytes; H_2O=control of water-injected eggs; (-EGF)= translation products which bound to membranes in the absence of EGF; (+EGF)=translation products which bound to membranes in the presence of 15 ug EGF. Arrows point to two (17,000 and 25,000 dalton) translation products specifically displaced from membranes by EGF. Below gel is densitometry tracing of an autoradiograph of the gel. Arrow points to the position of 17,000 dalton product.

In summary, the development of the humoral hypercalcemia of malignancy is a complex process which may involve the elaboration of one or more distinct tumor products. PTH-like factors are produced by a number of tumors and these factors can clearly act both on bone and kidney to bring about the resultant hypercalcemia. In addition, transforming growth factors which are produced by certain tumors have been shown to stimulate bone resorption _in vitro_ and these factors are also likely candidates for mediators of the humoral hypercalcemia of malignancy. The primary abnormality in this syndrome is an increase in osteoclastic bone resorption caused by a humoral factor. In patients with normal renal function, normal cellular mechanisms of regulation of serum calcium protect against hypercalcemia. However, in patients with cancer and increased renal tubular calcium reabsorption due to the possible effects of a PTH-like factor hypercalcemia may occur.

ACKNOWLEDGEMENTS

The authors acknowledge Drs. K. Ibbotson, S. D'Souza and G. Mundy of the University of Texas Health Science Center, San Antonio, Texas with whom these studies were performed.

REFERENCES

Breslau, N.A., McGuire, J.L., Zerwekh, J.E., Frenkel, E.P., and Pak, C.Y.C., 1974, Hypercalcemia associated with increased serum calcitriol levels in three patients with lymphoma, Annals Int Med 100:1.

Ibbotson, K.J., D'Souza, S.M., Ng, K.W., Osborne, C.K., Niall, M., Martin, T.J., and Mundy, G.R., 1983, Tumor derived growth factor increased bone resorption in a tumor associated with the humoral hypercalcemia of malignancy, Science 221:1292.

Ibbotson, K.J., Twardzik, D.R., D'Souza, S.M., Hargreaves, W.R., Todaro, G.J., and Mundy, G.R., 1985, Stimulation of bone resorption in vitro by synthetic transforming growth factor-alpha, Science 228:1007.

Klein, R.F., Strewler, G.J., Nissenson, R.A., Williams, R.D., and Leung, S.C., June, 1985, PTH-like protein from human carcinoma cells stimulates bone resorption, Abstract 145, 7th Annual Am Soc Bone and Min Res.

Mundy, G.R., Raisz, L.G., Cooper, R.A., Schechter, G.P., and Salmon, S.E., Evidence for the secretion of an osteoclast stimulating factor in myeloma. N Engl J Med 291:1041, 1974.

Mundy, G.R., Ibbotson, K.J., D'Souza, S.M., Simpson, E.L., Jacobs, J.W., and Martin, T.J., 1984, The hypercalcemia of cancer: Clinical implications and pathogenic mechanisms, New Engl J Med 310:1718.

Raisz, L.G., 1975, Epidermal growth factor resorbes bone _in vitro_, J Clin Invest 44:103.

Rodan, S.B., Insogna, K.L., Vignery, A. M-C., Stewart, A.F., Broadus, A.E., D'Souza, S.M., Bertolini, D.R., Mundy, G.R., and Rodan, G.A., 1983, Factors associated with humoral hypercalcemia of malignancy stimulate adenylate cyclase in osteoblastic cells, J Clin Invest 72:1511.

Simpson, E., Mundy, G.R., Ibbotson, K., D'Souza, S., Jacobs, J., 1983, Absence of parathyroid hormone messenger RNA in nonparathyroid tumors associated with hypercalcemia, N Engl J Med 309:325.

Stewart, A.F., Horst, R., Deftos, L.J., Cadman, E.C., Lang, R., and Broadus, A.E., 1980, Biochemical evaluation of patients with cancer associated with hypercalcemia: evidence for humoral and non-humoral groups, N Engl J Med 303:1377.

Stewart, A.F., Insogna, K.L., Goltzman, D., and Broadus, A.E., 1983, Identification of adenylate cyclase stimulating activity and cytochemical glucose 6 phosphate dehydrogenase stimulating activity in extracts of tumors from patients with humoral hypercalcemia of malignancy, PNAS 80:1454.

Strewler, G.J., Williams, R.D., and Nissenson, R.A., 1983, Human renal carcinoma cells produce hypercalcemia in the nude mouse and a novel protein recognized by parathyroid hormone receptors, J Clin Invest 71:769.

Tashjian, A.H., Voelkel, E.F., Lazzaro, M., Singer, F.R., Roberts, A.B., Derynck, R., Winkler, M.E., and Levine L., 1985, Alpha and Beta human transforming growth factors stimulate prostaglandin production and bone resorption in cultured mouse calvaria, Proc Natl Acad Sci 82:4535-4538.

MECHANISM OF THE HYPOCALCEMIC AGENT WR-2721 AND ITS ACUTE AND CHRONIC APPLICATION IN EU- AND HYPERPARATHYROIDISM

S. Hirschel-Scholz, J. Caverzasio and J.-P. Bonjour

Division of Pathophysiology, Department of Medicine
University Hospital
CH 1211 Geneva 4 (Switzerland)

Recent studies suggest that the organic phosphorothioate WR-2721 could be a useful drug in the treatment of hypercalcemia, particularly in relation with disorders of parathyroid hormone (PTH) homeostasis (1).

WR-2721 (S-,2-(3-aminopropylamino)-, ethylphosphorothioic acid) has been developed as a pharmacological agent because it provides selective protection of normal tissues against the toxicity of radiation and alkylating agents (2). In euparathyroid human subjects, this drug induces a rapid fall in calcemia (1). This acute hypocalcemic effect is associated with a decrease in the plasma level of PTH, suggesting that WR-2721 inhibits the production or release of PTH in subjects with normal parathyroid function (1).

Recently we have reported (3) that WR-2721 is also able to reduce an abnormal PTH secretion in a case of parathyroid carcinoma. Indeed, the intravenous infusion of a single dose (750 mg/m2) of WR-2721 was followed within 3 hours by a reduction in the plasma concentration of PTH by 60% which was accompanied by a diminution in urinary cyclic AMP excretion (3). This response was associated with a decrease in calcemia by about 0.45 mmol/l, a fall similar to that observed in euparathyroid patients (1).

The data so far obtained in human subjects (1,3) suggest that the major effect of WR-2721 on the plasma calcium (Ca) level is due to an inhibition of PTH secretion. However, other mechanisms could contribute to the fall in calcemia. Indeed WR-2721 contains a phosphate group which either bound to or liberated from the molecule extra- and/or intracellularly (4,5,6) might play some role in the hypocalcemic response. Furthermore, evidence for a direct inhibition of PTH secretion does not exclude additional interference with the action of the hormone on its target organs. Likewise, the drug could still exert a PTH-independent hypocalcemic effect by affecting bone and/or kidney Ca fluxes. Finally it has been reported (1) that the hypocalcemic effect of WR-2721 is associated with a fall in plasma magnesium (Mg). Thus, a change in Mg metabolism could also interfere with Ca homeostasis.

We have recently considered these various possible mechanisms by studying the effect of WR-2721 on Ca and Mg metabolism in intact and thyroparathyroidectomized (TPTX) rats. Our results which are described in details in another report (7) indicate that: 1) The phosphorylated and dephosphorylated form of WR-2721 induced an equal dose-dependent decrement in plasma Ca. 2) In intact rats a maximal hypocalcemic dose of WR-2721 reduced urine cyclic AMP from 70.5 $\pm$ 6.3 to 38.2 $\pm$ 3.1 pmol/ml GFR, a level comparable to that observed (35.9 $\pm$ 5.2 pmol/ml GFR) in TPTX rats. 3) WR-2721 given to TPTX rats did not significantly interfere with the calcemic effect of bovine PTH 1-34 infused at 2.5 IU/h. Likewise the action of PTH on the tubular reabsorption of Ca and inorganic phosphate (Pi), and urinary excretion of cyclic AMP was not altered by pretreating TPTX rats with WR-2721. 4) In TPTX rats made normocalcemic by low Pi diet, the hypocalcemic effect of WR-2721 was only about one-fourth of that observed in intact animals. However, it was associated with an increased urine Ca excretion per ml GFR, indicating a PTH-independent inhibitory effect on tubular Ca reabsorption. 5) In WR-2721 treated intact rats, prevention of hypomagnesemia by infusing MgCl2 did not reduce hypocalcemia.

Thus, the hypocalcemic effect of WR-2721 does not appear to be dependent upon the presence of a phosphate group in the molecule. Our findings do not suggest that there is a direct causal relationship between hypomagnesemia and hypocalcemia. Interestingly, the hypomagnesemic effect of WR-2721 is probably due, at least in part, to a PTH-independent inhibition of renal Mg reabsorption (S. Hirschel-Scholz et al., manuscript in preparation).

These results obtained in euparathyroid and TPTX rats sustain the notion that WR-2721 when administered acutely is a strong inhibitor of PTH secretion. In addition, WR-2721 also exerts a PTH-independent inhibitory effect on the tubular Ca reabsorption. This renal action could contribute to the hypocalcemic response, particularly in situations when renal Ca reabsorption is enhanced. It has been reported that WR-2721 added in vitro to bone tissue culture reduces bone resorption (8). Hence, it has been suggested that WR-2721 could also exert in vivo an inhibitory effect on bone resorption (8). Although our experiments in TPTX rats do not support this notion, they cannot exclude it. Other studies are necessary in order to clarify this important issue.

The acute decrease of PTH secretion in human beings (1,3) and animals (7,8) after one single dose of WR-2721 prompted us to test whether the application of WR-2721 over a period of several days could prevent the pathological consequences of chronic hyperparathyroidism. We have observed (9) in a rat model that WR-2721 was quite effective in preventing the nephrocalcinosis induced by secondary hyperparathyroidism. This preventive action could be elicited with doses which did not alter the food intake of the animals. Interestingly the same chronic administration of WR-2721 did not hinder the development of a nephrocalcinosis induced by 1,25-dihydroxyvitamin D3 in TPTX rats. These results suggest that the antinephrocalcinotic action of WR-2721 is directly related to its inhibitory effect on PTH secretion.

In conclusion, WR-2721 is a powerful hypocalcemic drug. It is able to block normal and excessive PTH secretion acutely. When chronically applied, it can prevent a PTH-induced nephrocalcinosis. These findings suggest that WR-2721 might be a potential drug in the medical treatment of hyperparathyroidism. Furthermore, other hypercalcemic conditions could benefit of its additional PTH-independent hypocalcemic action which appears to due, at least in part, to a direct inhibitory effect of the drug on tubular Ca reabsorption. Whether WR-2721 also exert a significant PTH-independent inhibitory effect on bone resorption in vivo remains to be clarified.

ACKNOWLEDGMENTS

We thank Mrs M.-Ch. Brandt for preparing this manuscript. This work was supported by the Swiss National Science Foundation (Grant 3.806.0.82), the Geneva League against Cancer and the Carlos & Elsie de Reuter Foundation. WR-2721 was kindly provided by the National Cancer Institute, Washington, USA.

REFERENCES

1. Glover, D., Riley, L., Carmichael, K., Spar, B., Glick, J., Kligerman, M., Agus, Z.S., Slatopolsky, E., Attie, M., and Goldfarb, S.: Hypocalcemia and inhibition of parathyroid hormone secretion after administration of WR-2721 (A radio-protective and chemoprotective agent). N.Engl.J.Med. 309: 1137 (1983).

2. Yuhas, J.M., Spellman, J.M., Culo, F.: The role of WR-2721 in radiotherapy and/or chemotherapy. Cancer Clin. Trials 3: 211 (1980).

3. Hirschel-Scholz, S., Jung, A., Fischer, J.A., Trechsel, U., and Bonjour, J.P.: Suppression of parathyroid secretion after administration of WR-2721 in a patient with parathyroid carcinoma. Clin. Endocrinol. 23: 313 (1985).

4. Yuhas, J.M.: Biological factors affecting the radioprotective efficiency of S-2-[3-Aminopropylamino] Ethylphosphorothioic Acid (WR-2721). LD50 (30) doses. Radiation Res. 44: 621 (1970).

5. Yuhas, J.M.: Protective drugs in cancer therapy: Optimal clinical testing and future directions. Radiation Oncol. Biol.Phys. 8: 513 (1982).

6. Durand, R.E.: Radioprotection by WR-2721 in vitro at low oxygen tensions: implications for its mechanisms of action. Br.J.Cancer 47: 387 (1983).

7. Hirschel-Scholz, S., Caverzasio, J., Bonjour, J.-P.: Inhibition of parathyroid hormone (PTH) secretion and PTH-independent diminution of tubular Ca reabsorption by WR-2721, a unique hypocalcemic agent. J.Clin.Invest. in press (1985).

8. Attie, M.F., Fallon, M.D., Spar, B., Wolf, J.S., Slatopolsky, E., Goldfarb, S.: Bone and parathyroid inhibitory effects of S-2(3-aminopropylamino)ethylphosphorothioic acid. Studies in experimental animals and cultured bone cells. J.Clin.Invest. 75: 1191 (1985).

9. Hirschel-Scholz, S., Caverzasio, J., Bonjour, J.-P.: Prevention by the organic thiophosphate WR-2721 of parathyroid-hormone (PTH)-induced nephrocalcinosis. Calcif.Tissue Int., in press (1985).

DETERMINANTS OF CIRCULATING LEVELS OF 1.25$(OH)_2D_3$ (CALCITRIOL) IN PRIMARY RY HYPERPARATHYROIDISM (PHPT)

Michel Paillard, Pascal Patron and Jean Paul Gardin

Laboratoire de Physiologie et de Physiopathologie Rénale et Electrolytique, Université Paris VII, Hôpital Louis Mourier 178 Rue des Renouillers, 92701 Colombes, France

INTRODUCTION

The circulating level of 1,25$(OH)_2D_3$ (calcitriol) which controls intestinal absorption of calcium, has been suggested to play a major role in the clinical presentation of primary hyperparathyroidism (PHPT) (Broadus et al., 1980). In a recent report (Broadus et al., 1980), abnormally high circulating values of calcitriol were found in the absorptive form of the disease, characterized by high intestinal absorption of calcium, hypercalciuria, high incidence of renal stones, whereas normal to high-normal values of calcitriol were observed in non-absorptive form of the disease, with normal intestinal absorption of calcium, normal calciuria and low incidence of renal stones. However, the differences in serum calcitriol values could not be explained on the basis of differences in any of the variables currently regarded as modulating the renal production of calcitriol (Broadus et al., 1980).

The present work was designed 1) to systematically study serum calcitriol and its known determinants in a large population of unselected patients with primary hyperparathyroidism ; 2) to attempt to establish a logical link between serum calcitriol levels and the factors modulating the renal production of calcitriol.

PATIENTS AND METHODS

Among 306 patients with further surgically confirmed primary hyperparathyroidism, and investigated in our unit between 1979 and 1984, 102 patients were randomly sampled. Nine patients were excluded from the study because they displayed overt bone disease with radiographic evidence of osteitis fibrosa cystica, generalized osteopenia with diffuse bone pain and serum alkaline phosphatases as high as five times the upper normal values. The investigation was carried out as previously described (Gardin and Paillard, 1984) after overnight fasting (the day before, meals were poor in calcium and drinking water was calcium free). Serum calcitriol was measured with the radioimmunoassay using rabbit antiserum and after calcitriol was extracted and the extract purified sequentially on Sephadex LH 20 and high performance liquid chromatography (Bouillon et al., 1980). Measurement of serum iPTH used a commercially available parathyroid RIA system (Saclay, France), in which all reagents were homogenous for the carboxy terminal fragment 53-84 of human synthetic PTH. Cyclic AMP was measured with a commercially available kit (Amersham) in urine and blood. Plasma cAMP was extracted from plasma with ethanol as previously described (Gardin and Paillard, 1984). Nephrogenous cyclic AMP (NcAMP) was calculated as previously described (Broadus et al., 1977). Serum 25(OH)D was measured with a competitive protein binding assay according to Preece et al. (1974). Glomerular filtration rate (GFR) was estimated from endogenous creatinine clearance.

Results in text, Figures and Table of the present work are means $\pm$ SD. The statistical significance between two groups was tested by the two-tailed test after the analysis of variance demonstrated a difference between groups.

RESULTS

Mean values of serum calcitriol and of factors currently known to control the renal production of calcitriol in the physiological state are shown in the Table.

The mean value of serum calcitriol was abnormally high in patients with PHPT and significantly higher than in normal subjects ($P < 0.01$). Mean for NcAMP was increased in patients and significantly higher than in normal subjects ($P < 0.001$). Serum Pi was lower and serum Ca was higher in patients than in normal subjects ($P < 0.001$). Means for 25(OH)D were identical in patients and normal subjects. In the whole patients with PHPT, no correlation was found between serum calcitriol on one hand, and either PTH activity (as reflected by NcAMP values), serum phosphorus, or serum calcium. However, a positive correlation was observed in the whole patients, between glomerular filtration rate (GFR) and serum calcitriol (Fig. 1). In addition, a negative correlation was

Table. Values of serum calcitriol and its determinants in patients with primary hyperparathyroidism.

	Calcitriol pM	NcAMP nmol per 100 ml GF	iPTH pg/ml	Serum Pi mM	Serum Ca mM	25(OH)D nM
PHPT	187	4.25	115	0.75	2.82	34
n=93	± 83*	±1.70	± 90	±0.15	±0.25	±28
Normal	101	1.28	<76	1.06	2.37	33
values	± 36	±0.34		±0.14	±0.10	± 9
n=41	(n=18)					
P<	0.01	0.001		0.001	0.001	

* Values are means ± SD.

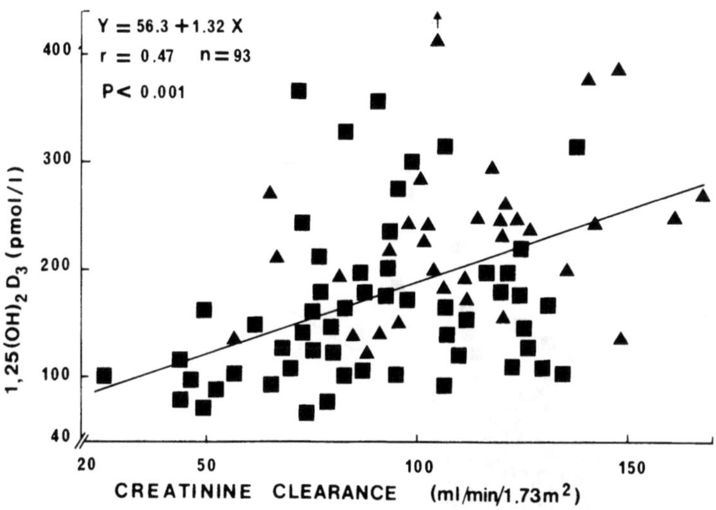

Fig. 1 : Relationship between creatinine clearance values and circulating 1,25 $(OH)_2D_3$ in patients with PHPT (▲, patients with renal stones; ■, with non specific symptoms).

noted between serum calcitriol (y, pmol/L) and age (x, years), y = 254 -1.48 x, r - 0.31, P<0.01. Finally, a negative correlation was observed in patients between glomerular filtration rate (y, ml/min.1.73 m^2) and age (x, years), y = 151 - 1.01 x, r - 0.58, P<0.001. The latter correlation had a slope and y intercept similar to those previously described in normal population (Rowe et al., 1976).

Since we found a correlation between serum calcitriol and GFR, which may be considered in steady state as an index of functional renal mass, this suggested that renal mass would be an important factor influencing the production of calcitriol. Thus we tested correlations with serum calcitriol values expressed as function of GFR (i.e., expressed per 100 ml GF) to minimize the role of renal mass as a source of variation in calcitriol values. We found a positive correlation between NcAMP (as index of PTH activity) and serum calcitriol/GFR (Fig. 2). In addition, we found a negative correlation between serum phosphorus and serum calcitriol/GFR (Fig. 3). No correlation was observed between serum calcium and serum calcitriol/GFR. These findings suggest that PTH activity and serum phosphorus to a lesser extent control the renal 1 alpha hydroxylase in PHPT, and hence circulating levels of calcitriol independent of variations in renal mass.

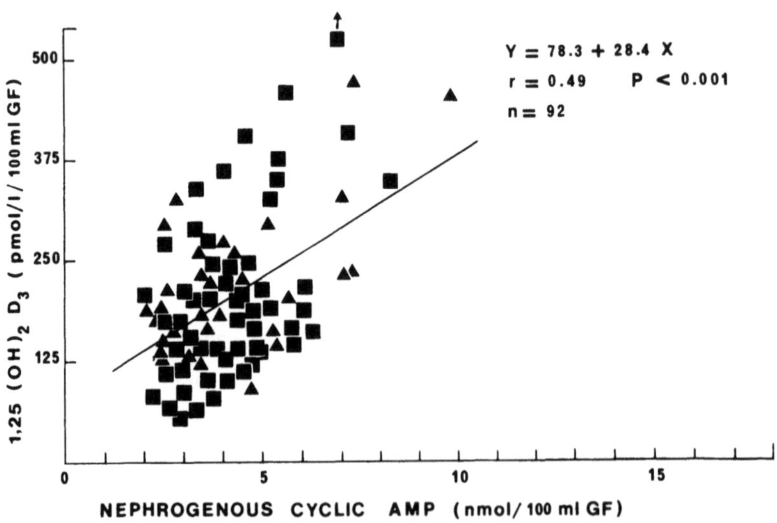

Fig. 2 : Relationship between nephrogenous cyclic AMP values and 1,25 $(OH)_2D_3$ values expressed per 100 ml glomerular filtrate in patients with PHPT (▲, patients with renal stones ; ■, with non specific symptoms.

DISCUSSION

The main results of the present study are the following : 1) Mean value of serum calcitriol is abnormally high in patients, indicating that the renal 1 alpha hydroxylase is stimulated in PHPT ; 2) PTH activity and serum phosphorus to a lesser extent (under conditions of sufficient reserve of vitamin D) are the primary determinants of serum calcitriol levels, independent of variations in renal mass ; 3) Renal mass is a secondary determinant of the production of calcitriol and hence of its circulating level, and could influence the clinical presentation of the disease.

The direct correlation between serum calcitriol and GFR observed in patients with PHPT is a new important finding. This suggests that functional renal mass is a major factor determining the circulating level of calcitriol. The correlations noted in the present work between serum calcitriol and age, and between GFR and age similar to that observed in normal subjects (Rowe et al., 1976) indicate that the decrease in circulating calcitriol with aging is secondary to the

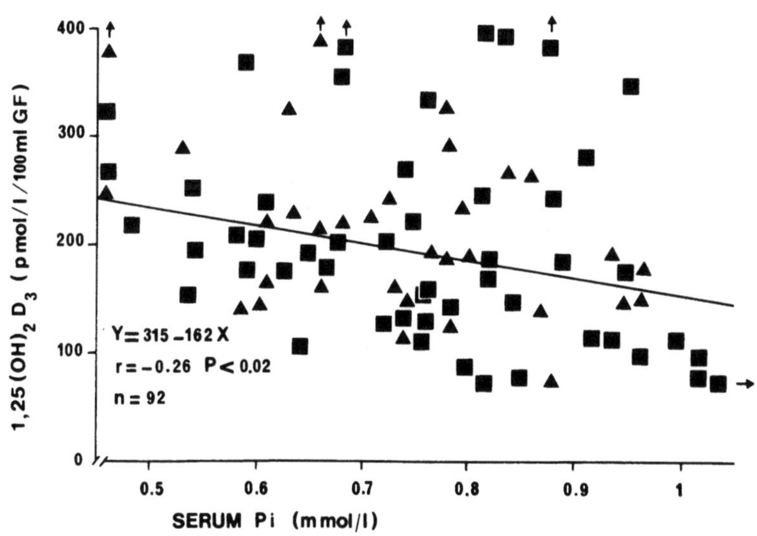

Fig. 3 : Relationship between serum phosphorus and circulating 1,25 $(OH)_2D_3$ expressed per 100 ml glomerular filtrate, in patients with PHPT (▲, patients with renal stones ; ■, with non specific symptoms).

progressive decline in renal mass. In support to this proposition is the existence of correlations previously found in normal population between serum calcitriol and age (Manolagas et al., 1983 ; Keh Sung Tsai et al., 1984), calcitriol and GFR (Keh Sung Tsai et al., 1984), and between incremental serum calcitriol after 24 hr infusion of hPTH fragment 1,34 and age or GFR (Keh Sung Tsai et al., 1984).

Moreover, PTH activity expressed as NcAMP and serum phosphorus to a lesser extent appear as the main determinants of serum calcitriol, independently of variations in renal mass, when a sufficient store of vitamin D is available. The existence of a direct correlation in patients with PHPT between PTH activity and calcitriol expressed per unit of renal mass is found for the first time to our knowledge. This is in agreement with experimental results showing that PTH stimulates 1 alpha hydroxylase activity, and that this effect is mediated by NcAMP (Kawashima and Kurokawa, 1983 ; Armbrecht et al., 1984). The absence of correlation noted in a previous study between serum calcitriol and renal excretion of cAMP in PHPT is not surprising since, in addition to the small number of patients studied, urinary cAMP was used instead of cAMP secreted by renal tubule as index of PTH activity (Kaplan et al., 1977). The well-known role of serum phosphorus as a determinant of serum calcitriol was confirmed in the present work from the inverse correlation between calcitriol expressed per unit of renal mass, and serum phosphorus in the patients.

The importance of renal mass in determining the circulating level of calcitriol, hence the clinical presentation of PHPT emerges when we examined the two subpopulations of patients of the present study, those with recurrent renal stones (33 patients) and those with non specific symptoms (60 patients). Indeed mean serum calcitriol values were dramatically elevated in patients with renal stones (222 $\pm$ 76 pmol/L) and high-normal or normal in patients with non specific symptoms (157 $\pm$ 73 pmol/L, $P < 0.001$). Moreover, in these two subpopulations of patients, means for NcAMP were similarly elevated (4.2 $\pm$ 1.8 vs. 4.2 $\pm$ 1.5 nmol/dl GF), and means for serum phosphorus were identical (0.73 $\pm$ 0.13 vs. 0.76 $\pm$ 0.15 mmol/L). Finally, in patients with renal stones, GFR was higher (113 $\pm$ 31 vs. 87 $\pm$ 28 ml/min, $P < 0.001$) and age was lower (41 $\pm$ 14 vs. 59 $\pm$ 15 years, $P < 0.001$). Thus in young patients with physiologically high functional renal mass, the chronic hypersecretion of PTH is responsible for the high level of calcitriol, with subsequent hypercalciuria and high incidence of renal stones. In older subjects with physiologically lower functional renal mass, the similar degree of PTH hypersecretion results in a normal or high-normal serum calcitriol, with normal calciuria and low incidence of renal stones.

In conclusion, under conditions of sufficient reserve of vitamin D, 1) PTH activity and serum phosphorus to a lesser extent, which control the renal 1 alpha hydroxylase, are the primary determinant of the circulating level of calcitriol, independent of variations in renal mass ; 2) renal mass is a major secondary determinant of serum calcitriol value, and could influence the clinical presentation of the disease ; 3) thus, young patients with normally high GFR values have abnormally high values of serum calcitriol and are at high risk for renal stone. Older patients with normally lower GFR values have high-normal or normal values of serum calcitriol and are at low risk for renal stones.

REFERENCES

Armbrecht, H. J., Wongsurawat, N., Zenser, T. V., and Davis, B.B. 1984, Effect of PTH and 1,25 $(OH)_2D_3$ on renal 25 (OH) D_3 metabolism, adenylate cyclase and protein kinase, Am. J. Physiol., 246:E102-E107.

Bouillon, R., De Moor, P., Baggiolini, E.G., and Uskokovic, M.R. 1980, A radioimmunoassay for 1,25 dihydroxycholecalciferol, Clin. Chem., 26:562-567.

Broadus, A. E., Mahaffey, J. E., Bartter, F. C., and Neer, R. M. 1977, Nephrogenous cyclic Adenosine Monophosphate as a parathyroid function test, J. Clin. Invest., 60:771-783.

Broadus, A. E., Horst, R. L., Lang, R., Littledike, E. T., and Rasmussen, H. 1980, The importance of circulating 1,25 dihydroxyvitamin D in the pathogenesis of hypercalcemia and renal-stone formation in primary hyperparathyroidism, New Engl. J. Med., 302:421-425.

Gardin, J. P., Paillard, M. 1984, Normocalcemic primary hyperparathyroidism. Resistance to PTH effect on tubular reabsorption of calcium, Mineral Electrolyte Metab., 10:301-308.

Kaplan, R. A., Haussler, M. R., Deftos, L. J., Bone, H., and Pak, C. Y. C. 1977, The role of 1 alpha 25 dihydroxyvitamin D in the mediation of intestinal hyperabsorption of calcium in primary hyperparathyroidism and absorptive hypercalciuria, J. Clin. Invest., 59:756-760.

Kawashima, H., and Kurokawa, K. 1983, Unique hormonal regulation of vitamin D metabolsm in the Mammalian kidney, Mineral Electrolyte Metab., 9:227-235.

Keh Sung Tsai, Heath III, H., Kumar, R., and Riggs, B.L. 1984, Impaired vitamin D metabolism with aging in women. Possible role in pathogenesis of senile osteoporosis, J. Clin. Invest., 73:1668-1672.

Manolagas, C., Culler, F. L., Howard, J. E., Brickman, A. S., and Deftos, J. 1983, The cytoreceptor assay for 1,25 dihydroxyvitamin D and its application to clinical studies, J. Clin. Endocrinol. Metab., 56: 751-760.

Preece, M.A., O'Riordan, J. L. H., Lawson, D. E. M., and Koticed, E. 1974, A competitive protein binding assay for 25 hydroxycholecalciferol and 25-hydroxyergocalciferol in serum, Clin. Chim. Acta, 54:235-242.

Rowe, J. W., Andres, R., Tobin, J.D., Norris, A.H., and Shock, N. W. 1976, The effect of age on creatinine clearance in man : a cross-sectional and longitudinal study, J. Geront., 31:155-163.

MECHANISM OF NEPHROCALCINOSIS IN PRIMARY HYPERPARATHYROIDISM

Ph. Jaeger*, W. Jones*, M. Kashgarian[+], G.V. Segre[++] and J.P. Hayslett*

Depts. of Int. Medicine and Pathology[+], Yale University, New Haven, CT, Endocrine Unit[++], Massachusetts General Hospital, Harvard University, Boston, Mass.

INTRODUCTION

Although nephrocalcinosis is a well-known complication of primary hyperparathyroidism, the mechanism by which calcium is deposited in the kidney is not understood. It is not known, for example, whether nephrocalcinosis is caused by hypercalcemia, hypophosphatemia or the product of calcium x phosphorus in plasma. In man it is not possible to regulate the independent variables that may be involved in the induction of nephrocalcinosis, and chronic animal models simulating primary hyperparathyroidism have not been developed. The present study was performed to examine the individual roles of Ca, P and the Ca x P product as causative factors in the induction of nephrocalcinosis, using an animal model of primary hyperparathyroidism in which each variable could be regulated by experimental design (1).

METHODS

Male Sprague-Dawley rats weighing an average of 250 g were used in all experiments. These animals underwent thyroparathyroidectomy (TPTX), thyroxine was replaced as L-thyroxin (levothyroid, Armour Pharm. Co., Kankakee, Ill, 2 µg thrice weekly) and parathyroid hormone was administered as the 1-34 fragment of bovine synthetic parathyroid hormone [bPTH(1-34)] (Beckman Instrument Company, Palo Alto, CA). To simulate primary hyperparathyroidism in man, PTH was administered for 6 days at 2.1 $U \cdot h^{-1}$ (i.e. three fold the estimated basal secretion rate of PTH) via osmotic minipumps (Alzet, Palo Alto, Ca, Model 2001) implanted subcutaneously in the interscapular fossa.

Group I: 10 TPTX rats received bPTH (1-34) in the absence of measures to modulate the hypercalcemic and hypophosphatemic response to PTH.

Group II: 4 TPTX rats received bPTH (1-34) during administration of a diphosphonate [dichloromethylene diphosphonic acid* (Cl_2MDP), 10 mg/kg/d, subcutaneously].

*Gift from the Proctor and Gamble Co., Cincinnati, Ohio.

Group III: 4 TPTX rats received bPTH (1-34) during administration of Cl_2MDP, administered at 45 mg/kg/d, subcutaneously.

Group IV: 5 TPTX rats received bPTH (1-34) during administration of Cl_2MDP (10 mg/kg/d, subcutaneously) and a Ca-free diet.

Group V: 5 TPTX rats received bPTH (1-34) while on a Ca-free diet. In addition, a Ca- and P- free diet was administered for 2 weeks preceding TPTX and PTH infusion.

Group VI: 12 TPTX rats received bPTH (1-34) during administration of Cl_2 MDP (10 mg/kd/d., subcutaneously) and a Ca-free diet. A Ca- and P- free diet was administered for 2 weeks preceding TPTX and PTH infusion.

Following the initiation of PTH infusion, blood was drawn daily and analyzed for Ca (2) and P (3). Results will be presented as the mean values recorded over the last 4 days of infusion. On the sixth day of infusion, animals were exsanguinated and blood samples analyzed for PTH (4). After sacrifice, the kidneys were removed and fixed in absolute ethanol for histological examination. Following paraffin embedding, thin sections were stained with Alizarin-red, to estimate calcium deposition, and with von Kossa stain to determine Ca-P deposition.

RESULTS

As shown in Figure 1, the plasma Ca levels were elevated above the upper limit of normal Ca range (2.8 mmol/l) in Groups I through V, but was reduced to the normal range in Group VI. Conversely, plasma P was lowered to a variable extent in Groups I through V and was normal in Group VI. These changes in Ca and P levels resulted in marked variations in the Ca x P product.

Nephrocalcinosis, when present, was always more severe in the renal medulla than in cortex, and calcium deposits were detected by both stains, suggesting that calcium was deposited in renal tissue predominantly as the phosphate salt. As seen in the lower panel of Figure 1, nephrocalcinosis was more prominent in Groups I and III with the highest levels of plasma Ca, and was absent in Group VI with normal calcium levels, despite elevated levels of the Ca x P product. Regression analysis of pool data, obtained from all 6 groups, showed a significant positive relationship between the nephrocalcinosis score (NC) and plasma Ca [$NC = 1.44\ Ca - 3.07$, $r = 0.79$, $p<0.001$ ($n = 28$)], but no correlation between the severity of nephrocalcinosis and either plasma P or the Ca x P product.

DISCUSSION

The infusion of the 1-34 fragment of bovine synthetic parathyroid hormone at three fold basal levels induced sustained hypercalcemia and hypophosphatemia, simulating primary hyperparathyroidism (Group I). After six days of infusion there were significant deposits of Ca-P in both the renal cortex and medulla. Experimental maneuvers which inhibited intestinal Ca absorption, and/or bone resorption, and/or increased the avidity of bone tissue for Ca, resulted in marked variations in each variable under study, namely plasma Ca and P, and the Ca x P product. The major finding of these experiments was the direct positive correlation between plasma Ca levels and the severity of nephrocalcinosis. There was no correlation, however, between renal calcium deposits and either P or the Ca x P product. These results suggest that in primary hyperparathyroidism soft tissue calcium

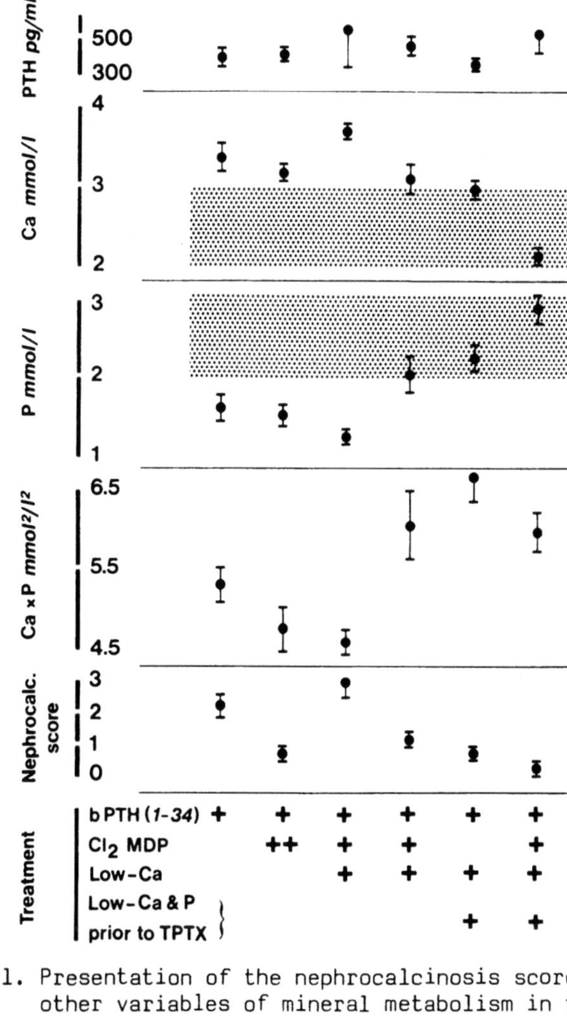

Fig. 1. Presentation of the nephrocalcinosis scores and other variables of mineral metabolism in the six experimental groups.
Values are means ± SEM.

deposition is related to hypercalcemia. Since nephrocalcinosis may also occur in phosphate-induced hyperparathyroidism (5), further studies will be required to evaluate the effect of variable levels of PTH in the mechansim of nephrocalcinosis.

REFERENCES

1. Jaeger Ph., Jones W., Kashgarian M., Baron R., Clemens T.L., Segre G.V. and Hayslett, J.P. Animal model of primary hyperparathyroidism. Submitted for publication, 1985.

2. Connerty H.V. and Briggs A.R. Determination of serum calcium by means of orthocresolphtalein complexone. Am. J. Clin. Pathol. 45:290, 1966.

3. Daly J.A. and Ertingshausen G. Direct method for determining inorganic phsophorus in serum with the "Centrifichem". Clin. Chem. 18:263, 1972.

4. Segre G.V., Tregear G.W. and Potts J.T., Jr. Development and application of sequence-specific radioimmunoassays for analysis of the metabolism of parathyroid hormone methods. Enzymol. 37:38, 1975.

5. Borle A.B. and Clark I. Effects of phosphate-induced hyperparathyroidism and parathyroidectomy on rat kidney calcium in vivo. Am. J. Physiol. 241:E136, 1981.

CHANGES IN SERUM CALCIUM AND PHOSPHORUS INDUCED BY CHRONIC RENAL FAILURE IN THE ABSENCE OF PARATHYROID HORMONE

Mariano Rodriguez, Victor Lorenzo, Arnold J. Felsenfeld, and Francisco Llach

Nephrology Section, Department of Medicine, University of Oklahoma Health Sciences Center and VA Medical Center Oklahoma City, Oklahoma 73104

INTRODUCTION

In the animal models of renal failure, despite the absence of parathyroid hormone (PTH), suppressed tubular reabsorption of phosphate is observed. Thus, processes independent of PTH result in enhancement of phosphate excretion even in chronic renal failure, as has been demonstrated in uremic parathyroidectomized dogs by Swenson et al (1).

This PTH-independent increase in phosphate excretion in the dog has been confirmed in rats with chronic renal failure by Caverzasio et al (2) who observed in thyroparathyroidectomized rats, a significant increase in fractional excretion of phosphate. In both sham and thyroparathyroidectomized rats, the influence of a reduced renal mass on the fractional excretion of phosphate could not be explained by changes in serum phosphate concentration; the decrease in tubular reabsorption of phosphate over glomerular filtration rate (TRP/GFR) was of the same magnitude in thyroparathyroidectomized rats as in those subjected to sham parathyroidectomy.

In addition, Kraus et al (3) in parathyroidectomized uremic rats, observed a decrease in phosphate excretion in response to low dietary phosphorus. All these observations suggest that in the absence of parathyroid hormone, the kidney is able to regulate phosphate excretion in response to changes in dietary phosphorus.

In the absence of parathyroid hormone, animals with renal failure are able to increase phosphate excretion; however, whether this adaptive mechanism is sufficient to maintain normal levels of plasma phosphorus and calcium remains to be proven.

The purpose of this study is to evaluate the effect of renal failure in plasma levels of calcium and phosphorus in the chronic parathyroidectomized rat.

MATERIALS AND METHODS

Eighteen male Wistar Munich rats weighing 300-400 gms underwent selective parathyroidectomy. Under a dissecting microscope, parathyroid glands were identified and removed, leaving the thyroid tissue intact.

After surgery, the rats were maintained on a low calcium diet (calcium 0.05%, phosphate 0.8%, vitamin D 2.5 U/mg) for a 17 day period.

Blood samples for calcium, phosphorus, and creatinine were obtained from the tail artery at 10 and 17 days; only 16 rats consistently showing serum calcium levels lower than 6 mg% were studied. On day 18, rats were fed a normal calcium diet (calcium 1.2%, phosphate 0.9%, vitamin D 850 U/mg) during the rest of the experimental period of 84 days; blood was again obtained at 32 and 41 days.

Surgical reduction of 5/6 of the renal mass was performed using a two-step procedure: 1) artery ligation of two main arteries of the left kidney (day 41), and 2) left nephrectomy five days later (day 46). Blood was obtained on days 56 and 101 after which the experiment was concluded.

Another group of 16 rats was used as controls (intact parathyroid glands). These rats underwent surgical reduction of renal mass as described above. After surgery, the rats were maintained on a normal calcium diet, and blood for calcium, phosphorus, and creatinine was obtained from the tail artery before and ten days after surgery.

Calcium was measured using an automated analyzer (Calcette Model 4008, Precision Systems) and phosphorus was determined with a specific kit (Stanbio TM, San Antonio, Texas).

Data were analyzed using analysis of variance and paired and unpaired t test. Results are expressed as mean ± SD.

RESULTS

After parathyroidectomy and while on a low calcium diet, the serum calcium decreased from 9.9±0.3 to 4.9±0.7 mg/dl, $p<0.001$, by the tenth day. The mean serum calcium remained low after 17 days on a low calcium diet (4.9±0.6 mg/dl). The decrease in serum calcium was associated with an increase in mean serum phosphorus from 6.1±0.4 to 12.4±2.0 mg/dl ($p<0.001$) at 10 days, and at 17 days, it remained elevated (13.9±2.0 mg/dl).

The change from a low calcium diet to a normal calcium diet produced a significant increase in the mean serum calcium concentration; after 15 days on a normal calcium diet, the mean calcium was 6.50±0.6 mg/dl as compared with 4.9±0.6 mg/dl during the low calcium diet ($p< 0.001$); no significant change was observed ten days later.

No significant change was observed in the mean serum phosphorus with the increase in dietary calcium. Thus, mean serum phosphorus concentration was 13.9±2 mg/dl in rats receiving a low calcium diet; and it was 13.1±1 and 11.1±2 mg/dl after a normal calcium diet.

As expected with renal mass reduction, the serum creatinine increased from 0.5±0.03 to 1.3±0.4 mg/dl in parathyroidectomized animals and to 1.1±0.1 mg/dl in controls. Renal failure produced significant changes in the levels of serum calcium and phosphorus. In parathyroidectomized rats, ten days after renal mass reduction, serum phosphorus decreased to 7.8±2 mg/dl, as compared with 11.1±2 mg/dl ($p< 0.001$) before renal failure. This relatively low level of serum phosphorus was maintained until the end of the experiment (7.9±3 mg/dl). In parathyroidectomized rats, the mean serum calcium concentration, 6.7±0.6 mg/dl, increased after ten days (day 56) of renal mass reduction to 9.4±0.8 mg/dl ($p<0.001$), a level which was maintained until the end of the experiment (9.9±0.9 mg/dl).

In control rats, the decrease in renal function resulted in a significant increase in the mean serum calcium concentration (10.0±0.6 vs 10.7±0.3 mg/dl, $p < 0.001$) and a decrease in the mean serum phosphorus (7.4±1.1 vs 6.6±0.3 mg/dl, $p < 0.005$). The serum calcium at the end of the experiment was 11.01±0.3 mg/dl, which was not different from that after 10 days of renal failure. The mean serum phosphorus concentration at the end of the experiment was 7.6±0.9 mg/dl.

DISCUSSION

In the present study, renal mass reduction in chronic parathyroidectomized rats produces a sustained decrease in serum phosphorus concentration as well as an increase in serum calcium concentration. The increase in serum calcium observed in the present study may be due to a decrease in glomerular filtration rate as has been previously shown in the dog (7). A less likely explanation is an increase in bone resorption resulting from chronic renal failure, leading to the increment in serum calcium. However, chronic renal failure usually is associated with hypocalcemia due to low levels of 1,25 dihydroxycholecalciferol and a decreased calcemic response to parathyroid hormone (5-7). Nevertheless, it should be appreciated that most of the data on hypocalcemia and renal failure comes from studies in humans or the experimental animal with intact parathyroid glands.

Most importantly, in the present study, chronic renal failure in the long term parathyroidectomized rat results in a decrease in serum phosphate despite a decrease in glomerular filtration rate and a dietary ingestion of 0.9% phosphorus. This was unexpected and is a new observation which is difficult to explain. Several hypothetical possibilities should be considered; an increase in fractional excretion of phosphorus independent of PTH may occur. Caverzasio et al (2) and Kraus et al (3) have demonstrated that the induction of renal failure in the rat produces an increase in fractional excretion of phosphorus in the presence or absence of parathyroid hormone. The presence of an intrinsic renal adaptive mechanism in response to a reduction in glomerular filtration rate has been postulated. However, such an adaptive renal response should have maintained the serum phosphorus at levels similar to those observed before the reduction of renal mass and does not fully explain the changes in serum concentrations of phosphate observed in the present study.

Recently, Wilson et al in our laboratory have shown in patients with early renal failure, a decrease in serum phosphorus concentrations which was associated with an increase in the fractional excretion of phosphorus (8). Furthermore, these patients had low levels of 1,25 dihydroxycholecalciferol and when treated with 0.25 μg of 1,25 dihydroxycholecalciferol twice a day, had an increase in serum phosphorus and a decrease in fractional excretion of phosphorus. Thus, another possible explanation for the decrease in serum phosphate after renal failure may be a decrease in the plasma levels of 1,25 dihydroxycholecalciferol.

The increase in serum calcium per se may have a specific effect on maximal tubular excretion of phosphorus; Bonjour et al (9) have shown that chronic elevation of plasma calcium is associated with a decrease in maximal tubular reabsorption over glomerular filtration rate in thyroparathyroidectomized rats. Since detailed measurements of tubular reabsorption of phosphorus with control urine determinations were not performed, we cannot be certain that the increase in urinary phosphate was the only explanation for the changes in serum phosphorus. Finally, it is also possible that with renal failure, a deposition of phosphate in bone and other extraskeletal sites may occur. A factor which cannot be excluded is the variation in the amount of phosphate ingested by our rats. However, all rats did have a similar weight gain before and after a reduction in renal mass. In addition, the increment in serum calcium increased in all rats, reflecting adequate dietary calcium intake.

In conclusion, the high serum phosphate concentration in chronic parathyroidectomized rats decreased significantly after renal mass reduction. A renal tubular adapatation in phosphate handling may be the most important factor responsible for this change. However, the levels of serum 1,25 dihydroxycholecalciferol, serum calcium, and deposition in bone and soft tissue may also play roles in the decrease in serum phosphorus.

REFERENCES

1. R.S. Swenson, J.R. Weisinger, J.L. Ruggeri, and G.M. Reaven, Evidence that Parathyroid Hormone is not Required for Phosphate Homeostasis in Renal Failure, Metabolism, 57:264 (1976).
2. J. Caverzasio, H.J. Gloor, H. Fleisch, and J.P. Bonjour, Parathyroid Hormone-Independent Adaptation of the Renal Handling of Phosphate in Response to Renal Mass Reduction, Kidney International, 21:471 (1982).
3. E. Kraus, G. Briefel, L. Cheng, B. Sacktor, and D. Spector, Phosphate Excretion in Uremic Rats: Effects of Parathyroidectomy and Phosphate Restriction, American Journal of Physiology, 248:F175 (abstract) (1985).
4. S.N. Tuma, and L.E. Mallette, Hypercalcemia after Nephrectomy in the Dog: Role of the Kidneys and Parathyroid Glands, Journal of Laboratory and Clinical Investigation, 102:213 (1983).
5. J.D. Kopple and J.W. Coburn, Metabolic Studies of Low Protein Diets in Uremia. II. Calcium, Phosphorus and Magnesium, Medicine, 52:597 (1973).
6. F. Llach, S.G. Massry, F.R. Singer, K. Kurokawa, J.H. Kaye, and J.W. Coburn, Skeletal Resistance of Endogenous Parathyroid Hormone in Patients with Early Renal Failure: A Possible Cause of Secondary Hyperparathyroidism, Journal of Clinical Endocrinology and Metabolism, 41:339-345 (1975).
7. E.B. Mawer, J. Backhouse, and C.M. Taylor, Failure of Formation of 1,25 Dihydroxycholecalciferol in Chronic Renal Insufficiency, Lancet 1:626 (1973).
8. L. Wilson, A. Felsenfeld, M.K. Drezner, and F. Llach, Altered Divalent Ion Metabolism in Early Renal Failure: Role of $1,25(OH)_2D$, Kidney International 27:565 (1985).
9. J.P. Bonjour and H. Fleisch, Calcium Supply and Renal Handling of Phosphate, Mineral and Electrolyte Metabolism 3:261 (1980).

MINERALS AND BLOOD PRESSURE

THEORETICAL MECHANISMS OF DIETARY CALCIUM'S ANTIHYPERTENSIVE ACTION

Richard Bukoski[+], Philip Lucas[*], Tilman Drüeke[*], and David McCarron†

†Division of Nephrology and Hypertension, Oregon Health Sciences University, Portland, OR 97201
*INSERM, U90, Hôpital Necker, Paris, France

SUMMARY

Theoretical mechanisms underlying dietary calcium's antihypertensive action are reviewed. Based upon known defects in the regulation of membrane Ca^{2+} transport and regulation of intracellular free Ca^{2+} concentration, we conclude that maneuvers that favorably modify calcium homeostasis such as dietary calcium supplementation or 1,25 $(OH)_2$ vitamin D_3 administration appear to favorably modify these defects. Consequent improvements in vascular smooth muscle function may mediate the reductions in blood pressure that follows chronic dietary calcium supplementation in experimental hypertension and in the clinical setting.

INTRODUCTION

Essential hypertension is a pervasive health problem in the United States - with conservative estimates set at 30-40 million individuals at risk. The personal and economic burden of the morbidity and mortality resulting from hypertension dictate that if even a small percentage of these patients favorably responded to a non-pharmacologic maneuver such as a dietary intervention, the potential benefits to our society might be profound (1).

New evidence for such a potential diet intervention has emerged from recent epidemiologic reports. These have demonstrated an inverse association between dietary calcium consumption and blood pressure, such that lower dietary calcium intake is associated is associated with higher blood pressure status or a greater risk of being hypertensive (2-10). These epidemiological observations assume more meaning in light of two recent reports of short-term, placebo-controlled trials in which oral calcium supplementation (1000 mg/day) lowered blood pressure in humans with mild essential hypertension (11,12). The results of these two studies are consistent with the data from two earlier preliminary trials (13,14). In addition, a genetic model of essential hypertension, the spontaneously hypertensive rat (SHR), also lowers its blood pressure when dietary calcium is increased (15-19). As discussed below, the SHR may be a particularly appropriate model for examining the mechanism of the effect of dietary calcium intake on blood pressure, since it possesses multiple cellular and organ-related defects to Ca^{2+} metabolism.

TABLE 1

Theoretical Mechanisms of Calcium's Antihypertensive Action

1. Increased extracellular Ca^{2+}: Induction of functional changes in VSM
2. Vitamin D suppression: Removal of vasoconstriction
3. Phosphate depletion: Impaired cardiovascular function
4. Naturesis: Intravascular volume depletion
5. Stabilizing membrane permeability to Na^+/K^+: Decreased intracellular Na^+
6. Correction of intracellular Ca^{2+} binding and/or transport: Reduction in intracellular Ca^{2+}

HYPOTHETICAL MECHANISMS

Serum Ca^{2+}

Based on the clinical and laboratory research to date, several hypotheses have been proposed to explain calcium's antihypertensive action (Table 1). First, any current interpretation of the theoretical mechanisms involved in the modulation of arterial pressures by dietary calcium must consider of the role of changes in extracellular ionized Ca^{2+} concentration. Low serum ionized calcium concentrations have been documented in both human (20,21) and experimental hypertension (16,22). In both our animal studies (16) and our human trial (11), the reduced serum Ca^{2+} concentration normalized when blood pressure improved following dietary calcium supplementation. Whether these minimal upward shifts in extracellular Ca^{2+} levels are a prerequisite for the induction of changes in arterial pressure and functional vascular properties or simply an index that a modification of calcium homeostasis has ocurred is unknown. The observation that serum Ca^{2+} normalizes promptly with supplementation (16) while attenuation of blood pressure may require significantly longer periods suggests that normalization of the serum Ca^{2+}, in and of itself, is not sufficient to improve cardiovascular regulation and lower blood pressure.

Vitamin D Suppression

A second hypothesis has postulated an antihypertensive effect of dietary calcium mediated by vitamin D (23,24). It has been proposed that $1,25(OH)_2$ vitamin D_3 acts as a vasoconstrictive ionophore that increases intracellular Ca^{2+}, VSM reactivity, vascular resistance and ultimately blood pressure. Resnick et al. (23) have shown that a subset of essential hypertensives (low renin) have elevated $1,25\ (OH)_2$ vitamin D_3 levels. They have hypothesized that increased Ca^{2+} intake lowers blood pressure by raising serum ionized Ca^{2+} concentrations which in turn lowers PTH and serum $1,25(OH)_2$ vitamin D_3. The attenuation of serum $1,25\ (OH)_2$ vitamin D_3, in effect, eliminates its tonic pressor action. While an attractive hypothesis, a primary ionophoric, vasoconstrictor action of $1,25\ (OH)_2$ vitamin D_3 has not been demonstrated to date either in vivo or in vitro. In fact, recent studies in cultures of chick skeletal muscle suggest the opposite. In those reports, $1,25(OH)_2$ vitamin D_3 increases the ability of skeletal muscle membranes to sequester Ca^{2+} (25,26). It is conceivable that $1,25\ (OH)_2$ vitamin D_3 acts <u>acutely</u> as an ionophore and <u>chronically</u>, as a trophic agent to increase the muscle cell's ability to regulate intracellular Ca^{2+} more efficiently. In the SHR, $1,25\ (OH)_2$, vitamin D_3 metabolism is abnormal. Basal levels are inappropriately suppressed in the face of increased PTH and decrease ionized Ca^{2+} values (27,28). Provocative maneuvers that normally stimulate $1,25\ (OH)_2$ vitamin D_3 production, fail to evoke a normal response in the SHR (28). If $1,25\ (OH)_2$ vitamin D_3 has a primary vascular action that is protective, then a defect in its regulation in the SHR may contribute to its altered vascular reactivity. More experimental evidence is needed in order to assess properly the role, if any, of 1,25 $(OH)_2$ vitamin D_3 in modulating blood pressure directly or through its modifying a direct antihypertensive effect of Ca^{2+}.

Phosphate Depletion

A third hypothesis has proposed that dietary calcium lowers blood pressure by phosphate depletion (17,29). This hypothesis evolved from studies that indicated that hyperabsorption of dietary calcium by the SHR's intestine, increased serum Ca^{2+} levels and ultimately enhanced urinary excretion of Ca^{2+} and PO_4. The resulting hypophosphatemia has been hypothesized to be deleteriously affecting cardiovascular performance (contractility), thereby lowering Blood pressure through a reduction in either cardiac output or peripheral resistance.

This hypothesis, however, is not fully supported by the data. First, the levels of Ca^{2+} fed the rats by Lau et al. (17) were excessively high (4.3% by weight), making comparative interpretation of the data difficult in terms of a normal (1%) or modestly supplemented (2%) Ca^{2+} intake which we have now documented will lower blood pressure (19). Second, while the balance studies reported by the authors of the phosphate hypothesis indicated that there is an increase in intestinal Ca^{2+} absorption in the older SHR, the very same report demonstrated that there is diminished intestinal Ca^{2+} transport in the 10-12 week old SHR (17). This latter finding argues strongly against intestinal hyperabsorption of Ca^{2+} leading to PO_4 depletion. Lau et al.'s transport studies in the 10-12 week old SHR are consistent with recent data of ours based upon studies of Ca^{2+} transport in isolated duodenal-jejunal segments using the modified Ussing chamber (17,19,28). In those experiments it was demonstrated that net active Ca^{2+} transport is virtually absent in the proximal small intestine of the 12-14 week old SHR. These latter findings concur with those of Shedl et al., (27) who also demonstrated reduced intestinal Ca^{2+} transport in the 16 week old SHR. Third, the hypothesized negative effect of mild hypophosphatemia on the SHR's cardiac performance has not been evaluated. Fourth, other reports that have demonstrated a pronounced effect of dietary calcium on blood pressure have failed to reveal evidence of significant hypophosphatemia (19). The role of hypophosphatemia in the antihypertensive action of dietary calcium remains unresolved at this time.

Natriuresis

A fourth hypothesis forwarded to explain the blood pressure lowering effect of dietary calcium involves a postulated reduction in Na^+ balance, intravascular volume and cardiac output. Several published observations, however, argue against this hypothesis (11,12,17,29). In two placebo-controlled trials of 1000 mg of calcium per day, the reduction in blood pressure was not associated with either a natriuresis or a change in net Na^+ balance (11,12). The latter possibility was specifically addressed in one of the trials (12). Furthermore, plasma volume and Na^+ balance have been measured in SHRs receiving Ca^{2+} supplemented diets (17,29). The finding that both an increase in plasma volume and a net retention of Na^+ occurs, argues strongly against the postulate that dietary calcium lowers Blood pressure through a natriuretic, volume depletion mechanism. In light of the well-established changes in systemic and regional hemodynamics that accompany the development of hypertension in the SHR, the likelihood that dietary calcium lowers blood pressure through a cardiac output or volume mechanism seems unlikely. Intravascular volume and cardiac output are either normal or reduced in this model. The SHR's abnormal peripheral resistance remains the primary hemodynamics alteration and therefore remains the parameter most likely effected by dietary calcium. While it seems likely that dietary calcium's antihypertensive action is expressed through some mechanism other than changes in Na^+ balance, thorough documentation of dietary calcium's effects on cardiac output, regional hemodynamics and intravascular volume are needed.

Membrane Stability

A fifth hypothesis to explain the antihypertensive action of calcium can be based upon the potential for dietary calcium to modify membrane permeability to Na^+ and K^+ in the VSM. By dampening membrane permeability and/or increasing the activity of Na^+/K^+ ATPase, intracellular Na^+ could be reduced and membrane potential increased. The former effect would theoretically promote maintenance of an optimal intracellular free Ca^{2+} concentration and lower VSM reactivity. Intracellular Na^+ is increased in both vascular and non-vascular tissue derived from humans (30) and experimental models of hypertension including the SHR (31). Two lines of indirect evidence suggest a that abnormal intracellular Na^+ content in hypertension may be linked to disordered Ca^{2+} metabolism and its correction. First, the short-circuit currents of mounted duodenal segments of the SHR are lower compared to the WKY (19). This observation is consistent with an abnormal increase of the intracellular Na^+ content of the enterocyte. Short-term 1,25 $(OH)_2$ vitamin D_3 administration to the rat tends to correct the short-circuit current measured in mounted duodenal segments of the SHR (28) at the same time that it improves Ca^{2+} transport in the duodenum. Whether this correction of short-circuit current reflects a normalization of intracellular Na^+ content is not known. Second, intracellular Na^+ content is elevated in RBCs of hypertensives, a finding that has been postulated to be secondary to a putative defect in the Na^+/K^+ pump (32). Recently we have reported that Ca^{2+} supplementation stimulates Na^+/K^+ ATPase activity in RBCs of hypertensives who respond to dietary calcium by lowering blood pressure, suggesting the potential in humans for dietary calcium to reduce the associated elevations of intracellular Na^+ through an effect on the Na^+/K^+ pump (33).

Intracellular Ca^{2+} Binding

A sixth hypothesis suggests that dietary calcium has an effect on vascular smooth muscle (VSM) that corrects an underlying defect in cellular Ca^{2+} metabolism (22). Our laboratory has recently provided evidence that favorable alterations in VSM properties, are associated with the blood pressure lowering effect of dietary calcium in the SHR (34,35). In addition, we observed an increase in basal Ca^{2+} influx in the absence of any change in basal tone of aortic tissue (35). The latter finding we interpreted as reflecting an increase in intracellular cell membrane Ca^{2+} binding capacity. We have postulated that this alteration results in a less reactive smooth muscle cell and presumably a reduction of the intracellular free Ca^{2+}.

Consistent with these observations and our hypothesis are data documenting defective cellular Ca^{2+} metabolism in the SHR and a resultant failure to maintain Ca^{2+} homeostasis (22). This hypothesis is supported by various observations. Extracellular ionized Ca^{2+} is lower in the SHR than in the WKY control (16). PTH levels are increased (16). Active Ca^{2+} transport by the intact duodenum (19,27,28) and by enterocytes isolated from the SHR is reduced (36). This is a pattern of organ-specific failure to regulate Ca^{2+} homeostasis. Extending these findings to the vasculature of the Ca^{2+} sensitive SHR, defective regulation of intracellular Ca^{2+} appears evident.

There is a diminished Ca^{2+} sequestering capacity in subcellular fractions of the SHR's VSM (37-42). These findings are consistent with the observations that intracellular Ca^{2+} is increased in platelets of both essential hypertensives (43) and the SHR (44). In parallel there is a decrease in the basal Ca^{2+} ATPase activity in erythrocytes of both hypertensive humans (33) and SHRs (36) when compared to their respective controls. As a unifying hypothesis one can propose that there is a defict of a specific intracellular Ca^{2+} binding protein in Ca^{2+} sensitive humans with essential hypertension and in the SHR that regulates the Ca-ATPase pump. This pump is central to the cell's, ability to actively transport Ca^{2+}, sequester Ca^{2+}, and ultimately maintain intracellular Ca^{2+} at some optimally lower level. Theoretically, by inducing a reversal of this process, i.e., stimulating the synthesis of membrane associated Ca^{2+} binding proteins, dietary calcium may stabilize the VSM membrane; lower intracellular free Ca^{2+}; reduce vascular reactivity; and lowers blood pressure.

Non-vascular Mechanism

As a final consideration, possible non-vascular mechanisms which might contribute to dietary calcium's lowering blood pressure should be entertained. The evidence summarized above appropriately focuses on the potential direct vascular mechanisms of dietary calcium. However, inherent in that assumption is that only factors involving vascular smooth muscle cells are modified. Several lines of evidence from human hypertension and experimental models suggest that non-vascular mechanisms of dietary calcium may contribute. The renal leak of calcium reported in both human (45) and experimental models (16,18) would theoretically reduce the quantity of retained calcium. In the SHR, the issue is further complicated by the animal's impairment of active duodenal Ca^{2+} transport (19,27,28), an obvious barrier to bioavailability. The transport mechanisms controlling intestinal absorption and renal reabsorption of calcium share a common dependency on the Ca^{2+} ATPase pump. Impairment of this pump's activity in both human and experimental forms of hypertension have been reported as noted above. This pump's regulation of cytosolic Ca^{2+} is critically dependent on both Ca^{2+} and calmodulin. This suggests the theoretical possibility that if the transport mechanisms of the duodenum and proximal nephron are also Ca^{2+} dependent then simply increased calcium exposure from the diet might yield greater bioavailability. This would theoretically occur via increasing cytosolic Ca^{2+}, activating calmodulin, thereby, stimulating Ca-ATPase regulated Ca^{2+} transport (46).

There are significant voids in our knowledge concerning the mechanism of the antihypertensive effect of Ca^{2+}, though substantive and remarkable progress has been made in an area of clinical and basic research that dates only from 1980 (45). However, the recent results from our laboratories which indicate that significant improvement in membrane Ca^{2+} transport and membrane stability occur in both vascular and duodenal tissue of the SHR following 1,25 $(OH)_2$ vitamin D_3 or Ca^{2+} supplementation. This observation suggests that an effect of dietary calcium on membrane cation transport and membrane stability may be one fundamental mechanism that underlies calcium's antihypertensive actions. Extensive future investigations are needed to confirm or refute this postulate.

Acknowledgements: We would like to thank Gloria Ellis for preparing this manuscript. Research support for our studies cited was funded by the National Dairy Board and administered in cooperation with the National Dairy Council.

REFERENCES

1. Hypertension: Is there a place for calcium? The Lancet i:359-361 (1986) (Editorial).
2. D. A. McCarron, C. D. Morris, and C. Cole, Dietary calcium in human hypertension, Science 217:267-269 (1982).
3. S. Ackley, E. Barrett-Connor, and L. Suarez, Dairy products calcium and blood pressure, Am J Clin Nutr 38:457-461 (1983).
4. D. A. McCarron, C. D. Morris, H. J. Henry, and J. L. Stanton, Blood pressure and nutrient intake in the United States, Science 224:1392-1398 (1984).
5. W. R. Harlan, A. L. Hull, R. L. Schmouder, J. R. Landis, F. E. Thompson, and F. A. Larkin, Blood pressure and nutrition in adults, Am J Epidemiol 120:17-28 (1984).
6. M. R. Garcia-Palmieri, R. Costas Jr., M. Cruz-Vidal, P. D. Sorlie, J. Tillotson, and R. J. Havlik, Milk consumption, calcium intake and decreased hypertension in Puerto Rico: Puerto Rico Heart Health Program Study, Hypertension 6:322-328 (1984).
7. W. R. Harlan, A. L. Hull, R. L. Schmouder, J. R. Landis, F. A. Larking and F. E. Thompson, High blood pressure in older Americans: The First National Health and Nutrition Examination Survey, Hypertension 6:802-809 (1985).
8. H. W. Gruchow, K. A. Sobocinski, and J. J. Barboriak, Alcohol, nutrient intake and hypertension in US adults, JAMA 253:1567-1570 (1985).
9. D. Reed, D. McGee, K. Yano, and J. Hankin, Diet, blood pressure and multicollinearity, Hypertension 7:405-411 (1985).
10. D. Kromhout, E. B. Bosschieter, and C. Coulander, Potassium, calcium, alcohol intake and blood pressure: The Zutphen Study, Am J Clin Nutr 41:1299-1304 (1985).
11. D. A. McCarron, and C. D. Morris, Blood pressure response to oral calcium in mild to moderate hypertension: A randomized, double-blind, placebo-controlled, corssover trial, Ann Intern Med 103:825-831, (1985).
12. F. Luft, G. R. Aronoff, R. S. Sloan, N. S. Fienberg, and M. H. Weinberger, Short term augmented Ca intake has no effect on Na homeostasis in man, Clin Pharmacol Ther in press, (1986).
13. N. E. Johnson, E. L. Smith, and J. L. Freudenheim, Effects of blood pressure of calcium supplementation of women, Am J Clin Nutr 42:12-17, (1985).
14. J. M. Belizan, J. Villar, O. Pineda, et al., Reduction of blood pressure with calcium supplementation in young adults, JAMA 49:1161-1165 (1983).
15. S. Ayachi, Increased dietary calcium lowers blood pressure in the spontaneously hypertensive rat, Metabolism 28:1234-1238 (1979).
16. D. A. McCarron, N. N. Yung, B. A. Ugoretz, and S. Krutzik, Disturbances of Ca metabolism in the spontaneously hypertensive rat, Hypertension (suppl 1) 3:I162-I167 (1981).
17. K. Lau, S. Chen, and B. Eby, Evidence for the role of PO_4 deficiency in antihypertensive action of a high-Ca diet, Am J Physiol 246:H324-H331 (1984).
18. K. Lau, J. Zikos, M. Spirnak, and B. Eby, evidence for an intestinal mechanism in hypercalciuria of spontaneously hypertensive rat, Am J Physiol 247:E625-633 (1984).
19. D. A. McCarron, P. A. Lucas, R. S. Schneidman, R. LaCour, and T. Drueke, Blood pressure development of the spontaneously hypertensive rat after concurrent manipulations of Ca^{2+} and Na^{+}: Relation to intestinal fluxes, J Clin Invest 76:1147-1154 (1985).
20. D. A. McCarron, Low serum concentrations of ionized calcium in patients with hypertension, New Engl J Med 307:226-228 (1982).

21. L. M. Resnick, J. H. Laragh, J. E. Sealey, and M. H. Alderman, Divalent cations in essential hypertension: Relations between serum ionized calcium, magnesium and plasma renin activity, New Engl J Med 309:888-891 (1983).
22. D. A. McCarron, Is calcium more important than sodium in the pathogenesis of essential hypertension? Hypertension 7:607-627 (1985).
23. L. M. Resnick, J. P. Nicholson, and J. H. Laragh JH, Calcium metabolism, blood pressure and salt intakes in essential hypertension, Circulation 70:II-1 (1984) (abstract).
24. M. R. Sowers, R. B. Wallace, and J. H. Lemke, The association of intakes of vitamin D and calcium with blood pressure among women, Am J Clin Nutr 42:135-142 (1985).
25. R. D. Boland, A. Norman, E. Ritz, and F. Haselbach, Presence of a 1,25-dihydroxy-vitamin D_3 receptor in chick skeletal muscle myoblasts, Biochem Biophys Res Comm 128:305 (1985).
26. A. R. Boland, S. Gallop, and R. L. Boland, Effects of vitamin D_3 on phosphate and calcium transport across and composition of skeletal muscle plasma cell membranes, Biochem Biophys Acta 733:264 (1983).
27. P. Shedl, D. L. Miller, J. M. Pape, R. L. Horst, and H. D. Wilson, Calcium and sodium transport and vitamin D metabolism in the spontaneously hypertensive rat, J Clin Invest 73:980-986 (1984).
28. P. A. Lucas, R. C. Brown, T. Drüeke, B. Lacour, J. A. Metz, and D. A. McCarron, Abnormal vitamin D metabolism and intestinal calcium transport in the SHR: Relation to bone calcium status, In press, JCI (1986).
29. J. F. E. Mann, J. Bommer, U. Ganten, and E. Ritz, Effects of parathyroidectomy and high dietary calcium intake in spontaneously hypertensive rat, Nephron (in press) (1986).
30. W. Zidek, H. Vetter, K. G. Dorst, H. Zumkley, and H. Losse, Intracellular Na^+ and Ca^{2+} activities in essential hypertension, Clin Sci 63:413-435 (1982).
31. W. Zidek, H. Vetter, H. Zumkley, and H. Losse, Intracellular cation activities and concentrations in spontaneously hypertensive and normotensive rats, Clin Sci 61:415-435 (1981).
32. M. P. Blaustein, Sodium ions, calcium ions, blood pressure regulation and hypertension: A reassessment and a hypothesis, Am J Physiol 232:C165-C173 (1977).
33. F. F. Vincenzi, C. D. Morris, L. B. Kinsel, M. Kenny, D. A. McCarron, Decreased in vitro Ca^{2+} ATP-ase activity in red blood cells of hypertensive subjects, Hypertension (in press) (1986).
34. R. D. Bukoski, and D. A. McCarron, Altered aortic reactivity and lowered blood pressure associated with high Ca^{2+} intake in the SHR, Personal communication (1986).
35. R. D. Bukoski, S. B. Plant, and D. A. McCarron, Aortic Ca^{2+} metabolism and blood pressure are altered by Ca^{2+} intake in the SHR, Personal communication (1986).
36. D. A. McCarron, P. Lucas, B. Lacour, and T. Drüeke, Ca^{2+} efflux rate constant (°KCa) in isolated SHR enterocytes, Abstract presented to the 18th annual meetings of the American Society of Nephrology (1985).
37. K. Aoki, Y. Yamashita, N. Tomita, K. Tazumi, and K. Hotta, ATPase activity and Ca binding ability of subcellular membranes of arterial smooth muscle in the spontaneously hypertensive rat, Jpn Heart J 15:180-181 (1974).
38. M. A. Devnyck, M. G. Pernollet, A. M. Nunez, and P. Meyer, Calcium binding alteration in plasma membrane from various tissues of spontaneously hypertensive rat, Clin Exp Hyp 3:797-807 (1981).
39. C. Y. Kwan, L. Belbeck, and E. E. Daniel, Abnormal biochemistry of vascular smooth muscle plasma membrane isolated from hypertensive rats, Molec Pharmacol 17:137-140 (1980).

40. C. Y. Kwan, and E. E. Daniel, Arterial muscle abnormalities of hydralizine-treated spontaneously hypertensive rats, Eur J Pharmacol 82:187-190 (1982).
41. S. Shibata, M. Kochii, and T. Taniguchi, Calcium fluxes and binding in the aortic smooth muscle from the spontaneously hypertensive rat, Blood Vessels 12:279-289 (1975).
42. R. C. Webb, and R. C. Bhalla, Altered calcium sequestration by subcellular fractions of vascular smooth muscle from spontaneously hypertensive rats, J Molec Cell Cardiol 8:651-661 (1976).
43. P. Erne, P. Bolli, E. Burgisser, and F. R. Buhler, Correlation of platelet calcium with blood pressure: Effect of antihypertensive therapy, N Engl J Med 310:1084-1088 (1984).
44. G. Bruschi, M. E. Bruschi, M. Caroppo, G. Orlandini, M. Spaggiari, and A. Caotatarta, Cytoplasmic free [Ca^{2+}] is increased in the platelets of spontaneously hypertensive rats and essential hypertensive patients, Clin Sci 68:179-184 (1985).
45. D. A. McCarron, P. Pingree, R. J. Rubin, S. M. Gaucher, M. Molitch, and S. Krutzik, Enhanced parathyroid function in essential hypertension: A homeostatic response to a urinary calcium leak, Hypertension 2:162-168 (1980).
46. J. A. Cox, M. Cornte, E. A. Stein, Activation of human erythrocyte Ca^{2+} - dependent Mg^{2+}-activated ATPase by calmodulin and calcium: Quantitative analysis, Proc Natl Acad Sci USA 79:4265-4269 (1982).

CALCIUM AS A MEDIATOR OF SALT SENSITIVE HYPERTENSION

Lawrence M. Resnick

Cardiovascular Center
The New York Hospital-Cornell Medical Center
New York, New York

BACKGROUND

Dietary salt intake has long been recognized as a factor in clinical and experimental hypertensive disease. However, when attention remains narrowly focused on salt and/or sodium alone, the mechanism(s) by which increases in dietary, extracellular, or intracellular sodium lead to increases in blood pressure, still remain unclear. Increased tissue and presumably intracellular sequestration of sodium was initially observed by Tobian (1), accounted for perhaps by now widely postulated circulating inhibitors of Na^+-K^+-ATPase, stimulated by the volume expansion of salt loading in salt sensitive hypertensive animals and humans (2). Linking calcium with sodium has led to a further understanding of salt-induced hypertension. Calcium is required as a final common mediator of smooth muscle contraction, and thus of peripheral vasoconstrictor tone, characteristically elevated in human hypertension (3). Parallel studies over the past three decades have also demonstrated the sodium dependence of cellular calcium uptake in arterial smooth muscle (4). This led to the much discussed hypothesis by which this sodium-calcium exchange mechanism may be involved in human hypertension (5).

Problems remain, however. First, putative Na^+-K^+-ATPase inhibitors have thus far eluded attempts at isolation. Second, other volume-stimulated factors such as auriculin, or atrial natriuretic factor, do not inhibit Na^+-K^+-ATPase, and are vasodilators (6). Third, even substances increasing with saline-volume expansion which are Na^+-K^+-ATPase inhibitors may not be chemically distinct molecules (7). Lastly, and clinically perhaps most relevant, current hypotheses thus far cannot account for the heterogeneity of human hypertension. Some hypertensives are sensitive to the pressor effects of increasing dietary salt, others are "salt-insensitive".

Our group has in the past few years begun to study calcium metabolism in human hypertension. Our approach has been a clinical-biochemical one, has emphasized the heterogeneity of hypertensive mechanisms, and has utilized the renin-angiotensin system to characterize different pathophysiologic subgroups among hypertensive populations (8). Using this approach, we have defined parallel deviations in calcium ionic and hormonal metabolism with plasma renin activity (9). Figure 1 demonstrates the relationship of serum magnesium and ionized calcium values to plasma renin activity in normotensive and essential hypertensive subjects. Magnesium is inversely related to

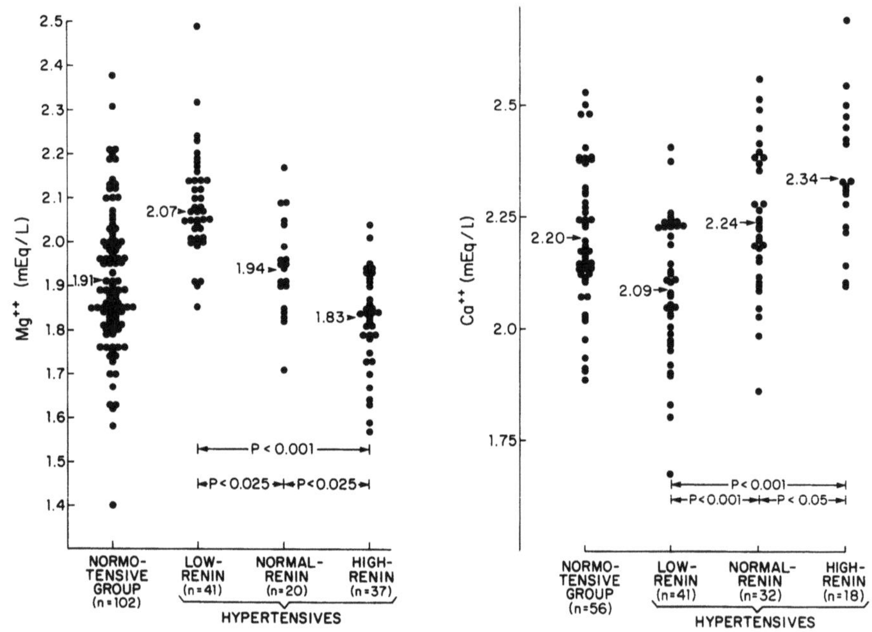

Figure 1: Relation of serum levels of magnesium (Mg^{++}) (left) and ionized calcium (Ca^{++}) (right) to plasma renin activity (PRA) in normotensive and essential hypertensive subjects (9).

plasma renin activity in essential hypertensive individuals, while ionized calcium is directly related to renin system activity. Of note, the range of values for both divalent cations are similar for hypertensive and normotensive control subjects, and thus if all hypertensives had been considered as a homogeneous group, no differences would have been observed in circulating levels of ionized calcium or of magnesium. Such small deviations in circulating ion levels "within normal limits" raise two questions: 1) are these deviations in circulating ion levels of biologic importance, rather than just of statistical significance? 2) what, if anything, do these ionic deviations have to do with the hypertensive process?

Table 1 demonstrates circulating levels of the calcium regulating hormones, calcitonin (CT) and 1,25 dihydroxyvitamin D (1,25D) in normotensive controls, and essential hypertensive subjects (10). Once again, deviations are observed in different renin subgroups of essential hypertensive individuals, which are consistent with those initially measured deviations in serum ionized calcium. Thus, the lower levels of CT and higher levels of 1,25D observed in e.g. the low renin hypertensive subgroup, are consistent with and tend to confirm the biologic significance of the lower serum ionized calcium levels measured in these subjects. Higher renin subjects, with higher levels of CT, and lower average levels of 1,25D, once again demonstrate hormonal levels appropriate for the higher average serum ionized calcium levels measured in these high renin hypertensive subjects. Altogether then, these coordinate ionic and hormonal deviations appear biologically as well as statistically significant.

TABLE 1: Calcium Regulating Hormones in Essential Hypertension

Group	PTH[a] (pg/ml)	CT (pg/ml)	1,25D (pg/ml)
Nl-BP (n=10)	244 ± 21	68.0 ± 6.5	53.9 ± 5.3
Lo-RH (n=18)	433** ± 38	56.4* ± 1.8	77.4* ± 5.4
Nl-RH (n=19)	318* ± 22	66.2 ± 4.0	57.2 ± 4.3
Hi-RH (n=11)	292 ± 37	†84.3* ± 3.9	†44.1 ± 4.9

* $p<0.05$ vs. Nl-BP † 0.001 vs. LoRH
** $p<0.005$

[a] abbreviations: PTH=parathyroid hormone, CT=calcitonin, 1,25D=1,25 dihydroxyvitamin D
Nl-BP=normotensive control subjects, Lo-RH, Nl-RH and Hi-RH= low, normal, and high renin hypertension

That these calcium metabolic deviations might be relevant to the pathophysiology of hypertension itself is also suggested by our earlier work. The blood pressure responses to oral calcium supplementation as well as to calcium channel blockade with nifedipine seem to depend on and can be predicted by the initial serum ionized calcium levels, and these skewed values return back towards average normotensive levels with either therapy (11,12). Furthermore, a linkage of clinical calcium metabolism with sodium-blood pressure effects has also emerged. Calcium supplementation in essential hypertensive humans lowered pressure most in those subjects with the highest average 24 hour urinary sodium excretion values ($r=-0.67$, $p<0.001$) (13). This has also been observed in the SHR model (14). Furthermore, the hypotensive response to calcium channel blockade with verapamil may be enhanced on high vs. low dietary salt for a given subject (15). Moreover, parathyroid hormone levels may depend on dietary sodium status in essential hypertensive subjects, especially in low renin individuals, who are more likely to be salt sensitive (10).

SALT LOADING IN ESSENTIAL HYPERTENSION

With these considerations in mind, we reasoned that if calcium metabolism was involved in the pathophysiology of salt-dependent hypertension in man, maneuvers which alter blood pressure such as dietary salt loading, should alter calcium metabolism as well. Furthermore, salt-induced changes in blood pressure ought to depend on and/or be predicted by changes in suitably measured calcium metabolic variables. Preliminary evidence in animal models suggest this might be the case (16). We therefore studied the short-term (1 week) and longer-term (1 month) effects of low and high dietary salt intakes in essential hypertensive subjects. Our results confirm a strong calcium-dependence of salt-sensitive hypertension in man, and suggest a practical approach to identifying patients for whom dietary salt restriction would be of benefit (17,18).

Unmedicated essential hypertensive subjects were hospitalized and were randomly allocated to metabolic balance diets containing either 10 or 200 mEq of sodium as sodium chloride each for one week's duration. Blood pressures were monitored four times daily in the supine and upright positions, and for comparison, the last eight readings on each salt intake were averaged. Fasting blood tests were obtained at the end of each diet and analyzed for plasma renin activity and calcium ionic and hormonal indices.

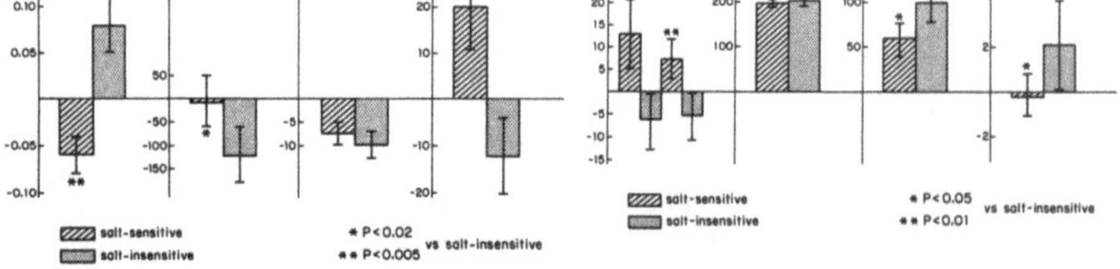

Figure 2: Effects of short-term dietary salt-loading on blood pressure and on serum and urinary ionized and hormonal indices of calcium metabolism (17).

High dietary salt intake was associated with elevated blood pressure in some (6/12) but not all the subjects studied. If all hypertensives had been considered as a homogeneous group, no significant changes in blood pressure would have been observed. However, when subjects were divided according to their blood pressure response to salt loading (greater or less than a 5% change in diastolic blood pressure on high vs. low dietary salt) into salt sensitive or salt-insensitive groups, significant differences emerged (Figure 2). For the same increment in urinary sodium excretion and in contrast to salt insensitive subjects, salt sensitive hypertensives had significantly blunted increases in urinary calcium and magnesium excretion, and demonstrated a significant suppression of serum ionized calcium associated with and confirmed by reciprocal elevations of serum 1,25D levels. There seems to be a close relationship between the salt-induced changes in blood pressure and alterations in calcium metabolic variables. Continuous relationships were observed between the ability of salt loading to change pressure and its ability to suppress serum ionized calcium and stimulate levels of 1,25D (Figure 3).

In attempting to translate these results to a more realistic, free-living hypertensive population, we studied the ionic, hormonal, and blood pressure effects of low and high dietary salt intakes randomly administered for one month each to out-patient essential hypertensive subjects on no concurrent medications. Qualitatively similar results were obtained. Once again, salt sensitive subjects on high vs. low dietary salt demonstrated blunted urinary calcium and magnesium excretory responses, a suppression of serum ionized calcium and a greater elevation of 1,25D levels, in comparison with salt insensitive subjects (Figure 4). Similarly, salt-induced changes in diastolic blood pressure

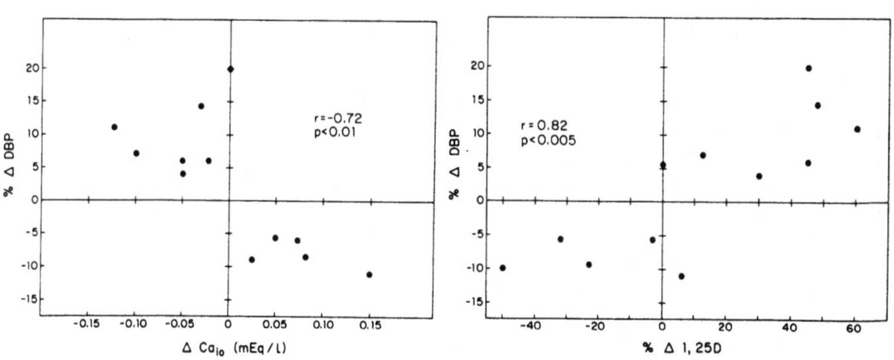

Figure 3: The relationship of high dietary salt-induced blood pressure effects (%ΔDBP) to salt-induced changes in calcium metabolism. ΔCa_{io} = salt-induced change in serum ionized calcium (left), %Δ1,25D = the salt-induced percent change in 1,25 dihydroxyvitamin D (right) (17).

were again closely related to salt-induced changes in both serum ionized calcium ($r=-0.90$, $p<0.001$) and levels of 1,25D ($r=0.74$, $p<0.01$). Interestingly, and consistent with our previous larger population survey (vide supra) the lower serum ionized calcium - higher 1,25D "profile" of the salt-sensitive subject was associated with lower average plasma renin activity (1.5±0.5 vs. 3.0±0.6 ng/ml/h, $p<0.02$) compared with salt insensitive subjects.

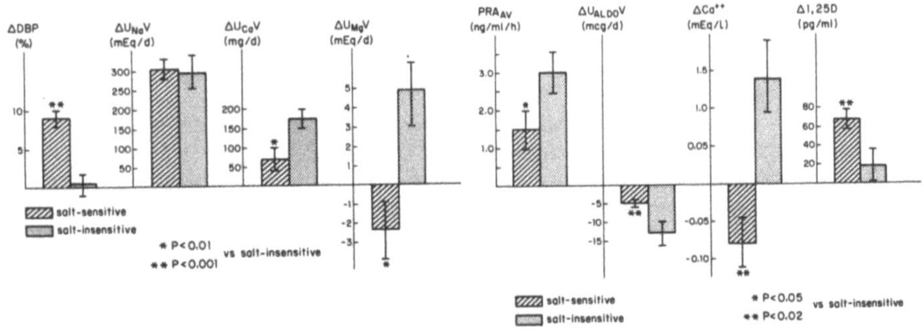

Figure 4: Effects of longer-term dietary salt loading on blood pressure and on serum and urinary ionic and hormonal indices of calcium metabolism (18).

CALCIUM - A MEDIATOR AND MARKER OF SALT SENSITIVITY

Altogether, these data suggest an intimate pathophysiologic linkage between mechanisms of salt sensitive hypertension and salt-induced alterations of calcium metabolism. Mechanistically, we hypothesize that the salt-induced suppression of serum ionized calcium in salt sensitive hypertensive subjects, represents intracellular accumulation of calcium from the extracellular space. In vascular smooth muscle, this "mirror image" elevation in intracellular calcium would induce vasoconstriction and mediate the pressure effects of salt in the salt-sensitive subject. The failure of parathyroid hormone to rise would also be explained if the fall in extracellular calcium was due to its intracellular transport. This hypothesis is supported by the interesting observation that when salt loading was performed in normotensive subjects no changes in pressure or in serum ionized calcium were noted (19). Our concept is also consistent with current postulated molecular mechanisms of sodium-related hypertension (5).

These data may have clinically useful as well as pathophysiologically important implications. Diagnostically, they suggest that appropriate analysis of renin system activity, of 1,25D, or more simply of serum ionized calcium, may help target those hypertensives most susceptible to the pressor effects of dietary salt, and thus allow us to more rationally individualize dietary recommendations in hypertension. Analysis of plasma renin activity would seem of greatest benefit in patients on low dietary salt intakes whereas in patients on a high dietary salt intake, serum ionized calcium measurement would have the greater ability to distinguish salt-sensitive from salt-insensitive subjects (Figure 5). Lastly, the provocative question arises, would oral calcium supplements, which we have already shown to lower blood pressure most in patients on high dietary salt intakes (13), be able to prevent the pressor effects of dietary salt loading? If so, calcium might be useful therapeutically and/or prophylactically as an alternative to dietary salt restriction in selected, salt-sensitive individuals. These exciting possibilities are the focus of current research.

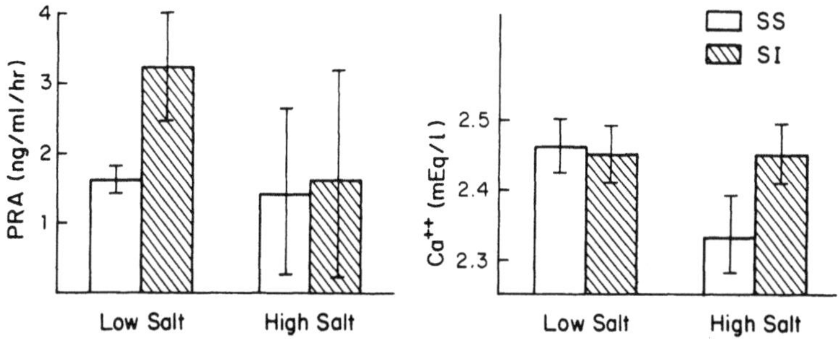

Figure 5: Plasma renin activity (PRA) and serum ionized calcium (Ca^{++}) as markers of salt sensitivity on low and high salt diets. Data displayed are for n=12. Differences in PRA on low salt and in Ca^{++} on high salt are significant, $p<0.05$.

REFERENCES

1. Tobian, L., and J.T. Binion, Tissue cations and water in artieral hypertension, Circulation. 5:754-758 (1952).
2. MacGregor, G.A. and H.E. deWardener, Is a circulating sodium transport inhibitor involved in the pathogenesis of essential hypertension? Clin. Ex. Hypertens. 3:81-5-830 (1981).
3. Resnick, L.M. and J.H. Laragh, The renin-angiotensin system in pregnancy, in: "Endocrinology of Pregnancy," F. Fuchs and A. Klopper, eds., Harper and Row (1983).
4. Reuter, H., M.P. Blaustein, G. Hauresler, Na-Ca exchange and tension development in arterial smooth muscle, Phil. Trans. R. Soc. Lond. B. 265:87-94 (1973).
5. Blaustein, M.P., Sodium ions, calcium ions, blood pressure regulation, and hypertension: a reassessment and a hypothesis, Am. J. Physiol. 232:C165-C173 (1977).

6. Kleinert, H.D., T. Maack, S.A. Atlas, A. Januszewicz, J.E. Sealey, and J.H. Laragh, Atrial natriuretic factor inhibits angiotensin-, norepinephrine-, and potassium-induced vascular contractility, Hypertension. 6:#2(Suppl I) 143-147 (1984).
7. Tamura, M., H. Kuwano, T. Kinoshite, and T. Inagami, Identification of linoleic and oleic acids as endogenous Na^+,K^+-ATPase inhibitors from acute volume-expanded hog plasma, J. Biol. Chem. 260:9672-9677 (1985).
8. Brunner, H.R., J.H. Laragh, L. Baer, et al., Essential hypertension: renin and aldosterone, heart attack and stroke, N. Engl. J. Med. 286:441-449 (1972).
9. Resnick, L.M., J.H. Laragh, J.E. Sealey, and M.H. Alderman, Divalent cations in essential hypertension. Relations between serum ionized calcium, magnesium, and plasma renin activity, N. Engl. J. Med. 309:888-891 (1983).
10. Resnick, L.M., and J.H. Laragh, Calcium regulating hormones and renin activity (PRA) are linked in human essential hypetension (EH), IX International Congress of Nephrology, June, 1984.
11. Resnick, L.M., J.E. Sealey, and J.H. Laragh, Short and Long-term oral calcium alters blood pressure (BP) in essential hypertension, Fed. Proc. 42(3):300 (1983).
12. Resnick, L.M., J.P. Nicholson, J.E. Sealey, and J.H Laragh, Acute and long-term effects of calcium channel blockade on divalent ions, blood pressure, and plasma renin activity, Clin. Res. 31(2):253A (1983).
13. Resnick, L.M., J.P. Nicholson, and J.H. Laragh, Outpatient therapy of essential hypertension with dietary calcium supplementation, J. Am. Coll. Card. 3:616 (1984).
14. Schneidman, R.J., S.A. Hann, and D.A. McCarron, Enchancement of Ca^{2+}. Antihypertensive action in the SHR by Dietary Na^+, IXth International Cong. Nephrology (Los Angeles), June, 1984.
15. Nicholson, J.P., L.M. Resnick, and J.H. Laragh, Sodium balance and the antihypertensive efficacy of verapamil. The Sixth Scientific Meeting of the Inter-American Society of Hypertension, Hypertension, (in press).
16. Kurtz, T.W., and R.C. Morris, Dietary NaCl as a determinant of disordered calcium metabolism in the Dahl salt-sensitive rat, Kidney Internat. 27:194 (1985).
17. Resnick, L.M., J.P. Nicholson, and J.H. Laragh, Alterations in calcium metabolism mediate dietary sodium sensitivity. Clin. Res, 33:608A (1985).
18. Resnick, L.M., J.P. Nicholson, and J.H. Laragh, Sodium sensitivity and calcium metabolism in essential hypertension, Circulation, 70:II-1 (1984).
19. McCarron, D.A., L.I. Rankin, W.M. Bennett, S. Krutzik, M.R. McClung, and F.C. Luft, Urinary calcium at extremes of sodium intake in normal man. Am. J. Nephrol. 1:84 (1981).

THE EFFECT OF CALCIUM ON RENIN, ALDOSTERONE AND PROSTAGLANDIN RELEASE

J.L. Nadler, and I. Antonipillai

University of Southern California Medical Center
Section of Endocrinology
2025 Zonal Avenue, Los Angeles, California 90033

INTRODUCTION

Evidence suggests that calcium (Ca^{2+}) has an important regulatory role in the control of systemic and renal vascular tone. Acute and chronic hypercalcemia can increase blood pressure (Bp) and reduce renal blood flow (1-3). In contrast, pharmacologic blockade of Ca^{2+} entry into vascular smooth muscle can improve many forms of hypertension (4). The mechanisms of Ca^{2+} induced changes in vascular tone are complex since Ca^{2+} can directly produce vascular smooth muscle constriction or mediate the pressor actions of vasopressin, norepinephrine and angiotensin II (AII) (5,6). In addition, other studies reviewed in this paper indicate that Ca^{2+} may regulate Bp through important interactions with the renin-angiotensin and prostaglandin (PG) systems.

RENIN SECRETION

Renin release is regulated by several mechanisms. These include the baroreceptor in the afferent arteriole, sodium fluxes sensed by the macula densa cells of the distal tubule, and signals from the sympathetic nervous system. Two second messengers for these events have been proposed. cAMP has a vasodilator action on vascular tissue and is a stimulatory signal for renin release mediated by the baroreceptor and sympathetic mechanisms (7). However, Ca^{2+} has been implicated as a renin inhibitor (8). We have performed a series of studies using rat renal cortical slices to confirm these recent reports and provide new evidence for role of extracellular Ca^{2+} entry, intracellular Ca^{2+} levels and calmodulin as regulators of renin release.

The renal cortical slices and incubation procedures have been previously described (9). Renin secretion in this preparation is stable during 5 consecutive 15 minute intervals. The slices are responsive to both stimulatory and inhibitory signals (Figure 1). Isoproterenol (10^{-6}M) produced a significant stimulation of renin release with a maximal change at 30 minutes (159 ± 10% of control, $p < 0.01$). At this same time, AII reduced renin release (69 ± 7%, $p < 0.01$).

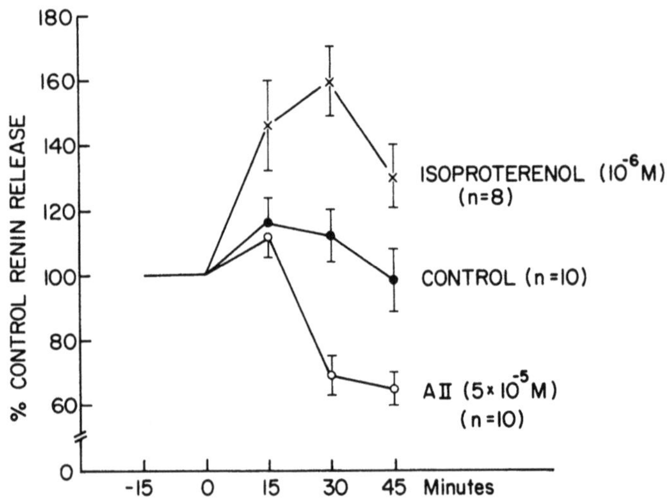

Fig.1. Renin release from rat cortical slices in control incubations (●—●) and after isoproterenol stimulation (X—X) or AII suppression (O—O). Baseline renin release is plotted as a percentage of the mean of -15min and zero time values. Effect of isoproterenol or AII is expressed as a percentage of the baseline after 15, 30 and 45 min. The values are the mean ± SE. The number of observations is shown in parentheses. Reprinted from Ref. #9 with permission of Endocrinology.

Additional experiments were performed to examine the role of Ca^{2+} in these responses. In these studies, normal Ca^{2+} concentration in the buffer (10^{-3}M) was replaced with low Ca^{2+} buffer (< 10^{-6}M) and 1mM EGTA. Exposure of the slices to low Ca^{2+} EGTA buffer did not alter isoproterenol stimulation of renin release (159 ± 10% in 10^{-3}M buffer vs 151 ± 21% in low Ca^{2+} buffer). In marked contrast, the low Ca^{2+} buffer blocked the effect of AII on renin and values were not different from control (100 ± 6%). These results indicate that extracellular Ca^{2+} plays little role in the effects of B adrenergic stimulation of renin release, while a reduction of Ca^{2+} blocks the inhibition of renin by AII.

To further examine the role of Ca^{2+} flux, the Ca^{2+} channel blocker nifedipine was added to the slices. Nifedipine produced a dose related increase in renin (Figure 2). At 10^{-5}M nifedipine increased renin to 177 ± 23% of baseline. These results suggest that reduction of Ca^{2+} entry into the juxtaglomerular cells is a potent stimulatory signal for renin release.

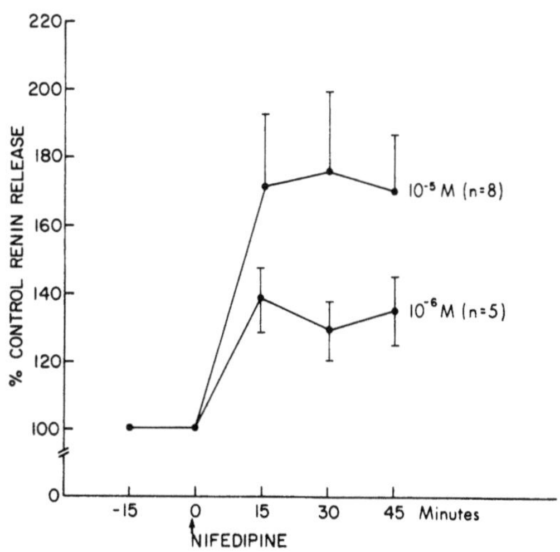

Fig.2. Dose-related stimulation of renin release by nifedipine. Reprinted from Reference #9 with permission of Endocrinology.

Similar to the results with nifedipine, an agent called TMB-8, which inhibits intracellular Ca^{2+} release also produced a significant dose dependent increase in renin release (154 ± 16%, 10^{-5}M and 207 ± 12%, 10^{-4}M). These additional studies indicate that intracellular Ca^{2+} levels are key signals for renin release.

Intracellular Ca^{2+} may act by altering the function of the Ca^{2+} binding protein Calmodulin. Therefore, we evaluated the effect of a specific calmodulin inhibitor, calmidizolium, on renin release. Calmidazolium (5×10^{-7}M) was a potent stimulator of renin release (160 ± 15%, $p < 0.01$). This indicates that Ca^{2+} interacts with calmodulin as a inhibitory signal for renin secretion.

In summary, these studies indicate that unlike most hormonal secretory systems, a reduction of Ca^{2+} flux or intracellular Ca^{2+} levels is a potent stimulus for renin secretion.

ALDOSTERONE SYNTHESIS

AII, potassium and ACTH are the primary regulators of aldosterone synthesis by the adrenal zona glomerulosa (10). ACTH acts primarily via a Ca^{2+} independent process (11) while AII and potassium stimulate aldosterone production by altering Ca^{2+} flux and phospholipid turnover (12). Studies in isolated glomerulosa cells have shown that verapamil can inhibit aldosterone synthesis induced by potassium and AII (13). Other evidence indicates that TMB-8 and calmodulin inhibitors can prevent AII induced aldosterone production (14). In man, nifedipine can increase plasma renin activity without altering aldosterone levels while Ca^{2+} channel blockade blunts the aldosterone stimulatory response to AII (15).

We recently completed a study designed to evaluate the role of Ca^{2+} flux in aldosterone production and Bp regulation in patients with primary aldosteronism (16). Ten hypertensive patients with well-documented primary aldosteronism due to an adrenal adenoma (n=5) or bilateral adrenal hyperplasia (n=5) were studied on a Clinical Research Center. All medications were withdrawn 3 weeks prior to the study and the Ca^{2+} blocker, nifedipine, was administered sublingually at a dosage of 20mg.

In all patients nifedipine lowered systolic and diastolic Bp to normal within 60 minutes (156 ± 6 to 124 ± 3mm Hg systolic and 104 ± 3 to 83 ± 3mm Hg diastolic, $p < 0.01$). In 6 of 10 patients the Bp remained normal for 6 hours after the sublingual dose. The pulse rate did not change (73 ± 3 vs 83 ± 4 beats min, $p > 0.1$). Plasma aldosterone levels were markedly reduced by nifedipine in patients with an adenoma or bilateral hyperplasia (Figure 3).

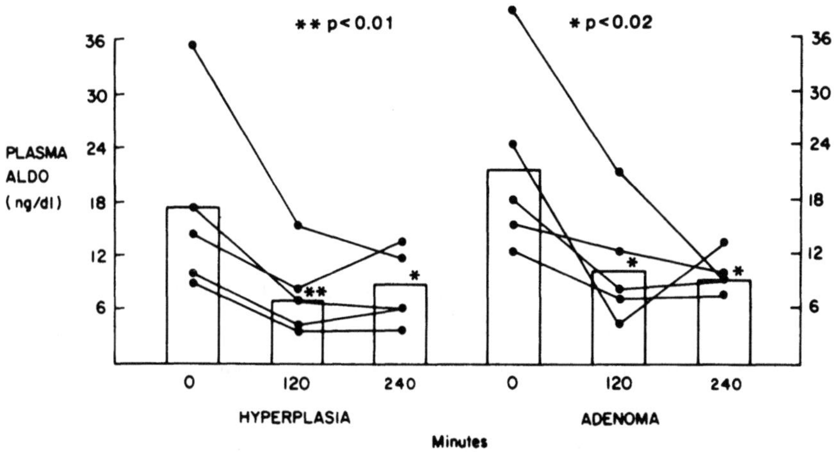

Fig.3. Supine PM plasma aldosterone at baseline, 120 and 240 minutes after nifedipine (20 mg sublingual) in 5 patients with hyperplasia (IHA) and 5 with an adenoma (APA). Bars represent mean plasma aldosterone levels for each group of PA subjects. The decrease in plasma aldosterone is significant in both IHA and APA ($p < 0.02$). Reprinted from Ref. #16 by permission of the Journal of Clinical Endocrinology and Metabolism.

TABLE I: Effect of 4 weeks of nifedipine therapy on morning supine plasma aldosterone, potassium, and blood pressure in six subjects with PA

Patient classification	Plasma aldosterone (ng/dl)		Potassium (mEq/liter)		Blood Pressure (mm Hg)	
	Baseline	4 weeks@	Baseline	4 weeks@	Baseline	4 weeks@
IHA	58	32	3.1	3.5	170/110	150/88
IHA	29	18	3.0	4.0	144/92	122/80
IHA	28	9	3.3	3.7	136/92	118/82
IHA	33	24	3.2	3.8	185/108	140/92
APA	75	21	2.6	3.5	188/112	142/88
APA	54	17	3.0	3.7	150/96	132/80
Mean ± SE	46±8	20±3	3.0±0.1	3.7±0.1	162±9/102±4	134±5/85±2
	P < 0.02		P < 0.01		P < 0.01	
						P < 0.001

@Values at 4 weeks represent the mean of pooled results from 3 and 4 week visits

Reprinted from reference #16 with permission from the Journal of Clinical Endocrinology and Metabolism.

Nifedipine therapy was continued (30-60mg/day) for 1 month in 6 patients to further assess the chronic effects of Ca^{2+} blockade on Bp and aldosterone synthesis in primary aldosteronism. Mean values are shown in Table 1. Plasma aldosterone in the 4 patients with hyperplasia and 2 with an adenoma was reduced (46 ± 8 to 20 ± 3ng/dl (0900 supine) $p < 0.02$) and serum potassium normalized without supplementation (3.0 ± 0.1 to 3.7 ± 0.1 mEq/liter, $p < 0.01$). Bp was normalized and was similar to the reduced Bp values seen after the first nifedipine dose suggesting a lack of tachyphylaxis. These results suggest that Ca^{2+} channel blockers may provide an effective medical therapy for patients with primary aldosteronism by lowering Bp and normalizing plasma potassium and aldosterone concentrations.

In summary, much evidence in vitro and in man indicates that Ca^{2+} is the primary messenger for potassium and AII mediated aldosterone synthesis. These studies suggest that Ca^{2+}, required for steroidal synthesis, arises from both an extracellular pool via Ca^{2+} channels and an intracellular source through activation of the phosphotidalinositol pathway.

Ca^{2+} - PG MODULATING SYSTEM

Considerable evidence suggests that vasodilatory PG's play a key role as protective modulators of systemic and renal blood flow during states of increased pressor activity (17-19). Studies in animals and man indicate that AII and norepinephrine stimulate PGE_2 and/or PGI_2 release (20,21) while cyclooxygenase blockade produces decrements in renal function during states of ischemia, heart failure or sodium depletion (22-24).

Ca^{2+} can directly activate membrane phospholipases to release arachidonic acid from membrane phospholipids which can then be converted into products including PGE_2 and/or PGI_2. This is supported by in vitro data showing that the Ca^{2+} ionophore A23187 stimulates PGE_2 synthesis in the renal medulla and PGI_2 production in endothelial cells (25,26). However, the interaction of Ca^{2+} and PG's in man has not been clearly shown.

We designed a study to investigate 1) the effect of mild and moderate hypercalcemia on PGE_2 and PGI_2 formation, 2) whether PGE_2 and/or PGI_2 can modulate the systemic and renal vasoconstrictive actions of Ca^{2+}. The study protocol and methods have been previously described

(27). In brief, 12 healthy subjects were admitted to a Clinical Research Center and placed on a fixed sodium and potassium diet. After a 3 hour control infusion (dextrose in water), they received on different days, a low dose of Ca^{2+} gluconate (2mg/kg) or a pressor or high dose Ca^{2+} infusion (8mg/kg). The infusions were given over 3 hours using a constant infusion pump. To evaluate the functional significance of PG's released by Ca^{2+}, the low dose Ca^{2+} infusion was repeated during cyclooxygenase inhibition with indomethacin or ibuprophen. In addition, to clarify the role of Ca^{2+} entry and α adrenergic activation in the vascular and PG responses to Ca^{2+}, some subjects received the Ca^{2+} channel blocker, nifedipine (20mg) or the α antagonist, prazosin, prior to high Ca^{2+} infusion. PG's were measured in 3 hour urines collected during the control and experimental infusions. Urinary PGE_2 and the stable PGI_2 metabolite, 6 keto $PGF_{1\alpha}$ were measured via our published RIA methods (20,27). During some infusions renal blood flow was determined using a paraaminohippurate clearance technique.

The effect of the low dose Ca^{2+} infusion alone or with cyclooxygenase inhibition on mean Bp, renal blood flow and urinary PGE_2 excretion is shown in Table II. The low dose Ca^{2+} infusion did not alter mean Bp, or renal blood flow. However, this same infusion given during indomethacin or ibuprophen pretreatment now increased Bp and reduced renal blood flow. This suggests that vasodilatory PG release may be an important modulator of Ca^{2+} vasoconstrictive effects. PGE_2 release was not enhanced by the Ca^{2+} infusion indicating that it is not a modulator of Ca^{2+} vasoconstrictive actions.

In contrast, Figure 4 shows that PGI_2 is stimulated by the low dose and high dose Ca^{2+} infusion in a dose related fashion. These results suggest that PGI_2 is the vasodilatory PG modifying the systemic and renal vasoconstrictive effects of Ca^{2+}. Indomethacin alone does not alter Bp or renal blood flow in healthy subjects. However, as shown in Table 1 and Figure 4 indomethacin given with a low dose Ca^{2+} infusion prevents the rise in PGI_2 and now produces changes in vascular tone, further supporting the role of PGI_2 release in preventing Ca^{2+} induced vasoconstriction.

Since Ca^{2+} can induce catecholamine release which may activate α receptors to stimulate PGI_2 (28), we studied whether α blockade with prazosin (5mg) would alter the Ca^{2+} induced 6 keto $PGF_{1\alpha}$ release. Also, to clarify whether increases in extracellular Ca^{2+} stimulates PGI_2 directly, nifedipine was administered prior to some high Ca^{2+} infusions.

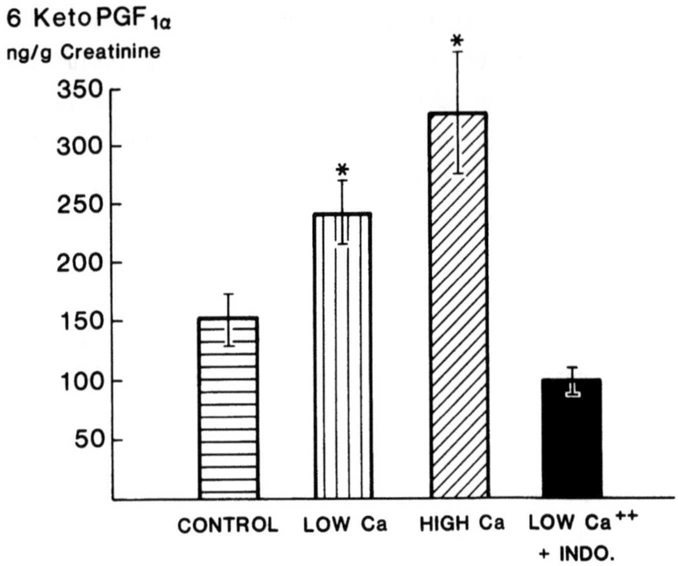

Fig. 4. Stimulation of 6 keto $PGF_{1\alpha}$ excretion by low dose and high dose Ca^{2+} infusions (⋆ $p < 0.02$ Low Ca and $p < 0.01$ high Ca vs control). Indomethacin pretreatment prevented the Ca^{2+} induced rise in 6 keto $PGF_{1\alpha}$.

TABLE II: Effect of low-dose Ca^{2+} infusion alone and with indomethacin on mean Bp, RBF, and PGE_2 excretion

Infusion	Mean Bp (mm Hg)	RBF (ml/min/1.73m²)	PGE_2 ng/g creatinine
Control	82 ± 2	1210 ± 100	269 ± 26
Low-dose Ca^{2+}	84 ± 2	1180 ± 80	150 ± 36[a]
Low-dose Ca^{2+} + indomethacin	94 ± 3[a]	860 ± 60[a]	91 ± 20[b]

Values are mean ± SE
[a]$p<0.02$ vs. control
[b]$p<0.01$ vs. control

Prazosin, given prior to the Ca^{2+} infusions did not alter the Ca^{2+} induced rise in mean Bp or 6 keto $PGF_{1\alpha}$ (159 ± 21 to 258 ± 23ng/gram creatinine, $p < 0.02$). In marked contrast, nifedipine completely prevented both the rise in mean Bp produced by the high Ca^{2+} infusion (87 ± 2 vs 82 ± 2mm Hg, p = NS) and the stimulation of 6 keto $PGF_{1\alpha}$ (170 ± 24 vs 210 ± 46ng/gram creatinine). Nifedipine given alone on a different day did not alter basal 6 keto $PGF_{1\alpha}$ excretion (158 ± 30 vs 182 ± 38). These results suggest that Ca^{2+} induced increases in PGI_2 are not through α receptor activation but are mediated via direct Ca^{2+} entry.

These results suggest that increases in extracellular Ca^{2+} can produce vasoconstriction which, if unopposed, can increase Bp and reduce renal blood flow. However, Ca^{2+} also stimulates PGI_2 production through the release of arachidonic acid. PGI_2 then acts as a vasodilator which modulates Ca^{2+} vasoconstrictive actions and maintains vascular homeostasis.

Therefore, Ca^{2+} has complex actions in the vascular system that can produce changes in Bp. Ca^{2+} is an inhibitory signal for renin release, thus Ca^{2+} blockers can enhance renin secretion. Ca^{2+} is also the second messenger for aldosterone secretion induced by potassium and AII. This provides additional rationale for the use of Ca^{2+} blockers in states of aldosterone excess and hypertension. Finally, vascular and renal PGI_2 synthesis is stimulated by Ca^{2+} and the PGI_2 released plays an important role as modulator of Ca^{2+} vascular effects.

Based on our results and those by other investigators, changes in the renin-angiotensin and prostaglandin systems should be evaluated when using dietary or pharmacologic methods to treat hypertension that can alter Ca^{2+} metabolism.

ACKNOWLEDGEMENTS

Dr. Nadler is the recipient of a "Clinical Investigator Award" (HL#1496) and Dr. Antonipillai is the recipient of a "New Investigator Award" (AM#34245) from the NIH. This research was also supported in part by grants from the NIH (HL#2112 and GRCG #43).

The authors thank Dr. Richard Horton for helpful advice in these projects and J. Yamamoto, B.S. for her excellent technical assistance. We also thank Denise Walters for her excellent secretarial assistance.

REFERENCES

1. J.M. Earll, N.A. Kurtzman and R.H. Moser: Hypercalcemia and hypertension. Ann Int. Med 64:378 (1966).

2. C. Marone, C. Beretta-Piccoli and P. Weidman: Acute hypercalcemic hypertension in man: role of hemodynamics, catecholamines and renin. Kidney Int 20:92 (1980).

3. H.D. Humes, I. Ichikawa, J.Z. Troy and B.M. Brenner: Evidence for a parathyroid hormone dependent influence of calcium on the glomerular ultrafiltration coefficient. J. Clin Invest 61:32 (1978).

4. O.L. Pedersen: Calcium blockade in arterial hypertension. Hypertension 5:(Suppl II) 74 (1983).

5. J.P. Goldberg, R.W. Schrier: Effect of calcium membrane blockers on in vivo vasoconstrictor properties of norepinephrine, angiotensin II and vasopressin. Min Elect Metab 10:178 (1984).

6. R. Deth and C. VanBreeman: Relative contributions of Ca^{2+} influx and cellular Ca^{2+} release during drug induced activation of the rabbit aorta. Pflugers Arch 348:13 (1974).

7. T.K. Keeton and W.B. Campbell: The pharmacologic alteration of renin release. Pharmacol Rev 32:81 (1981).

8. P.C. Churchill: Second messengers in renin secretion. Am J. Physiol 249:F175 (1985).

9. I. Antonipillai and R. Horton: Role of extra and intracellular calcium and calmodulin in renin release from rat kidney. Endocrinology 117:601 (1985).

10. I.A. Reid and W.F. Ganong: Control of aldosterone secretion in: "Hypertension: Pathophysiology and Treatment. J. Genest, E. Koiw and O. Kuchel eds. McGraw-Hill, New York (1977).

11. F.K. Fujita, G. Aguilera, K.J. Catt: The role of cyclic AMP in

aldosterone production by isolated zona glomerulosa cells. J Biol Chem 254:8567 (1979).

12. R. Foster, M.V. Lobo, H. Rasmussen, E.T. Marusic: Calcium: its role in the mechanism of action of angiotensin II and potassium in aldosterone production Endocrinology 109:2196 (1981).

13. J.L. Fakunding and K. Catt: Dependence of aldosterone stimulation in adrenal glomerulosa cells on calcium uptake: effects of lanthanum and verapamil. Endocrinology 107:1345 (1980).

14. M.E. Elliot, R.C. Alexander and T.L. Goodfriend: Aspects of angiotensin action in the adrenal: key roles for calcium and phosphatidylinositol. Hypertension (Suppl II) 4:52 (1982).

15. G.P. Guthrie, Jr., R.G. McAllister, Jr., and T.A. Kotchen: Effects of intravenous and oral verapamil upon pressor and adrenal steroidogenic responses in normal man. J. Clin Endocrinol Metab 57:339 (1983).

16. J.L. Nadler, W. Hsueh and R. Horton: Therapeutic effect of calcium channel blockade in primary aldosteronism. J Clin Endocrinol Metab. 60:896 (1985).

17. J. McGiff, K. Crowshaw, N. Terragno and A. Lonigro: Renal prostaglandins possible regulators of the renal actions of pressor hormones. Nature 227:1255 (1970).

18. J. Aiken and J. Vane: Intrarenal prostaglandin release attenuates the renal vasoconstrictor activity of angiotensin. J Pharmacol Exp Ther 184:678 (1973).

19. P. Negus, R. Tannen and M. Dunn: Indomethacin potentiates the vasoconstrictor actions of angiotensin II in normal man. Prostaglandins 12:175 (1976).

20. J. Nadler, R.Zipser, R. Coleman and R. Horton: Stimulation of renal prostaglandins by pressor hormones in man: comparison of PGE_2 and PGI_2. J Clin Endocrinol Metab 56:1260 (1983).

21. L. Levine and M. Moskowitz: α and β stimulation of arachidonic acid metabolism in cells in culture. Proc Natl. Acad. Sci USA 76:6632 (1979).

22. R. Zipser, J. Hoefs, P. Speckart, P. Zia and R. Horton. Prostaglandins: modulators of renal function and pressor resistance in chronic liver disease. J Clin Endocrinol Metab 48:895 (1979).

23. V. Dzau, M. Packer, L. Lilly, S. Swartz, N. Hollenberg and G. Williams: Prostaglandins in severe congestive heart failure: relations to activation of the renin angiotensin system and hyponatremia. N Eng J Med 310:347 (1984).

24. D. Cline and J.S. Stoff: Renal syndromes associated with nonsteroidal anti-inflammatory drugs. N Engl J Med 310:563 (1984).

25. B. Weksler, C. Ley and E. Jaffee: Stimulation of endothelial cell prostacyclin production by thrombin, trypsin and the ionophore A23187. J Clin Invest. 62:923 (1978).

26. H.R. Knapp, O. Oelz, L.J. Roberts, B.J. Sweetman, J.A. Oates and P.W. Reed: Ionophores stimulate prostaglandin and thromboxane biosynthesis. Proc. Natl Acad Sci USA 74:4251 (1977).

27. J.L. Nadler, M. McKay, V. Campese and R. Horton: Evidence that prostacyclin modulates the vascular actions of calcium in man. J. Clin Invest 77: (1986, in press).

28. J.L. Nadler, R. Zipser and R. Horton: The effect of adrenergic stimulation on urinary prostaglandin E_2 and 6 keto $PGF_{1\alpha}$ in man. Prostaglandins 26:519 (1983).

DISTURBED PHOSPHATE METABOLISM IN SPONTANEOUSLY HYPERTENSIVE RATS

R.J.M. Bindels and C.H. van Os

Department of Physiology, University of Nijmegen
P.O.Box 9101
6500 HB Nijmegen. The Netherlands

INTRODUCTION

In spontaneously hypertensive rats (SHR) Ca^{2+} metabolism is abnormal when compared to normotensive Wistar-Kyoto rats (WKY). Disturbances in Ca^{2+} metabolism include a slight decrease in serum ionized Ca^{2+} concentration (McCarron et al., 1981; Wright and Rankin, 1982), hypercalciuria (McCarron et al., 1981; Bindels et al., 1984) and increased serum parathyroid hormone (PTH) levels (McCarron et al., 1981; Anderson et al., 1983). Finally, a high Ca^{2+} diet lowers the blood pressure in SHR (McCarron et al., 1981) after 18 weeks, proving the functional importance of the disturbed Ca^{2+} homeostasis in SHR.

Disturbances in inorganic phosphate (Pi) metabolism have also been reported in SHR. Recent studies demonstrate hypophosphatemia (Bloomquist et al., 1982) and hypophosphaturia in SHR (Hsu et al., 1984; Bindels et al., 1984). Moreover, in normotensive subjects Pi concentrations were inversely related to blood pressure (Ljunghall and Hedstrand, 1977), suggesting that decreased serum Pi levels may be associated with increased blood pressure.

Mg^{2+} homeostasis has not been studied extensively in SHR, although a relationship between plasma Mg^{2+} and blood pressure has been proven (Altura et al., 1984; Resnick et al., 1984). A recent study reported decreased intracellular free Mg^{2+} in erythrocytes of humans with essential hypertension (Resnick et al., 1984). Mg^{2+} deficiency in rats is associated with elevation in blood pressure (Altura et al., 1984). Moreover, variation in dietary Mg^{2+} affected the course of hypertension in SHR (Berthelot and Esposito, 1983).

The aim of the present investigation was therefore, to characterize Ca^{2+}, Mg^{2+} and Pi metabolism at the same time in one group of animals. Serum and 24-hour urine were analyzed in WKY and SHR from 6 till 23 weeks of age. Finally, in view of a consistent lower Pi excretion, the Na^+-dependent Pi transport in renal brush-border membranes of SHR was studied.

MATERIALS AND METHODS

At 5 weeks of age 5 male WKY and 5 age-matched SHR were individually housed in metabolic cages. 24-Hour urine collections were obtained at

1 week intervals. Serum samples were taken from WKY and SHR of 6, 12 and 20 weeks of age. Systolic blood pressure was measured indirectly from the tail artery of prewarmed rats using a tail cuff and a pneumatic pulse transducer. Ca^{2+}, Mg^{2+} and Pi concentrations in serum and acidified urine were determined by colorimetry with commercially available kits (Lancer, St.Louis).

Renal brush-border membranes were isolated from WKY and SHR of 6, 12 and 20 weeks of age with two subsequent Mg^{2+} precipitation steps according to Biber et al. (1981). The uptake of labelled Pi and D-glucose was measured by a rapid filtration technique using 0.45 µm filters (Schleicher and Schull, GmbH, Dassel). Incubation was started by adding 5 µl of membrane suspension to 45 µl of incubation medium containing 0.01 to 1 mM Pi or 0.1 to 3.0 mM D-glucose, 100 mM NaCl, 100 mM Mannitol, 1 mM NaN_3, buffered with 20 mM Hepes/Tris at pH 7.4 (final concentration). The uptake was quenched by addition of 1.1 ml of an ice-cold stop solution. All incubations were carried out in triplicate at 25 °C. The stop solution contained 100 mM KCl, 100 mM Mannitol, 1 mM NaN_3, 10 mM Na_2HAsO_4 or 1 mM phloridzin buffered with 20 mM Hepes/Tris at pH 7.4. The initial rates are calculated from Pi uptake values at 15 s minus uptake values at 5 s in the presence of 100 mM NaCl and corrected for Na^+-independent uptake determined in the presence of 100 mM KCl instead of NaCl between the same time intervals.

All results were evaluated statistically by unpaired t-test or analysis of variance followed by contrast analysis.

RESULTS

WKY and SHR were kept in metabolic cages from 6 weeks of age up to 23 weeks. From each individual rat body weight and blood pressure were measured (Table 1). There was no significant difference in body weight between the two strains ($F=2.2$, $p>0.14$). Blood pressure in SHR was significantly ($F=418$, $p<0.0001$) elevated compared with the age-matched WKY at all ages studied.

Serum data of 6, 12 and 20 weeks old WKY and SHR are summarized in Table 2. Serum Pi was significantly lower ($F=34$, $p<0.001$) in SHR from 6 weeks on, whereas serum Ca^{2+} and Mg^{2+} were similar in both strains ($F=1.3$, $p>0.26$). Urinary Pi, Mg^{2+} and Ca^{2+} excretion in WKY and SHR were also measured and the results are shown in Fig. 1. Urinary Pi excretion in SHR was significantly lower than in the age-matched WKY from 6 through 23 weeks of age. In contrast, urinary Mg^{2+} excretion was significantly higher in SHR throughout the whole period of observation. Finally, from 16 weeks onward SHR excrete significantly more Ca^{2+} than the age-matched WKY.

Table 1. Body weight and blood pressure as a function of age in WKY and SHR

Age (weeks)		8	15	22
Body weight (gr)	WKY	228 ± 2	337 ± 4	377 ± 4
	SHR	229 ± 6	330 ± 3	370 ± 4
Blood pressure (mmHg)	WKY	121 ± 9	131 ± 3	128 ± 4
	SHR	180 ± 7*	221 ± 4*	222 ± 4*

Values are mean ± SEM. Number of rats used is 5 for each group. *, significant different ($p< 0.001$) between SHR and age-matched WKY (analysis of variance).

Table 2. Serum Pi, Ca^{2+} and Mg^{2+} as a function of age in WKY and SHR

Age (weeks)		6	12	20
Pi (mM)	WKY	3.62 ± 0.04	2.97 ± 0.03	2.23 ± 0.07
	SHR	3.21 ± 0.14*	2.62 ± 0.07*	1.90 ± 0.10*
Mg^{2+} (mM)	WKY	0.96 ± 0.03	0.97 ± 0.05	1.07 ± 0.03
	SHR	0.84 ± 0.03	1.02 ± 0.02	1.01 ± 0.03
Ca^{2+} (mM)	WKY	2.59 ± 0.07	2.34 ± 0.02	2.39 ± 0.02
	SHR	2.40 ± 0.03	2.45 ± 0.03	2.35 ± 0.03

Values are mean ± SEM. Number of rats used is 5 for each group. *, significant different ($p < 0.01$) between SHR and age-matched WKY (analysis of variance).

Initial rates of Na^+-dependent Pi transport in BBMV from kidneys of WKY and SHR were studied in the range of 0.03-1.0 mM Pi in the incubation medium. The data of Na^+-dependent Pi transport in BBMV from 6, 12 and 20 week old rats were fitted to Michaelis-Menten kinetics, yielding the kinetic parameters summarized in Table 3. Na^+-dependent Pi uptake was increased in SHR compared to the age-matched WKY for all Pi concentrations and all ages studied. Therefore, the maximal transport capacity, V_{max}, was significantly higher in SHR at all ages ($F= 8$, $p< 0.01$). In contrast, the apparent affinity constant, K_m, is similar in both strains and at all ages studied ($F = 0.3$, $p> 0.7$). Moreover, there were no significant differences in D-glucose uptake nor in intravesicular volume between the two strains (data not shown).

In Fig. 2 the serum Pi concentration is plotted against the V_{max} of the Na^+-dependent Pi transport observed in WKY and SHR at 6, 12 and 20 weeks of age. The shift of the SHR data to the left points towards an increasing need for Pi reabsorption in SHR as an attempt to reach a normal blood Pi level. Therefore, Fig. 2 supports the view of a primary hypophosphatemia in SHR, followed by adaptation of the kidney in the direction of increased Pi retention.

Table 3. Kinetic parameters of Na^+-dependent Pi uptake in brush-border membranes of kidney cortex from WKY and SHR

Age (weeks)		6	12	20
V_{max} (nmol/min mg/prot)	WKY	11.2 ± 0.7	9.6 ± 0.9	8.5 ± 0.6
	SHR	12.3 ± 1.1	11.6 ± 1.1	10.4 ± 0.7
K_m (μM)	WKY	62 ± 3	57 ± 9	61 ± 5
	SHR	64 ± 9	64 ± 9	61 ± 4

Values are mean ± SEM. Number of rats used is 5 for each group.

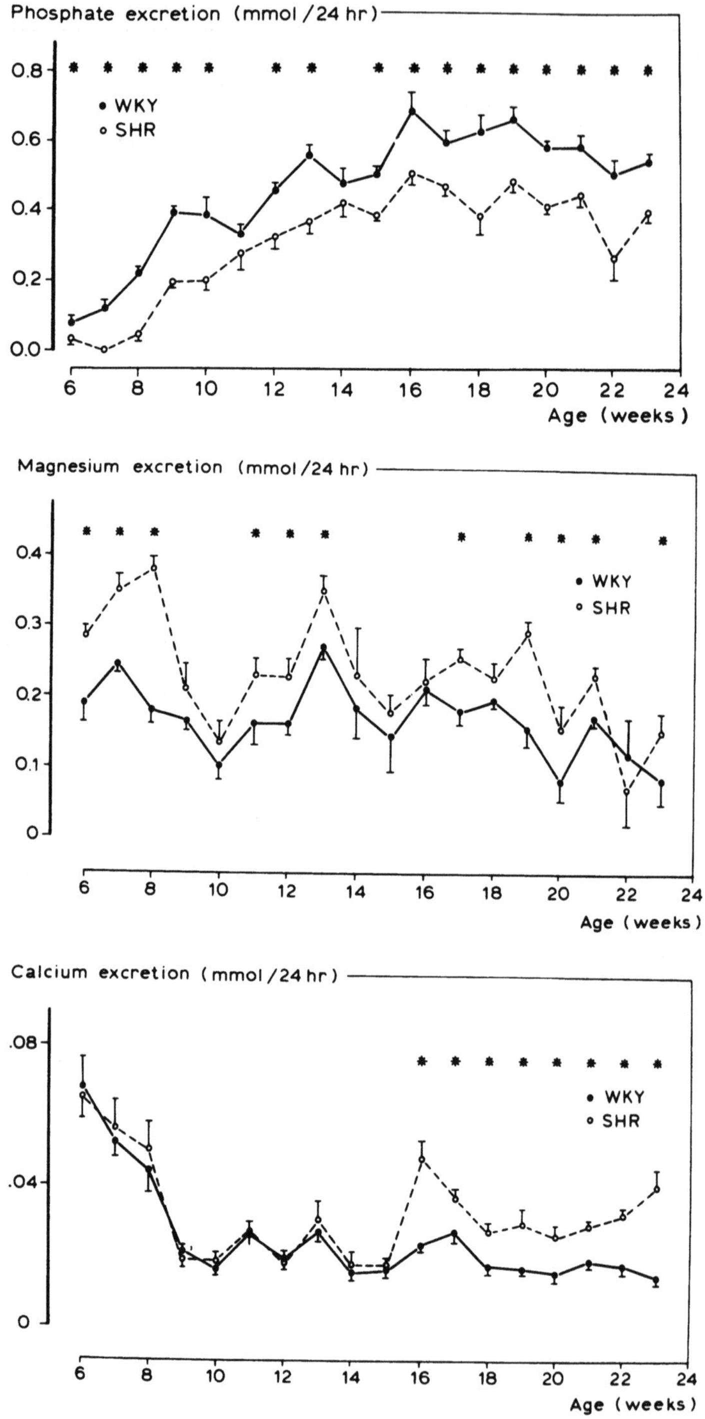

Fig. 1. Urinary Pi, Mg^{2+} and Ca^{2+} excretion (mmol/24 h) as a function of age in WKY and SHR placed in metabolic cages. Values represent mean ± SEM. *, significant different (p> 0.05) between SHR and age-matched WKY (unpaired t-test).

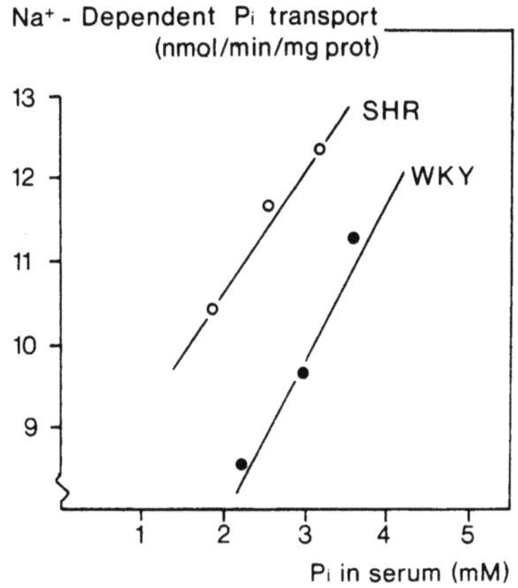

Fig. 2. Dependence of the V_{max} of Na^+-dependent Pi uptake in renal BBMV on serum Pi concentration in WKY and SHR of 6, 12 and 20 weeks of age. The values are taken from Tables 2 and 3.

CONCLUSIONS

The results of the present study indicate that Pi, Mg^{2+} and Ca^{2+} homeostasis in SHR are disturbed, including hypophosphatemia, hypophosphaturia and hypermagnesiuria from 6 weeks onward, followed at 16 weeks by a hypercalciuria. These results are in favour of a mean disturbed Pi metabolism in SHR.

Our finding of a primary hypophosphatemia and an adaptive hypophosphaturia in SHR are consistent with a recent study describing decreased renal Pi clearance and an increased renal tubular Pi reabsorption in SHR (Hsu et al., 1984). In addition, we found an increase in the Na^+-dependent Pi transport capacity in isolated renal cortical BBMV from SHR which parallels the hypophosphaturia. In view of these findings it is unlikely that hypophosphatemia in SHR is due to a defect in Pi handling by the kidney but more likely the result for instance of decreased intestinal absorption or altered cellular or bone Pi metabolism.

Factors leading to hypermagnesiuria in SHR are not explored in this study. However, hypermagnesiuria and hypercalciuria in SHR are not likely to be related because hypermagnesiuria preceeded hypercalciuria by at least 10 weeks. Hypermagnesiuria is compatible with hypophosphatemia, since Pi depletion in man and rat leads to marked hypermagnesiuria (Lee et al., 1981).

Three major causes of hypercalciuria have been recognized, including resorptive hypercalciuria, absorptive hypercalciuria and reduced renal reabsorption (renal leak). There is indirect evidence in favour of a renal Ca^{2+} leak as the cause of hypercalciuria in hypertension (McCarron et al., 1981; Strazzullo et al., 1983). Hypercalciuria in SHR may simply reflect a progressive abnormality in the renal handling of Ca^{2+} secondary to the development of hypertension. Hypercalciuria developed in SHR after 16 weeks (McCarron et al., 1981), long after hypertension in SHR is established. Reports on intestinal Ca^{2+} transport in SHR are difficult to interprete, since higher levels (Toraason and Wright, 1981; Lau et al., 1984), unaltered (Stern et al., 1984) and lower levels (Schedl et al., 1984) of Ca^{2+} transport across the duodenum of aging SHR have been reported. Also Pi depletion causes a reduction in tubular Ca^{2+} reabsorption and hypercalciuria (Sutton, 1983). Since disorders in Ca^{2+} excretion can only be understood with complete reference to all organs involved in the Ca^{2+} homeostasis, more conclusive experiments are needed to delineate the possible cause of hypercalciuria.

REFERENCES

Anderson, S.A., Grades, J.R., Ellison, D.H., McCarron, D.A., 1983, Calcium balance and parathyroid hormone in the spontaneously hypertensive rat, Hypertension, 5 (suppl I): 159.

Biber, J., Steiger, B., Haase, W., and Murer, H., 1981, A high yield preparation for rat kidney brush border membranes. Different behaviour of lysosomal markers, Biochim.Biophys.Acta, 647: 169.

Bindels, R.J.M., Geertsen, J.A.M., van Os, C.H., and Slegers, J.F.G., 1984, Renal calcium and phosphate handling in the spontaneously hypertensive rat and normotensive Wistar-Kyoto control rat, in: Progress in Clin. and Biol.Res., Vol. 168, "Epithelial calcium and phosphate transport: Molecular and cellular aspects", F.Bronner and M.Peterlik, eds., Alan Riss, New York.

Bloomquist, E.I., El-Bermani, A.W.I., Yaney, G.C., and Karr-Dullien, V., 1983, Hypophosphatemia in infant and adult spontaneously hypertensive rats, Clin.Exp.Hypertension, A5(3): 447.

Hsu, C.H., Chen, P.S., and Caldwell, R.M., 1984, Renal phosphate excretion in spontaneously hypertensive rats, Kidney Int., 25: 789.

Lau, K., Zikos, D., Spirnak, J., and Eby, B., 1984, Evidence for an intestinal mechanism in hypercalciuria of spontaneously hypertensive rats, Amer.J.Physiol., 247: E625.

Lee, D.B.N., Brautbar, N., and Kleeman, C.R., 1918, Disorders in phosphate metabolism, in: "Disorders in mineral metabolism. Pathophysiology of calcium, phosphate and magnesium", Vol. III, F.Bronner and J.W.Coburn, eds., Academic Press, New York.

Ljunghall, S., and Hedstrand, S., 1977, Serum Phosphate inversely related to blood pressure, Brit.Med.J., 1: 553.

McCarron, D.A., Yung, B.A., Ugorets, B.A., and Krutzik, S., 1981, Disturbances of calcium metabolism in the spontaneously hypertensive rat, Hypertension, 3 (suppl I): I162.

Stern, N., Lee, D.B.N., Sillis, V., Beck, F.W.J., Deftos, L., Manolagas, S.C., and Sowers, J.R., 1984, Effects of high calcium intake on blood pressure and calcium metabolism in young SHR, hypertension, 6: 639.

Strazzullo, P., Nunziata, V., Cirrillo, M., Giannattasio, R., Ferrara, L.A., Mattioli, P.L., and Mancini, M., 1983, Abnormalities of calcium metabolism in essential hypertension, Clin.Sci., 65: 137.

Sutton, R.A.L., 1983, Disorders of renal calcium excretion. Kidney Int., 23: 665.

Toraason, M.A., and Wright, G.L., 1981, Transport of calcium by duodenum of spontaneously hypertensive rat, Amer.J.Physiol. 241: G344.

Wright, G.L., and Runkin, G.O., 1982, Concentration of ionic and total calcium in four models of hypertension, Amer.J.Physiol., 243: H365.

PARTICIPATION OF VARIOUS BRAIN NUCLEI IN THE ALTERED TIME COURSE OF GENETIC HYPERTENSION IN SHR DEPENDENT ON CHANGES IN CALCIUM METABOLISM

H. Geiger, M. Palkovits,[*] U. Bahner, G. Schmid, and A. Heidland

Medical Clinic of the University of Wuerzburg
Josef-Schneider-Str. 2, D-8700 Würzburg
[*] National Institute of Mental Health, Bethesda, MD

INTRODUCTION

In recent theories concerning the pathogenesis of essential hypertension, abnormalities of calcium metabolism play an important part. The mechanisms involved in Ca-mediated blood pressure control are incompletely understood. An epidemiological survey of Kesteloot and Geboers in the Belgian army (1) showed a significant relationship between serum calcium, urinary calcium, and blood pressure in male participants. Results of animal studies demonstrated that restriction of calcium increases, and supplementation with calcium lowers the blood pressure of spontaneously hypertensive rats (2, 3, 4, 5). Based on the data of various studies made of the dietary calcium intake of normotensive and hypertensive individuals (6, 7, 8, 9, 10, 11) it is suggested that inadequate calcium consumption may be a previously unrecognized factor in the development of hypertension.

Besides the calcium status the parathyroid hormone activity plays an important role in the development of high blood pressure. Hypertension is common in hyperparathyreoidism but at present the mechanism of parathyroid hypertension is not clear (11,12). Parathyroid hormone is a potent vasoactive peptide (13).

Parathyreoidectomy (PTX) performed after weaning delayed systolic blood pressure (SBP) increase (14, 15). These investigations allude to a potential participation of calcium status and parathyroid hormone activity in the pathogenesis of genetic hypertension. The mechanisms in particular with regard to central blood pressure regulation are still unknown.

Saavedra, Grobecker and Axelrod (16) described changes in central cathecholaminergic neurons in the spontaneously hypertensive rats and Schmid and coworkers (17) found altered cAMP-content in specific brain areas of SHR. These results suggest that cAMP-dependent processes are involved in the regulatory mechanisms of central blood pressure control. For this reason we examined the influence of parathyroidectomy and different calcium supplementation on the cAMP-concentration in specific brain areas of SHR.

METHODS

In our experiment four week-old male SHR received two basic diets: a synthetic rat diet deficient in calcium (0.5 % Ca) or a high-calcium diet of 5 % calcium. Half of them underwent parathyroidectomy (PTX). Four groups were studied:

1) PTX-SHR with low Ca-diet (PTX-lCa)
2) sham-operated SHR with low Ca-diet (SO-lCa)
3) PTX-SHR with high Ca-diet (PTX-hCa)
4) sham-operated SHR with high Ca-diet (SO-hCa).

After a 4-week treatment period the animals were sacrified by means of a focused microwave irradiation of the brain. The different brain nuclei were removed by the micro-punch technique of Palkovits (18) and worked up by the method described by Schmid (17).

Cyclic AMP was measured radioimmunologically using a commercial Kit. The statistical evaluation of results was carried out by the Wilcoxon-U-Test for unpaired observations.

RESULTS

The average body-weight of the 4-week-old rats was 65 g (Figure 1). After the 4-week-treatment period the body weight amounted to about 160 g in three groups. Only the PTX-rats with low calcium diet showed a significant lower body weight with 136 g. Apparently the lack of PTH coincidently with an extreme reduced calcium supplementation leads to a distinct unfavourable effect on the development of body weight and affects adversely the normal physiological maturation of the animals.

The initial blood pressure was 112 mmHg (Figure 2). After the 4-week-diet regime a highly significant difference in blood pressure has evolved both in the PTX and in the SO-group. The rats fed with a low calcium diet display a hypertension already after 4 weeks, whereas the animals with high calcium supplementation show normal blood pressure. If you compare PTX with SO-rats with respect to the same calcium content of the chow there is no significant difference.

The findings of the present study indicate that variations in calcium status influence the development of high blood pressure in genetic hypertensive rats. In young SHR, a low calcium diet enhanced the time course of genetic hypertension. A high calcium supplementation significantly decreases high blood pressure and that confirms earlier works obtained with calcium enriched diets that lessened but did not prevent genetic hypertension in SHR.

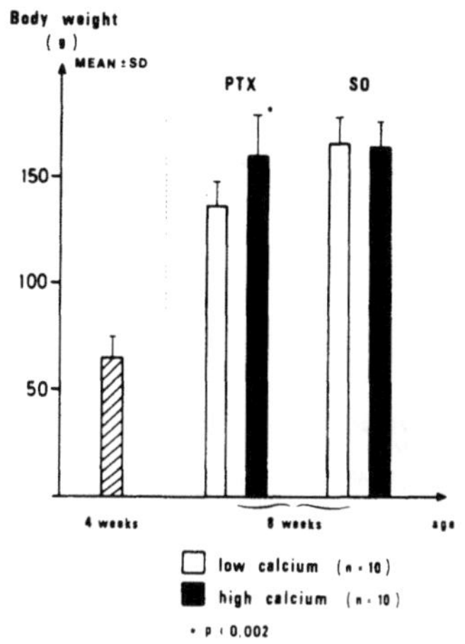

Figure 1 Body Weight of SHR After PTX Or SO

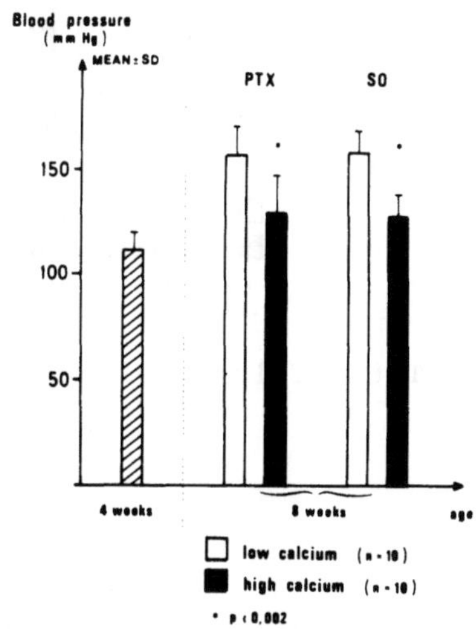

Figure 2 Blood Pressure of SHR-PTX and SHR-SO with Different Calcium Diet

The mean serum total calcium was determined as 3.00 mmol/l at the beginning of the study (Figure 3). After the 4-week-treatment period, the calcium levels showed significant differences. Experience the variable calcium alimentation led to different calcium levels both in PTX and in SO-rats. The PTX-rats fed with a high calcium diet are able to attain with nearly 4.00 mmol/l high serum total calcium concentration despite the lack of PTH. There were no differences between parathyroidectomized and sham-operated rats receiving the diet with high calcium content. In contrast parathyroidectomized rats with low calcium feeding showed a marked decrease of serum calcium to 2.32 mmol/l.

The main question of our study was to clear up whether central mechanism participate in the modulated time course of hypertension in SHR in dependence on altered calcium supplementation and on changes in parathyroid hormone status. For this reason we determined the cAMP-concentration in specific brain areas (Table 1).

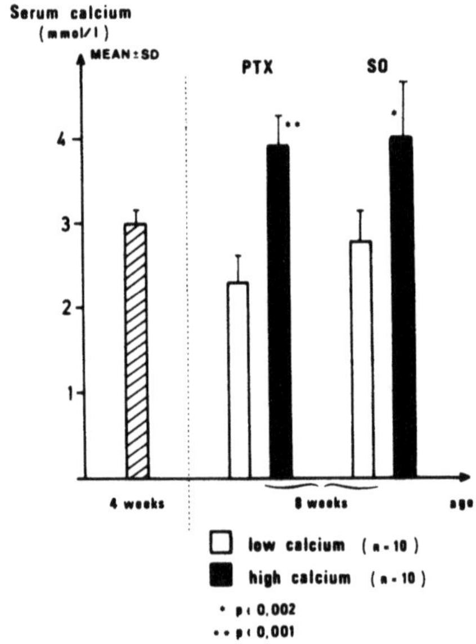

Figure 3 Serum Total Calcium of SHR After PTX Or SO

Table 1 cAMP Concentration in Specific Brain Areas

	Cortex cinguli	Cortex parietalis	Nucleus caudatus	Amygdala	Hippocampus	Central grey	NTS A_2-Cellg.	A_1-Cell group	Locus coeruleus
PTX - l Ca SO - l Ca	-	p <0,05	-	-	p <0,01	-	-	-	p <0,05
PTX - h Ca SO - h Ca	-	p <0,01	-	-	-	p<0,05	-	-	
PTX - l Ca PTX - h Ca	p<0,05	p <0,05	-	-	p<0,05	p<0,01	-	-	p <0,001
SO - l Ca SO - h Ca	-	p <0,05	-	-	p<0,05	-	p <0,01	-	p <0,001

Our results show that different calcium diet did not cause significant changes of cAMP content in A 1-cell group, caudate nucleus and Amygdala and that a lack of PTH didn't alter cAMP-concentration in NTS with a A 2-cell-group, caudate nucleus, Amygdala and cingulate cortex.

PTX-rats fed with different calcium diet have different cAMP-concentrations in cingulate cortex, parietal cortex, hippocampus, central grey and especially locus coeruleus. Sham-operated rats with variable calcium supplementation show a different cAMP content in parietal cortex, hippocampus, NTS with A 2-cell-group and pronounced in locus coeruleus.

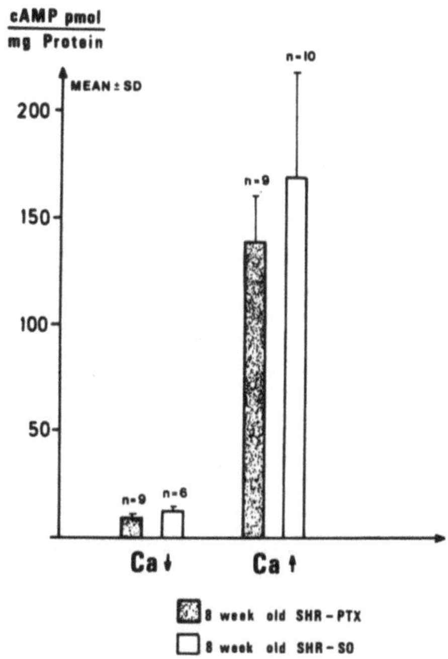

Figure 4 cAMP-Content of Locus Coeruleus

Rats fed with a low calcium diet had different cAMP-concentrations in parietal cortex, hippocampus and locus coeruleus when they differed in PTH-status. Rats with a calcium enriched diet showed various cAMP-contents only in parietal cortex and central grey comparing sham-operated with parathyroidectomized animals.

There were significant elevated cAMP-concentrations in the locus coeruleus of all animals with high calcium diet irrespective of PTH-status, that means of all rats with normotensive blood pressure levels in our study (Figure 4).

After a critical examination of our results we conclude that an altered calcium and PTH status may cause changes in the cAMP-content of specific brain areas. Both neuroanatomical and neurophysiological studies suggest the idea that brain nuclei showing significant cAMP alterations in SHR might represent secondary or modulatory areas capable to affect baroreceptor reflex function. Our results let us presume that the effects on blood pressure which have been induced by an altered calcium and PTH-metabolism are associated with changes of central blood pressure regulating processes. Under these conditions the locus coeruleus may play an important role.

The locus coeruleus represents the most important noradrenergic brain area. Containing nearly half of all norepinephrine-synthesizing neurons it is origine of many catecholaminergic neurons to various brain regions. Electrophysiological investigations have shown that the norepinephrine neurons of the locus coeruleus act as an inhibitor in almost every affected brain structure.

In previous investigations we found a highly significant increase (164 %) of the cAMP content in this cerebral nuclei of SHR which contributes to a participation of this brain area in the pathogenesis of genetic hypertension.

Thus, afferent neurons from brain areas which exhibit an altered activation in SHR could contribute to an increased activity and consequently to elevated cAMP levels in the locus coeruleus (e. g. afferent neurons from hypothalamus, limbic structures). On the other hand the altered activation of the locus coeruleus could effect the activity to other brain areas via efferent connections.

The efferent connections to the locus coeruleus constitute a major ascending pathway designated the dorsal noradrenergic bundle (Figure 5).

The dorsal noradrenergic bundle traverses the midbrain tegmentum with central grey, and ascends to cortical structures as cingulate cortex and neocortex. Besides there are descending branches to the hippocampus region and to the nucleus dorsalis of the vagal nerve (X), which represents a connection to the baroreceptor reflex arc. There remains to be mentioned that all designated brain regions show altered cAMP-concentrations in our experiment. Based on the data of our previous investigations we discuss the hypothesis that elevated cAMP-concentrations in the locus coeruleus are not primarily associated with the development of hypertension, but that they reflect a new balance in the blood pressure regulating system of the SHR.

Studies of Elam and coworkers (19) recently have shown that cardiovascular vagal afferents can participate in the regulation of the locus coeruleus neurons in the rat. Their data strengthen the concept that brain norepinephrine neurons in the locus coeruleus are subject to control by peripheral blood volume receptors.

On condition of a high calcium diet the locus coeruleus may work as a central structure with depressor function, as it were a counteracting or modulating structure of cell groups with pressor (blood elevating) function.

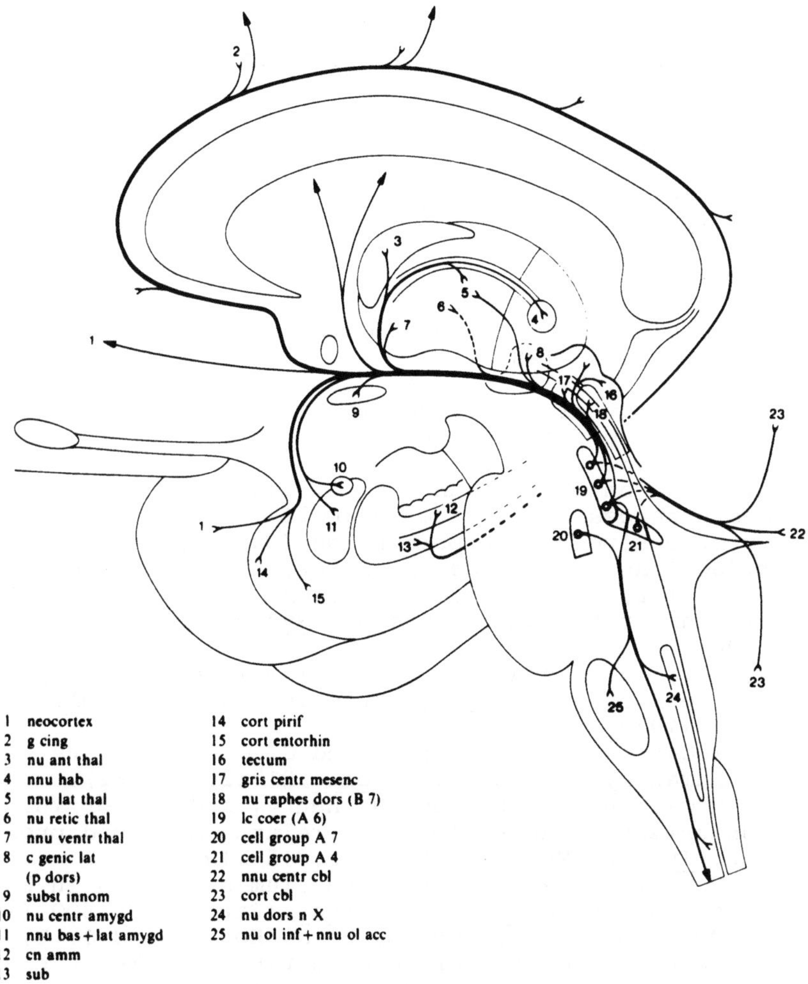

The noradrenergic system I: connections of the locus coeruleus

Figure 5
Adopted from R. NIEUWENHUYS, J. VOOGD, C. VAN HUIJZEN: "The Human Central Nervous System". A Synopsis and Atlas. Springer Verlag Berlin 1978.

SUMMARY

1. Variations in calcium alimentation influence the development of high blood pressure in genetic hypertensive rats.
2. Different calcium feeding and changes in parathyroid hormone status cause altered cAMP-content in specific brain areas.
3. All animals with high calcium diet show significant elevated cAMP-content in the locus coeruleus, irrespective of parathyroid status. These findings support the hypothesis that elevated cAMP-concentrations reflect an increased depressor activity of the locus coeruleus.

ACKNOWLEDGEMENT

We thank Miss Ch. Ullrich and Miss H. Greb for skillful technical assistance.

REFERENCES

1. H. Kesteloot and J. Geboers, Calcium and blood pressure, Lancet 1:813 (1982).
2. R. Schleiffer, F. Pernot, A. Berthelot, and A. Gairard, The calcium dietary level modulates the development of genetic hypertension in rats, C.R. Soc. Biol. (Paris) 177 (4):481 (1983).
3. D. A. McCarron, N. N. Yung, B. A. Ugoretz, and S. Krutzik, Disturbances of calcium metabolism in the spontaneously hypertensive rat, Hypertension (Supp I) 3:162 (1981).
4. J. M. Belizan, O. Pineda, E. Sainz, L. A. Menendez, and J. Villar, Rise of blood pressure in calcium-deprived pregnant rats, Am. J. Obstet. Gynecol. (2) 141:163 (1981).
5. S. Ayachi, Increased dietary calcium lowers blood pressure in the spontaneously hypertensive rat, Metabolism 28: 1234 (1979).
6. D. A. McCarron, C. D. Morris, and C. Cole, Dietary calcium in human hypertension, Science 217:267 (1982).
7. D. A. McCarron, J. Stanton, H. Henry, and C. Morris, Assessment of nutritional correlates of blood pressure, Ann. Intern. Med. (Part 2) 98:715 (1983).
8. D. A. McCarron, Calcium and magnesium nutrition in human hypertension, Ann. Intern. Med. (Part 2) 98:800 (1983).
9. D. A. McCarron, C. D. Morris, H. J. Henry, and J. L. Stanton, Blood pressure and nutrient intake in the United States, Science 224:1392 (1984).
10. S. Ackley, E. Barrett-Connor, and L. Suarez, Dairy products, calcium, and blood pressure, Am. J. Clin. Nutr. 38:457 (1983).
11. A. K. Sangal and D. G. Beevers, Parathyroid hypertension, Br. Med. J. 286:498 (1983).
12. E. L. Bradley and J. O. Wells, Primary hyperparathyroidism and hypertension, Am. Surg. 49:569 (1983).
13. D. A. McCarron, D. H. Ellison, and S. Anderson, Vasodilation mediated by human PTH 1-34 in the spontaneously hypertensive rat, Am. J. Physiol. 246:96 (1984).
14. J. F. Mann and E. Ritz, Effects of parathyroidectomy on blood pressure development in eucalcaemic, spontaneously hypertensive rats, Proc. Eur. Dial. Transplant. Assoc. 20:220 (1983).

15. J. Mann, M. Rambausek, P. Klooker, A. Schömig, D. Ganten, and E. Ritz, Hinweis auf einen permissiven Effekt von Parathormon in der Blutdruckentwicklung von spontan hypertonen Ratten, Verhandlungen der Deutschen Gesellschaft für Innere Medizin 88:776 (1982).
16. J. M. Saavedra, H. Grobecker, and J. Axelrod, Changes in central catecholaminergic neurons in the spontaneously (genetic) hypertensive rat, Circ. Res. 42:529 (1978).
17. G. Schmid, M. Palkovits, U. Bahner, and A. Heidland, Altered cyclic AMP levels in specific cardiovascular brain centers of spontaneously hypertensive rats (SHR), Klin. Wochenschr. 58:1091 (1980).
18. M. Palkovits, Isolated removal of hypothalamic or other brain nuclei of the rat, Brain Res. 59:449 (1973).
19. M. Elam, T. Yao, T. H. Svensson, and P. Thoren, Regulation of locus coeruleus neurons and splanchni, sympathetic nerves by cardiovascular afferents, Brain Res. 290:281 (1984).

15. J. Manny M. Rombauseck, F. Kleckner, A. Schömig, D. Gratzl, and E. Fitz, Hinweis auf einen [illegible] Effekt von Katecholaminen in der Nieren[illegible]entwicklung [illegible] hypertonen Ratten, Verhandlungen der Deutschen Gesellschaft für Innere Medizin [illegible] (19[illegible]).
16. [illegible]

DISSOCIATION BETWEEN THE HYPOTENSIVE EFFECT OF THIAZIDES AND PLASMA DIVALENT CATIONS

Daniel C. Batlle, Wu-Chang Yang,
Alisa Von Riotte and Salim K. Mujais

Department of Medicine
West Side VA Hospital and
University of Illinois College
of Medicine at Chicago
Chicago, IL

INTRODUCTION

Alterations in divalent cation metabolism are increasingly being sought in experimental and human hypertension (1-11). The finding of a low ionizable plasma calcium in some hypertensive patients (2,4) has lead to the exploration of the effect of calcium intake on the level of arterial blood pressure (6-11). This has been done by either dietary survey of large populations groups or by dietary supplementation in both experimental and human hypertension (6-12). Derangements of extracellular ionizable calcium and magnesium concentrations in hypertensive patients have been related to the renin profile (4). Ionizable serum calcium levels were found to be lower in patients with low renin hypertension and higher in patients with high renin hypertension than in those with normal renin hypertension (4).

Thiazide diuretics are effective hypotensive agents in most forms of primary and secondary hypertension. They are also known to significantly influence the balance of divalent cations in normal subjects and in a variety of pathologic states (13-25). These effects on cation metabolism could influence arterial pressure on their own right. The present study was undertaken to examine whether there is a relationship between the hypotensive response to thiazide therapy and the changes in plasma divalent cations that these agents may induce in primary hypertension. We reasoned that part of the hypotensive effect of thiazides could involve an increase in plasma ionizable calcium. In addition, we wanted to examine if the increase in plasma renin activity associated with the administration of this type of diuretic agents correlates with a rise in plasma calcium. A thiazide-induced fall in blood pressure if associated with an increase in ionizable calcium would support (but not prove) the notion that changes in the extracellular concentration of this cation are involved in blood pressure regulation (1,2,4). A lack of an effect of thiazides on ionizable plasma calcium, on the other hand, would strongly suggest that volume-dependent control of hypertension does not involve changes in extracellular calcium. This issue is also important in view of recent data suggesting that primary hyperaldosteronism, the prototype of low-renin volume-dependent hypertension, is associated with low ionizable plasma Ca levels which increase after cure of the disease (11).

PATIENTS AND METHODS

Thirty patients with essential hypertension followed at the Hypertension Clinic of the University of Illinois Hospital at Chicago were the subject of the present study. There were twenty-four women and six men with a population mean age of 47.8±2.2 years. Their racial distributions was as follows: 23 blacks, 5 caucasians and 2 hispanics. All were either untreated for hypertension or their previous antihypertensive therapy was discontinued at least three weeks before entry into the study. None of the patients had any disease besides hypertension that could affect divalent cation metabolism. All patients had diastolic blood pressure values higher than 95 mmHg on at least five clinic visits, and none had evidence of secondary forms of hypertension. All patients were on a regular diet except that moderate salt restriction was advised. After selection, patients were entered into a placebo phase of two weeks during which determination of baseline clinical and laboratory parameters was undertaken. Subsequently, patients were started on a single daily dose of 50 mg of hydrochlorothiazide for eight weeks. The drug was provided free and compliance was assessed by patient interview and pill count. Patients were seen at one, two, four and eight weeks after initiation of therapy at which times clinical and laboratory measurements were obtained on five occasions at approximately 3 minute intervals. The majority of the thirty patients complied with all the return appointment schedules. Each patient was seen by the same nurse practitioner at the initial and all subsequent visits. Blood pressure was obtained in the sitting position using a cuff appropriate to arm size. Body weight was measured at each visit. Each patient gave his or her informed consent for entry into the study, according to a protocol approved by the Institutional Review Board of the University of Illinois College of Medicine at Chicago.

The following laboratory tests were obtained during each clinic visit: plasma creatinine, sodium, potassium, chloride, osmolarity, magnesium, phosphate, total calcium, and ionizable calcium. In an additional group of sixteen patients, plasma renin activity and plasma electrolytes were determined approximately one week prior to hydrochlorothiazide and after approximately 2 months of continued therapy with this diuretic.

Plasma sodium and potassium were measured using a flame photometer with lithium as the internal standard (International Instruments, model IL64). Plasma chloride was determined by Chloridometry (Corning Model 920M). Creatinine was assayed by a Beckman Analyzer II). Total calcium was measured using Corning 940 titrator and ionizable calcium was determined using a Orion Ionized Calcium electrode (SS-20). Magnesium was measured by atomic absorption utilizing a Varian-100. Phosphate was determined using an Autoanalyzer Model II. Osmolality was measured with a Wescor Vapor Pressure Osmometer. Plasma renin activity was measured with radioimmunoassay (E.R. Squibb and Sons, Princeton, N.Y.).

Statistical analysis was performed on a IBM PC/XT computer using the software program ABSTAT (Co). The statistical significance was assessed by analysis of variance (ANOVA) and by paired T test. Correlations were done by linear regression analysis and analysis of covariance. Results are reported as mean ± SEM.

RESULTS

Effects of thiazides on blood pressure. In the majority of patients blood pressure was mildly or moderately elevated. The mean pretreatment blood pressure of the 30 patients was 157.1±2.7 and 100.5±1.1 mmHg for systolic and diastolic, respectively. Thiazide administration led to a significant decrease in both systolic and diastolic blood pressure which was already apparent by the end of the first week of therapy and sustained throughout the period of observation (figure 1, left panel). At the end of the first week on thiazide

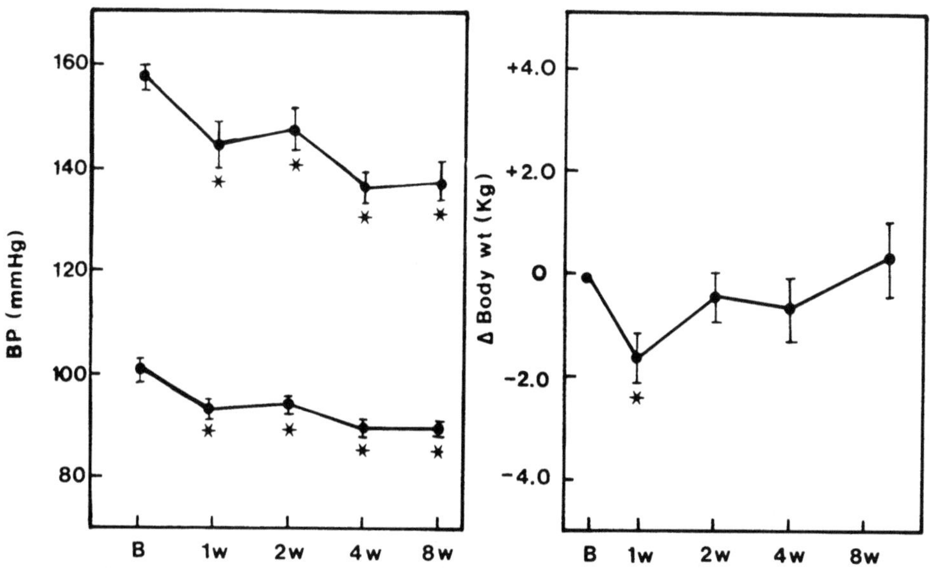

Fig. 1. The effect of thiazides on systolic and diastolic blood pressure (left panel) and weight (right panel) over a period of 8 weeks.

therapy blood pressure fell to 144±4.5/92±1.2 mmHg, in association with a significant decrease in total body weight (1.6±0.5 kg). The decrement in blood pressure correlated directly with the decrease in body weight ($r=0.48$, $p<0.05$ and $r=0.48$, $p<0.05$ for systolic and diastolic blood pressure respectively). By the second week of thiazide administration, total body weight began to increase returning to baseline by the end of the second month (figure 1, right panel). A slight but significant further reduction in blood pressure occurred at the end of one month of treatment and was sustained at the end of two months despite of return of body weight to that observed prior to treatment (figure 1).

Effect of thiazides on renal function and plasma potassium. Pretreatment and follow-up values of plasma creatinine and electrolytes are summarized in Table I. Plasma creatinine rose slightly but significantly. Thiazide administration had the expected effect of inducing a significant and sustained fall in plasma potassium. Plasma sodium and plasma osmolality were not significantly affected while plasma chloride was significantly reduced in most patients probably reflecting the development of diuretic-induced hypochloremic metabolic alkalosis.

Effect of thiazide on divalent ions. Pretreatment total plasma calcium was normal in all patients and was not significantly changed by diuretic therapy (Table 1). Individual values for plasma ionizable calcium before and during thiazide administration are depicted in Figure 2. Administration of hydrochlorothiazide had no significant effect on ionizable plasma calcium at any interval during the study period. Plasma magnesium values tended to decrease slightly during diuretic therapy but this decrement did not attain statistical significance by analysis of variance (figure 3). By paired t test analysis a significant fall in plasma Mg was noted only in the second week of therapy. Similarly, no significant change was observed in the plasma levels of inorganic phosphate.

Since examination of mean values may mask a significant relationship between individual variables, we examined the correlations between each individual variable before and during therapy. There was no significant correlation between pre-treatment ionizable calcium and either pre-treatment systolic or diastolic blood pressure (r=0.13, NS and r=0.21, NS respectively). Likewise, the level of either systolic or diastolic blood pressure achieved during thiazide therapy did not correlate with the corresponding ionizable plasma calcium levels (r=0.15 and 0.03, respectively. The decrement in either systolic or diastolic blood pressure (i.e. ΔBP) at all time intervals studied also did not correlate with the change in ionizable calcium (i.e. ΔCa^{++}) at those intervals. Likewise, no correlation was found between the decrement in blood pressure and changes in plasma magnesium.

Effect on plasma renin activity. In an additional subset of sixteen patients, plasma renin activity was measured before and approximately 8 weeks following continuous thiazide therapy. The diuretic resulted in a 3-fold increase in PRA (from 1.7±0.4 to 5.4±1.1 ng/ml/hr $p<0.001$). There was no correlation between PRA and plasma ionizable calcium values obtained prior to thiazide therapy (r=0.40, NS). Likewise, no correlation was found between those variables during therapy (r=0.44, NS). When the data obtained before and during thiazide administration was pooled for linear regression analysis, no significant correlation between plasma ionizable calcium and plasma renin activity was found (r=0.26). Furthermore, there was no significant correlation between the increment in PRA (Δ PRA) and the change in ionizable plasma calcium (ΔCa^{++}). The correlation coefficient between these two variables (r=0.51), however, might have achieved statistical significance if a larger number of patients had been studied. No significant correlations were found between PRA and plasma magnesium either before (r=0.30) or during thiazide therapy (r=0.34).

DISCUSSION

This study shows that the hypotensive action of thiazide diuretics in patients with primary hypertension is not associated with an increase in ionizable plasma calcium. A role for ionizable plasma calcium in the pathogenesis of hypertension has been recently suggested (1-5). Clinical and experimental observations supporting a role of extracellular calcium

Table I: Biochemical Data

	Creat mg/dl	Total Ca mEq/l	Ionized Ca mEq/l	Mg mEq/l	PO_4 mmol/l	Na mEq/l	K mEq/l	Cl mEq/l
Baseline	0.95 ±0.03	4.48 ±0.05	2.41 ±0.04	1.53 ±0.03	1.13 ±0.05	140.6 ±0.4	4.17 ±0.06	100.4 ±0.8
1 wk	1.06* ±0.05	4.60 ±0.05	2.43 ±0.05	1.45 ±0.05	1.13 ±0.05	139.7* ±0.4	3.70*** ±0.08	97.3* ±1.1
2 wk	0.99 ±0.05	4.61 ±0.08	2.48 ±0.04	1.40** ±0.05	1.16 ±0.05	139.8 ±0.5	3.69*** ±0.10	98.8 ±1.2
4 wk	1.04** ±0.04	4.50 ±0.08	2.48 ±0.04	1.47 ±0.03	1.15 ±0.04	140.0 ±0.5	3.74*** ±0.08	96.4*** ±0.7
8 wk	0.99 ±0.04	4.58 ±0.05	2.46 ±0.03	1.43 ±0.07	1.06 ±0.04	140.3 ±0.6	3.76*** ±0.10	97.9 ±0.9

*$P<0.05$, **$p<0.01$, ***$P<0.001$ as compared with baseline value.

in the pathogenesis of primary hypertension include the following: decreased levels of ionizable plasma calcium (2,4), increased parathyroid hormone activity and reduced serum phosphorus concentration (5), increased urinary calcium excretion (5), epidemiologic data showing that decreased dietary calcium intake increases the risk of hypertension and cardiovasular disease (6-8) and lowering of blood pressure by calcium supplementation in spontaneously hypertensive rats (9,10) and in normotensive human subjects (12).

On the basis of these observations we reasoned that lowering of blood pressure by thiazide diuretics could be associated with an increase in plasma ionizable calcium. This type of diuretics could increase ionizable plasma calcium by reducing renal calcium excretion and, perhaps, by increasing plasma renin activity in association with reduction in extracellular volume and lowering of blood pressure. While the hypocalciuric effect of thiazides is well established (13-17), the net effect of these diuretics on calcium balance, however, is less well defined. Balance studies have shown that after more than 1 year on the drugs, fecal calcium increased so that net calcium balance was not influenced by renal calcium retention (18). Other studies in patients with idiopathic hypercalciuria, however, revealed no change in calcium absorption despite of a fall in urinary calcium excretion (19,20). The effect of thiazide diuretics on plasma calcium in normal individuals is also not well defined. Some studies in normal subjects treated with thiazide diuretics revealed an increase in total plasma calcium and a rise in serum protein concentration such that plasma calcium corrected for the change in protein did not rise significantly (15,18). Other studies in normal subjects, however, have shown a significant rise in both total and ionizable plasma calcium during prolonged thiazide adminstration (21). A rise in plasma calcium concentration in response to larger doses of oral thiazide was noted by Koppel et al (22) in patients with absent renal function, a finding which suggested an extrarenal effect of thiazides on calcium metabolism. In addition, a transient increase in plasma ionizable calcium after an acute infusion of chlorothiazide was noted by Popovtzer et al (23). This effect was rapid and required the presence of either endogenous or exogenous parathyroid hormone suggesting enhancement of the hypercalcemic effect of PTH on bone by chlorothiazide (23). In summary, thiazides have the potential to increase plasma calcium by causing renal calcium retention and by extrarenal actions related to potentiation of the calcemic effect of PTH.

A rise in plasma calcium, however, does not need to occur in normal subjects on chronic thiazide therapy. A compensatory decrease in net intestinal calcium absorption, inhibition of 1-25-$(OH)_2$-vitamin D synthesis and reduced PTH secretion secondary to a transient increase in plasma calcium might prevent a sustained elevation in plasma calcium unless there is a marked negative calcium balance to begin with (i.e. idiophatic hypercalciuria). Our findings clearly negate any significant effect of thiazide on ionizable plasma calcium in patients with primary hypertention (figure 2). The lack of effect of thiazides on ionizable plasma calcium therefore does not support the contention that alterations of extracellular calcium are involved in the pathogenesis of volume-dependent hypertension. This hypothesis has been proposed by Resnick and Laragh based on their findings of low ionizable calcium in patients with low-renin hypertension and in patients with primary hyperaldosteronism (4,11). In the scheme proposed by these authors a negative calcium balance in these two types of hypertension may result from renal hypercalciuria resulting from volume expansion or mineralocorticoid excess (11). In agreement with this notion, these authors showed that dietary sodium loading resulted in a fall in ionized calcium

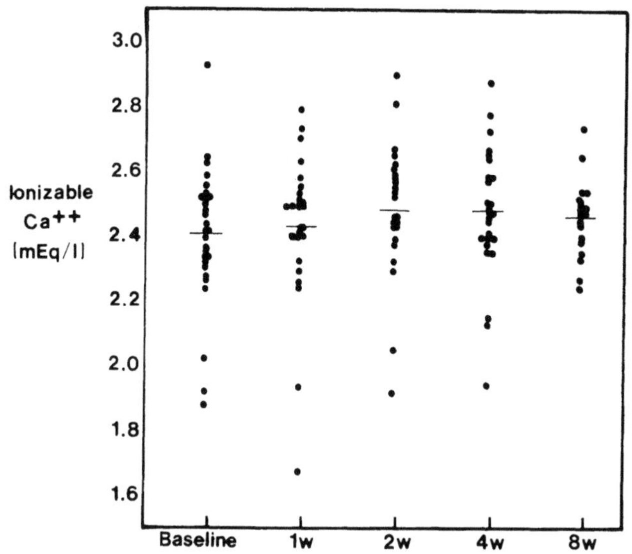
3.0
2.8
2.6
2.4
2.2
2.0
1.8
1.6
Ionizable
Ca++
(mEq/l)
Baseline
1w
2w
4w
8w

Fig. 2

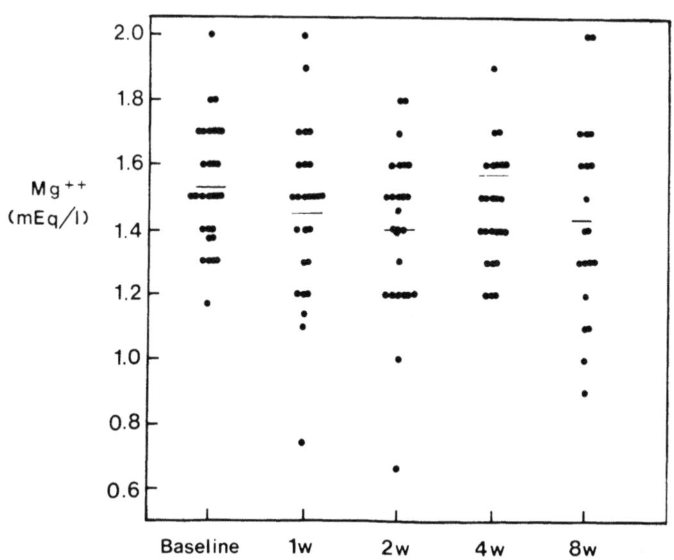
2.0
1.8
1.6
1.4
1.2
1.0
0.8
0.6
Mg ++
(mEq/l)
Baseline
1w
2w
4w
8w

Fig. 3

and elevation of blood pressure in patients with low-renin hypertension (24). If this hypothesis were correct lowering of blood pressure during thiazide therapy should have the exact opposite effect on plasma Ca than dietary salt loading. Control of hypertension by thiazides should also result in a rise in ionizable calcium much in the same way than that reported after control of volume-dependent hypertension by removal of aldosterone producing adenomas (11). Thiazides, however, corrected the volume-dependent mechanism that appeared to sustain the elevated blood pressure of our patients without affecting plasma ionizable calcium. Moreover, a rise in ionizable calcium did not occur despite the predictable effect of thiazides of increasing plasma renin activity. Our data therefore negate a role for extracellular calcium either as a mediator or as a marker of hypertension responsive to volume removal by thiazides. Our data also does not support the contention that plasma ionizable calcium correlates with plasma renin activity in patients with primary hypertension. A significant correlation between plasma magnesium and plasma renin activity was also not found in the present study. It may be argued that a correlation between plasma renin activity and these divalent cations was not found in our hypertensive patients because a full spectrum of renin values was not a feature of the population studied. We feel, however, that the marked rise in plasma renin activity that occurred during thiazide administration should have uncovered a significant correlation with plasma cations if a relationship existed.

In summary, the notion that alterations of calcium metabolism mediate the hypertensive effect of volume expansion is not supported by this study. Our data clearly indicates that the mechanism whereby thiazide diuretics lower blood pressure in hypertensive patients is independent of extracellular calcium levels. This finding does not negate a role for intracellular calcium in blood pressure regulation. An impressive fall in free intracellular calcium has been reported in association with lowering of arterial blood pressure by diuretics and other antihypertensive agents (25). While alterations in the cell's storage and mobilization of calcium may result in increased peripheral resistance and elevated blood pressure our data indicates that extracellular calcium levels cannot be used as a marker of a basic defect in cell transport of calcium underlying the pathogenesis of primary hypertension.

REFERENCES

1. D. A. McCarron, Calcium in the pathogenesis and therapy of human hypertension, Am J Med. 78:27-34 (1985).
2. D.A. McCarron, Low serum concentrations of ionized calcium in patients with hypertension, N Engl J Med, 307:226-228 (1982).
3. P. Strazzullo, V. Nunziata, M. Cirillo, et al, Abnormalities of calcium metabolism in essential hypertension, Clin Sci, 65: 137-141 (1983).
4. L. M. Resnick, J.H. Laragh, J. E. Sealy, M.H. Alderman, Divalent cations in essential hypertension, N Engl J Med, 309:888-891 (1983).
5. D.A. McCarron, P. Pingree, R.J. Rubin, et al, Enchanced parathyroid function in essential hypertension; a homeostatic response to a urinary calcium leak, Hypertension, 2:162-168 (1980).
6. D.A. McCarron, C.D. Morris, C. Cole, Dietary calcium in human hypertension, Science, 217:267-269 (1982).
7. S. Ackley, E. Barrett-Conner, L. Suarez, Dairy products, calcium and blood pressure, Am J Clin Nutr 38:457-461 (1983).

8. F. W. Stitt, M.D. Crawford, D. G. Clayton, et al, Clinical and biochemical indicators of cardiovascular disease among men living in hard and soft water areas, Lancet, 1:122-126 (1973).
9. D. A. McCarron, N. N. Yung, B.A. Ugoretz, et al, Disturbances of calcium metabolism in the spontaneously hypertensive rat, Hypertension 3:1162-1167 (1981).
10. McCarron DA, Blood pressure and calcium balance in the Wistar-Kyoto rat. Life Sci, 30:683-689 (1982).
11. L. M. Resnick, J. H. Laragh, Calcium metabolism and parathyroid function in primary aldosteronism, Am J Med, 78:385-390 (1985).
12. J. Belizan, J. Villar, et al, Reduction of blood presure with calcium supplementation in young adults, JAMA, 249:1161-1165 (1983)
13. B. A. Lamberg, B. Kuhlback, Effect of chlorothiazide and hydrochlorothiazide on the excretion of calcium in urine, Scand J Lab Clin Invest, 11:351-360 (1959).
14 A. M. Parfitt, The acute effects of mersalyl chlorothiazide and mannitol on the renal excretion of calcium and other ions in man, Clin Sci, 36: 267-282 (1936).
15. A. S. Brickman, S. G. Massry, J. W. Coburn, Changes in serum and urinary caclium during treatment with hydrochlorothiazide: Studies on mechanisms, J Clin Invest, 51:945-954 (1972).
16. L.S. Costanzo, I. M. Weiner, On the hypocalciuric action of chlorothiazide, J Clin Invest, 54:628-637 (1974).
17. L. S. Costanzo, E. W. Windhager, Calcium and sodium transport by the distal convoluted tubule of the rat, Am J Physiol, 235:F492-F506 (1978).
18. E. R. Yendt, R.J.A. Gagne, M. Cohanim, The effects of thiazides in idiophatic hypercalciuria, Am J Med Sci, 251:449-460 (1966).
19. U. Ehrig, J.E. Harrison, D.R. Wilson, Effect of long-term thiazide therapy on intestinal calcium absorption in patients with recurrent renal calculi, Metabolism, 23:139-149 (1974).
20. D.E. Barilla, R. Tolentino, R.A. Kaplan, et al, Selective effects of thiazide on intestinal absorption of calcium in absorptive and renal hypercalciurias. Metabolsim, 27:125-130 (1978).
21. R.M. Stote, L.H. Smith, D.W. Wilson, et al, Hydrochlorothiazide effects on serum calcium and immunoreactive parathyroid hormone concentrations: Studies in normal subjects, Ann Intern Med, 77: 587-594 (1972).
22. M. H. Koppel, S. G. Massry, J.H. Shinaberger, et al, Thiazide-induced rise in serum calcium and magensium in patients on maintainance hemodialysis, Ann Intern Med, 72: 895-901 (1970).
23. M. M. Popvtzer, V. L. Subryan, A.C. Alfrey, et al, The acute effect of chlorothiazide on serum-ionized calcium. Evidence for a parathyroid hormone-dependent mechanism, J Clin Invest, 55:1295-1302 (1975).
24. L. M. Resnick, J.P. Nicholson, J.H. Laragh, Calcium metabolism, blood pressure, and salt intake in essential hypertension (abstr), Circulation, 70 (suppl II):1 (1984).
25. P. Erne, P. Bolli,, et al, Correlation of platelet calcium with blood pressure: Effect of antihypertensive therapy, N Engl J Med 310:1084-1088 (1984).

ACKNOWLEDGMENTS

This study was supported, in part, by a grant from the Chicago Heart Association and a grant from Smith Kline and Fench Laboratories. The authors wish to express their appreciation to the nurse practitioners, Mrs. Susan Fontana, and Mrs. Theresa Grosskof and for their invaluable assistance in the completion of this study.

NEPHROLITHIASIS

URINE CITRATE AND CALCIUM IN CALCIUM NEPHROLITHIASIS

Joan H Parks and Fredric L Coe

Renal Program
University of Chicago
Chicago, IL

INTRODUCTION

Our particular interest concerns urine citrate in idiopathic calcium oxalate stone disease. Only 3 previous (1,2,6) studies have included both sexes; they disagree as to the level of urine citrate, and the reduction of citrate in patients compared to normal, although both find that patients excrete less citrate than normals. They used different citrate methods. In one of the two, the numbers of women were small: 7 patients and 15 normals. In four other studies, of only men, patients were found to excrete between 97 and 63 percent as much citrate as normals (3,4,5,6,7).

Urine citrate is of interest because it could reduce the likelihood of calcium oxalate crystallization in at least 2 ways. Citrate is calcium binding and accounts for over half of the bound calcium in human urine (8). Calcium binding reduces the free ionic calcium in urine, and therefore the calcium oxalate supersaturation available for crystal nucleation and growth (9). Citrate reduced the growth rate of preformed calcium oxalate monohydrate seed crystals (10), and could therefore inhibit the growth of small crystals in urine into clinically significant stones. In principle, to the extent that low urine citrate is a risk factor for calcium oxalate stones, low urine citrate should be more characteristic of patients who form stones than of normal people.

We report, here, urine citrate levels in a large group of men and women with and without calcium oxalate stones, and compare citrate levels to those of other materials important in stone disease such as calcium, oxalate, and uric acid. We find that low urine citrate is present in both men and women with calcium oxalate stones, and multivariate examination of the data identifies low citrate as a leading difference between stone formers and normals. In addition, we find a gender difference, in that citrate levels in urine of women with stones are more consistently low compared to normal women than urine citrate of men with stones compared to normal men.

METHODS

330 patients, 95 of them women, were compared to 60 normal people, 27 of them women. The patients formed calcium oxalate stones. All were studied as outpatients. 24 hour urines were collected while eating a free choice diet, using thymol as a preservative; in general patients each

collected three urines, normals one or two. Mean values of all measurements for all collections in each subject were analysed, so each person is represented once in group averages. Blood samples were drawn at the end of each urine collection. All subjects were normocalcemic, and had normal serum creatinine levels. No patient had stones due to renal tubular acidosis, intestinal disease, or a systemic disease.

Citrate was measured by the lyase method (11). Calcium and magnesium were by atomic absorption, creatinine and phosphorus by autoanalyser, uric acid by the uricase method (12), oxalate by zinc reduction (13), supersaturation by the concentration product method (14), and sodium and potassium by flame photometry.

Ionic concentrations and formation of salts in urine were calculated using an iterative computer program (15) that considers 62 simultaneous interactions between urine ions.

RESULTS

Gender strongly affected the differences between patients and normals (Table 1). Among women, patients excreted less citrate, more calcium, more oxalate and more uric acid than normals, and their urine was more supersaturated with respect to calcium oxalate monohydrate as indicated by a higher value of the CPR. Among men with stones, calcium, and oxalate excretion were above normal, but citrate and uric acid excretions, and CPR all were normal. Absolute excretion rates of men were higher than women for calcium, oxalate and uric acid, but because urine volumes of men patients were higher than those of normal men or women, or women with stones, supersaturation with respect to calcium oxalate was not increased.

TABLE 1: SELECTED URINARY MEASUREMENTS

MEASUREMENT	MEN NORMAL	STONES	WOMEN NORMAL	STONES
Volume mL/24 hr	1293±84	1626±40**,#	1372±136	1370±72
Weight Kg	77±2	81±1	62±2	66±2*
Calcium mg/24 hr	181±14	254±7***	125±12	206±9***
Oxalate mg/24 hr	33±2	39±1**	25±1	29±1*
Citrate mg/24 hr	547±31	516±15	729±47	551±24**
Uric acid mg/24 hr	699±33	740±11	520±21	557±13*

*, differs from normal, same sex, $p<.05$; **, $p<.02$; ***, $p<.002$. All values are mean ±SEM; #, differs from women, same group, $p<.02$. Excretion rates of calcium, oxalate, uric acid, sodium, magnesium, and phosphorus by men exceeded those of corresponding women, $p<.01$ for all comparisons. Urine calcium and citrate excretion of women and men patients exceeded normal ($p<.01$) when expressed per kg body weight, gram creatinine, or liter of creatinine clearance (not shown).

Multivariate discriminant analysis was used to determine which of these abnormalities were independent, and which were merely statistical covariates; in other words, which abnormalities contributed independently to distinguishing between stone formers and normals. For both sexes, urine calcium and citrate concentrations were independent predictors of the stone forming state; CPR in women, but not men, was an additional predictor.

The principal physical chemical consequence of the high calcium excretion and lower citrate excretion in women and men with stones is to increase the ionic calcium in urine (Table 2). As a consequence of the increased urine calcium ion concentration, supersaturation with respect to calcium oxalate and brushite, indicated by their relative supersaturation (RSR) - the ratios of the concentration of dissolved calcium oxalate or brushite in urine to their respective solubilities - is higher in all men, and women with stones than in normal women.

The effects of citrate alone (Table 3) are to lower ionic calcium and supersaturation at any level of total urine calcium. However, in the women, whom we have chosen for illustration, urine citrate must by increased by three fold to lower urine calcium ion and RSR values to normal levels, given the hypercalciuria of patients.

TABLE 2: CALCULATED URINE IONS

MEASUREMENT	MEN NORMAL	STONES	WOMEN NORMAL	STONES
$[Ca^{++}]$ mM/L	1.71	1.83	0.866	1.827
[citrate] mM/L	0.439	0.391	0.727	0.404
[Ca Citrate] mM/L	1.103	1.114	1.17	1.27
$[Ox^{--}]$ mM/L	0.109	0.099]	0.108	0.093
[Ca Ox] uM/L	43.3	44.4	25.3	43.8
RSR, CaOx	7.03	7.20	4.10	7.11
RSR, Brushite	2.00	2.03	0.70	1.62

RSR, ratio of salt in solution to the solubility of salt.

TABLE 3: EFFECTS OF INCREASING OR DECREASING CITRATE IN WOMEN

CITRATE mM/L	NORMALS $[Ca^{++}]$	RSR(caox)	RSR(br)	STONES $[Ca^{++}]$	RSR(caox)	RSR(br)
0	1.61	6.0	1.27	2.64	8.62	2.29
2.87*	1.016	4.59	.818	1.827	7.11	1.62
3.80$	.866	4.10	.70	1.60	6.56	1.43
5	-	-	-	1.34	5.84	1.20
7.5	-	-	-	.943	4.49	.84

*, actual value for stone patients; $, actual value for normal women

DISCUSSION

Our results indicate that reduced urine citrate excretion is not a consistent feature of men with calcium oxalate stones. Men with stones excrete as much citrate as normal men. However, to the extent that a major role of citrate is as a ligand for calcium, a normal urine citrate excretion in men with stones, who are hypercalciuric, is relatively low for the calcium levels in the urine. The result is an increase in urine calcium ion level, and supersaturation that is proportional to the increase in calcium concentration. Men with stones have higher urine volumes than normal men, perhaps because of habitual water drinking, so their hypercalciuria does not express itself in proportional increases in urine calcium concentration.

Women with stones are also hypercalciuric, but have no greater urine volume than normal women; their urine calcium concentration is increased in proportion to their hypercalciuria. They also excrete less citrate than normal women, and have urine free calcium ion levels equal to those of men, and corresponding degrees of supersaturation. If they excreted as much citrate as normal women, their hypercalciuria would still cause considerable supersaturation; they would need a threefold increase of citrate to offset their hypercalciuria.

Normal women are the people who stand out from the rest. They have lower urine calcium excretion than men, and women with stones, and higher citrate excretion rates than women with stones; their urine is much less supersaturated. It is reasonable that women are, as a group, not as prone to stone disease as men.

Our results agree well with those of Hodgkinson (2), who found almost exactly the same relationships between normals and stone forming patients as we, with respect to urine citrate excretion (Table, 4). Our findings do not agree with Welshman et al (5), or Schwille (6), in finding no significant reduction of urine citrate in men with stones; we all agree in finding a reduction in women with stones, and that women excrete more citrate than men.

TABLE 4: SUMMARY OF PREVIOUS STUDIES

REF	MEN NORMAL	STONES	P/N	WOMEN NORMAL	STONES	P/N
1*@	96+37(12)	372+52(5)	.75	495+28(2)	527+142(2)	1.06
2#h	556±53(14)	421±39(26)	.75	716±58(15)	457±83(7)	.63
3#o	619±35(60)	599±34(60)	.97	653±71(16)	-	
4o	528±52(30)	440±46(44)	.83	706±91(17)	-	
5$h	370±21(59)	236±17(117)	.63	552±38(49)	221+33(47)	.40
6$h	429(21)	306(25)	.71	611(18)	338(20)	.55
7#@	481+30(15)	304+43(15)	.63	-	-	
p#o	547±31(33)	516±15(235)	.94	729±47(27)	551±24(95)	.76

*, Colorimetric Method; #, alkaline pyridine; $ citrate lyase; @, fixed diets; h, hospital diet; o, outpatients eating free choice diet; P/N, ratio of patient to normal citrate excretion rate; p, present study.

REFERENCES

1. Conway, N.S., Maitland, A.I.L., Rennie, J.B.: The urinary citrate excretion in patients with renal calculi. Br J Urol 21:30-38, 1949.
2. Hodgkinson, A.: Citric acid excretion in normal adults and in patients with renal calculus. Clin Sci 23:203-212, 1962.
3. Robertson, W.G., Peacock, M., Nordin, B.E.C.: Activity products in stone-forming and non-stone-forming urine. Clin Sci 34:579-594, 1968.
4. Elliot, J.S., Ribeiro, M.E.: The urinary excretion of citric, hippuric, and lactic acid in normal adults and in patients with calcium oxalate urinary calculus disease. Invest Urol 10:102-106, 1972.
5. Welshman, S.G. and McGeown, M.G.: Urinary citrate excretion in stone-formers and normal controls. Br J Urol 48:7-11,1976.
6. Schwille, P.O., Scholz, D., Paulus, M., Engelhardt, W., Sigel, A.: Citrate in daily and fasting urine; results of controls, patients with recurrent idiopathic calcium urolithiasis, and primary hyperparathyroidism. Invest Urol 16:457-462, 1979.
7. Rudman, D., Kutner, M.H., Redd, S.C.II, Waters, W.C.IV, Gerron, G.G., Beier, J.: Hypocitraturia in calcium nephrolithiasis. J Clin Endo & Metab 55:1052-1057, 1982.
8. Walser, M.: Dissociation constants for complexes of citrate with sodium, potassium, calcium and magnesium ions. J Phys Chem 65:159-169, 1961.
9. Weber, D.V., Coe, F.L., Parks, J.H., and Tembe, V.: Urinary saturation measurements in calcium nephrolithiasis. Ann Int Med 90:180-184, 1979.
10. Meyer, J.L., Smith L.H.: Growth of calcium oxalate crystals II. Inhibition by natural urinary crystal growth inhibitors. Invest Urol 13:36-39, 1975.
11. Moellering H., Gruber W.: Determination of of citrate with citrate lyase. Anal Biochem 17:369-376, 1966.
12. Coe, F.L.: Uric acid and calcium oxalate nephrolithiasis. Kid Int 24:392-403, 1983.
13. Hodgkinson A., Williams A.: An improved colorimetric procedure for urine oxalate. Clin Chim Acta 36:127-132, 1972.
14. Pak, C.Y.C., Holt, K.: Nucleation and growth of brushite and calcium oxalate in urine of stone-formers. Metab 25:665-673, 1976.
15 Finlayson, B.: Calcium stones: some physical and clinical aspects in calcium metabolism in renal failure and nephrolithiasis. D.S. Davis, ed., Wiley and Sons, New York, Ch 10, p 337-382, 1977.

INFLUENCE OF OXALATE ON CALCIUM OXALATE CRYSTALS FORMATION IN URINE

M. Olmer*, Y. Berland*, M. Grandvuillemin*, L. Madsen**, and R. Boistelle**

*Service de Nephrologie (Pr M. Olmer)
Hopital de la Conception, 144, rue st-Pierre, 13005 Marseille
**CRMC2, Campus Luminy, Case 913, 13228 Marseille Cedex 9

Among the different mechanisms involved in calcium oxalate nephrolithiasis, idiopathic hypercalciuria is well documented and is of interest for therapy (1.2.3.4.). The role of hyperoxaluria in the genesis of these stones is suggested in some studies (5.6).

In this report we provide both experimental and clinical studies evaluating the significance of oxaluria in calcium oxalate stone.

METHODS

Experimental Study

Urine from several subjects without history of nephrolithiasis or urinary tract infection were collected before lunch time. In each sample, the pH (pH meter ISIS 20 000), calcium concentration, oxalate concentration (gas chromatography) (7), citrate (enzymatic method) sulfate, ammonia, magnesium, phosphorus (spectrophotometer), sodium, potassium, and chloride (Technicon autoanalyser).

Immediately after collection, urine samples maintained at 37°C were centrifuged at 3500 revolutions/minute during 10 minutes, then passed through a millipore filter 0.20 µm. Two samples of 25 ml of urine were used. To sample A, calcium chloride ($CaCl_2 - H_2O$) 1.10^{-1} moles was added to increase calcium concentration. In sample B oxalate concentration was increased by adding oxalic acid ($H_2C_2O_4 - 2H_2O$) 1.10^{-2} moles. Mixing of these 2 samples produced a sample C whose pH maintained between 5.5 and 6.5. In this range of pH, the rate of crystalline growth and the solubility of calcium oxalate monohydrate are hardly influenced by the pH value (8).

Solution C was kept at 37°C, without shaking and was observed in light microscopy several times during the first twenty-four hours, then daily during the ten following days in order to note the appearance of calcium oxalate crystals and to follow their progression. Crystals were identified by their habit, light properties and by an x-ray study. The two phases observed were always calcium oxalate, monohydrate or dihydrate.

The initial supersaturation β of calcium oxalate is defined as following:

$\beta = a\ (Ca^{++})\ .a(Ox^{--})\ /\ Ksp.$, ratio of activity product of calcium and oxalate and the thermodynamic solubility product of calcium oxalate monohydrate. For simplification, supersaturation has always been calculated with respect to this only stable phase and by using a program taking into account all the complexes present in urine, i.e. 50 complexes and 12 constituents. Ionic activity determination has been carried out by an interactive method (9). If $\beta > 1$, the crystalline growth or a new crystal formation. When $\beta=1$, the solution is saturated and for $\beta<1$, the solution is undersaturated.

Clinical Study

66 recurrent calcium stone formers, 47 males and 19 females, mean age 39 ± 11 years were included in the study. For two days they were fed a 1 g calcium, 120 mg oxalate and 6 g sodium chloride diet per day. With this regimen all the patients presented an hypercalcuria (24 hr calciuria > 0.1 mmol/kg). In the subsequent four days they ingested a low calcium diet, 400 mg/day and a low oxalate diet, 40to 70 mg/day. Sodium chloride regimen was 6g/day. After each period the different components of urine were determined and calcium oxalate supersaturation calculated.

RESULTS

Experimental Study

The range of formation of calcium oxalate crystals can be defined by the supersaturation β as a function of oxalate concentration(Fig 1). The space is divided into 2 areas. Calcium oxalate monohydrate or dihydrate crystals form above the curve and crystal formation is all the more favoured as we get away from the curve. Below the curve, no crystallization appears either because urine is undersaturated ($\beta < 1$) or despite being supersaturated ($\beta <1$) urine remains in a metastable state. It seems that for a given oxalate concentration corresponds a minimal supersaturation β necessary to form crystals. When oxalate concentration increases, the supersaturation required for crystallization decreases. A critical oxalate concentration exist ~0.16 mmol/l above which no calcium oxalate formation is observed even if the supersaturation is important.

No simple relation allows to link supersaturation to calcium concentration. Low calcium concentration may coexist with crystal production whereas high concentration may not. In fact, the crystalline growth depends on urinary oxalate concentration.

Clinical Study

Table 1 shows mean values of calciuria, oxaluria, calcium and oxalate concentrations of the 66 related hypercalciuric patients. When calcium oxalate-urinary supersaturation of these patients is evaluated by referral to the diagram obtained in vitro, linking β and oxalate concentration , we observe that the urine of 29 patients (group I) is in a metastable phase; 37 patients (group II) have urine in zone of crystalline formation. The calciuria is equivalent for both groups; in contrast group I presents a significantly lower oxaluria ($p <0.05$) and an oxalate concentration inferior to the critical concentration defined in vitro (Table 2).

After the period of oxalate and calcium restriction diet, calciuria decreased significantly in both groups I and II, to 6.30 ± 2.6 mmol/24 h ($p < 0.001$) and 6.10 ± 2.2 mmol/24 h ($p < 0.001$) respectively. However, variation of the supersaturation β allows a division of each group in 2 populations. In group I, 21 patients (group IA) have urine that remains in the metastable zone (Fig.2a). The oxalate concentration of these patients remains below the critical concentration. Eight patients (group IB) have a supersaturation that increases and migrates into the homogeneous nucleation zone (Fig. 2b). The mean urinary oxalate concentration and the 24-hour oxaluria increased to 0.11 ± 0.04 mmol/1 and 0.21 ± 0.11 mmol/24 h ($p < 0.05$) respectively, during the first diet period compared to 0.33 ± 0.20 mmol/1)p <0.05) and 0.67 ± 0.47 mmol/24 h ($p < 0.05$) after the calcium restricted regimen. The 24-hour diuresis is equivalent in this group between the two periods (Table 3). In group II, 19 patients (group IIC) have urine that remains with an elevated supersaturation (Fig. 2C). During the two diet periods of the study these patients had an elevated oxaluria 0.58 ± 0.44 mmol/24 h, then 0.51 ± 0.23mmol/24 h and an oxalate concentration greater than the critical concentration 0.37 ± 0.26 mmol/1, then 0.32 ± 0.15 mmol/1 (Table 3). In contrast 18 patients (group IID) have urine that becomes metastable (Fig 2d). A non significant decrease of oxaluria was noted between the two dieting periods. Diuresis also increased non significantly. A joint variation of these two parameters was responsible for a significant decrease of oxalate concentration 0.22 ± 0.08 mmol/1, then 0.15 ± 0.04 mmol/1 ($p < 0.001$) (Table 3).

Discussion

An elevated supersaturation is the prinicipal factor required for the formation of large crystals in a solution such as urine. We have established, in vitro, the range of supersaturation necessary to obtain calcium oxalate crystals and have shown that crystal formation is closely linked to the urinary oxalate concentration. The predominate role of oxalate is due to the presence in urine of a higher calcium concentration than oxalate concentration. In solution, the most important complex between calcium and oxalate is CaC_2O_4; this complex corresponds to the following thermodynamic equilibrium:

$$\frac{[\ Ca\ C_2O_4\ ]}{a(Ca^{2+}).\ a(C_2O_4^{2-})} = Kcom$$

Any increase of calcium concentration is followed by an equivalent decrease of ionized oxalate by the formation of soluble complexes. Thus calcium oxalate supersaturation rises slightly. In contrast, when oxalate concentration increases, ionized oxalate rises but ionized calcium varies minimally. Therefore calcium oxalate supersaturation increases by a factor corresponding to the increase of oxalate concentration.

In the presence of a weak oxalate concentration, the calcium oxalate supersaturation decreases rapidly to values which do not allow a homogeneous nucleation. In the range of weak oxalate concentrations, an elevation of calcium concentration is followed by a slight increase of the calcium oxalate supersaturation for the same reasons considered above. In contrast, amorphous phosphate supersaturation increases rapidly. Thus, the frequency of the phosphate nucleation becomes greater than that of calcium oxalate. This explains why the metastable zone of calcium oxalate extends when oxalate concentration decreases.

Table 1: Mean values of calciuria, oxaluria, urinary calcium and oxalate concentrations of 66 hypercalciuric stone formers.

24 HOUR URINE		URINE CONCENTRATION	
CALCIUM (mM)	OXALATE (mM)	CALCIUM (mM)	OXALATE (mM)
10.6 ± 3.4	0.39 ± 0.33	6.1 ± 2.5	0.22 ± 0.18

66 HYPERCALCIURIC PATIENTS

Table 2: Mean values of calciuria, oxaluria, urinary oxalate concentration and supersaturation in group I and group II patients.

PATIENTS	CALCIUM (mmol|24h)	OXALATE (mmol|24h)	URINARY OXALATE CONCENTRATION (mM)	SUPERSATURATION β
GROUP I n=29	10.80 ± 4.30	0.24 ± 0.16	0.11 ± 0.04	8.90 ± 3.10
GROUP II n=37	10.50 ± 2.40	0.50 ± 0.39	0.30 ± 0.21	28.6 ± 20.8
STUDENT t TEST	N.S.	$p < 0.05$	$p < 0.05$	$p < 0.001$

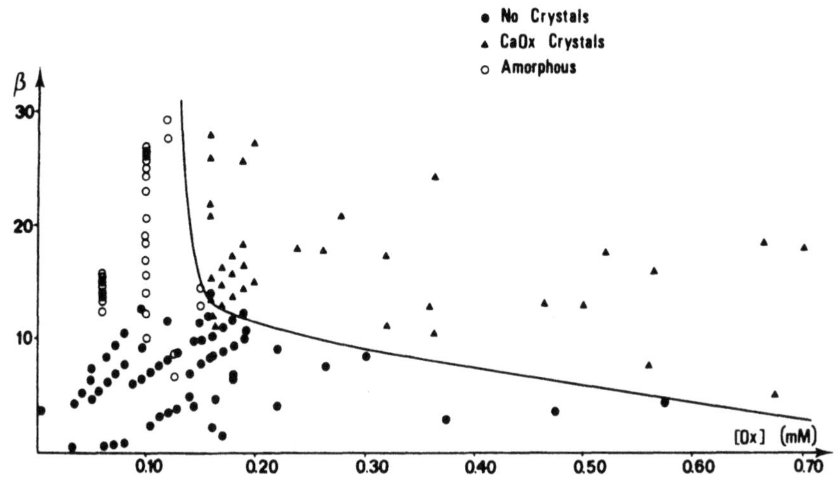

Figure 1: The nucleation fields of calcium oxalate in urine as a function of oxalate concentration and supersaturation β .

Table 3: Mean values of calciuria, oxaluria, urinary calcium and oxalate concentration, urine volume before (1) and after (2) calcium and oxalate restricted diet, in different subgroups IA, IB, IIC, IID.

		CALCIUM mM|24h	OXALATE mM|24h	[Ca] mM	[Ox] mM	URINE VOLUME(1)
GROUP I	A1	10.95 ± 4.80	0.26 ± 0.17	4.70 ± 1.70	0.11 ± 0.04	2.40 ± 0.77
	A2	*** 6.40 ± 2.70	0.31 ± 0.14	*** 2.90 ± 1.50	0.12 ± 0.04	2.40 ± 0.78
	B1	10.50 ± 2.90	0.21 ± 0.11	6.30 ± 2.30	0.11 ± 0.04	1.76 ± 0.42
	B2	*** 5.95 ± 2.40	* 0.67 ± 0.47	*** 3.50 ± 1.70	* 0.33 ± 0.20	1.88 ± 0.57
GROUP II	C1	10.00 ± 2.30	0.58 ± 0.44	6.95 ± 1.80	0.37 ± 0.26	1.51 ± 0.45
	C2	*** 6.50 ± 2.50	0.51 ± 0.23	*** 4.10 ± 1.40	0.32 ± 0.15	1.66 ± 0.60
	D1	10.90 ± 2.50	0.42 ± 0.32	6.80 ± 3.30	0.22 ± 0.08	1.90 ± 0.84
	D2	*** 5.70 ± 1.80	0.32 ± 0.12	*** 2.70 ± 0.90	*** 0.15 ± 0.04	2.26 ± 0.58

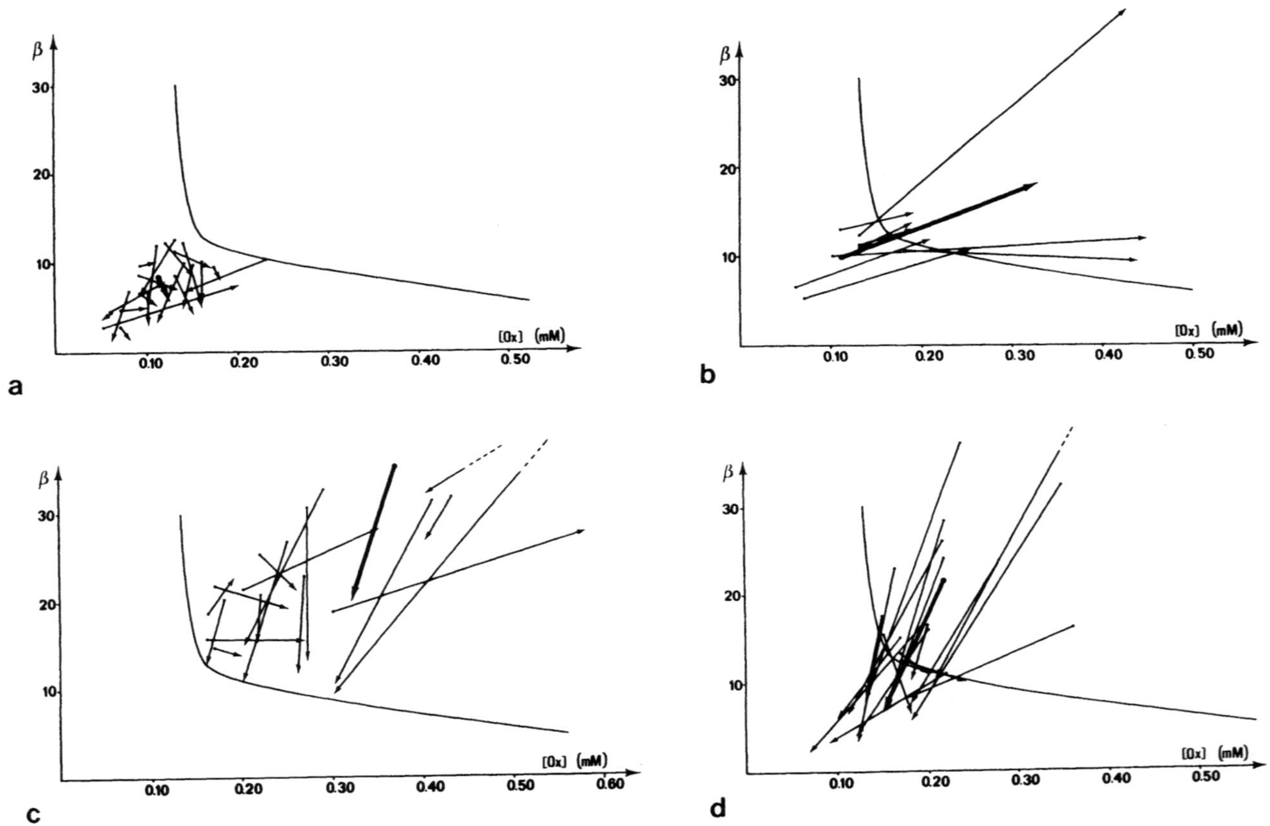

Figure 2: Evolution of the supersaturation β in the 66 hypercalcuric patients after calcium and oxalate restricted diet. Four subgroups can be individualized Ia(a), IB(b), IIC(c), IID(d). The arrows join supersaturation (β) before and after restricted diet. The thick arrow represent a mean vector.

When the calcium oxalate lithogen risk is evaluated with respect to urinary supersaturation for the 66 hypercalciuric patients, we find two populations that cannot be distinguished by their calciuria. The first one corresponds to patients with a high lithogen risk as their urine has an elevated supersaturation favouring a homogeneous nucleation. These patients have an increased oxaluria. The second population comprises low lithogen risk patients. Their urine are metastable so that, only a heterogeneous nucleation may eventually appear. These patients have a low oxaluria.

When the two populations followed a calcium and oxalate restricted diet, an equivalent decrease of calciuria was noted for both. However, urinary calcium oxalate supersaturation progressed in a variable way and was closely linked to the progression of oxaluria. Thus, in patients of the group IB hyperoxaluria observed after low calcium diet may be related to insufficiently oxalate restricted diet responsible for an increase of oxalate intestinal absorption (10,11) taking into account the absence of urine volume variation (12). Patients from group IIC conserved an elevated oxaluria after calcium restriction. These high lithogen risk patients might have an impairment of the transmembrane transport of oxalate in intestinal as well as in renal cells (13). Group IID has a decrease of the supersaturation which might be related to the fall of oxaluria. An impairment of intestinal cellular transport of oxalate may be suspected in these patients. They respond satisfactorily to oxalate restriction.

Conclusion

In urine, calcium oxalate supersaturation β is closely related to oxalate concentration. Supersaturation better than calcium or oxalate concentration is helpful to predict the risk of recurrent calcium oxalate lithiasis. In nephrolithiasis patients with hypercalciuria, low calcium diet induces different oxaluria pathways responsible for a not univocal therapy. Each individualized group may be a guide for pre-venting recurrent lithiasis.

REFERENCES

1. Coe, F.L: Treated and untreated recurrent calcium nephrolithiasis in patients with idiopathic hypercalciuria, hyperuricouria in metabolic disorder. Ann. Int. Med. 87:404 (1977).

2. Pak, C.Y.C.; Kaplan, R.; Bone, H. , Townsend, J.; Waters, O.: A simple test for diagnosis of absorption and renal hypercalciurias. N. Engl. J. Med. 292:497 (1975).

3. Pak, C.Y.C: Kidney stones: Various forms and treatment. Nephron. 23:142 (1979).

4. Yendt, Cohanim, E.R.: Prevention of calcium stones with thiazides, Kidney Int. 13:397-409, (1978).

5. Robertson, W.G.; Peacock, M.; Nordin, B.E.C.: Activity products in stone-forming and non-stone-forming urine. Clin. Sci. 34:579 (1968).

6. Robertson, W.G.; Scurr, O.S.; Bridge, C.M.: Factors influencing the crystallisation of calcium oxalate in urine-critique. J. Crystal Growth. 53:182 (1981).

7. Calaf, R.; Marie, P.; Ayavou, T.; Reynaud, J.: Etude critique du dosage de l'acide oxalique urinaire par chromatographie en phase gazeuse. Annals Bio. Clin. 40:673 (1982).

8. Meyer, J.L.: Nucleation kinetics in the calcium oxalate sodium urate monohydrate system. Invest. Urol. 19:197 (1981).

9. Lundager Madsen, H.D.: Aspects physicochimiques de la lithiase urinaire. Nephrologie. 5:151 (1984).

10. Hodgkinson, A.: Evidence of increased oxalate absorption in patients with calcium-containing renal stones. Clin. Sci. and Mol. Med. 54:291 (1978).

11. Marangella, M.; Fruttero, B.; Bruno, M.; Linari, F.: Hyper oxaluria in idiopathic calcium stone disease, further evidence of intestinal hyperabsorption of oxalate. Clin. Sci. 63:381 (1982).

12. Zarembsk, P.M.; Hodgkinson, A.: Some factors influencing the urinary excretion of oxalic acid in man. Clinica Chimica. Acta. 25:1 (1969).

13. Baggio, B., Gambaro, G.; Marchini, F.; Cicerello, E.; Borsatti, A: Raised transmembrane oxalate flux in red blood cells in idiopathic calcium oxalate nephrolithiasis. The Lancet. i:12 (1984).

THERAPEUTIC MODALITIES FOR THE PREVENTION OF NEPHROLITHIASIS

Wolfgang Tschöpe, Reinhold Horsch, and Eberhard Ritz

Department Internal Medicine and Urology
University of Heidelberg, Germany, FRG

For a critical analysis of recurrent calcium nephrolithiasis, it is useful to view the problem in broader phylogenetic perspective. Ridding the organism of Ca, Pi and oxalate is a formidable task. Lower species, e.g. bryozoa or tunicates pack their excreta, including calcium, into excretory cells of the cuticule or the gastrointestinal tract from which they are shed into the ambient sea water (1). In insects, calcium is normally excreted in the form of microconcrements via the Malpighian tubules, i.e. their aglomerular excretory ducts. The work of Wessing et al. (2) demonstrated that Malpighian tubule fluid contains numerous small microliths, 5 µm across, with concentric lamellae containing Ca, Pi or oxalate within a mucopolysaccharide matrix. Such phylogenetic examples illustrate the difficulty of excreting Ca and oxalate without incurring the risk of nephrolithiasis. Viewed against this background it is remarkable that mammalians acquired the ability to excrete supersaturated urine without formation of concrements. Indeed it is surprising that nephrolithiasis, i.e. breakdown of protective systems against crystal growth in the mammalian excretory ducts, does not occur more frequently.

Protein intake and Ca nephrolithiasis

Much can be learned on the causes of the recent stone epidemic and on the consequent therapeutic strategies to follow by analyzing the problem in historical perspective. Nephrolithiasis was well recognized when the first autopsies were performed at the time of Morgagni (3). In addition, Thomas Willis in his "liber de urinis" recognized earth alkali as a component of renal stones, implicating Ca and/or Mg in the genesis of nephrolithiasis (4). Thus, it was not lack of expertise that explains the failure to observe calcium nephrolithiasis in non-obstructed non-infected kidneys until the beginning of the 19th century when well kept records of London Hospitals demonstrate the appearance of calcium nephrolithiasis (5). The conclusion that such de novo appearance must have been due to affluence and altered dietary habits is supported by the experience of the St. Georg Hospital in Leipzig (6) where autopsy incidence of nephrolithiasis rose continuously from the 2nd to the 6th decade of the 20th century only to decline markedly during the two world wars (7). Robertson and others suggested a correlation between high consumption of animal protein within a given population on the one hand and high incidence of nephrolithiasis on the other hand (8). This raises the obvious question whether individuals who suffer from recurrent calcium nephrolithiasis consume even more protein, in specific animal protein, than non-stone forming individuals. In a previous study (table 1) we found no difference of body weight, urinary sodium or urea excretion (as indices of dietary intake) between normal individuals and untreated recurrent calcium stone formers.

Table 1 Dietary indices in recurrent Ca stone formers (ref.8)

	Normal population (n=238)	stone formers (n=84)	
Weight (kg)	74 (55-117)	74 (61-107)	
Cholesterol (mmol/l)	4.97 (2.79-9.66)	5.46 (2.62-8.52)	p 0.05
UV_{Na} (mmol/24 h)	201 (55-570)	216 (56-460)	
UV_{urea} (mmol/24 h)	423 (127-785)	435 (135-755)	

Similarly, more recent data of Power (9) and of our group (to be published) show similar daily intake of energy and protein, more specifically animal protein, in stone formers and in matched controls (table 2).

Table 2 Daily intake in recurrent Ca stone formers (ref. 9)

	Cases	Controls
Energy (kJ)	11833 ± 2900	11374 ± 2527
Protein (g)	88 ± 18	85 ± 18
Animal protein (g)	55 ± 15	53 ± 13

We must therefore conclude that stone formers do not differ from non stone formers by higher protein intake but by intrinsic susceptibility to stone formation. It may still be valid to advise them to restrict animal protein. Unfortunately, this obvious conclusion from historical and epidemiological studies has never been put to controlled practical tests.

Water intake and nephrolithiasis

Epidemiological studies show an influence of fluid intake on nephrolithiasis; for instance stone prevalence is higher in professions with notoriously lower water intake, e.g. airline pilots and conductors; in populations with low as opposed to high fluid intake, e.g. in Ashkenazim vs Sephardim in Israel; and stone incidence is higher in the hot season. The latter observation may also be due, however, to more intensive sun exposure, since a correlation exists between vitamin D status and calciuria in the general population (8) as well as in recurrent calcium stone formers. In recurrent stone formers we found a correlation between UV_{Ca} and 25(OH) D_3 levels of r=0.3 (to be published) in good agreement with findings of Berlin et al. (10).

In view of the above recurrent calcium stone formers are widely given advice to increase fluid intake, but scientific evidence for the efficacy of this measure is scarce. This may be due, in part, to unsatisfactory patient compliance. In one study (11) it was found that patients, given advice to increase urine output, when followed over 4 years increased urine volume by no more than +10%; no significant effect on stone recurrence could be demonstrated. In a more recent study, Ulman et al. (12) randomly allocated 38 recurrent calcium stone formers to either water diuresis or 50 mg hydrochlorothiazide and found a similar decrease of stone recurrence rate with both modalities. Although a placebo group was not included, this is the best indirect evidence available that suggests that water diuresis is indeed effective. The beneficial action of water diuresis on lithogenesis has even been questioned, since diuresis may diminish both inhibitory activity and lithogenic ion concentrations; however, more recent studies of Pak (13) indeed suggest a beneficial action on activity product ratios. Obviously, diuresis may also be efficacious by alternative actions e.g. via diminished particle retention time etc.

Drug intervention in nephrolithiasis

Magnesium, orthophosphate, citrate, allopurinol and thiazides have been widely used in the treatment of recurrent calcium nephrolithiasis.

With respect to magnesium salts, the Uppsala group (14) showed that administration of magnesium hydroxide increases urinary magnesium and citrate excretion rates and reduces stone recurrence. However, in later controlled trials, despite even higher doses of magnesium hydroxyde, no effect was observed by Ettinger (15) and by Wilson (16). So the issue remains open.

With respect to phosphate, on the basis of uncontrolled observations, Thomas (17) and Smith (18) emphasized the safety and efficacy of orthophosphate treatment. Therapeutic "success" was reported in 91% of 150 patients with idiopathic calcium nephrolithiasis (18). A beneficial effect was confirmed in two more recent controlled studies. Both Wilson (19) and Robertson (20) found a significant reduction of stone recurrence under treatment with orthophosphate. While orthophosphate is undoubtedly efficacious, and while systemic metabolic side effects, e.g. parathyroid stimulation, are negligible or non-existent, local complications in the urinary tract may occur in the presence of urinary tract infection and/or alkaline pH. This reduces the clinical safety of this treatment modality which has not become very popular in Europe.

Ever since the seminal paper of Lamberg in 1958 documenting the hypocalciuric action of thiazides (21) and the subsequent report of Yendt and Cohanim documenting reduced calcium stone formation (22), thiazides have generated considerable research effort. All extant therapeutic trials have been criticised by Churchill who argued with some justification that failure to randomly allocate patients in most trials failed to consider statistical regression to the mean and that cointervention, particularly diet and fluid therapy, influenced outcome (23). Certainly, non-random control groups, as in some studies (24) do not protect against selection factors.

Some small controlled studies suggested no effect of thiazides but statistical power (beta-error) was clearly inadequate. Brocks et al. (24) found 0.09 stone recurrences/patient year vs 0.4 before treatment in patients on thiazides (n=33), while the respective figures were 0.11 vs 0.7 in placebo treated patients (n=9).

On the other hand, the groups from Chicago (25), Toronto (26), Uppsala (2/) and Heidelberg (28) found that more than 85% of patients on thiazides remained free of recurrence after 3-5 years. Preliminary results of adequately large controlled trials (20) also suggest the efficacy of thiazides.

Despite some lingering biostatistical doubts one can conclude that thiazides are effective. However, difficult questions remain relating to risk/benefit considerations of intervention with thiazides or orthophosphate.

Longterm treatment with thiazides is not without potentially hazardous longterm side effects. The role of hypokalemia has been suggested in the genesis of polymorphous ventricular arrhythmia (29), electrical instability after myocardial infarction (30) and elevated blood pressure (31). The risk of hypokalemia may conceivable be diminished by the concomitant use of potassium-sparing diuretics with thiazides. According to the study of Costanzo (32) amiloride causes additional increment of tubular calcium reabsorption while diminishing urinary potassium excretion. Since triamteren, a water-soluble pteridin derivative, may cause stones on its own (32) it should be used with circumspection, or not at all, as a potassium-sparing diuretic in stone formers.

Thiazides may also cause hyperlipemia and, more embarrassingly, impaired glucose tolerance (34). Such glucose intolerance will, even in the absence of frank chemical diabetes, cause macroangiopathic complications of similar magnitude as frank chemical diabetes mellitus. Finally, in the MRC trial, 18% of male patients became impotent under thiazides (35). On the other hand, stone recurrence has recently lost some of its previous drama with the present availability of lithotripsy. Table 3 summarizes our experience in Heidelberg, covering 500 treatment sessions in 452 patients with X-ray-positive stones. 70% of the patients had calcium stones.

Table 3 Experience with lithotripsy in Heidelberg (31.7.85)

510 sessions in	452 patients (100%)
necessary interventions:	
auxilliary measures	97 patients (22%)
surgery	7 patients (2%)
Complications:	
colic	25%
fever (without preexistent UTI)	15% of patients
(with preexistent UTI)	30% of patients

The procedure was very well tolerated indeed. Auxiliary measures e.g. ureteral catheterism, sling or percutaneous nephrostomy, were necessary in no more than 22% of patients and subsequent surgery in a total of 2.5%. Complications included fever, predominantly when preexistent urinary tract infection was present, although we saw no single case of urosepsis. Ureteral obstruction from sludge ("Steinstraßen") has remarkably decreased in incidence after stones in excess of 2.5 cm across are no longer subjected to lithotripsy.

Three months after intervention, over 85% of patients were free of residual stones by X-ray. It must be admitted that longterm studies are necessary to prove that small nuclei are not left within the pelvis which might serve as a nidus and increase the risk of subsequent recurrence.

Given the availability of such non-invasive procedure to adequately deal with stone recurrence, it will be imperative in the future to carefully balance risk and benefit of thiazides (or orthophosphate) for longterm prophylaxis against repetitive lithotripsy. While controlled information is not present available, the majority of patients in our experience presently prefer eventual lithotripsy to inconvenience and potential hazards of prolonged drug therapy. The introduction of this revolutionary new technique certainly forces us to review established procedures of the past, but only time and carefully conducted studies will give definite scientific information on the relative roles of drug prophylaxis versus repetitive lithotripsy.

References

1. Siewing, R., 1985, Lehrbuch der Zoologie, G. Fischer, Stuttgart

2. Wessing, A., and Eichelberg, D., 1975, Ultrastructural aspects of transport and accumulation of substances in the Malpighian tubules, in: Fortschritte der Zoologie, A. Wessing, ed., G. Fischer, Stuttgart

3. Ruysch, F., 1691, Observationum anatomico-chirurgicarum centuria, Amsterdam, p 73

4. Willis, Th., 1682, Liber de urinis, in: Opera ommnia, Amsterdam, p 152 ff

5. Andersen, D.A., 1972, Environmental factors in the aetiology of urolithiasis. Urinary Calculi. Int. Symp. Renal Stone Res., Madrid, Karger Verlag, Basel

6. Schumann, H.-J., 1963, Die Häufigkeit der Urolithiasis im Sektionsgut des Pathologisches Institutes St. Georg, Leipzig, Zbl. All. Pathol. 105: 88

7. Robertson, W.G., Peacock, M., Hodgkinson, A., 1979, Dietary changes and the incidence of urinary calculi in the U.K. between 1958 and 1976, J. Chron. Dis. 32: 469

8. Tschöpe, W., Schellenberg, B., and Ritz, E., 1980, Plasma phosphate and urinary calcium in recurrent stone formers. Effect of age, blood pressure and diet, Mineral Electrolyte Metab. 4: 237

9. Power, Ch., and Nelson, M., 1984, Diet and renal stones: a case-control study, in: Urinary Stone. Proceedings of the Second International Urinary Stone Conference, Signapore, 1983, R.L. Ryall, J.G. Brockis, V.R. Marshall, B. Finlayson eds., Churchill Livingstone, Melbourne

10. Berlin, T., Holmberg, I., Björkhem, I., 1985, Is 25-hydroxyvitamin D_3 in importance for development of hypercalciuria in renal stone formers? in: Vitamin D. A chemical, biochemical and clinical update. Walter de Gruyter & Co., Berlin, New York

11. Ljunghall, S., Backman, U., Danielson, B.G., Fellström, B., Johansson, G., and Wikström, B., 1979, Prophylactic treatment of renal calcium stones. Experiences with diatary advice, cellulose phosphate and thiazides. Scand. J. of Urol. and Nephrol. , Suppl. 32: 239

12. Ulmann, A., and Clavel, J., 1984, Is 24 h calciuria of predictive value for calcium stone recurrence? Abstracts 5th Int. Symp. Urolithiasis, Garmisch-Partenkirchen

13. Sakhaee, K., Zerwekh, J.E., Pak, C.Y.C., 1981 , Objective evidence for the beneficial effect of a high fluid intake in the management of nephrolithiasis, in: Urolithiasis. Clinical and Basic Reseach, L.H. Smith, W.G. Robertson, and B. Finlayson, eds., Plenum Press, New York and London, 1981

14. Johansson, G., Backman, U., Danielson, B.G., Fellström, B., Ljunghall, S., and Wikström, B., 1981,Prophylactic treatment with magnesium hydroxide in renal stone disease, in: Urolithiasis. Clinical and Basic Research. L. Smith, W. Robertson, and B. Finlayson, eds., Plenum Press New York and London, 1981

15. Ettinger, B., 1984, Controlled studies in stone prophylaxis, 5th Int. Symp. Urolithiasis, Garmisch-Partenkirchen

16. Wilson, D., 1984, Comparison of medical treatment for the prevention of recurrent calcium nephrolithiasis, 5th Int. Symp. Urolithiasis, Garmisch-Partenkirchen

17. Thomas, W.C. Jr., 1971, Effectiveness and mode of action of orthophosphate in patients with calcareous renal calculi, Trans. Amer. Clin. Climat. Assoc. 83: 113

18. Smith, L.H., Thomas Jr., W.C., and Arnaud, C.D., 1973, Orthophosphate therapy in calcium renal lithiasis. Urinary calculi. Int. Symp. Renal Stone Res., Madrid 1972, Karger, Basel

19. Wilson, D.R., Strauss, A.L., Manual, M.A., 1984, Comparison of medical treatments for the prevention of recurrent calcium nephrolithiasis, Urol. Res. 12: 39

20. Robertson, W.G., Peacock, M., Selby, P.L., Williams, R.E., Clark, P., Chisholm, G.D., Hargreave, T.B., Rose, M.B., Wilkinson, H., Hammonds, J.C., and Gall, W.J., 1984, A multicentre trial to evaluate three treatments for recurrent idiopathic calcium stone disease - a preliminary report. Urol. Res. 12: 48

21. Lamberg, B.A., Kulbeck, B., 1959, Effect of chlorothiazide on the excretion of calcium in urine, Scand. J. Clin.Lab. Invest. 11: 351

22. Yendt, E.R., Gagne, R.J.A., Cohanim, M.,1966, The effect of thiazides in idiopathic hypercalciuria. Am.J.Med.Sci. 251: 449

23. Churchill, D.N., 1984, Appraisal of methodology in studies of either thiazide or orthophosphate therapy for recurrent calcium urolithiasis, Urol. Res. 12: 469

24. Brocks, P., 1981, Do thiazides prevent recurrent idiopathic renal calcium stones? Lancet 2: 124

25. Coe, F.L., 1977, Treated and untreated recurrent calcium nephrolithiasis in patients with idiopathic hypercalciuria, hyperuricosuria or no metabolic disorder, Ann. Int. Med. 87: 404

26. Yendt, E.R., Cohanim, M., 1978, Prevention of calcium stones with thiazides, Kidney Int. 13: 197

27. Ljunghall, S., Backman, U., Danielson, B.G., Fellström,B., Johansson, G., and Wikström, B., 1980, Prophylactic treatment of renal calcium stones. Experiences with dietary advice, cellulose phosphate and thiazides, Scand. J. Urol. and Nephrol., Suppl. 53

28. Tschöpe, W., Ritz, E., and Knebel, L., 1981, Calciumhaushalt während Diuretika-Applikation, in: Diuretika II. Diuretika-Symposium Paris, Krück, F., and A. Schrey eds., Dr. C. Wolf und Sohn, München

29. Holland, O.B., Nixon, J.V., Kuhnert, R.N., 1981, Diuretics-induced ventricular ectopic activity, Am. J. Med. 70: 762

30. Ramsay, 1984, Diuretic use, serum potassium and ventricular fibrillation in patients with myocardial infarction, Proc. Brit. Pharmacol. Soc. (abstract)

31. Kaplan, N.M., Carnegu, A., Raskin, Ph., Heller, J.A., Simmons, M., 1985, Potassium supplementation in hypertensive patients with diuretic-induced hypokalemia, New Engl.J.Med. 312: 746

32. Costanzo, L.S.,1984, Comparison of calcium and sodium transport in early and late rat distal tubules: effect of amiloride, Am. J. Physiol. 246: F945

33. Ettinger, B., Oldryd, N.O., Sörgel, F., 1981, Renal calculi in triamterene users, in: Urolithiasis. Clinical and Basic Research, L.H. Smith, W.G. Robertson, B. Finlayson, eds., Plenum Press, New York and London, 1981

34. Murphy, M.B., Kohner, E., Lewis, P.J., Schumer, B., and Dollery, C.T., 1982, Glucose intolerance in hypertensive patients treated with diuretics; a fourteen-year follow-up, Lancet 11: 1293

35. Report of Medical Research Council Working Party on Mild to Moderate Hypertension; Adverse reactions to bendrofluazide and propranolol for the treatment of mild hypertension. Lancet 1981, II: 539

MISCELLANEOUS

FACTORS CONTROLLING THE INTRACELLULAR CONCENTRATION OF ORTHOPHOSPHATE (Pi) IN MAMMALIAN CELLS

A. Bevington, G.J. Kemp, and R.G.G. Russell

Department of Human Metabolism & Clinical Biochemistry
University of Sheffield Medical School
Beech Hill Road
Sheffield
S10 2RX
UK

INTRODUCTION

It is ironic that while considerable effort and ingenuity have been expended over the past decade in measuring the low cytoplasmic concentrations of calcium (10^{-6} to 10^{-7} mole/l) in mammalian cells, relatively little attention has been paid to measurement of intracellular orthophosphate (Pi) (10^{-2} to 10^{-3} mole/l) and the factors controlling it. While it is commonly assumed that the biochemical basis for tissue damage during phosphate deprivation is a decrease in the intracellular concentration of Pi,[1] there are a number of other important processes in which changes in intracellular concentrations of Pi have been implicated; notably regulation of anaerobic glycolysis,[2] oxidative phosphorylation[3] and basal metabolic rate;[4] and modulation of calcium-activated cellular processes such as exocytosis.[5]

In spite of this, progress in defining the role of intracellular Pi in cellular regulation has been slow, because of the problems involved in measuring it selectively in the presence of organic phosphates and extracellular Pi, and because of ignorance of the factors controlling its intracellular concentration.

In determining these controlling factors, a clear distinction must be drawn between processes controlling the intracellular concentration of Pi at steady-state (when the concentrations of Pi and organic phosphates in all intracellular compartments, and in extracellular fluid, are constant); and processes leading to transient changes in intracellular Pi (for example a net flux of Pi across the plasma membrane, or net degradation of organic phosphates to Pi).

To date, the only detailed model which has been proposed for the cytoplasmic concentration of Pi at steady-state is the passive distribution model for the simple case of the human erythrocyte,[6] in which only two compartments, extracellular fluid and cytoplasm, are present. If the molal concentrations (in moles per litre of cell water or plasma water) of a passively transported mono-anion such as chloride (Cl^-) are measured, the cytoplasmic/plasma concentration ratio r is less than unity, as the resting membrane potential of the erythrocyte (10mV inside negative) tends to repel anions from the cytoplasm.[7]

If Pi, which is a mixture of $H_2PO_4^-$ and HPO_4^{2-} ions, also distributes passively across the cell membrane, the cytoplasmic/plasma concentration ratio should be equal to r for $H_2PO_4^-$ and r^2 for HPO_4^{2-}. Hence the corresponding molal concentration ratio R, for total Pi, is given by

$$R = (r + m\,r^2)/(1 + m) \qquad (1)$$

where m is the concentration ratio $HPO_4^{2-}/H_2PO_4^-$ at extracellular pH. If R and r are expressed in more convenient molar units (moles per litre of cells or plasma) the numerator of this expression becomes $(r + (0.93/0.72)mr^2)$, as erythrocytes are 72%, and plasma 93%, water by volume.

Consequently, this simple model predicts that, at a given pH, the cytoplasmic concentration of Pi should be directly proportional to that in extracellular fluid or plasma, and should be R times this extracellular concentration.

METHODS FOR MEASUREMENT OF INTRACELLULAR Pi

Even though extraction of Pi from cell suspensions or freeze-clamped tissue with perchloric acid causes negligible hydrolysis of the more concentrated organic phosphates[8,9] (with the possible exception of phosphocreatine),[10,11] subsequent colorimetric assay of the Pi with acidified molybdate may cause more serious hydrolysis.[12] Fortunately, rapid separation of Pi from the organic phosphates, for example by extraction with organic solvents,[6,9,13] reduces this problem to negligible levels; and procedures now exist for assay of Pi even in the presence of a large excess of phosphocreatine.[12]

An important problem with these methods is that they do not distinguish between the various sub-cellular compartments, although these assays, coupled with rapid sub-cellular fractionation, have been used to study the distribution of Pi between the cytoplasmic and particulate fractions of human leucocytes and blood platelets.[14] Improvements in histological methods, using precipitation of heavy metal salts of Pi, have allowed qualitative changes in the sub-cellular distribution of Pi to be visualised in rat pancreatic islets,[15] but hydrolysis of organic phosphates during fixation of the sample could also contribute to the observed deposits of Pi.[15]

In principle, the cytoplasmic concentration of Pi can be calculated selectively from the glyceraldehyde-3-phosphate dehydrogenase/ phosphoglycerate kinase equilibrium.[16] However, because of imprecision in the equilibrium constants for these reactions, only a rough estimate can be obtained.[9]

Chemical extraction and assay of Pi can be eliminated if the tissue sample is sufficiently large (at least hundreds of mg wet weight) to be studied by the non-destructive technique of phosphorus-31 nuclear magnetic resonance (^{31}P NMR).[17-19] The Pi pool detected is thought to be mainly cytoplasmic[3,20] and, if ^{31}P NMR measurements and chemical extraction can be performed on the same sample, bound or "NMR-invisible" Pi pools, which may be sequestered in intracellular organelles, can be estimated by difference.[3,10,21] However, in perfused rat heart, intra-mitochondrial Pi may also be detectable by NMR as a partly resolved Pi resonance;[22] whereas in cat biceps, the discrepancy between NMR and freeze-clamping methods is thought to arise from hydrolysis of phosphocreatine during freezing of the tissue.[11]

INTRACELLULAR Pi AT STEADY STATE

Mammalian Erythrocytes

As these cells lack sub-cellular organelles, their total chemically extractable Pi could be a simple model for cytoplasmic Pi in multi-compartment cells. Unfortunately, early estimates of the intracellular concentration of Pi in rapidly washed human erythrocytes[6,23] were close to the concentrations predicted from the passive distribution model (equation 1); whereas estimates of cytoplasmic Pi in multi-compartment cells are considerably higher than expected from passive distribution (see below), indicating that erythrocytes may handle Pi in a way which differs fundamentally from other cells. However, if the concentration of Pi in erythrocytes is calculated from the difference between the concentrations of Pi in the whole suspension and in plasma, washing of the cells can be avoided, thus removing a possible source of artifact. Under these conditions, higher cellular concentrations of Pi have been reported.[9,24] For example, normal human erythrocytes, in autologous plasma under 95% O_2/5% CO_2 at 37°C, maintain a concentration of Pi which is 50 - 100% higher than predicted from passive distribution.[9]

^{31}P NMR spectra of mammalian erythrocytes show a Pi resonance which is incompletely resolved from one of the two resonances of 2,3-diphosphoglycerate,[25] and part of this Pi is probably extracellular. Even so, spectra of suspensions of rabbit erythrocytes and spectra of perchloric acid extracts of the same suspensions, show similar proportions of the cytoplasmic phosphorus metabolites, including Pi,[25] indicating that, in the intact cell, Pi is not bound significantly to membranes or proteins.

Multi-compartment Cells

Measurements of chemically extractable Pi in Ehrlich ascites tumour cells,[8] human blood platelets and leucocytes[14] and freeze-clamped skeletal muscle from dog[26] detected a concentration of Pi in the range 2 to 17 mmole/l when extracellular Pi was in the range 1 to 5 mmole/l. Superficially, this implies active accumulation of Pi against an electrochemical gradient (bearing in mind the tendency of the membrane potential to exclude Pi from the cytoplasm). An obvious problem in interpreting these results is that a significant fraction of the cellular Pi could be located in subcellular organelles, such as mitochondria, in a thermodynamically inactive form[3,10,21] so that the free aqueous Pi concentration in the cytoplasm could be similar to that predicted from passive distribution in response to the membrane potential (equation 1). However, ^{32}Pi transport studies in renal and intestinal brush border membrane[27,28] have clearly demonstrated an active uptake mechanism for Pi, utilising the free-energy of the trans-membrane sodium gradient. Similar mechanisms have been proposed in Ehrlich ascites cells,[8] mouse fibroblasts transformed with SV-40[29] and cell line L1210,[30] although, in Ehrlich cells, the nature of the Pi transporter may be more complex than this.[31]

The existence of such sodium-linked mechanisms does not prove that the steady-state concentration of cytoplasmic Pi is maintained significantly higher than the value predicted from passive distribution. For example, if the plasma membrane also has a high passive permeability to Pi (as seen in the basolateral membrane of renal epithelial cells[27]), this could abolish the putative trans-membrane Pi gradient arising from active transport. Nevertheless, in human flexor digitorum superficialis muscles at rest, the sarcoplasmic concentration of Pi has been estimated, by ^{31}P NMR, to be approximately 2.5 mmole/kg wet weight or 3.5 mmole/l of cell water,[10] in normal adults whose plasma concentration of Pi is around 1 mmole/l. As the membrane potential of human skeletal muscle is about 80-90 mV inside negative,[32] from equation 1 the sarcoplasmic concentration of Pi here should be only about 10^{-5} mole/l if passive distribution is the sole contributor. This large discrepancy between observed and predicted values strongly suggests that these muscles maintain a high concentration of Pi by active transport, although this may not be a universal feature of skeletal muscle as, in cat biceps at rest, Pi is almost undetectable by ^{31}P NMR.[11]

Effect of Extracellular Pi

From the passive distribution model (equation 1), the cytoplasmic concentration of Pi would be expected to be directly proportional to that in extracellular fluid. Similarly, for a constant degree of coupling between the plasma membrane gradients of sodium and Pi, direct proportionality might be expected even if Pi were transported actively by a mechanism like that in kidney.[27] Undoubtedly, marked changes in the intracellular concentration of Pi can be imposed upon normal human erythrocytes when incubated at 37°C and pH 7.4 in simple salt solutions containing extreme extracellular concentrations of Pi (0.1 or 5 mmole/l). A new steady-state concentration of intracellular Pi is established after about an hour, and this can be used to study the effects of abnormal concentrations of intracellular Pi on organic phosphate metabolism.[33-35]

Erythrocytes drawn from patients with similar extreme concentrations of Pi in their plasma, also show marked alterations in intracellular Pi;[9] but it is still unclear how the cells behave at intermediate concentrations of extracellular Pi (0.5 to 2.5 mmole/l). In patients with vitamin D-resistant hypophosphataemic rickets (VDRR), who had plasma concentrations of Pi around 0.5 mmole/l, erythrocyte Pi decreased directly in proportion to plasma Pi.[6] Similarly, the total cellular Pi detectable in the myocardium[36] and skeletal muscle[37] of phosphorus-deprived rats was reported to decrease in direct proportion to the extracellular concentration of Pi, although a non-linear relationship between cellular and extracellular Pi could also be consistent with the data.[36,37] In contrast, patients with Paget's disease of bone who were made hyperphosphataemic with the bisphosphonate drug EHDP[38] showed disproportionately small increases in intracellular Pi in their erythrocytes, and negligible increases in intracellular Pi in their leucocytes and blood platelets. In erythrocytes, this did not arise from a change in membrane potential, pH or cell water content, as chloride distribution between cells and plasma was unaffected.[39]

In a recent study to confirm and extend these findings, over the plasma Pi range 0.5 to 2.5 mmole/l, (A. Bevington, K.I. Mundy, A.J.P. Yates, J.A. Kanis, and R.G.G. Russell, unpublished observations) erythrocyte Pi was measured in patients with Paget's disease before and after treatment with EHDP, in normal adults and in patients with VDRR. In agreement with the earlier reports,[38,39] the cell/plasma Pi distribution ratio decreased as plasma Pi increased, whereas the values calculated from the chloride distribution (using equation 1) showed no dependence on plasma Pi. This effect is unlikely to be an artifact of the patients' disease or therapy as, in normal erythrocytes incubated to a steady-state in vitro in autologous plasma (1.00 $\pm$ SD 0.16 mmole Pi/l) or Pi-supplemented plasma (2.27 $\pm$ SD 0.19 mmole Pi/l), the corresponding molar cell/plasma concentration ratios for Pi declined from 0.57 $\pm$ SD 0.12 to 0.34 $\pm$ SD 0.05 ($p<0.001$).

This effect was also observed for sarcoplasmic Pi by ^{31}P NMR in flexor digitorum superficialis muscles of patients with these disorders([40] and D.J. Taylor, B. Rajagopalan, and G.K. Radda, unpublished observations). Similarly, measurements of total cellular Pi in freeze-clamped skeletal muscle of hypophosphataemic dogs,[26] and in leucocytes and blood platelets of patients with a variety of hypo- and hyper-phosphataemic disorders,[38] showed the cell/plasma Pi distribution ratio decreasing as plasma Pi rose although, in part, this could have arisen because non-cytoplasmic Pi was bound in a form which only slowly attained a new steady-state with extracellular and cytoplasmic Pi.

As the plasma membranes of erythrocytes[9] and myocytes[41] are permeable to Pi, these results suggest that, under certain circumstances, these cells can buffer or regulate their steady-state cytoplasmic concentration of Pi when the extracellular concentration is perturbed. This may be of clinical importance as, if it is still to be accepted that cell damage in severe hypophosphataemia arises from loss of cytoplasmic Pi, with concomitant depletion of organic phosphates,[1] it is important to know under what circumstances, and in which cells, this apparent regulation occurs.

Apart from protection against depletion of intracellular Pi, regulation of the concentration of Pi in intracellular pools may have other advantages. For example, when the concentration of Pi in erythrocytes is forced to change, the concentration ratio adenosine 5'-triphosphate/adenosine 5'-diphosphate (ATP/ADP) is unaltered, so that the phosphorylation potential ATP/(ADP x Pi) is disturbed.[35] As the free energy of hydrolysis of ATP is a function of the phosphorylation potential,[21] cellular energy metabolism may be impaired via alterations in the mass-action ratios of ATPase reactions[35] if cytoplasmic Pi is not controlled.

TRANSIENT PHENOMENA

Just as the steady-state concentration of cytoplasmic calcium is normally maintained at a low value and increases in response to external stimuli (such as occupation of cell-surface receptors), so cytoplasmic Pi also shows transient changes in its concentration in response to stimulation of the cell. However, unlike calcium, the biological functions and consequences of these changes are still largely unknown.

In principle, transient changes in the cytoplasmic concentration of Pi can be attributed to two main sources; firstly a change in the kinetics of Pi transport across the plasma membrane or of Pi release from intracellular pools;[15,42,43] and secondly generation of Pi within the cell from organic phosphates. A frequently studied example of the latter is the net flux of phosphorus from phosphocreatine to Pi in exercising skeletal muscle,[10,11,44] but even here, processes of the first type may be occurring. For example, exercising human leg muscles release considerable quantities of Pi into extracellular fluid,[41] and there is indirect evidence, from ^{31}P NMR of human flexor digitorum superficialis, of a net flux of Pi into non-sarcoplasmic pools during recovery from exercise, as the sarcoplasmic Pi resonance transiently disappears.[10]

Clinically, an important example of cellular concentrations of Pi changing because of fluxes in or from the organic phosphate pools, occurs in pH disorders. In cells which utilise glycolysis, such as erythrocytes, it is well-established that increasing the pH stimulates the glycolytic flux and leads to accumulation of sugar phosphates.[45] It has therefore been postulated[46] that, in humans, intracellular alkalosis leads to a shift of phosphorus from intracellular Pi to sugar phosphates in glycolytic cells (notably skeletal muscle) and this, in turn, depletes extracellular Pi, leading to hypophosphataemia.

The human erythrocyte is a convenient glycolytic cell in which to test this hypothesis, and a decrease in cellular Pi with increasing pH has already been demonstrated.[45] In part, this may be an effect on steady-state distribution of Pi across the cell membrane, as erythrocyte membrane potential becomes more negative as pH rises.[7] Nevertheless, from 90 min incubations at 37°C in autologous plasma, with added HCl or NaOH, (G.J. Kemp, A. Bevington and R.G.G. Russell, unpublished observations) it has been shown that the cellular concentration of Pi declines too steeply with increasing pH to be accounted for solely by changes in membrane potential. The observed plot of mol<u>ar</u> cell/plasma concentration ratio for Pi (ordinate) versus plasma pH (abscissa) over the pH range 7.1 to 7.9, has a slope of - 0.85 ± SE 0.15, compared with - 0.42 ± SE 0.10 ($p<0.05$) calculated from the chloride ratios r using equation 1.

In these cells, 2,3-diphosphoglycerate accumulated markedly over a period of 24h at higher pH, in agreement with measurements made in clinical alkalosis;[45] and at 45% haematocrit, a significant loss of Pi from plasma was detectable after only 90 min, suggesting that this process in cells which utilise glycolytic pathways may contribute to hypophosphataemia in vivo.

CONCLUSIONS

Even in the simple case of the mammalian erythrocyte, the steady-state concentration of Pi in the cytoplasm cannot always be explained solely by passive distribution of Pi across the plasma membrane. In multi-compartment cells, the discrepancy may be much larger and, in some, the cytoplasmic concentration of Pi at steady-state may be regulated in face of changes in the extracellular concentration of Pi. This apparent regulation implies that the absolute concentration of cytoplasmic Pi is important for normal cell function; but the biological significance of transient changes in intracellular concentrations of Pi, in response to external signals, has still to be determined.

ACKNOWLEDGEMENTS

AB was supported by a J.G. Graves Medical Research Fellowship from the University of Sheffield; and GJK by the Special Trustees of the United Sheffield Hospitals.

REFERENCES

1. B.D. Hettleman, R.L. Sabina, and M.K. Drezner, Defective adenosine triphosphate synthesis : An explanation for skeletal muscle dysfunction in phosphate-deficient mice, J. Clin. Invest. 72:582 (1983).
2. E. Racker, "Mechanisms in Bioenergetics", Academic Press, New York and London (1965).
3. B.D. Ross, D.M. Freeman, and L. Chan, Phosphorus metabolites by NMR, in: "Phosphate and Mineral Metabolism" (Vol. 178, in Advances in Experimental Medicine and Biology), Proc. 6th Int. Workshop on Phosphate and other Minerals, Verona 1983, S.G. Massry, G. Maschio, and E. Ritz, eds., Plenum Press, New York (1984).

4. L. Sestoft, Is the relationship between the plasma concentration of inorganic phosphate and the rate of oxygen consumption of significance in regulating energy metabolism in mammals? Scand. J. Clin. Lab. Invest. 39:191 (1979).
5. J.E. Campillo, A.S. Luyckx, and P.J. Lefebvre, Effect of phosphate on the arginine-induced insulin release by the isolated perfused rat pancreas, in: "Phosphate and Minerals in Health and Disease" (Vol. 128, in Advances in Experimental Medicine and Biology), Proc. 4th Int. Workshop on Phosphate and other Minerals, Strasbourg 1979, S.G. Massry, E. Ritz, and H. Jahn, eds., Plenum Press, New York (1980).
6. H.S. Tenenhouse and C.R. Scriver, Orthophosphate transport in the erythrocyte of normal subjects and of patients with X-linked hypophosphatemia, J. Clin. Invest. 55:644 (1975).
7. S.B. Hladky and T.J. Rink, pH equilibrium across the red cell membrane, in: "Membrane Transport in Red Cells", J.C. Ellory and V.L. Lew, eds., Academic Press, London (1977).
8. J.W. Bowen and C. Levinson, Phosphate concentration and transport in Ehrlich ascites tumor cells : effect of sodium, J. Cell. Physiol. 110:149 (1982).
9. A. Challa, A. Bevington, C.M. Angier, A.J. Asbury, C.J. Preston, and R.G.G. Russell, A technique for the measurement of orthophosphate in human erythrocytes, and some studies of its determinants, Clin. Sci. 69: In press (1985).
10. D.J. Taylor, P.J. Bore, P. Styles, D.G. Gadian, and G.K. Radda, Bioenergetics of intact human muscle. A ^{31}P nuclear magnetic resonance study, Mol. Biol. Med. 1:77 (1983).
11. R.A. Meyer, T.R. Brown, and M.J. Kushmerick, Phosphorus nuclear magnetic resonance of fast- and slow-twitch muscle, Am. J. Physiol. 248:C279 (1985).
12. R.J.M. Lane and N.J. Watmough, Measurement of phosphate concentration in the presence of very labile esters, Anal. Biochem. 140:424 (1984).
13. R.J. Hall, An improved method for the micro-determination of inorganic phosphate in small volumes of biological fluids, J. Med. Lab. Technol. 20:97 (1963).
14. A. Noorwali, C.J. Preston, A. Challa, and R.G.G. Russell, Measurement of intracellular inorganic phosphate in human blood red cells, leucocytes and platelets, in: "Regulation of Phosphate and Mineral Metabolism" (Vol. 151, in Advances in Experimental Medicine and Biology), Proc. 5th Int. Workshop on Phosphate and other Minerals, New York 1981, S.G. Massry, J.M. Letteri, and E. Ritz, eds., Plenum Press, New York (1982).
15. R.W. Dudek, A.F. Boyne and M. Freinkel, Quick-freeze fixation and freeze-drying of isolated rat pancreatic islets. Application to the ultrastructural localization of inorganic phosphate in the pancreatic beta cell, J. Histochem. Cytochem. 29:321 (1981).
16. M. Stubbs, R.L. Veech, and H.A. Krebs, Control of the redox state of the nicotinamide-adenine dinucleotide couple in rat liver cytoplasm, Biochem. J.126:59 (1972).
17. D.G. Gadian and G.K. Radda, NMR studies of tissue metabolism, Ann. Rev. Biochem. 50:69 (1981).
18. G.K. Radda, P.J. Bore, and B. Rajagopalan, Clinical aspects of ^{31}P NMR spectroscopy, Brit. Med. Bull. 40:155 (1984).
19. A.A. Maudsley, S.J. Hilal, H.E. Simon, and S. Wittekoek, In vivo MR

spectroscopic imaging with P-31 : work in progress, Radiology 153:745 (1984).
20. I.A. Bailey, S.R. Williams, G.K. Radda, and D.G. Gadian, Activity of phosphorylase in total global ischaemia in the rat heart. A phosphorus-31 nuclear-magnetic-resonance study, Biochem. J. 196:171 (1981).
21. D. Freeman, S. Bartlett, G. Radda, and B. Ross, Energetics of sodium transport in the kidney. Saturation transfer ^{31}P NMR, Biochim. Biophys. Acta 762:325 (1983).
22. P.B. Garlick, T.R. Brown, R.H. Sullivan, and K. Ugurbil, Observation of a second phosphate pool in the perfused heart by ^{31}P NMR; is this the mitochondrial phosphate? J. Mol. Cell. Cardiol. 15:855 (1983).
23. G.R. Bartlett, Human red cell glycolytic intermediates, J. Biol. Chem., 234:449 (1959).
24. J. Goodman and A.N. Bessman, Effect of hemodialysis on red cell organic and inorganic phosphates, Am. J. Med. Sci. 270:447 (1975).
25. R.J. Labotka, T. Glonek, M.A. Hruby, and G.R. Honig, Phosphorus-31 spectroscopic determinations of the phosphorus metabolite profiles of blood components : erythrocytes, reticulocytes, and platelets, Biochem. Med.15:311 (1976).
26. J.P. Knochel, R. Haller, and E. Ferguson, Selective phosphorus deficiency in the hyperalimented hypophosphatemic dog and phosphorylation potentials in the muscle cell, in:"Phosphate and Minerals in Health and Disease" (Vol. 128, in Advances in Experimental Medicine and Biology), Proc. 4th Int. Workshop on Phosphate and other Minerals, Strasbourg 1979, S.G. Massry, E. Ritz, and H. Jahn, eds., Plenum Press, New York (1980).
27. M.R. Hammerman and S.J. Schwab, Phosphate transport in the kidney, studies with isolated membrane vesicles, in: "Epithelial Calcium and Phosphate Transport : Molecular and Cellular Aspects", (Vol. 168, in Progress in Clinical and Biological Research), F. Bronner and M. Peterlik, eds., Alan R. Liss Inc., New York (1984).
28. G. Karsenty, B. Lacour, A. Ulmann, E. Pierandréi, and T. Drüeke, Early effects of vitamin D metabolites on phosphate fluxes in isolated rat enterocytes, Am. J. Physiol. 248:G40 (1985).
29. R.T. Hamilton and M. Nilsen-Hamilton, Transport of phosphate in membrane vesicles from mouse fibroblasts transformed by Simian Virus 40, J. Biol. Chem. 253:8247 (1978).
30. G.B. Henderson and E.M. Zevely, Intracellular phosphate and its possible role as an exchange anion for active transport of methotrexate in L1210 cells, Biochem. Biophys. Res. Commun. 104:474 (1982).
31. J.P. Wehrle and P.L. Pedersen, Characteristics of phosphate uptake by Ehrlich ascites tumor cells, J. Biol. Chem. 257:9698 (1982).
32. D. Elmqvist, W.W. Hofmann, J. Kugelberg, and D.M.J. Quastel, An electrophysiological investigation of neuromuscular transmission in myasthenia gravis, J. Physiol. (Lond.) 174:417 (1964).
33. M.A. Lichtman and D.R. Miller, Erythrocyte glycolysis, 2,3-diphosphoglycerate and adenosine triphosphate concentration in uremic subjects : relationship to extracellular phosphate concentration, J. Lab. Clin. Med. 76:267 (1970).
34. C.H. Wallas, Metabolic studies on the erythrocyte from patients with chronic renal disease on haemodialysis. II ATP metabolism, Brit. J. Haematol. 27:145 (1974).

35. A. Bevington, A.J. Asbury, C.J. Preston, and R.G.G. Russell, Phosphate metabolism in erythrocytes of critically ill patients, Clin. Sci. 69: In press (1985).
36. N. Brautbar, R. Baczynski, C. Carpenter, S. Moser, P. Geiger, P. Finander, and S.G. Massry, Impaired energy metabolism in rat myocardium during phosphate depletion, Am. J. Physiol. 242:F699 (1982).
37. N. Brautbar, C. Carpenter, R. Baczynski, R. Kohan, and S.G. Massry, Impaired energy metabolism in skeletal muscle during phosphate depletion, Kid. Int. 24:53 (1983).
38. C.J. Preston, A. Noorwali, A. Challa, A.D. Paterson, D.J. Beard, J.A. Kanis, and R.G.G. Russell, Intracellular inorganic phosphate and ATP levels in human blood erythrocytes, leucocytes and platelets in normal subjects and in diseases associated with altered phosphate metabolism, in:"Regulation of Phosphate and Mineral Metabolism" (Vol. 151, in Advances in Experimental Medicine and Biology), S.G. Massry, J.M. Letteri, and E. Ritz, eds., Plenum Press, New York (1982).
39. A. Bevington, C.J. Preston, A. Challa, A. Noorwali, J.A. Kanis, and R.G.G. Russell, Why does sodium etidronate (EHDP) change the distribution of orthophosphate (Pi) across the red cell? Calcif. Tiss. Int. 35 (Supplement) : A37 (1983).
40. R. Smith, R.J. Newman, G.K. Radda, M. Stokes, and A. Young, Hypophosphataemic osteomalacia and myopathy : studies with nuclear magnetic resonance spectroscopy, Clin. Sci. 67:505 (1984).
41. D. Sorlie, K. Myhre, O.D. Saugstad, and K.-E. Giercksky, Release of hypoxanthine and phosphate from exercising human legs with and without arterial insufficiency, Acta Med. Scand. 211:281 (1982).
42. P. Jirounek, G.J. Jones, W. Pralong, M. Rouiller, and R.W. Straub, Involvement of Ca ions with phosphate movements in rabbit vagus nerve fibres, J. Physiol. (Lond). 328:52P (1982).
43. G. Medina and J.A. Illingworth, Some hormonal effects on myocardial phosphate efflux, Biochem. J. 224:153 (1984).
44. R.H.T. Edwards, M.J. Dawson, D.R. Wilkie, R.E. Gordon, and D. Shaw, Clinical use of nuclear magnetic resonance in the investigation of myopathy, Lancet I:725 (1982).
45. A.J. Grimes, "Human Red Cell Metabolism", Blackwell Scientific Publications, Oxford (1980).
46. J.P. Knochel, Hypophosphatemia, West. J. Med. 134:15 (1981).

PHOSPHATE TRANSFER AND ^{31}P NUCLEAR MAGNETIC RESONANCE (NMR) SPECTROSCOPY PRE AND POST HEMOFILTRATION (HF)

T. Haas, Ph. Jehenson, D. Hillion, S. Tran Dinh, E. de Viel, J. Verrier, and A. Syrota

CHI Poissy: Dialysis Dept., 78303 Poissy, France
and SHFJ Dept. Biol. CEA, 91406 Orsay, France

INTRODUCTION

Plasma phosphate levels off during hemofiltration and returns to basal values after hemofiltration (1, 2, 3, 4, 5). Plasma phosphate kinetics cannot be related to delayed mass transfer. A phosphate transfer to the plasma compartment occurs, probably from a cellular reservoir. Phosphorus-31 nuclear magnetic resonance (^{31}P-NMR) spectroscopy was used in order to specify this transfer, its effect on the energy metabolism of the skeleral muscle cell. Furthermore the influence of an acute variation of the kalemia on this transfer was investigated.

PATIENTS AND METHODS

Four patients with chronic renal failure treated by hemofiltration (HF) for over 6 months agreed on participating to this study. Informed consent was obtained from all of them. Hemofiltration sessions : capillary polysulfone hemofilters (F60 FRESENIUS) were used for all sessions, ultrafiltration (UF) and reinjection (RI) were automatically controlled by a gravimetric generator (Haemoprocessor 40005, SARTORIUS). The exchanged ultrafiltrate volume was 1l UF/2.5 kg of body weight at each session. The RI fluid composition was as follows (mmol/l) : Na 145, Ca 2, Mg 0.5, glucose 11, bicarbonate 28, potassium 4 (study A) or 2 (study B). The patient's body weight was kept constant during the HF session. The desired dry body weight could be obtained by a pure ultrafiltration of one hour or more, if necessary, after the end of HF. The plasma kinetics of phosphate, calcium, creatinine, sodium and potassium were obtained from ultrafiltrate samples taken every 15 minutes during and after the HF session. Samples from the total ultrafiltrate volume of the session allowed calculation of the removed quantity of each of these solutes. Blood gases were measured while NMR spectra were recorded.

Nuclear magnetic resonance

The ^{31}P NMR spectra were obtained at 2 Tesla with a 3 cm diameter surface-coil placed on the muscles of the flexor compartment of the forearm. The 53 cm bore superconducting magnet was built by the "Service Technique et Instrumental des Particules Elémentaires" of the Saclay Nuclear Research Center (Gif-sur-Yvette). A CXP 90 Bruker console with an Aspect 2000 computer was used for acquisition and handling of date. Either 60 or 200 free induction decays (FID) were acquired for each spectrum with a 90° flip angle and a 1.5 s pulse interval.

Table 1 : Body weight, UF volume and UF rate (mean±SD)

	Study A (n=4)	Study B n=4)	P
UF Liters	24.2±1.6	24.9±1.9	NS
UF rate ml/min	132±12	129±14	NS
Body weight Kg	66.2±8.6	65.6±7.5	NS

Table 2 : Phosphatemia (PO4) and kalemia (K) before (b) and after (a) HF, and phosphate removed (mPO4) per HF (mean±SD)

		Study A (n=4)	Study B (n=4)	P
PO4 mmol/l	b	1.76±0.43	1.59±0.39	NS
	a	0.92±0.22	1.02±0.27	NS
K mmol/l	b	6.0±0.2	5.8 ±0.2	NS
	a	4.8±0.2	4.2 ±0.3	<0.05
mPO4 mmol		26.7±7.9	29.9±6.4	NS

The summed FID was multiplied by an exponential corresponding to 10-Hz line broadening before Fourier transformation. Signal frequencies are reported as chemical shifts in parts per million (PP m) relative to the phosphocreatine (PCr) signal. The chemical shift (d) of the inorganic phosphate (Pi) peak was used to estimate intracellular pH using the following formula : $pH = 6.72 + \log (d - 3.27)/(5.69 - d)$.

The spectra were recorded just before, in the middle of, at the end of and one hour after the end of the HF session. The extracorporeal circuit was bypassed and the arteriovenous fistula needles were heparinized whilst the spectra were recorded during the HF session. The extracorporeal circuit was then reconnected to the patient in order to go on with the session until the next recording of spectra.

Statistical analysis

Mean, standard deviation, Student's test (with 5% significance : $p < 0.05$) and linear regression analysis were used in the statistical analysis of data.

Table 3 : Blood pH, pO_2, pCO_2 and plasma bicarbonate (mean ± SD)

		pH	pO_2 mmHg	pCO_2 mmHg	Bicarbonate mmol/l
Before HF	A	7.33	116 ± 10	38 ± 2	20.6 ± 2.9
	B	7.34	114 ± 12	37 ± 3	20.1 ± 2.5
Mid HF	A	7.33	114 ± 11	39 ± 3	23.5 ± 1.4
	B	7.39	120 ± 8	39 ± 1	23.9 ± 1.2
End of HF	A	7.40	110 ± 4	38 ± 2	24.2 ± 0.3
	B	7.41	110 ± 2	39 ± 2	25.5 ± 0.8
1 hour post HF	A	7.40	110 ± 5	36 ± 3	23.8 ± 0.5
	B	7.40	105 ± 9	39 ± 2	24.5 ± 0.2

Table 4 : Blood pH (pH_B) and intracellular pH (pH_i) (mean)

	Before HF	Mid HF	Post HF	One hour post HF
pH_B A	7.33	7.38	7.40	7.41
pH_i A	7.08	7.10	7.07	7.11
pH_B B	7.34	7.39	7.41	7.40
pH_i B	7.06	7.09	7.10	7.08

RESULTS

- Studies A and B were performed in the same conditions of mean body weight before HF, of mean UF volume and of mean UF flow (table 1).
- The phosphatemia and kalemia before and after HF as well as the phosphate mass removed during the session are listed in table 2.
- The mean values of blood pH, pO_2, pCO_2 and bicarbonate during studies A and B are listed in table 3. There never is any significant difference between the two studies for any parameter.
- The NMR spectra of a patient and a normal subject in the basal state and those of two patients before, in the middle of, at the end of and one hour after HF are shown in figures 1 and 2, respectively.
- The mean values of blood pH and intracellular pH are given in table 4.
- The phosphate mass removed from the extracellular compartment and that coming from the intracellular compartment was calculated for both studies A and B by assuming the mean extracellular volume to be 20% of the patient's body weight. The results (table 5) show a significantly larger phosphate transfer to the extracellular compartment with a RI fluid containing 2 mmol/l potassium (study B) rather than 4 mmol/l potassium (study A).
- In order to verify the latter result a complementary study was performed on 13 patients routinely treated by hemofiltration. These patients were hemofiltered with a RI fluid containing 0 and then 3,5 mol/l potassium at one week interval.
- The intra and extracellular phosphate and potassium balances were calculated by taking the weight loss during HF into account. The results are given in table 6.

Table 5 : Phosphate removed : Total (ΔmPi), from extra cellular (Δ mPi_{EC}) and intracellular (Δ mPi_{IC}) compartments (mean $\pm$ SD)

	Δ m Pi mmol	Δ m Pi_{EC} mmol	Δ mPi_{IC} mmol
Study A (n=4)	26.7 $\pm$ 7.9	11.4 $\pm$ 4.8	15.4 $\pm$ 3.3
Study B (n=4)	29.9 $\pm$ 11.4	7.6 $\pm$ 4.3	22.3 $\pm$ 7.8
P	NS	NS	< 0.05

DISCUSSION

NMR spectra in the basal state (before HF). The 8 recorded spectra do not differ from those of normal subjects. The presence of sugar phosphates and phosphodiesters is inconstant. Intracellular pH (study A : 7.08 ; study B : 7.06) is similar to that of normal subjects (pH=7.05) even though blood pH is low (study A : 7.33 ; study B : 7.34).

NMR spectra during and after HF. The spectra do not significantly vary during the HF session in studies A and B. The sugar phosphates either appear or disappear. The measured phosphocreatine to ATP (PCr/ATP) and inorganic phosphate to ATP (Pi/ATP) ratios show no major variations. However, even though no statistically significant variation is observed due to the small variations and the small number of patients, the fact that the various patients show variations in the same direction leads to a few comments. The PCr/ATP ratio decrease by 5 to 26% (mean 12.5%)

Table 6 : Second study : mean results with potassium free reinjection fluid (K = 0) and potassium rich reinjection fluid (K = 3,5)

Pi_b and K_b = Phosphatemia and Kalemia before HF ; Pi_a and K_a = Phosphatemia and Kalemia after HF ; ΔmPi and ΔmK = Phosphate and Potassium mass removed during HF ; ΔmPi_{EC} and ΔmK_{EC} = Phosphate and Potassium removed from extra cellular space ; ΔmPi_{IC} and ΔmK_{IC} = Phosphate and Potassium removed from intra cellular space.

	K = 0 mmol/1	K = 3,5 mmol/1	P
Pi_b mmol/1	1.97	2.08	NS
Pi_a mmol/1	1.04	0.94	NS
K_b mmol/1	5.3	5.7	NS
K_a mmol/1	3.7	4.9	< 0.001
ΔmPi mmol	35.8	32.4	NS
ΔmK mmol	97.4	42.3	< 0.001
ΔmPi_{EC} mmol	14.4	16.4	NS
ΔmK_{EC} mmol	31.8	24.6	< 0.005
ΔmPi_{IC} mmol	21.4	16.0	< 0.03
ΔmK_{IC}	65.6	17.7	< 0.001

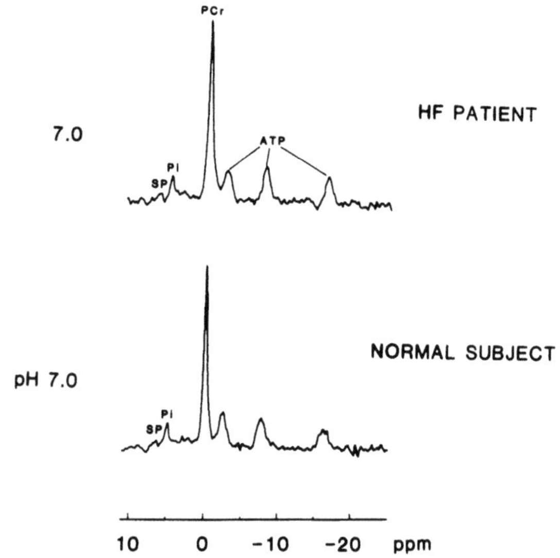

Figure 1 : NMR SPECTRA IN ONE PATIENT BEFORE HF AND IN NORMAL SUBJECT

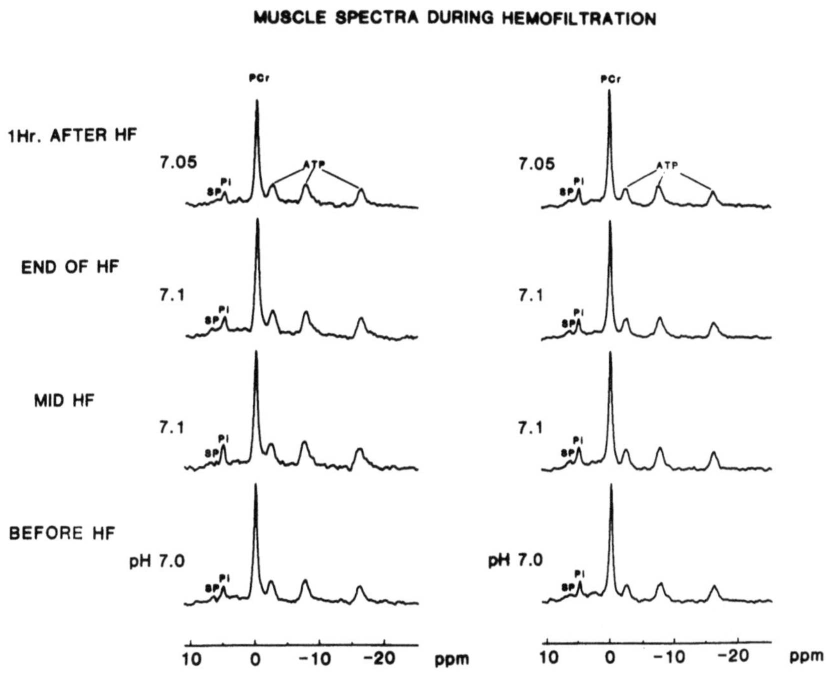

Figure 2 : NMR SPECTRA PRE, PER AND POST HF IN TWO PATIENTS

during the sessions. This is in contradiction with the increase in the PCr/ATP ratio found by Kokko (6) during dialysis. The Pi/ATP ratio does not vary in two patients and decreases in the other two (5% and 39%) in study B. It decreases in two patients (25% and 34%) and increases in the other two (7% and 42%) in study A. On the whole, the decrease in the PCr/ATP ratio during HF is similar to that of a light muscular exercise. The absence of a parallel increase in the Pi/ATP ratio in 6 cases out of 8 (the ratio increases in 2 cases) suggests an exit of Pi from the cell. The variations towards alkalinization of the intracellular pH during HF are also slight, so the metabolic pathways, the ionic transfers during HF, and the capacity of the cell to buffer an acute alkalinization result in an intracellular pH stability.

Pi transfer from intra to extra cellular space. The measured quantity of phosphate removed (Δm Pi) is not significantly different with the potassium poor RI fluid than with the potassium rich one. Nevertheless the Pi flux (Δ m Pi_{IC}) towards the extracellular space is larger in study B than in study A ($p < 0.05$). This is verified by the complementary study on 13 patients where the effect of a potassium-free RI fluid is compared to that of a RI fluid containing 3.5 mmol/l potassium and where the intra and extracellular balances of potassium are measured together with those of phosphate. The Pi fluxes (Δm Pi_{IC}, $p < 0.03$) and the K fluxes (Δ m K_{IC}, $p < 0.001$) both significantly differ for the two cases. A linear relationship can be established between the Pi and the K fluxes :

$$\Delta m\ Pi_{IC} = 0.119 \times \Delta m\ K_{IC} + 13.58\ (r = 0.457\ ;\ p < 0.02).$$

This relationship could be explained by 2 hypotheses :

1) The cell hyperpolarisation due to the decrease in extracellular potassium favours the Pi exit.

2) A Pi-K co-transport exists at the cell membrane level.

CONCLUSION

The NMR study before during and after HF shows that HF does not appreciably affect the muscle cell's energy metabolism and intracellular pH. Nevertheless, in our experience, NMR did not allow to specify phosphate transfer from the cell to extracellular compartment. Further studies are necessary in order to explain the relationship between K and Pi fluxes.

REFERENCES

1. T. Haas, G. Dongradi, F. Villeboeuf, E. de Viel, J.F. Fournier and D. Duruy, Plasma kinetics of small molecules during and after hemofiltration : decrease in hemofiltration efficiency related to increase in ultrafiltration rate. Clin. Nephrol., 19:193, 1983
2. G. Dongradi, T. Hass, F. Villeboeuf, F. Khazine, J.P. Fendler, High efficiency hemofiltration (ultrafiltration of more than 250 ml/min for two hours) in eight uremic patients, in : Proceedings of the European Dialysis and Transplant Association., 18:176, 1981
3. G. Dongradi, T. Hass, F. Villeboeuf, E. de Viel, Intrabody phosphate transfer, Artif. Organs. 7A:58, 1983
4. H. Sugisaki, M. Onohara, T. Kunitomo, Dynamic behavior of plasma phosphate in chronic dialysis patients, Trans. Am. Soc. Artif. Intern. Organs., 28:302, 1982
5. T. Kunitomo, M. Onohara, S. Nishiumi, F. Inagaki, H. Sugisaki, Phosphate in chronic dialysis (HD) patients, Artif. Organs., 7A:43, 1983
6. J. Kokko, Applications of nuclear magnetic resonance (NMR) in nephrology, in : Phosphate and Mineral Metabolism., 178:465, 1983

REGULATION OF RENAL Na-K-ATPase: EFFECTS OF ALDOSTERONE IN PHOSPHATE DEPLETION

Salim K. Mujais

Department of Medicine
Northwestern University Medical School
Chicago, IL

INTRODUCTION

Phosphate depletion is associated with multiple abnormalities in renal metabolism particularly the induction of depletion of intracellular adenine triphosphate nucleotides (ATP) (1-3). These nucleotides play a key role in renal transport since they are the substrates for multiple transport enzymes crucial to renal function, namely the transport ATPases: Na-K-ATPase, Ca-ATPase and H^+-ATPase. Of these Na-K-ATPase consumes most of the energy required during renal transport operation and the activity of the enzyme is dependent on the intracellular level of the high energy nucleotide (4). Thus, a decrease in intracellular ATP, if critical, may lead to a reduction in the activity of the enzyme.

The close relationship between the enzyme and its substrate nucleotide is best illustrated in the cortical collecting tubule where the enzyme is subject to hormonal modulation by aldosterone (5,6). To achieve its transport effect at this nephron site, aldosterone induces a series of consorted changes in the cellular proteins and metabolism all orchestrated to enhance sodium reabsorption and potassium and hydrogen ion excretion. These changes include an increase in the luminal membrane permeability to sodium, a stimulation of ATP synthesis (7), and, if the presence of the hormone is sustained, an enhancement in Na-K-ATPase activity (6). Increased sodium entry across the luminal membrane will raise intracellular sodium concentration which is the main regulator of Na-K-ATPase activity (4). An enhancement of the latter will return intracellular sodium concentration to normal and preserve cellular integrity and volume. In the process the consumption of ATP by the enzyme is increased. It is clear then that aldosterone has a series of consorted effects on cellular processes all directed to handle an increased rate of sodium transport. The hypothesis that was tested in the present work was whether the increase in enzyme activity observed with chronic aldosterone administration (6) is dependent on the availability of augmented levels of intracellular ATP. We reasoned that prolonged phosphate depletion will lead to a curtailment of nucleotide synthesis that may be reflected in a blunting of the effect of the hormone on the enzyme.

METHODS

Male albino Sprague-Dawley rats weighing 100-110 g were fed a phosphate deficient diet (see Materials for content of diets), and subsequently assigned to one of two groups at 7, 11, and 15 weeks after the start of phosphate deprivation.

Group I rats underwent a sham operation and constitute the control phosphate-depleted group. In group II rats, osmotic minipumps (ALZET model 2001, Alza Co, Palo Alto, CA) containing aldosterone in concentrations calculated to deliver 5 µg/100g/day, were implanted subcutaneously in the interscapular area. These rats constitute the aldo-treated phosphate-depleted group.

Rats maintained on a regular phosphate diet were weight matched to 15 week phosphate-depleted rats and allocated to parallel groups respectively labelled control phosphate-replete group, and aldo-treated phosphate-replete group. Except for duration of dietary modification and diet content, all animals were studied in an identical fashion.

Metabolic balance study

Animals of Groups I and II were individually placed in metabolic cages (Nalge Co, Rochester, NY) for two separate periods of 24 hours (5th and 6th days after start of hormone administration). On the first day they were allowed to acclimate to the cage. The urinary excretion of electrolytes was determined during the 2nd day (6th day).

ATPase Assay

Na-K-ATPase was determined in nephron segments obtained on the 7th day after initiation of hormone administration in Groups I and II. The procedures for tubule microdissection and Na-K-ATPase assay have been reported before (5,6), and will be described only briefly here: The animals were killed by exsanguination from the abdominal aorta, and the left kidney was perfused with a collagenase containing solution. Tubules were dissected in the cold under stereomicroscopic observation in a medium containing (in mM): NaCl, 137; KCl, 5; $MgSO_4$, 0.8; Na_2HPO_4, 0.44; $MgCl_2$, 1; $CaCl_2$, 0.25; Tris-HCl, 10; pH 7.4. The tubules were individually transferred to a concave bacteriological slide and photographed to determine their length. Nephron segments were identified on the basis of morphologic and topographic characteristics during dissection, and their identification was verified under high power microscopy during photography.

ATPase activity was measured after a two-step permeabilization procedure, by incubation of tubule segments for 15 minutes at 37°C in a 2-µl droplet of the following solutions: For determination of total ATPase activity the incubation medium contained (in mM): NaCl, 50; KCl, 5; $MgCl_2$, 10; EGTA, 1; Tris-HCl, 100; Na_2ATP (grade II, vanadate-free, Sigma Chemical Co.), 10; and ($-^{32}P$)ATP (New England Nuclear, MA) in tracer amounts (5 nCi/µl). For determination of Mg-dependent ATPase activity, NaCl and KCl were omitted, Tris-HCl was 150 mM and 1 mM ouabain was added. Phosphate liberated by the hydrolysis of ($-^{32}P$) ATP was separated by filtration through a Millipore filter (Millipore Continental Water Systems, Bedford, MA) after absorption of the unhydrolyzed nucleotide on activated charcoal, and the radioactivity in the eluate was counted in a liquid scintillation spectrophotometer (Beckman, model LS100C, Brea, CA).

Total and Mg-dependent ATPase activities were each determined on five replicate samples from individual animals, and were expressed as picomoles inorganic phosphate liberated per millimeter of tubule length per hour.

Sodium-and potassium-dependent, ouabain-inhibitable ATPase was taken as the difference between the means of each group of measurements, and thus represents a single data point in each animal; the results show mean values of these determinations in all animals within a given experimental group. To minimize the variability between experiments, one appropriate control and one experimental animal were studied simultaneously.

Analytical Methods

Plasma sodium and potassium were measured using a flame photometer with lithium as the internal standard (International Instruments, model IL-64, Lexington, MA). Chloride was determined by chloridometry (Corning, model 920M, Medfield, MA). Blood urea nitrogen (BUN) was assayed using a Beckman Analyser II. Total calcium was measured using Corning 940 titrator. Phosphate was determined using an Autoanalyzer Model II. Osmolality was measured with a Wescor Vapor Pressure Osmometer.

Materials

The low phosphorous diet (ICN 902206, ICN Nutritional Biochemicals, Cleveland, OH) had the following composition: Blood fibrin-bovine washed 6x, 20; sucrose 66% vegetable oil hydrogenated 10%; sodium chloride 1.01%; potassium chloride 1.53%, calcium carbonate 1.01%; magnesium sulfate 0.3%.

Statistical Methods

Statistical analysis was done using Student's t-test for nonpaired samples. Results are reported as mean ± standard error (SE).

RESULTS

Effects of phosphate deprivation

Prolonged phosphate deprivation significantly blunted body growth. At eight weeks, phosphate deprived rats had barely attained 50 per cent of the weight of their litter mate given a normal phosphate diet. In addition, their food intake paralleled their weight in that they were consuming less food than both their age controls and weight controls. Several metabolic abnormalities characteristic of phosphate deprivation were identified in phosphate deprived rats (figure 1). In general these abnormalities were related in their magnitude to the duration of phosphate deprivation. Plasma phosphate was significantly reduced by eight weeks of phosphate deprivation, and further significant reductions were observed between twelve and sixteen weeks ($p<0.05$). This hypophosphatemia was associated with complete absence of phosphate from the urine reflecting the renal adaptive mechanisms to conserve phosphate. Hypercalcemia was observed at eight weeks and worsened progressively: calcium levels being higher at 16 weeks than at eight weeks ($p<0.01$). This hypercalcemia was paralled by a hypercalciuria (figure 1) that also worsened with time being significantly greater at 12 and 16 weeks than at eight ($p<0.01$, $p<0.01$ respectively). Blood urea nitrogen was not different in phosphate deprived rats at any time period (8 week: 23 ± 4; 12 week: 19 ± 2; 16 week: 19 ± 3 mg/dl) from that in phosphate replete weight matched controls (18 ± 1 mg/dl). In the face of lower food intake of phosphate deprived rats, this normal BUN level can be construed as evidence for decreased renal function. Plasma sodium and plasma osmolality were normal in phosphate deprived rats (plasma sodium 8 week: 141 ± 0.6; 12 week: 142 ± 1; 16 week: 142 ± 0.4 mEq/l ; plasma osmolality 8 week: 300±2; 12 week: 305±3; 16 week: 311±2 mosml/kg). Plasma chloride was normal at eight and 12 weeks (figure 1). Hypercloremia was, however, detected after 16 weeks of phosphate deprivation. Mild

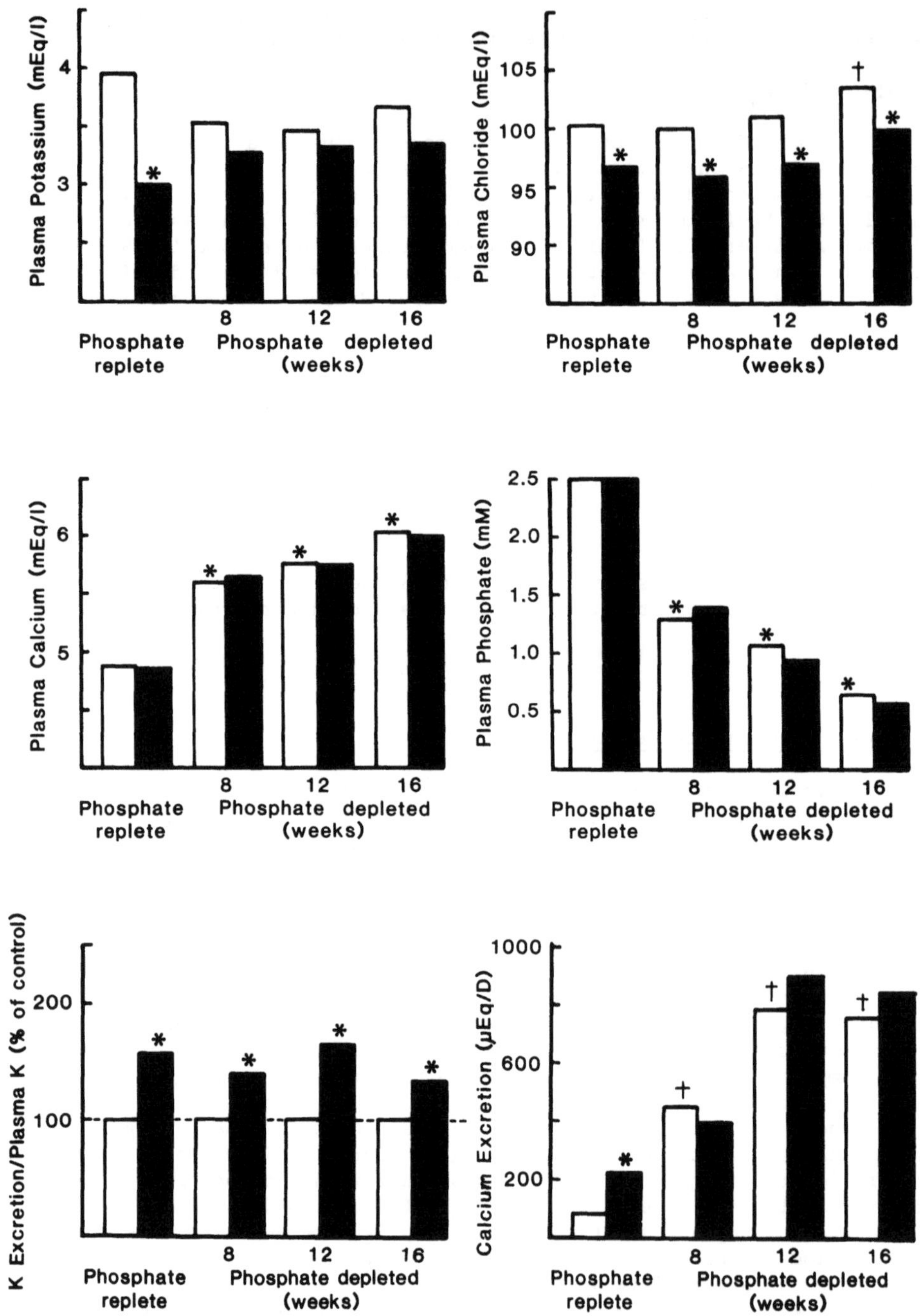

Figure 1. Metabolic balance study

hypokalemia was present in phosphate deprived rats at all follow up intervals despite normal potassium content in the diet.

Na-K-ATPase activity measured in the cortical collecting tubule was similar in 16 week phosphate deprived rats and their weight matched controls (figure 2).

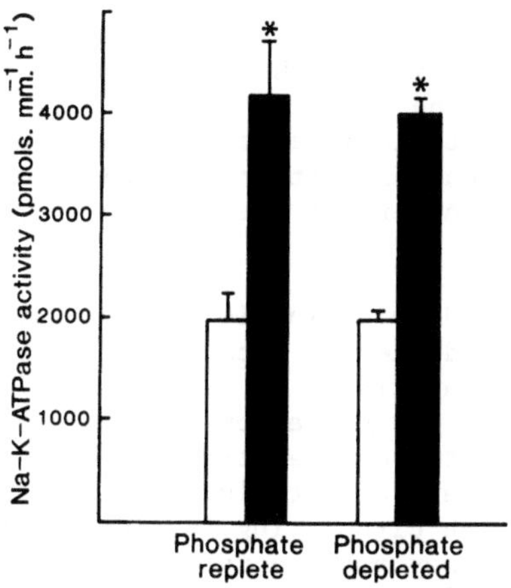

Figure 2. Na-K-ATPase activity in cortical collecting tubules from phosphate-replete and 16 week phosphate-depleted rats. Aldosterone (5 μg/100 g/day for 7 days) stimulated enzyme activity in both groups. (* $p<0.01$).

Effects of aldosterone

The effects of aldosterone on the various metabolic parameters are illustrated in figure 1. In phosphate replete weight matched controls, aldosterone led to a decrease in BUN (from 18 ± 1 to 12 ± 1, $p<0.001$), hypokalemia (fig 1), hypochloremia (fig 1), and hypercalciuria (fig 1). Aldosterone administration had no effect on plasma calcium or plasma phosphate. The hypokalemic effect of aldosterone was significantly blunted in phosphate deprived rats (figure 1). Because of the overlap and variability of urinary potassium excretion, the ratio of urine potassium excretion to plasma potassium was used as an index of the kaliuretic effect of aldosterone. The values of this index obtained in animals not receiving aldosterone were considered as controls. Figure 1 shows that the kaliuretic effect of aldosterone was still present in phosphate deprived animals at all follow up intervals despite the fact that the effect was not statistically evident when plasma potassium and urinary potassium excretion were examined seperately. The hypochloremic effect of aldosterone, a reflection of the expected metabolic alkalosis of mineralocorticoid excess, was evident in all groups. In 16 week deprived rats, the hypochloremic effect was of equal magnitude as in the other groups, but because of the higher baseline, the post aldosterone chloride values were higher in this group than in the others ($p<0.05$). Aldosterone had no effect on plasma calcium, plasma phosphate or BUN in the phosphate deprived groups. The hypercalciuric effect of aldosterone was not evident in the phosphate deprived rats probably overshadowed by the massive hypercalciuria of phosphate deprivation itself.

Aldosterone administration led to a doubling of Na-K-ATPase activity in the cortical collecting tubule of phosphate replete rats (fig 2). A doubling of enzyme activity was also observed in 16 week phosphate deprived rats that was indistinguishable from that observed in phosphate replete controls.

DISCUSSION

The present study was designed to explore the renal transport and metabolic effects of aldosterone in an animal model of phosphate depletion. In particular, the interaction of the hormone and renal Na-K-ATPase was evaluated as an index of the coupling between cell metabolism and function in the kidney. The results indicate that while some of the transport consequences of aldosterone are either blunted or overshadowed in phosphate depletion, the interaction between the hormone and Na-K-ATPase is unaffected.

Prolonged phosphate deprivation in growing rats was used to assure the development of renal ATP depletion because of the efficiency of the mechanisms operative in the kidney to prevent a fall in cellular inorganic phosphate, a prerequisite for depletion of high energy nucleotides (1,3). These mechanisms include the unique structure of the kidney cell and its ability for phosphate uptake at both luminal and basolateral membranes. Another factor is the high circulating levels of vitamin D observed in phosphate depletion that may enhance uptake by the tubule cell, thus helping maintain intracellular phosphate levels against the falling extracellular phosphate concentration. Finally, the rich intracellular stores of phosphate in renal mitochondria will buffer any rapid changes in cellular phosphate content. Indeed, while extracellular phosphate levels fall after one week of dietary phosphate deprivation, there is little change in renal phosphate until two weeks of phosphate deprivation (1). The fall in ATP levels is even more delayed being detectable only after six weeks of phosphate deprivation. This time lag reflects the dependence of ATP synthesis on the rich mitochondrial stores of phosphate that fall below critical levels only with prolonged deprivation (1).

In the present study, evaluation of the effects of aldosterone was done after 8 to 16 weeks of phosphate deprivation, a time period when cellular ATP content is expected to be reduced (1). Many of the manifestations of phosphate deprivation were manifest and quite advanced in this animal model including hypophosphatemia, hypercalcemia, and hypercalciuria. Na-K-ATPase activity in the cortical collecting tubule, however, was normal and was stimulated by aldosterone administration in a fashion indistinguishable from that in phosphate replete rats. Several explanations for this observation can be forwarded. The level of cellular ATP even after prolonged phosphate deprivation may still be adequate for basal Na-K-ATPase activity. This explanation would require a reevaluation of the kinetic relationship between the enzyme and its substrate with chronic alterations in substrate levels. The strict dependence of the enzyme on the level of intracellular ATP has been defined mostly with acute changes in cellular ATP. Chronic changes in the level of the nucleotide may be associated with an alteration in the kinetics of the enzyme. A second explanation is that under conditions of cellular energy shortage and sparsity of high energy nucleotides, ATP may be recruited from other energy requiring processes to fuel the operation of the Na-K-pump. A third possibility is that although enzyme activity as measured in vitro is normal, in vivo many of these enzyme units may be in a dormant state and actual enzyme work would be reduced (4). This requires that synthesis of enzyme units be independent of the energy supply of the cell, a postulate consistent with the observation that enzyme activity increases with uninephrectomy in another model of cellular energy shortage, hypothyroidism (4).

The intactness of the enzymatic response to aldosterone also requires explanation. It may imply that the increase in Na-K-ATPase activity induced by aldosterone is independent of the increase in cellular ATP and can occur in its absence. The hormone induces a series of consorted effects in the cell directed towards one goal: enhanced sodium reabsorption. It is not known whether these effects are all strictly interdependent. It has been proposed that the increase in Na-K-ATPase activity following aldosterone administration is secondary to the increase in sodium entry from the luminal side, since the increase in enzyme activity could be blocked by the sodium channel blocker amiloride (8). This proposition has been challenged on many accounts. It is based on the acute stimulation of enzyme activity in cortical collecting tubule by aldosterone, an observation that has been difficult to reproduce in both cortical collecting tubules (9) and collecting tubule analogues, the anuran urinary bladder (10). Furthermore, when enhanced enzyme synthesis was measured after prolonged exposure to the hormone, it was found to occur independent of sodium transport which was blocked by amiloride (11). Similarly, the increase in Na-K-ATPase activity may occur independently of an increase in intracellular ATP and even in the face of cellular ATP depletion. Again, the increase in enzyme activity as measured in vitro does not imply that the new enzyme units are activated and a larger proportion may be in a dormant state. It is highly unlikely that aldosterone actually increases the level of cellular ATP since this would require utilization of intracellular phosphate that is expected to be severely diminished below the level required for oxidative phosphorylation.

In summary, the present study has explored the effects of aldosterone in an animal model of phosphate depletion. We have shown that the activity of Na-K-ATPase activity in cortical collecting tubules in its basal state, and the enzyme increase with aldosterone remain intact despite depletion of intracellular ATP secondary to prolonged phosphate depletion.

ACKNOWLEDGEMENT

The present study was supported in part by a grant from the National Kidney Foundation of Illinois and the Chicago Heart Association. The author would like to thank Ms Susan Kauffman and Mr Edward Cooney for their technical assistance.

REFERENCES

1. K. Kurokawa, Phosphate depletion and renal cell metabolism, in: "Nephrology," R. R. Robinson, ed, Springer-Verlag, N.Y. (1984).
2. S. Sabatini, Clinical spectrum of phosphate depletion and its effects on urinary acidification, in: "Nephrology," R. R. Robinson, ed, Springer-Verlag, N.Y. (1984).
3. Z. S. Agus and R. E. Garrick, Effect of phosphate depletion on renal tubular transport, in: "Nephrology," R. R. Robinson, ed, Springer-Verlag, N.Y. (1984).
4. A. I. Katz, Renal Na-K-ATPase: Its role in tubular sodium and potassium transport. Am. J. Physiol. 242:F207 (1982).
5. S. K. Mujais, M. Chekal, W. J. Jones, J. P. Hayslett, and A. I. Katz, Regulation of renal Na-K-ATPase in the rat. Role of the natural mineralo-and glucocorticoid hormones. J. Clin. Invest. 73:13 (1984).
6. S. K. Mujais, M. A. Chekal, W. J. Jones, J. P. Hayslett, and A. I.

Katz.. Modulation of renal Na-K-ATPase by aldosterone. Effect of high physiologic levels on enzyme activity in isolated rat and rabbit tubules. J Clin Invest 76:170 (1985).
7. N. Cortas, , E. Abras, and M. Walser. Aldosterone response in the turtle bladder is associated with an increase in ATP. Am. J. Physiol. 245:F512 (1983).
8. K. J. Petty, J. P. Kokko, and D. Marver, Secondary effect of aldosterone on Na-K-ATPase activity in the rabbit cortical collecting tubule. J Clin Invest 68:1514 (1981).
9. S. K. Mujais, M. A. Chekal, S.-M. K. Lee, and A. I. Katz, Relationship between adrenal steroids and renal Na-K-ATPase: Effect of short-term hormone administration on the rat cortical collecting tubule. Pflugers Archiv Eur J Physiol 402:48 (1984).
10. J. H. Hill, N. Cortas, and M. Walser, Aldosterone action and sodium- and potassium-activated adenosine triphosphatase in toad bladder. J Clin Invest 52:185 (1973).
11. K. Geering, M. Grardet, C. Bron, J. P. Kraehenbuhl, and B. C. Rossier, Hormonal regulation of (Na,K)-ATPase biosynthesis in the toad bladder: Effect of aldosterone and 3,5,3'-triiodo-l-thyronine. J Biol Chem 257:10338 (1982).

EFFECTS OF THE IMPAIRMENT OF RENAL FUNCTION AND OF THE pH OF GASTRIC SECRETION ON THE EFFICACY OF $Al(OH)_3$ TO REDUCE SERUM INORGANIC PHOSPHORUS

G. Barsotti, M. Lazzeri, A. Polloni, E. Morelli,
E. Giovannetti, S. Lupetti, A. Cupisti, L. Dani,
and S. Giovannetti

Glinica Medica 1°, University of Pisa, Italy

INTRODUCTION

In spite of the demonstration that the Al of the antacids is absorbed in the gastrointestinal tract (1, 2, 3, 4) and that its retention may cause severe adverse effects in patients with chronic renal failure (CRF) (5, 6, 7, 8), the Al-containing phosphate binders continue to be widely used to correct hyperphosphatemia.

The present investigation has been performed to assess the role of CRF and that of the gastric secretion pH on the efficiency of colloidal solid $Al(OH)_3$ in reducing phosphatemia (sPi).

PATIENTS AND METHODS

For both the "in vitro" and the "in vivo" experiments tablets of ALUCOL(R) were used in order to avoid discrepancies in the efficiency to bind phosphate related to different sources of $Al(OH)_3$. Each tablet contains 500 mg of colloidal $Al(OH)_3$, 400 mg of saccarose and 80 mg of talc.

For the measurements of the efficiency of $Al(OH)_3$ to bind PO_4 "in vitro" in different conditions, solutions of monosodium PO_4 were used whose pH was adjusted to wanted value by adding NaOH or HCl. The tablets were powdered in a mortar and added to the solutions of PO_4 and shaked in a Dubnoff shaker at 37°C for the wanted time.

The amounts of Pi bound by $Al(OH)_3$ was calculated by difference between its original concentration in the solution and the concentration found in the supernatant fluid after centrifugation at 1,500 g for 10 minutes.

The experiments in normal volunteers (8 cases) were performed on students and doctors of the staff of the renal unit of this Clinic.

The 16 patients with CRF were on conservative therapy and followed a conventional low-protein diet supplying daily, per Kg body weight, approximately 0.6 g of protein and 12 mg of Pi. CRF (serum creatinine between 5 and 11 mg/dl) was due to glomerulonephritis in 10 cases, to pyelonephritis in 4 and to polycistic kidney disease in 2.

Gastroscopy with aspiration of gastric secretion was performed in all these patients and 4 mucosa samples were taken for histology. The diagnosis of atrophic gastritis (made in 6 of the 16 patients studied) was based on a pH value of the gastric juice higher that 5 (normal pH levels: about 2) and on the histology findings.

The "in vivo" experiments were performed in fasting conditions and consisted in: the accurate chewing and the ingestion within a few minutes of 8 tablets of ALUCOL(R) (4 g of solid colloidal $Al(OH)_3$); the withdrawal of a blood sample immediately before the ingestion of ALUCOL(R) and, subsequently, every hour for the following 3 hours; the collection of the urine of the 3 hours preceeding the ingestion of ALUCOL(R) and of those of the subsequent 3 hours. Collection of urine was made on 1.0 ml of concentrated HCl.

The concentrations of Pi in the water solutions, in serum and in urine were measured according to Ziversmith and Davis and the standard auto-analyzer methodology was employed for the measurement of creatinine in serum. The pH values of gastric juice and those of the solutions of Pi were measured with a Orion pH-meter.

The Student's T test was employed for the statistical analysis.

RESULTS

The "in vitro" experiments showed that the amounts of Pi that can be bound by $Al(OH)_3$ at pH_2 increases with increasing values of the Pi concentration. In the experimental conditions indicated in Fig.1-left the removal of Pi was complete at the lower concentrations used (10 and 16 mg/dl of Pi) but was partial at the higher concentrations (from 25 to 40 mg/dl of Pi).

These experiments showed also that the amounts of Pi that can be removed in 1.0 hour by 1.0 g of $Al(OH)_3$ increases as the H+ concentration increases. In the experimental conditions we used (Fig.1-center) the maximum removal (31.5 mg) occurred at pH 2.

Also the speed of binding is increased by the low pH levels (Fig. 1-right).

The time of contact between $Al(OH)_3$ and Pi is also important: the binding of Pi increases with time (Fig.1-right) both at pH 2 and at pH 7.

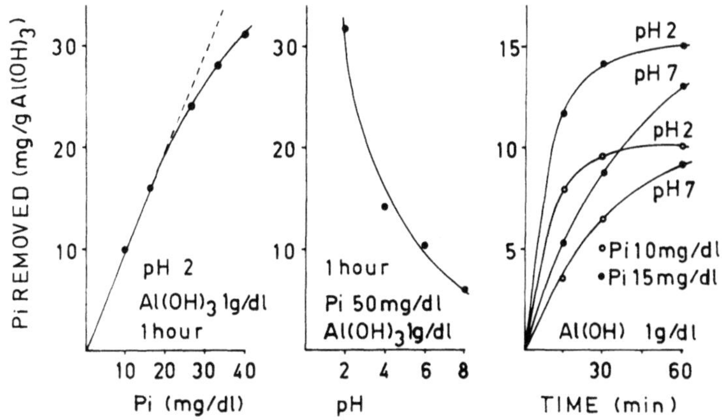

Fig.1. (1-left) The amounts of Pi removed by 1 g $Al(OH)_3$ plotted against the Pi concentration; (1-center) the amounts of Pi removed at different pH of the medium; (1-right) the speed of binding of Pi at different pH values.

The "in vivo" experiments showed that the ingestion of 4 g of colloidal solid $Al(OH)_3$ is followed by a fall of sPi that reaches its maximum after 1 hour ($p<0.001$). The fall is not significantly different, as percentage, in normal volunteers and in patients with CRF (12.4±3.1 per cent and 11.9±3.4 per cent, respectively) (Fig.2).

However, while in normal volunteers the sPi levels rebounded rapidly (from a mean of 3.21±0.24 mg/dl, after one hour) and reached the previous levels within the following 2 hours (previous mean levels 3.75±0.25 mg/dl as compared to 3.68±0.3 mg/dl 3 hours after the ingestion of ALUCOL(R)) in the uremic patients the sPi levels were still significantly lower ($p<0.02$) 3 hours after the ingestion of ALUCOL(R): previous levels 5.74±1.69 mg/dl, after 1 hour 5.0±1.62 mg/dl, and after 3 hours 5.32±1.49 mg/dl (Fig.2).

The urinary output of Pi of the 3 hours following the ingestion of ALUCOL(R) (mg 56.9±15.5) was significantly lower ($p<0.01$) than that of the previous 3 hours (mg 129.0±42.8) in normal volunteers (Fig.3). In the uremic patients, instead, no significant changes occurred in the urinary output of Pi: 83.0±30.1 mg before ALUCOL(R) and 81.3±25.6 after (Fig.3).

The pre-medication with cimetidine (400 mg by mouth 3 hours before the ingestion of ALUCOL(R) prevented almost completely the fall of sPi in normal volunteers (Fig.4) and, likewise, no significant fall of sPi occurred in uremic patients with atrophic gastritis and hypochloridria (Fig.5).

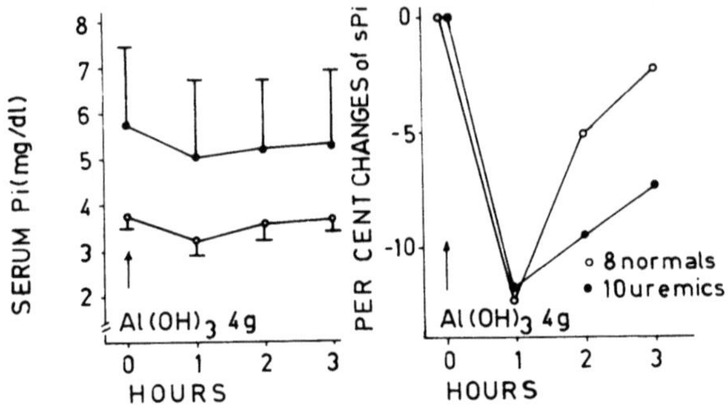

Fig.2. The changes of sPi occurring in normal volunteers and in patients with chronic renal failure, following the ingestion of 4 g of $Al(OH)_3$.

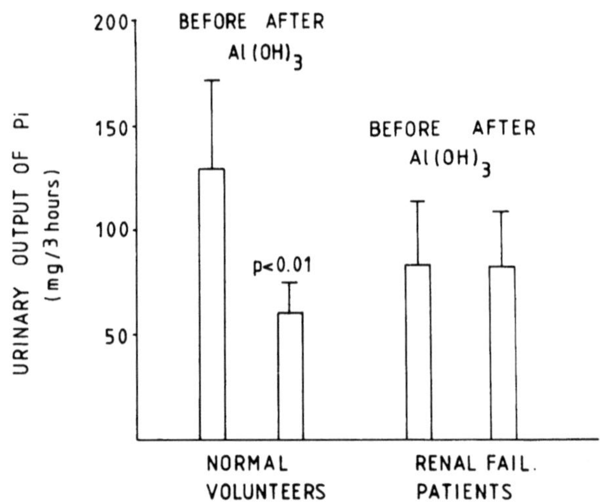

Fig.3. The changes in the urinary output of Pi occurring in normal persons and in patients with chronic renal failure in the 3 hours before and in those following the ingestion of 4 g of $Al(OH)_3$.

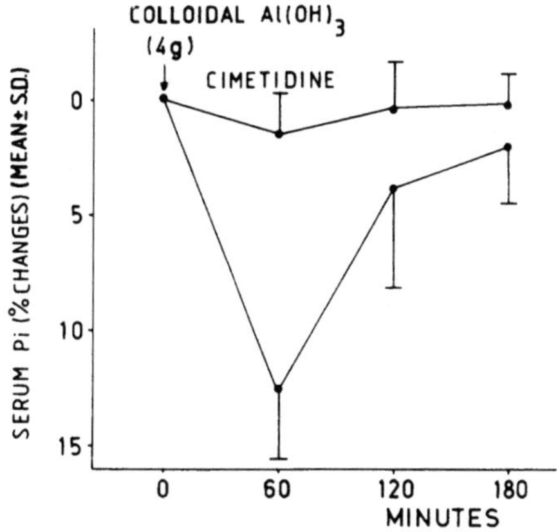

Fig.4. The changes of sPi following the ingestion of $Al(OH)_3$ by normal volunteers who had ingested cimetidine 3 hours before compared with the effect without cimetidine.

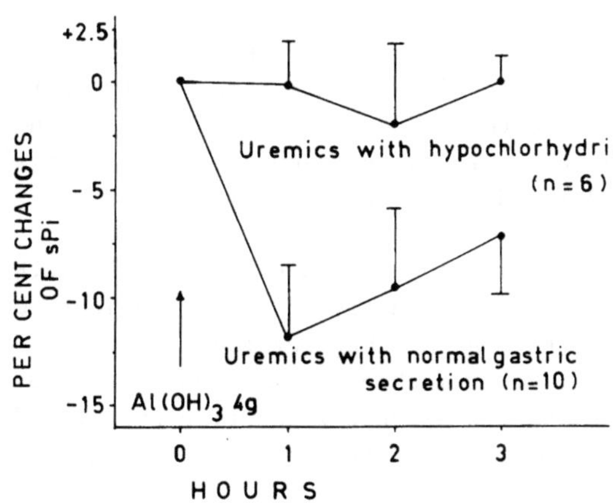

Fig.5. The changes of sPi in patients with chronic renal failure and atrophic gastritis compared with those occurring in patients with normal gastric secretion.

DISCUSSION

The results of the "in vitro" experiments and the different handling of Pi by diseased kidneys as compared with normal subjects, seem to explain the different effects of the ingestion of colloidal solid $Al(OH)_3$ on sPi in the different conditions studied.

The "in vitro" binding of PO_4 by $Al(OH)_3$ is higher and faster at low pH levels. Likewise, the fall of sPi (indicating the removal of Pi from extracellular fluid) is very rapid and much more relevant in subjects (normal and uremic) with normal gastric acid secretion than in uremic patients with hypochlorhydria or in persons whose gastric secretion had been inhibited with cimetidine.

The observation that the maximum fall of sPi occurs within 1 hour after the ingestion of $Al(OH)_3$ indicated that binding of PO_4 occurs largely in the stomach. This is also suggested by our unpublished observations with gastric-coated $Al(OH)_3$ that failed to induce any fall of sPi in the following 24 hours.

The rapid fall of sPi following the removal of Pi from gastric juice, implies the assumption that the Pi of the extracellular fluid moves rapidly and in large amounts into the gastric cavity where it is bound by $Al(OH)_3$. It is difficult to conceive such a shift and seems reasonable to postulate that other mechanism(s) of action contribute to the fall of sPi induced by $Al(OH)_3$. The possibility cannot be ruled out that the absorbed Al binds Pi in the extracellular fluid causing its precipitation in tissues and thus lowering of sPi. Our unpublished observations support such hypothesis.

The late behaviour of sPi has been found to be different in normal volunteers and in patients with chronic renal failure: in normal subjects sPi rebounded rapidly after 1 hour and, 3 hours after the ingestion of ALUCOL(R) the previous levels were reached. In patients with CRF the sPi levels remained low during the same period of time.

During the 3 hours following the ingestion of ALUCOL(R) the urinary output of Pi decreased significantly in normal subjects causing a recovery of Pi of 72 mg that may account for the rebound of sPi. It is obviously possible that other mechanisms may contribute to cause it. Patients with CRF continue to excrete with urine amounts of Pi not significantly different from those of the previous period and no Pi is recovered.

The practical implications of these observations are that the patients with atrophic gastritis and those taking inhibitors of the gastric secretion should not be treated with $Al(OH)_3$ to lower sPi. Atrophic gastritis is common in CRF (10) as well as other forms of gastric and duodenal diseases requiring the use of the inhibitors of the gastric secretion. This makes very frequent the risk of giving uselessly a potentially toxic drug.

REFERENCES

1. G.M.Berlyne, D. Pezt, J. Ben-Ari, J.Weinberger, M.Stein, G.R.Gilmore, R.Levine, Hyperaluminemia from aluminium resins in renal failure, Lancet ii:494 (1970).
2. W.D.Kaenny, A.P. Hegg, A.C.Alfrey, Gastrointestinal absorption of aluminium from aluminium containing antacids, New Engl.J.Med.296: 1389 (1977).
3. A.D. Cumming, J.F. Cowie, R.J. Winney, Factors influencing the plasma Aluminium concentration in hemodialysed patients, Dialysis and Transpl. 13: 692 (1984).
4. O.Knoll, H.Kellinghans, M.P.Bertam, H. Zumkley, U.Graefe, Gastrointestinal absorption of aluminium in chronic renal insufficiency, Contr. Nephrol. 38: 24 (1984).
5. F.P.Masselot, J.P. Adhemar, M.C. Jaudon,D.Kleinknecht, A.Galli, Reversible dialysis encephalopathy:role of aluminium containing gels, Lancet ii: 1386 (1978).
6. M.Poisson, R. Mashaly, B. Lebkiri, Dialysis encepalopathy:recovery after interruption of aluminium intake, Brit.Med. J. 2: 1610 (1978).
7. S.P.Andreoli, J.M. Bergstein, D.J.Sherrard, Aluminium intoxication from aluminium-containing phosphate binders in children with azotemia not undergoing dialysis, New Engl.J.Med.310, 1079 (1984).
8. G. Cournot-Witmer, Aluminium osteodystrophy, Contr.Nephrol. 38:59 (1984).
9. D.B.Ziversmith, A.K.Davis, Microdetermination of plasma phospholipids by trichloroacetic acid precipitation, J.Lab.Clin.Med. 35:155 (1950).
10. G.Barsotti, M.Lazzeri, A.Polloni, A.Giannoni, C.Cristofano,E.Morelli, Hypochloridria and serum gastrin levels in chronic renal failure, in: "Abstract Book - Int. Congress of Nephrology" pag.341 A, Los Angeles 1984.

EFFECTS OF DIFFERENT INSULIN ADMINISTRATION MODALITIES ON VITAMIN D METABOLISM OF INSULIN-DEPENDENT DIABETIC PATIENTS

Claude Colette, Louis Monnier, Jacques Arnal,
Jean-Louis Selam, and Jacques Mirouze

Department of Metabolism, Lapeyronie Hospital
Montpellier, France

INTRODUCTION

Several studies have established that diabetes mellitus is usually associated with abnormalities in bone mineral metabolism[1,2,3], calcium homeostasis[4,5,6], and vitamin D metabolism[7,8,9]. A reduced production of 1,25 dihydroxyvitamin D3, the biologically active metabolite of vitamin D3 is the most common finding in experimentally induced diabetic rats[7,9,10], or in young insulin-dependent (type 1) diabetic patients[11]. However the relationships between the diabetic state and the 1,25 dihydroxyvitamin D metabolism remain controversial[12]. The results of in vitro studies seem to indicate that insulin can modulate the production of the 1,25 dihydroxy-vitamin D[8]. It has been recently demonstrated that diabetic rats suffering from minimal reduction in insulin secretion (type 2) had no abnormalities in vitamin D metabolism[13]. However short term, i.e. less than two weeks, improvements of diabetic control under continuous subcutaneous insulin infusion (CSii) in insulin-dependent diabetic patients have failed to either normalize[14] or modify[5] circulating 1,25 dihydroxyvitamin D levels. Further-more in 65 per cent hepatectomized dogs, it has been observed that the hepatic production of 25 hydroxyvitamin D was only impaired when the hor-monal supply arising from the pancreas was suppressed by a spleno pancreatico caval shunt[15]. For all these reasons we were led to study the effects of both prolonged tight diabetic control and insulin delivery through portal routes on vitamin D metabolism in insulin-dependent diabetic patients. These two objectives were reached by using the continuous intra peritoneal infusion technique[16] and the results obtained were compared to those observed under systemic insulin deliveries using either CSii or conventional treatments with single or multiple daily insulin injections.

SUBJECTS AND METHODS

Twenty-eight insulin-requiring diabetic patients (10 males and 18 females) were included after approval by the Ethical Committee of the hospital and after an informed consent had been obtained from each subject. The patients were 20 to 53 year-old (mean ± SD = 35 ± 11 years). In order to avoid any interference from other factors on vitamin D metabolism, the patients selected for the study received no medication other than their insulin at the time of the study. All had been free of any intercurrent disease and administration of either dietary preparations or medications containing vitamin D or analogues. The investigations were performed from November to March in order to reduce at a minimum the effects of sunlight exposure on plasma vitamin D metabolites. In all patients renal function, estimated from blood urea, plasma creatinine levels and tests for urinary protein were found to be normal.

The patients were divided into 3 groups according to whether they were treated for several months or years with conventional treatment (group I, n=9), continuous subcutaneous insulin infusion (group II, n=8), or continuous intraperitoneal insulin infusion (group III, n=11). Additional details on the clinical status of the patients of each group are given in table 1.

Techniques of insulin delivery

The conventional treatment was conducted by the administration of one or two subcutaneous insulin injections per day. The once-daily regimen used a single injection of lente insulin (Lente, Novo Industri, Denmark) before breakfast, while the twice-daily regimen used a mixture of regular (Actrapid, Novo Industri, Denmark) and lente insulin (Monotard, Novo Industri, Denmark) before breakfast and dinner.

The continuous subcutaneous insulin (CSii) was delivered with either portable syringe (Microjet Infuser, Ames Co, USA) or peristaltic (Promedos, Siemens AG, FRG) battery powered pumps. Neutral porcine regular insulin (Actrapid) or U40 acidic porcine regular insulin (CS21, Hoechst AG, FRG) were respectively used with the syringe or peristaltic pumps. Insulin was delivered through a subcutaneous catheter that was implanted in the subcutaneous tissue of the lateral abdominal wall according to the technique previously described by Pickup[17].

The continuous intraperitoneal insulin infusion (CPii) was delivered with a peristaltic Promedos pump (Siemens AG) the 30 ml reservoir being filled with U40 porcine regular insulin (CS21, Hoechst AG).The insulin infusion was made via a polyethylene catheter that was passed through the umbilicus into the peritoneal cavity under local anesthesia. The implantation procedure used a technique which has been previously described[16], and permitted long term ambulatory insulin infusions since the average half life of the catheter is 16 months[18].

Table 1. Clinical status of the 3 groups of investigated patients. Results are given as mean ± sem

Groups	I[a]	II[b]	III[c]
No of patients tested	9	8	11
Age (years)	39 ± 3	35 ± 6	33 ± 2
Percentage of ideal body weight	102 ± 3	102 ± 6	96 ± 3
Diabetes duration (years)	26 ± 3	16 ± 4	20 ± 2
Duration of the pump treatment (months)		12.5±4.9	10.4±1.8

[a]Patients treated with conventional insulin treatment.
[b]Patients treated with continuous subcutaneous insulin infusion (CSii)
[c]Patients treated with continuous intraperitoneal insulin infusion (CPii)

Analytical procedures

The degree of the diabetic control was estimated from the measurement of total glycosylated hemoglobin (HbA_1) by using a chromotographic separation through prepacked microcolumns (Isolab system).

Plasma 25 hydroxyvitamin D (25-OH D, ng/ml) and 24,25 dihydroxyvitamin D (24,25 $(OH)_2D$, ng/ml) were measured by competitive protein binding assay procedures using normal diluted rat serum as binding protein[19]. A lipid extract of plasma was chromatographied on silica gel (Serva, 50-100 mesh) columns (0.5 x 6 cm) in 7 ml of diethyl ether light petroleum (1:2, v/v). The 25-OH D was eluted by 7 ml of diethyl ether and the 24,25 $(OH)_2$ D was separated by high pressure liquid chromatography (HPLC, Waters).

Plasma 1,25 dihydroxyvitamin D (1,25-$(OH)_2$ D, pg/ml) was determined by using the cytosolic fraction of intestinal mucosa from chicks as binding protein[20], after proper separation by high pressure liquid chromatography (HPLC, Waters).

Plasma immunoreactive parathyroid hormone (iPTH) was measured by using a carboxy terminal antibody.

Plasma calcium was measured by atomic absorption spectroscopy.

Plasma phosphorus and creatinine were measured by routine methods.

Statistical analysis

The comparisons between the different groups were made by the Student's t test. The different groups of investigated diabetic patients were compared to a control group of 30 non diabetic subjects matched for age, sex, bodyweight, dietary intakes and sunlight exposure. Correlations between plasma 25-OH D and 24,25 $(OH)_2$ D were evaluated by calculating the r of Pearson.

RESULTS

The results are shown in table 2. Plasma 25-OH D levels were significantly diminished in patients treated with either conventional insulin treatment ($P<0.02$) or CSii ($P<0.01$), i.e. with subcutaneous routes compared to those treated with CPii. Plasma 25-OH D concentrations in patients under CPii were similar to that observed in control subjects, but were lower in patients submitted to conventional treatment ($P<0.005$) or CSii ($P<0.001$) when compared to controls. As for 25-OH D, plasma 24,25 $(OH)_2$ D levels were lower in diabetic patients treated with conventional treatment ($P<0.01$) or CSii ($P<0.05$) than in diabetic patients under CPii. Plasma 24,25 $(OH)_2$ D were significantly decreased in diabetic patients treated with conventional insulin therapy ($P<0.05$), but not in those treated with CSii when compared with controls. In the diabetic group considered as a whole (n=28) a positive relationship was noted between plasma 24,25 $(OH)_2$ D (y, ng/ml) and plasma 25-OH D (x, ng/ml) : $y = 0.074x + 0.54$, $r = 0.52$, $P<0.01$. As indicated in table 2, plasma 1,25$(OH)_2$ D was slightly lower in the diabetic patients treated with conventional insulin administration than in control subjects and in diabetic patients treated with either CSii or CPii, but the differences were not statistically significant.

The diabetic control, estimated from HbA1 percentages, was better in the diabetic patients under CPii than in the two other groups (table 2). In all patients treated with CPii, HbA_1 percentages were less than 10 per cent and most of them (7 out of 11) were tightly controlled, i.e. HbA_1 within the normal range in non diabetic subjects : 6 to 8 per cent.

Table 2. Laboratory data in the 3 groups of diabetic patients and in 30 non diabetic subjects (control group)
Results are given as mean ± sem

Groups	I[a]	II[b]	III[c]	Control
No of subjects tested	9	8	11	30
HbA_1 (per cent)	10.2±0.6 ****d	9.6±0.7 ***d	7.2±0.4	
Plasma 25-OH D (ng/ml)	+e 12.2±1.1 **d	++e 9.7±1.4 ***d	21.2±3.0	17.0±0.9
Plasma 24,25 $(OH)_2$ D (ng/ml)	+e 1.1±0.2 ***d	1.3±0.4 *d	2.5±0.4	1.7±0.1
Plasma 1,25 $(OH)_2$ D (pg/ml)	35.5±4.8	42.0±6.4	49.9±4.9	43.0±2.0
Plasma creatinine (mg/dl)	1.19±0.08	1.22±0.18	1.07±0.04	
Plasma calcium (mg/dl)	9.3±0.2	9.1±0.1	9.1±0.1	
Plasma phosphorus (mg/dl)	4.1±0.1	3.6±0.3	3.2±0.3	
Plasma iPTH (mu/ml)	2.3±0.1	2.4±0.3	2.6±0.2	

a Patients treated with conventional insulin treatment
b Patients treated with CSii
c Patients treated with CPii
d Statistical comparisons between CPii v.s. CSii or conventional treatment are indicated by asterisks : * P<0.05, ** P<0.02, *** P<0.01, **** P<0.001
e Statistical comparisons between control subjects v.s. CSii or conventional treatment are indicated by the following symbols : + P<0.005, ++ P<0.001

No significant differences were observed for the other parameters : plasma calcium, phosphorus, creatinine and iPTH.

In order to test whether the differences of 25-OH D levels between the patients treated via subcutaneous and intraperitoneal routes were due to differences in diabetic control or to differences in the site of insulin delivery, plasma 25-OH D levels in the 11 patients treated with CPii were compared to that calculated by averaging the 25-OH D values of 8 patients treated via subcutaneous routes (4 under CSii and 4 under conventional treatment) and matched for diabetic control with those treated with CPii (HbA_1 less than 10 per cent). As shown in table 3, plasma 25-OH D and 24,25 $(OH)_2$ D were lower in patients treated via subcutaneous routes than in those treated with CPii.

Table 3. Comparison of plasma vitamin D metabolites in the 11 patients treated with CPii (group I) and in the 8 patients treated with subcutaneous routes (group II) but matched for diabetic control with those treated with CPii. Results are given as mean ± sem

Groups	I	II
No of patients tested	11	8
Plasma 25-OH D (ng/ml)	21.2 ± 3.0 **a	10.4 ± 1.5
Plasma 24,25 $(OH)_2$ D (ng/ml)	2.5 ± 0.4 ***a	0.8 ± 0.1
Plasma 1,25 $(OH)_2$ D (pg/ml)	49.9 ± 4.9	37.7 ± 5.5

a Statistical comparisons between groups I and II are indicated by asterisks : ** $P<0.02$, *** $P<0.005$

DISCUSSION

The present results indicate that the circulating levels of 25-OH D were found normal in tightly-controlled diabetic patients treated with continuous intraperitoneal insulin infusion. On the contrary plasma 25-OH D were decreased in patients treated via subcutaneous routes and exhibiting fair but less than ideal diabetic control. As the differences between subcutaneous and intraperitoneal insulin administrations persisted when the patients who were submitted to either type of insulin delivery, were matched for diabetic control, it seems therefore that the intraperitoneal infusion per se and not the degree of diabetic control was responsible for the normalization of plasma 25-OH D levels.

In the present study all patients were submitted to similar dietary intakes, using foods which were not fortified with vitamin D and all investigations were made in the late fall or in winter. For that reason the differences observed between the intraperitoneal and subcutaneous insulin administrations can be interpreted as differences in 25-OH D production by the liver. Insulin treatments through subcutaneous routes, i.e. with systemic insulin deliveries in fairly-controlled diabetics are usually associated to normal or even excessive peripheral insulin concentrations[21]. However, and although no clear demonstration has never been provided, one can suspect that subcutaneous insulin administration results in low delivery of insulin in the portal vein to the liver. On the contrary during CPii treatments one part of the intraperitoneally infused insulin is probably delivered into the portal stream and therefore to the liver[22]. For these reasons, the differences in plasma 25-OH D levels between intraperitoneally and subcutaneously treated patients can be explained by a direct stimulatory effect of insulin on the hepatic 25 hydroxylation of vitamin D.

Several works have been previously devoted to the investigation of vitamin D metabolites in experimental diabetic rats, and no significant modifications in plasma 25-OH D levels were usually observed[9,10,13]. However it must be noted that the investigations were made a few days or weeks after the animals had been rendered diabetic. As any modification in 25-OH D metabolism requires several weeks or months to result in significant changes of circulating levels of 25-OH D, it seems therefore that the experimental diabetic rat model is not useful for studying the effects of insulin on 25-OH D levels. The same remarks can be made for the studies of plasma 25-OH D in recently diagnosed or in young insulin treated diabetic patients[11]. Therefore the discrepancies between our results and those reported by other authors[11,14] are certainly due to the fact that our patients had a long history of diabetes and had been treated with insulin for approximately 20 years at the time of the investigation.

Concerning the dihydroxylated metabolites of vitamin D, our results show that plasma levels of 24,25 $(OH)_2$ D, a non or poorly active metabolite of vitamin D exhibited similar changes than did plasma 25-OH D. Such results and the observation of a positive relationship between these two metabolites seem to indicate that the modifications of plasma 24,25 $(OH)_2$ D concentrations were secondary to that observed for 25-OH D metabolism. The plasma levels of 1,25 $(OH)_2$ D, the biologically active metabolite of vitamin D, were normal and similar whatever was the type of insulin treatment. Therefore it appears that in chronically insulin-treated diabetic patients, the 1,25 $(OH)_2$ D production was regulated at a normal level even when the production of its precursor, the 25-OH D was depressed. These results are in agreement with previous works which have reported that circulating levels of 1,25 $(OH)_2$ D are diminished in severely insulin-deficient rats[7,9,10], but are restored to normal in chronically insulin-treated diabetic patients[5], or in non insulin-dependent diabetic states[13] which are characterized by the persistance of an endogenous insulin secretion by the pancreas.

In conclusion, insulin seems to play a role in the regulation of the hepatic 25 hydroxylase activity. In insulin-dependent diabetic patients, intraperitoneal insulin infusions result in normal circulating 25 OH D levels while subcutaneous, i.e. systemic routes, do not. However such an abnormality has probably no noticeable adverse effects in diabetic patients treated through subcutaneous routes since the 1,25 $(OH)_2$ D appears to be synthesized at a normal level.

REFERENCES

1. M. E. Levin, V. C. Boisseau and L. V. Avioli, Effects of diabetes mellitus on bone mass in juvenile and adult onset diabetes, N. Engl. J. Med. 294:241 (1976).
2. P. Mc. Nair, S. Madsbad, C. Christiansen, M. S. Christensen, O. K. Faber, C. Binder and I. Transbøl, Bone loss in diabetes : Effects of metabolic state, Diabetologia. 17:283 (1979).
3. A. L. Rosenbloom, D. C. Lezotte, F. T. Weber, J. Gudat, D. R. Heller, M. L. Weber, S. Klein and B. B. Kennedy, Diminution of bone mass in childhood diabetes, Diabetes. 26:1052 (1977).
4. L. Monnier, C. Colette, L. Aguirre, C. Sany and J. Mirouze, Intestinal and renal handling of calcium in human diabetes mellitus : influence of acute oral glucose loading and diabetic control, Eur. J. Clin. Invest. 8:225 (1978).
5. J. M. Gertner, W. V. Tamborlane, R. L. Horst, R. S. Sheswin, P. Felig and M. Genel, Mineral metabolism in diabetes mellitus : changes accompanying treatment with a portable subcutaneous insulin infusion system, J. Clin. Endocrinol. Metab. 50:862 (1980).

6. P. Mc Nair, N. Fogh-Andersen, S. Madsbad and M. S. Christensen, Decreased serum concentration of ionized calcium in insulin dependent human diabetes mellitus, Eur. J. Clin. Invest. 13:267 (1983).
7. L. E. Schneider, H.P. Schedl, T. Mc Cain and M. R. Haussler, Experimental diabetes reduces circulating 1,25 dihydroxyvitamin D in the rat, Science 196:1452 (1977).
8. N. Wongsurawat, H. J. Armbrecht, T. V. Zenser, B. B. Davis, M. L. Thomas and L. R. Forte, 1,25-dihydroxyvitamin D3 and 24,25-dihydroxyvitamin D3 production by isolated renal slices is modulated by diabetes and insulin in the rat. Diabetes. 32:302 (1983).
9. S. Hough, A. Fausto, O. K. Dong Jo, S. Young and L. V. Avioli, Vitamin D metabolism in the chronic streptozotocin-induced diabetic rat, Clin. Res. 29:540 A (1981).
10. E. M. Spencer, M. Khalil and O. Tobiassen, The effect of diabetes on the renal activity of 25 OH D3-1α hydroxylase in rats, Diabetes. 28:367 A (1979).
11. T. E. Frazer, N. H. White, S. Hough, J. V. Santiago, B. R. Mc Gee, G. Bryce, J. Mallon and L. V. Avioli, Alterations in circulating vitamin D metabolites in the young insulin-dependent diabetic, J. Clin. Endocrinol. Metab. 53:1154 (1981).
12. H. Heath III, P. W. Lambert, F. J. Service and S. B. Arnaud, Calcium homeostasis in diabetes mellitus, J. Clin. Endocrinol. Metab. 49:462 (1979).
13. J. Levy, S. L. Teitelbaum, J. R. Gavin III, A. Fausto, H. Kurose and L. V. Avioli, Bone calcification and calcium homeostasis in rats with non-insulin dependent diabetes induced by streptozotocin, Diabetes, 34:365 (1985).
14. J. Gertner, R. Horst and W. Tamborlane, Mineral metabolism and vitamin D status in juvenile diabetics (JOD) : changes following normalization of plasma glucose with a portable infusion pump, Diabetes. 28:354 A (1979).
15. C. Colette, B. Saint Aubert, D. Piccolboni, P. C. Andriguetto, F. Quijano, C. Astre, H. Joyeux and L. Monnier, Production of labelled 25 hydroxyvitamin D from ^{14}C vitamin D as a marker of liver function after 65 per cent hepatectomy in dogs, Nutr. Res. 2:31 (1982).
16. J. L. Selam, A. Slingeneyer, B. Hedon, P. Mares, J. J. Beraud and J. Mirouze, Long-term ambulatory peritoneal insulin infusion of brittle diabetes with portable pumps : comparison with intravenous and subcutaneous routes, Diabetes Care. 6:105 (1983).
17. J. C. Pickup, H. Keen, J. A. Parsons and K. G. M. M. Alberti, Continuous subcutaneous insulin infusion : an approach to achieving normoglycemia, Br.Med. J. 1:204 (1978).
18. J. L. Selam, P. Giraud, J. Mirouze, S. Saeidi, B. Hedon, A. Slingeneyer, H. Lapinski and C. Humeau, Continuous peritoneal insulin infusion with portable pumps : factors affecting the operating life of the chronic catheter, Diabetes Care. 8:34 (1985).
19. M. A. Preece, J. L. H. O'Riordan, D. E. M. Lawson and E. Kodiceck, A competitive protein-binding assay for 25 hydroxycholecalciferol and 25 hydroxyergocalciferol in serum, Clin. Chim. Acta. 54:235 (1974).
20. J. A. Eisman, A. J. Hamstra, B. E. Kream, H. F. De Luca, A sensitive, precise and convenient method for determination of 1,25 dihydroxyvitamin D in human plasma. Arch. Biochem. Biophys. 176:235 (1976).

21. R. Nosadini, G. A. Noy, M. Natrass, K. G. M. M. Alberti, D. G. Johnston, P. D. Home and H. Orskov, The metabolic and hormonal response to acute normoglycemia in type 1 (insulin-dependent) diabetes : Studies with a glucose controlled insulin infusion system (artificial endocrine pancreas).Diabetologia. 23:220 (1982).
22. D. S. Schade, R. P. Eaton, T. Davis, F. Akiya, E. Phinney, R. Kubica, E. A. Vaughn and P. W. Day, The kinetics of peritoneal insulin absorption, Metabolism. 30:149 (1981).

INFLUENCE OF INORGANIC PHOSPHATE ON GLUCOSE-INDUCED INSULIN RELEASE IN VITRO

J.I. Osuna, M. Castillo, E. Rodriguez, J.E. Campillo and C. Osorio

Departamento de Fisiologia y Bioquimica
Facultad de Medicina
Universidad de Granada, Granada, Spain

SUMMARY

The effect of various extracellular phosphate concentrations on glucose induced insulin secretion has been investigated in two experimental systems the isolated perfused rat pancreas and the isolated rat pancreatic islets incubated in vitro. In both systems, total insulin secretion in response to 16.7 mmol/l glucose increased as extracellular phosphate concentration increased from 0 to 3.6 mmol/. In perfusion experiments insulin secretion was significantly reduced at 4.8 mmol/l phosphate. In incubation experiments insulin secretion was inhibited only at 9.6 mmol/l phosphate concentration. When IBMX 1 mmol/l was used in incubation experiments glucose induced insulin secretion increased as extracellular phosphate increased from 0 to 3.6 mmol/l, but was also inhibited at 4.8 mmol/l.

INTRODUCTION

It is known that glucose influence the fluxes and content of several ions (Cl^-, H^+, Na^+, K^+, Ca^{2+}) in pancreatic islets (SEHLIN, J., 1980; WOLLHEIN, C.B. and SHARP, G.W.G., 1981).

FREINKEL et al. reported that glucose also caused a transient increase in inorganic phosphate release from pancreatic islets (FREINKEL, N. et al., 1974). Histochemical and microprobe examination of beta cells show an accumulation of inorganic phosphate adjacent to the inner face of the plasma membrane and nucleolus, which is lost during glucose stimulation (FREINKEL, N. et al., 1978). On the other hand several reports have shown that extracellular phosphate influences insulin release in response to various stimuli both in vivo and in vitro (CAMPILLO et al., 1977; CAMPILLO et al., 1979; HELLMAN and ANDERSSON, 1978; KIKUCHI et al., 1979). The intimate mechanism whereby phosphate influences insulin secretion remain to be investigated. The present work was designed to study the effect of various extracellular phosphate ($H_2PO_4^-$) concentrations on the glucose-induced insulin release by two different experimental systems: the isolated perfused rat pancreas; and the isolated rat pancreatic islets incubated in vitro. Therefore, we have studied the influence of two agents that modify the intracellular Ca^{2+} as 3-isobutyl-1-methyl-xanthine (IBMX) and Mg^{2+}.

MATERIAL AND METHODS

Overnight fasted male Wistar rats (200-300 g) were used. Pancreases were isolated and perfused as previously described (CAMPILLO et al., 1978). Pancreatic islets were isolated by the collagenase (Worthington 169 U/mg) technique, (LACY and KOSTIANOWSKY, 1967). The perfusate and incubation medium were a modified Krebs-Henseleit bicarbonate buffer containing in mmol/l: Glucose 2.7; Na^+ 145; Cl^- 126; K^+ 5.0; Mg^{2+} 0.7 and 1.4; Ca^{2+} 1.0; HCO_3^- 25; SO_4^{2-} 0.7 and H_2PO_4 0, 1.2; 2.4; 3.6; 4.8; 7.0 and 9.6 for each experimental condition. Whenever the phosphate concentration (H_2PO_4K) was changed, the K^+ concentration was maintained constant by appropiate alterations in KCl. The medium warmed at 37ºC and continuously gassed with O_2 and CO_2(95:5) was supplemented by 0.5 g/100 ml bovine albumin (Armour), and 2 g/100 ml dextran MW 70000 (Ibys, Madrid. Spain). The incubation medium contained also 500 U/ml Trasylol (Bayer).

In perfusion experiments, after a 40 minute prestimulation period, glucose was infused into the circuit immediately above the pancreas through a side arm perfusion pump (0.11 ml/min). The final glucose concentration reached in the perfusate was 16.7 mmol/l. In incubation experiments, 100 islets were preincubated for 30 minutes in basal medium and then harvested in groups of 10 of approximately the same diameter, transferred to vials containing 2 ml of medium with 2.7 or 16.7 mmol/l glucose and incubated for 45 minutes under agitation at 90 cycles/min. In every experiment the phosphate concentration was the same throughout the preincubation and incubation periods.

Perfusate (2 ml) and incubation (1 ml) samples were collected into chilled tubes and immediately frozen at -20ºC until assayed. Immunoreactive insulin (IRI) was measured using a double antibody procedure (HALES and RANDLE, 1963) with rat insulin (Novo) as standard. In perfusion experiments, insulin secretion rates were calculated by multiplying the concentration measured in the respective samples by the flow rate (2 ml/min). Total release during the stimulation period (30 minutes) was calculated by planimetry of the individual perfusion profiles.

All results are expressed as mean ± SEM. Statistical analysis was performed using the Student t test for non paired data.

RESULTS

The insulin release in response to 16.7 mmol/l glucose by the perfused pancreas at the various extracellular phosphate concentrations was biphasic (FIGURE 1). Insulin secretion during first phase release (ng/5 min) was significantly increased at 1.2(58.4 ± 10.7); 2.4(49.1 ± 8.7) and 3.6(65.1 ± 6.5) and 4.8 mmol/l (17.4 ± 4.8) phosphate. Insulin secretion during second phase release (ng/25 min) increased as phosphate concentration increased from 0(96.7 ± 19.2) to 3.6 mmol/l (315.0 ± 49.5) but it was significantly reduced at 4.8 mmol/l (76.1 ± 32.0) (Figure 2). Basal insulin release, assessed by estimating the insulin release in the last 5 minutes of the prestimulation period, was not significantly affected by changes in the phosphate concentration of the medium.

Figure 3 shows the effect of various extracellular phosphate concentrations on the insulin released by the isolated islets incubated in the presence of 2.7 or 16.7 mmol/l glucose. Insulin secretion in response to 2.7 mmol/l glucose increased as phosphate concentration increased in the incubation medium from 0 to 4.8 mmol/l; thereafter the insulin release was reduced despite the increased phosphate concentration in the medium. The maximun insulin release was observed at 4.8 mmol/l phosphate instead of 3.6 mmol/l as described in perfusion experiments. 16.7 mmol/l glucose-stimulated insulin release was significantly higher than basal release at all phosphate concentrations except at 9.6 mmol/l.

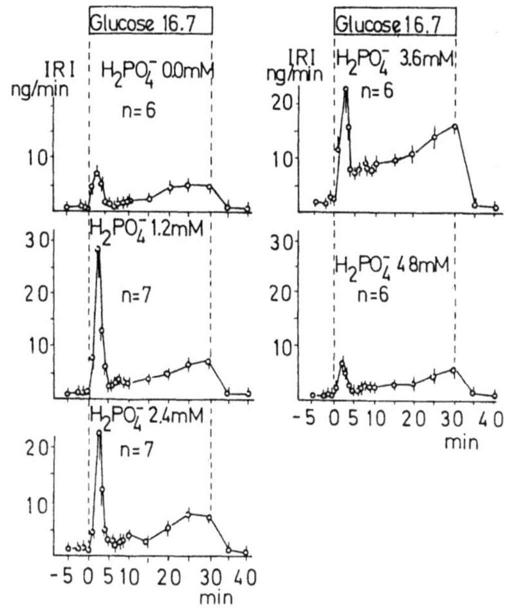

FIGURE 1: Insulin (IRI) release in response to the infusion of 16.7 mmol/l glucose at the various extracellular phosphate ($H_2PO_4^-$) concentrations.

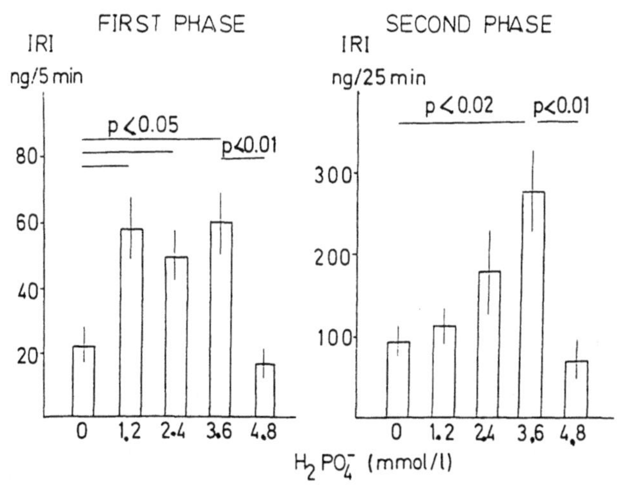

FIGURE 2 : Total insulin (IRI) released during first and second phases of 16.7 mmol/l glucose-induced insulin release at the various extracellular phosphate ($H_2PO_4^-$) concentrations.

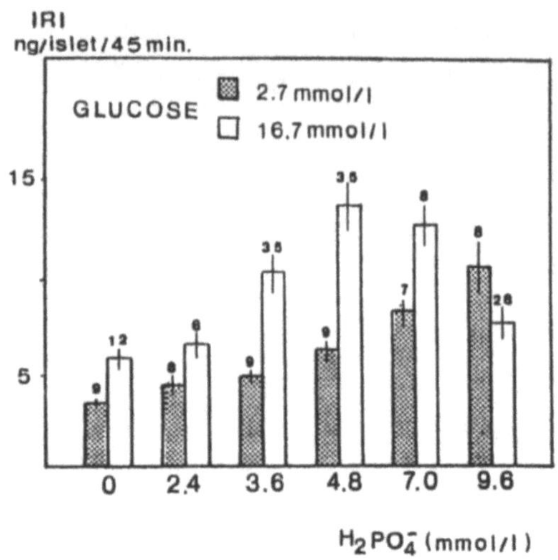

FIGURE 3: Insulin (IRI) release by isolated islets incubated with 2.7 or 16.7 mmol/l glucose at the various extracellular phosphate ($H_2PO_4^-$) concentrations.

Figure 4 shows the effect of 1.0 mmol/l IBMX on the insulin released by the isolated islets incubated in the presence of 2.7 or 16.7 mmol/l glucose. At the various phosphate concentrations tested basal insulin release was not affected by the presence of IBMX. In the presence of 1.0 mmol/l IBMX stimulated insulin secretion was significantly higher at 3.6 mmol/l phosphate and markedly inhibited at 4.8 and 9.6 mmol/l phosphate.

Figure 5 shows the effect of 1.4 mmol/l Mg^{2+} on the insulin released by the isolated islets incubated in the presence of 2.7 or 16.7 mmol/l glucose. At various phosphate concentrations tested basal insulin release was not significantly affected at 1.4 mmol/l Mg^{2+}. Insulin release induced by glucose 16.7 mmol/l was significantly higher at 9.6 mmol/l extracellular phosphate. At 0, 3.6 and 4.8 mmol/l extracellular phosphate no significant diferences were observed in the insulin release induced by glucose.

DISCUSSION

Although several reports have shown the effect of phosphate omission (CAMPILLO et al., 1978 and CAMPILLO et al., 1979) and hypophosphatemia (MARSHALL; BANASIAN and KALKHOFF, 1978) on glucose-induced insulin release, data were lacking on the effect of various extracelular phosphate concentrations on the beta cell response to glucose. These results are in agreement with previous reports (KIKUCHI et al., 1979) in islets perfused in the presence of 5 mmol/l phosphate and the inhibitory effect observed in the isolated islets at 9.6 mmol/l phosphate agree with that reported by HELLMAN and ANDERSON (1978), in islets microdissected from ob/ob mice incubated in the presence of 10 mmol/l phosphate.

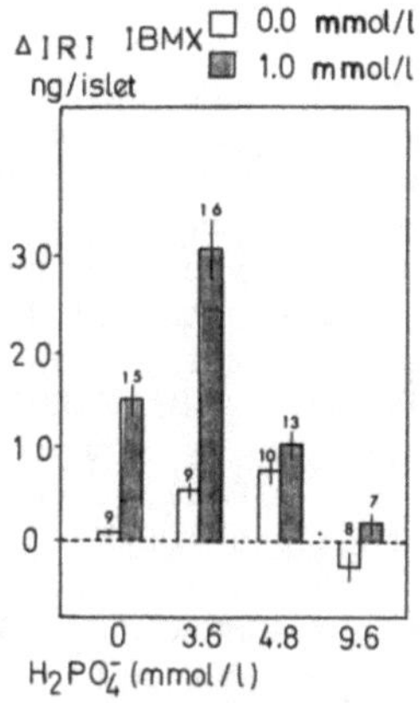

FIGURE 4: Insulin (IRI) release by isolated islets incubated with 2.7 or 16.7 mmol/l glucose and 1.0 mmol/l IBMX at the various extracellular phosphate ($H_2PO_4^-$) concentrations.

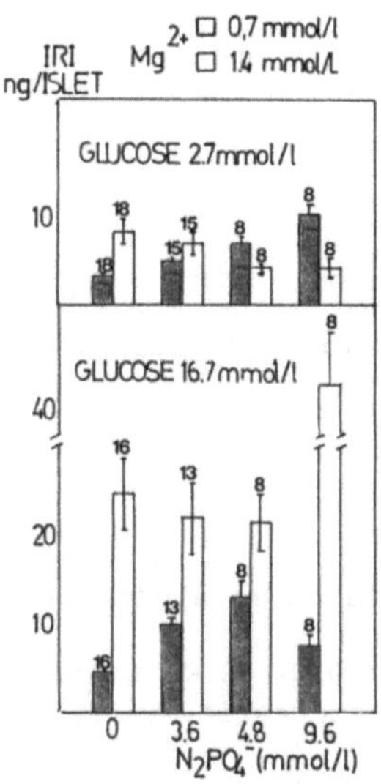

FIGURE 5: Insulin (IRI) release by isolated islets incubated with 2.7 or 16.7 mmol/l glucose and 1.4 mmol/l Mg^{2+} at the various extracellular phosphate ($H_2PO_4^-$) concentrations.

The mechanism whereby phosphate influences insulin secretion remain to be established. As in other cells and tissues, however, a relationship between phosphate and calcium movements in beta cell could be suggested (RASMUSSEN and GOODMAN, 1977; WOLLHEIN and SHARP, 1981). There is evidence to suggest that a rise in cytosolic-free calcium concentration is the common link in the stimulus-secretion coupling in various secretory cells (WOLLHEIN and SHARP, 1981). It is still unclear to what extent intracellular redistribution of calcium or a change in calcium movement across plasma membranes contribute to the regulation of cytosolic free calcium in the beta cell exposed to glucose (MALAISSE et al. 1979). Phosphate could be involved in either cases.

In fact, it has been demonstrated that calcium accumulation in pancreatic islets is increased in the presence of high phosphate concentrations (HELLMAN and ANDERSSON, 1978; KIKUCHI et al., 1979). Most, if not all, of the calcium is taken up by intracellular organelles: secretory granules (RAZARROLA et al., 1976), endoplasmic reticulum (BOROWITZ and MATTHEWS 1980) and mitochondria (DEAN, 1973). It has been suggested that mitochondria could represent the major sites for calcium accumulation in beta cells (SUGDEN and ASHCROFT, 1978) and other cells (BORLE, 1973). Phosphate stimulates calcium net uptake into a Lanthanum-non displaceable pool (HELLMAN and ANDERSSON, 1978) and phosphate constitutes a permeant anion for calcium uptake by mitochondria in beta cells (SUGDEN and ASHCROFT, 1978). Omission of phosphate could reduce the calcium accumulation by mitochondria. Phosphate greatly enhances calcium accumulation by mitochondria, at concentrations up to 1 mmol/l, but reduces it at higher concentrations (SUGDEN and ASHCROFT, 1978).

The shift to the left with respect to phosphate concentration determined by the presence of IBMX or to the right, with 1.4 mmol/l Mg^{2+}, agrees with the hypothesis that changes in extracellular phosphate concentration modified calcium movements and insulin secretion in the beta cells.

Further experiments are underway to study the intimate mechanism (s) of such an effect.

Acknowledgements

The authors wish to thank Mrs. M.J. Ayala for her technical assistance and A. Quintana for typewriting.

REFERENCES

1.- BORLE; A.S. Calcium metabolism at the cellular level. Fed. Proc. 32, 1944-1950, 1973.
2.- BOROWITZ, J.L. and MATTHEWS, E.K. Calcium exchangeability in subcellular fractions of pancreatic islet cells. J. Cell. Sci., 41, 233-243, 1980.
3.- BUKOWIECKI, L. and FREINKEL, N. Relationship between efflux of ionic calcium and phosphorus during excitation of pancreatic islets with glucose. Biochim. Biophys. Res. Commun. 63, 870-874, 1976.
4.- CAMPILLO, J.E., LUYCKX, A.S., TORRES, M.D. and LEFEBVRE, P.J. Effect of phosphate omission on arginine-induced insulin and glucagon release by the isolated perfused rat pancreas. FEBS. Letts. 34, 141-143, 1977.
5.- CAMPILLO, J.E., LUYCKX, A.S., TORRES, M.D. and LEFEBVRE, P.J. Effect of various concentrations of calcium on arginine induced insulin and glucagon release in vitro. Rev. Esp. Physiol., 34, 191-198, 1978.

6.- CAMPILLO, J.E., CASTILLO, M., VALDIVIA, M. and RODRIGUEZ, E. Effect of phosphate omission on glucose induced insulin release in vitro. Experientia, 35, 1260-1261, 1979.
7.- CAMPILLO, J.E., OSUNA, J.I. PAGES, M.T., BLAZQUEZ; A., CASTILLO, M. and OSORIO, C. Effect of inorganic phosphate on basal and glucose induced insulin release in vitro. Influence of magnesium concentration. Diabetologia, 21, 256-257 (Abstract) (1981).
8.- DEAN, P.M. Ultrastructural morphometry of the pancreatic B-cells. Diabetologia, 9, 115-119, 1973.
9.- SEHLIN, J. Univalent ions in islet function. Horm. Metab. Res. 10 (Suppl), 73-80, 1980.
10.-FREINKEL, N., PEDLEY, K.C., WOODING, P. and DAWSON, R.M.C. Localization of inorganic phosphate in the pancreatic B-cell and its less on glucose stimulation. Science, 201, 1124-1126, 1978.
11.-HALES, C.N. and RANDLE; P.J. Immunoassay of insulin with insulin antibody precipitate. Biochem. J., 88, 137-146, 1963.
12.-HELLMAN, B. and ANDERSSON, T. Calcium and pancreatic B-cell function. 4. Evidence that glucose and phosphate stimulate calcium-45 incorporation in to different intracellular pools. Biochim. Biophys. Acta, 241, 423-491, 1978.
13.- KIKUCHI, M., WOLLHEIM, C.B., SIEGEL, E.G., RENOLD, A.E. and SHARP, G.W.G. Biphasic insulin release in rat islets of Langerhans and the role of intracellular calcium stores. Endocrinology, 105, 1013-1019, 1979.
14.- LACY, P.E. and KOSTIANOWSKY, M. Methods for the isolation of intact islets of Langerhans from the rat pancreas. Diabetes, 16, 35-39, 1967.
15.- MALAISSE, W.J., HUTTON, J.C., KAIWAZU, S., HERCHUELZ, A., VALVERDE, I. and SENER, A. The stimulus secretion coupling of glucose-induced insulin release. XXV. The links between metabolic and cationic events. Diabetologia, 16, 325-330, 1979.
16.- MARSHALL, W.P., BANASIAK, M.F. and KALKHOFF, P.K. Effects of phosphate deprivation on carbohydrate metabolism. Horm. Met. Res. 10, 369-375, 1978.
17.- RASMUSSEN, H., GOODMAN, D.R.P. Relationship between calcium and cyclic nucleotides in cell activation. Physiol. Rev. 57, 421-509, 1977.
18.- RAVAZZOLA; M., MALAISSE-LAGAE, F., AMHEROTM, M., MALAISSE; W.J. and ORCI, L. Patterns of calcium localization in pancreatic endocrine cells. J. Cell Sci., 27, 107-117, 1976.
19.- FREINKEL, N., EL YOUNSI, C., BONNAR, J. and DAWSON, M.C. Rapid transient efflux of phosphate ions from pancreatic islets as an early action of insulin secretagogues. J. Clin. Invest. 54, 1179-1180, 1974.
20.- SUGDEN, M.C. and ASHCROFT, S.J.H. Effects of phosphoenolpyruvate, other glycolitic intermediates and methylxantines on calcium uptake by a mitochondrial fraction from rat pancreatic islets. Diabetologia, 15, 173-180, 1978.
21.-WOLLHEIM, C.B. and SHARP, G.W.G. Regulation of insulin release by calcium. Physiol. Rev. 61, 914-973, 1981.

MECHANISM OF INHIBITION OF GLYCOLYSIS BY VANADATE

Julio E. Benabe[+], Luis A. Echegoyen* and Manuel Martínez-Maldonado[+]

*From the Department of Chemistry, University of Miami, Miami, Florida, [+]The Department of Medicine Veterans Administration Hospital and University of Puerto Rico School of Medicine, San Juan, Puerto Rico 00936

INTRODUCTION

Vanadate is a well known inhibitor of the Na^+, K^+-ATPase (1,2). In addition to the effects on the sodium pump, vanadium compounds have been shown to inhibit acid phosphatase, alkaline phosphatase and adenylate kinase (3,4) as well as several of the enzymes in the glycolytic pathway (5), including glyceraldehyde-3-posphate dehydrogenase (6) phosphoglucomutase and phosphoglyceromutase (7-10). Since vanadate can adopt a stable trigonal bipyramidal structure which resembles phosphate, it could inhibit enzyme activity by replacing phosphate as a substrate in transfer or release reactions and lead to the formation of unstable analogues (5-7). This hypothesis, nevertheless, may not completely explain the inhibitory effect of vanadate on glycolytic enzymes. An alternative mechanism may be the sudden influx of extracellular calcium into cells by inhibition of Na^+, K^+-ATPase (11,12) or Ca^{++}-ATPase (11, 12). This would decrease the cell requirements for ATP (thus, decreasing the rate of glycolysis) by reduction in the activity of the ATPases or inhibition of phophofructokinase, pyruvate kinase or pyruvate carboxylase in the glycolytic pathway (13, 14).

Furthermore it has been shown that intracellular vanadate is readily reduced by cytoplasmic glutathione and after several hours, essentially all the vanadium is present as vanadyl (vanadium IV) complexed to proteins or small molecules (15). Thus another mechanism for inhibition of glycolysis could be a fall in cellular levels of glutathione thereby making the cell susceptible to oxydizing injury. Alternatively or concomitantly it could lead to direct oxidation of the enzymes of the glycolytic pathway.

We performed the following experiments using ^{31}P NMR to examine the possible mechanism for inhibition of glycolysis by vanadate in erythrocytes (RBC).

MATERIAL AND METHODS

Fresh human blood from normal donors was drawn into a heparin containing tube. The blood was placed in centrifuge tubes and the cells separated from the plasma. The cells were washed with 5% Dextrose, lactated Ringers's solution thrice at room temperature (26°C). In the experiments with ethacrynic acid the incubations were done using a phosphate buffer of monosodium phosphate-sodium phosphate dibasic at pH: 7.4. After washing, the cells were recentrifuged and the compounds studied, added. A JEOL FX-90Q NMR instrument with a spectral window of 3,000 Hz was used with a 10mm insert multinuclear probe operating at 36.4 MHz for ^{31}P. All samples were analyzed at room temperature under complete proton decoupling conditions. The pulse width was 50 pulse and the pulse delay was 3 seconds. 8K data points were used and chemical shifts were determined against an external 85% H_3PO_4 standard. On an average, 19,000 scans were accumulated. Cells incubated for 48 hrs. were kept at room temperature for 24 hrs. and scanned for 16 to 24 hrs.

For the experiments with ethacrynic acid a Varian FT-80 NMR [from the University of Miami] was used with a spectral window of 2,000 Hz operating at 29.95 MHz for ^{31}P. the incubations were done with a PO_4 buffer pH:7.41.

Na_3VO_4 was purchased from Alfa Products (Danvers, MA). $VOSO_4$ was obtained from Fisher's Co. (Pittsburgh, PA), iodoacetamide from Sigma (St. Louis, Mo.), nifedipine from Pfizer Pharmaceutical (New York, NY) and sodium edecrin (ethacrynic acid) from Merck Sharp & Dohme (West Point, PA). Lactated Ringers 5% Dextrose used to wash the cells contained sodium: 130 mEq/L, potassium 4mEq/L, calcium 3mEq/L, chloride 109 mEq/L, lactated 38 mEq/L, approximately 524 mOsm/L, pH: 5.0.

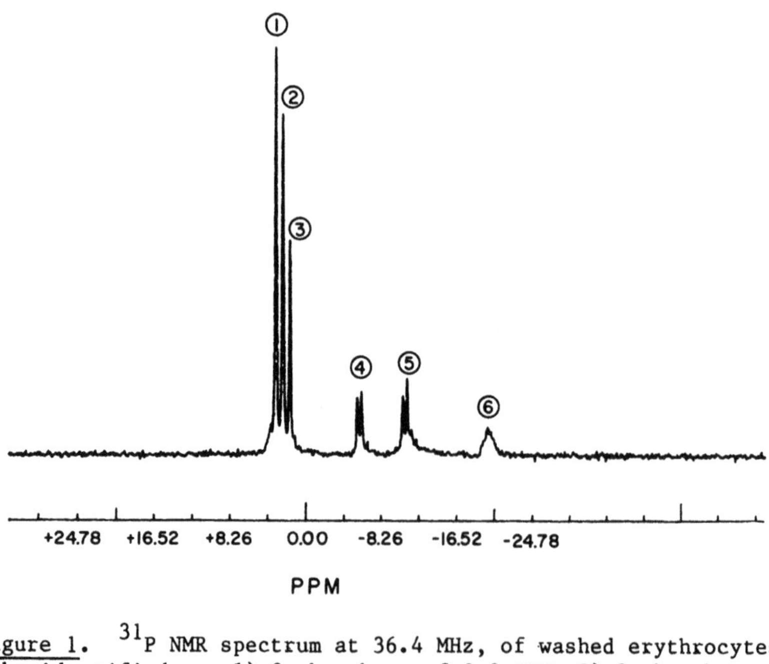

Figure 1. ^{31}P NMR spectrum at 36.4 MHz, of washed erythrocytes. The peaks identified are 1) 3-phosphate of 2,3-DPG 2) 2-phosphate of 2,3-DPG 3) inorganic phosphate (Pi) 4) gamma-phosphate of ATP 5) alpha-phosphate of ATP 6) beta-phosphate of ATP. The sample was scanned for 26 hrs and 31,604 accumulations were obtained. Chemicals shifts were determined against an external 85% H_3PO_4 standard.

RESULTS

The NMR spectrum for normal packed human blood can be seen in Fig 1. The characteristic chemical shifts for the 2,3-diphosphoglycerate (2,3-DPG) 2-phosphate and 3-phosphate, the inorganic orthophosphate Pi and the gamma, alpha, and beta phosphates from ATP are similar to those previously reported in the literature (16-18).

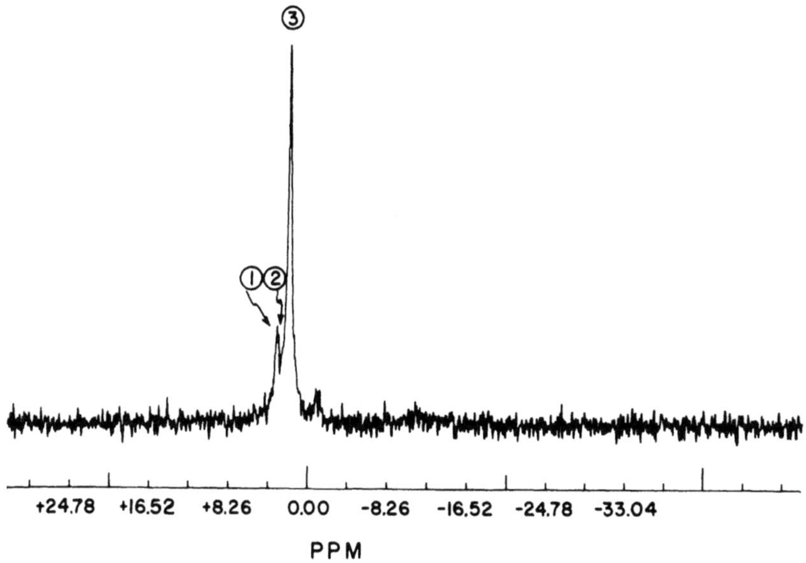

Figure 2. ^{31}P NMR spectrum at 36.4 MHz of washed RBC incubated with Na_3VO_4 10^{-2}M. The sample was scanned for 6 hrs with 8,570 accumulations. The cells had been exposed to Na_3VO_4 for less than one hour prior to scanning; 3) corresponds to Pi.

Figure 2 shows the changes in the ^{31}P spectrum after incubation with Na_3VO_4 (10^{-2}M). The signal for the ATP phosphates, as well as for the 2-phosphate and 3-phosphate of the 2,3-DPG, disappear while the signal for Pi is markedly enhanced.

Since the changes observed in the ^{31}P spectrum after incubation could be non-specific or related to inhibition of the Na^+, K^+, -ATPase, an experiment was done using ouabain (10^{-2}M). As can be seen in Fig. 3 the ^{31}P spectrum did not change from normal, except for a modest decrease in the Pi signal.

An alternative way in which vanadate could inhibit glycolysis is by causing an increase of calcium influx. Nevertheless, as shown in Fig. 4, equimolar concentrations of nifedipine, a calcium channels blocker, did not prevent vanadate induced changes in the ^{31}P spectrum.

In addition, the incubation of cells in extremely high concentrations of calcium (10mM) only caused a decrease in the signal for Pi (Fig. 5).

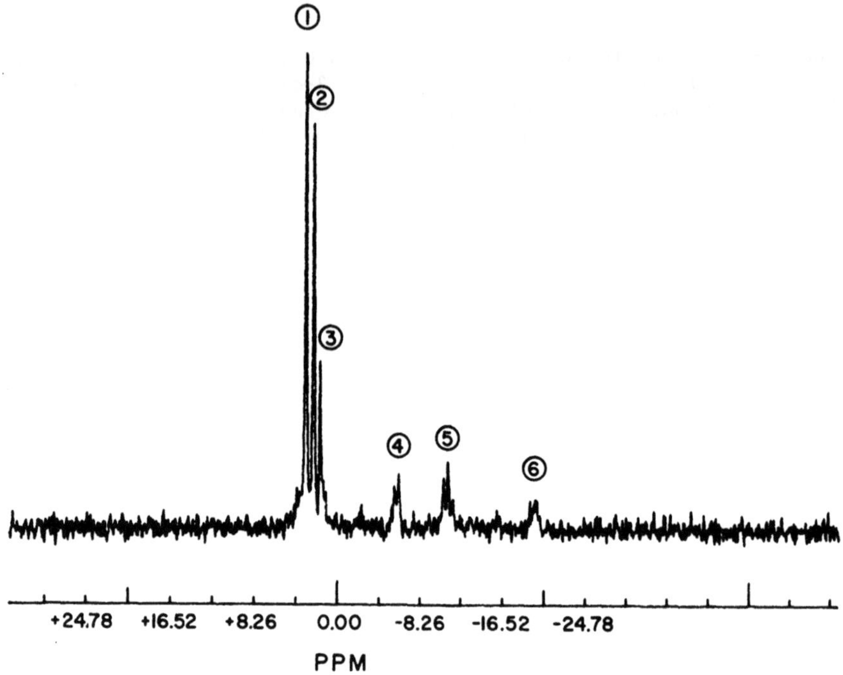

Figure. 3 ^{31}P NMR spectrum at 36.4 MHz of washed RBC incubated with ouabain 10^{-2}M. The cells were incubated with ouabain for 20 hrs. before scanning. They were scanned for 2 hrs. and 3,400 accumulations were obtained. The labelled peaks correspond to those in Fig. 1.

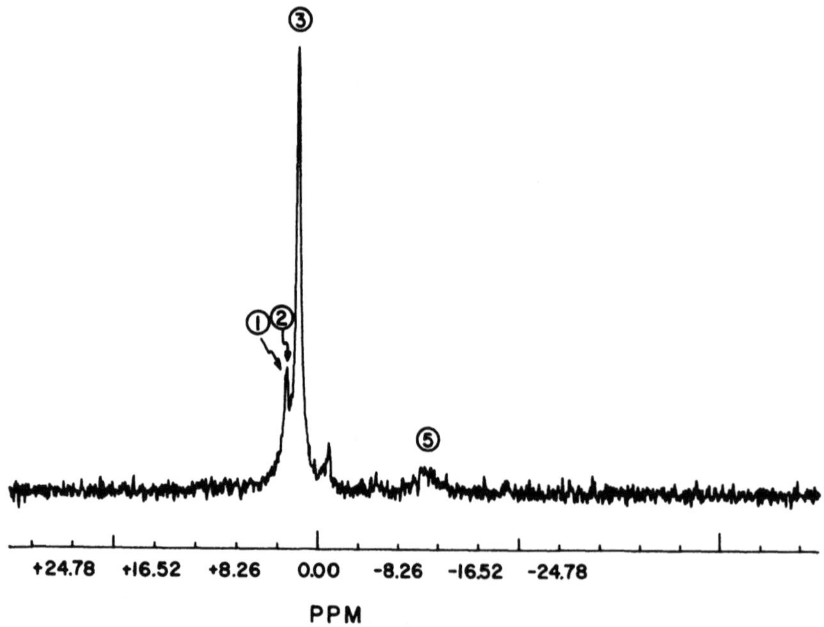

Figure. 4. ^{31}P NMR spectrum at 36.4 MHz of washed RBC incubated with Na_3VO_4 10^{-2}M and nifedipine 10^{-2}M. The cells were scanned for 26 hrs. and 31,560 accumulations were obtained.

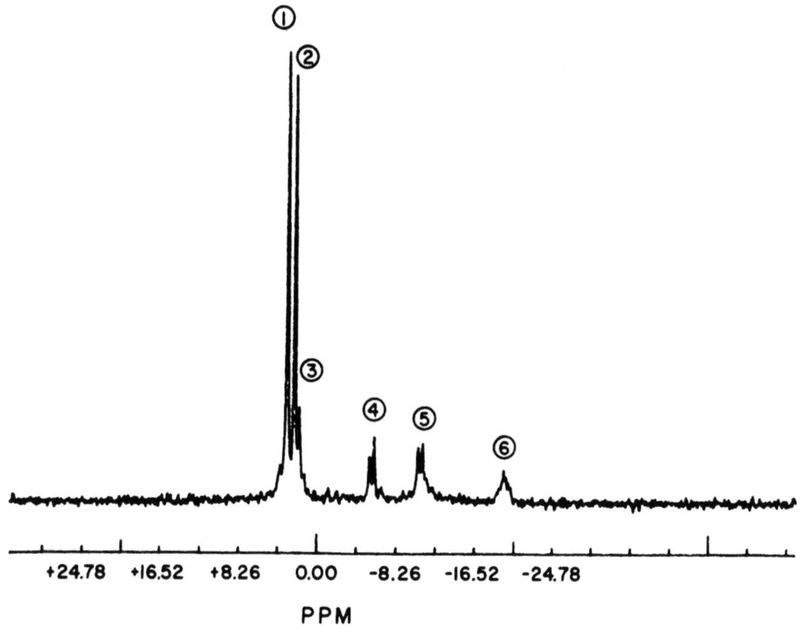

Figure. 5. ^{31}P NMR spectrum at 36.4 MHz of washed RBC incubated with 10mM $CaCl_2.2H_2O$, and scanned for 12 hrs. Around 15,272 accumulations were obtained.

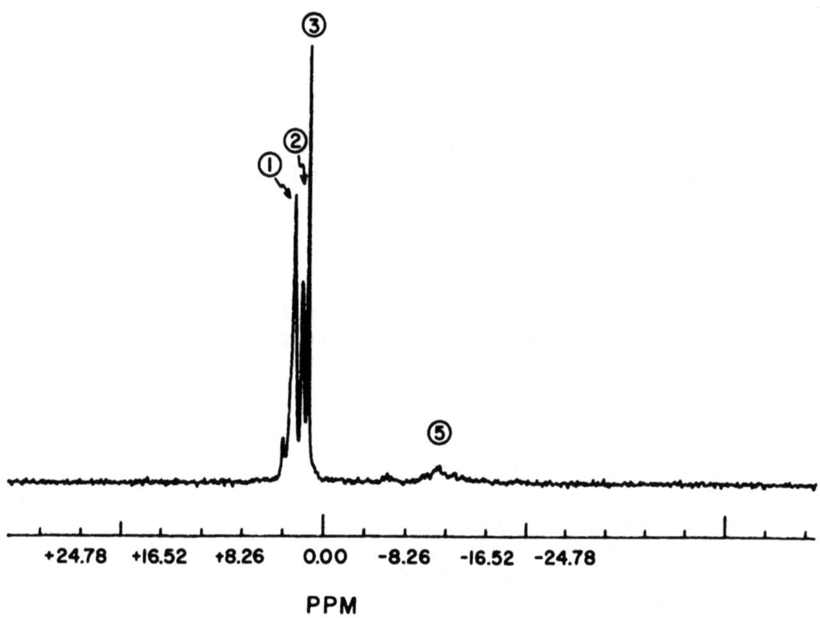

Figure 6. ^{31}P NMR spectrum at 36.4 MHz of washed RBC incubated with iodoacetamide $10^{-2}M$; 23,062 accumulations were obtained during 15 hrs. of scanning. Peaks 1, 2 and 3 correspond to those described in Fig. 1.

Because of the possibility that vanadate may interfere with glycolysis by inhibition of glyceraldehyde 3-phosphate dehydrogenase we compared the effects of iodoacetamide (10^{-4} M), an inhibitor of this enzyme, to those of vanadate. Fig. 6 shows the ^{31}P spectrum of RBC's incubated with iodoacetamide. A decrease in the signal of DPG, 2-phosphate and 3-phosphate, an enhancement of the Pi signal and disappearance of ATP are similar to those observed with vanadate.

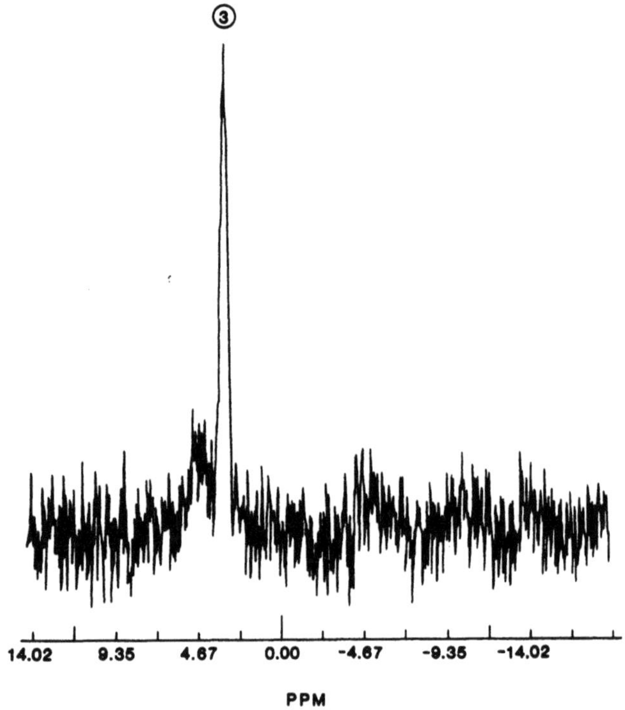

Figure 7. ^{31}P NMR spectrum at 29.95 MHz. of washed RBC incubated with ethacrynic acid 1.6 X 10^{-5} M in PO_4 buffer pH 7.4. The cells were scanned overnight for 10-12 hrs.

In addition, as shown in Fig. 7, ethacrynic acid (1.6 X 10^{-5}M) which inhibits glycolysis by a mechanism similar to iodoacetamide (19), enhanced the Pi signal while the 2,3-DPG and ATP signals disappeared.

Experiments were also done at lower, and 10^{-6}M, vanadate concentrations. The results can be seen in Fig. 8. Incubations with 10^{-6}M vanadate lasted for 48 hrs (Fig. 8).

To assure that a time dependent vanadate independent change was not taking place red cells were incubated for 48 hrs. without vanadate. As shown in Fig. 9 the changes in 2,3-DPG and Pi were distinct from those induced by vanadate. Finally, the incubation of cells with reduced form of vanadate (as the vanadyl sulfate) did not alter the ^{31}P spectrum (Fig. 10).

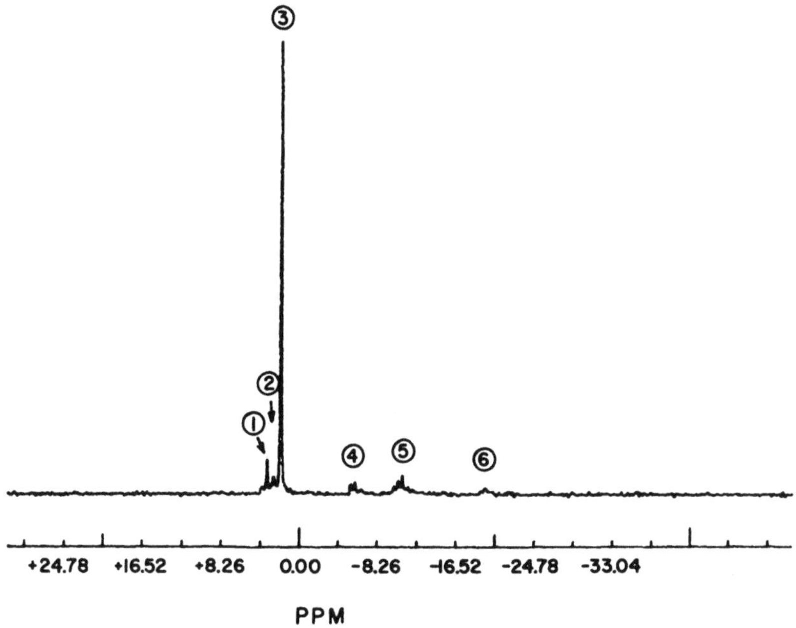

Figure 8. ^{31}P NMR spectrum at 36.4 MHz of washed RBC incubated with Na_3VO_4 10^6 M. The cell were incubated with Na_3VO_4 15 hrs prior to scan. They were scanned for 33 hrs. and 39,714 accumulations obtained.

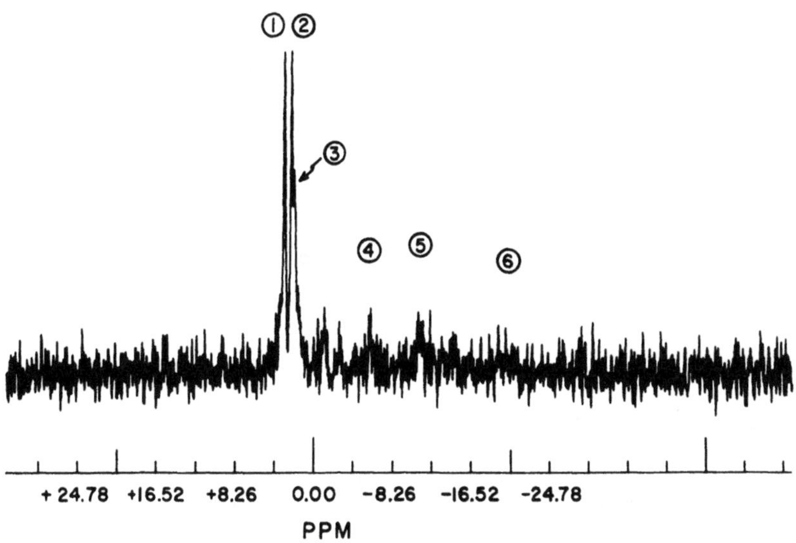

Figure. 9. ^{31}P NMR spectrum at 36.4 MHz of washed RBC after 48 hrs.

DISCUSSION

The NMR traces obtained in intact erythrocytes in these studies were identical to those obtained by Moon and Richards (20) in the original study of ^{31}P NMR of intact erythrocytes done in 1973. Furthermore, our results are similar to those of Henderson et al (16), who found that 2,3-DPG decreased with time, but that the addition of the substrates inosine and pyruvate to aged erythrocytes, lead to an increase in DPG. Our results are consistent in that erythrocytes that were incubated with dextrose and lactate showed a decrease in DPG and ATP and an increase in Pi after 48 hours (Fig. 9). Nevertheless, these changes were markedly accelerated and accentuated in the presence of vanadate at 10^{-6} M (Fig 8). The changes in the ^{31}P spectra of erythrocytes incubated with vanadate seem to be specific for the oxidized form of vanadium. The reduced from, vanadyl, did not elicit significant changes.

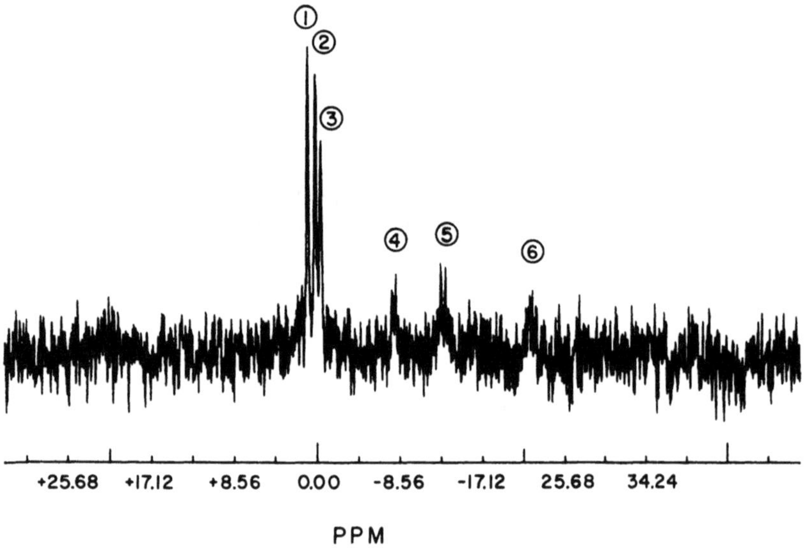

Figure 10. ^{31}P NMR spectrum at 36.4 MHz of washed RBC incubated with $VOSO_4$ 10^{-2} M. The cells were scanned for 23 hrs. and 27,669 accumulations were obtained. The numbered peaks correspond to those in Fig. 1.

The micromolar concentrations of vanadium used in the present studies are compatible with normal physiological and/or biochemical regulatory action of vanadium, since such concentrations may be present in muscle as well as in kidney (20). In recent studies, plasma concentrations of elemental vanadium in normal subjects ranging from 0.27 g/L to 3.94 g/L have been reported (21, 22). Since the concentration of vanadate inside erythrocytes may be 12.5 fold higher than extracellular vanadate (23) it is therefore possible that intracellular micromolar concentrations may occur. In the presence of 10^{-2} M vanadate marked enhancement of the Pi signal and disappearance of DPG and ATP were found at 6 hrs. as shown in Fig. 2.

The effects seen with vanadate are unlikely a consequence of inhibition of Na^+, K^+-ATPase since incubations with ouabain, another inhibitor of the ATPase failed to show similar changes (Fig. 3) Nifedipine, a slow calcium channels blocker which has been shown to inhibit vanadate-induced renal vasoconstriction which is dependent on calcium influx (11, 24-26), failed to prevent the effects of vanadate in RBC. In addition, the presence of very high concentrations of extracellular calcium failed to elicit changes similar to those of vanadate (Fig. 5). In this experiment the only effect was a marked decrease in Pi signal intensity consistent with entry of calcium to cells and formation of Ca-Pi complexes.

Iodoacetamide and ethacrynic acid induced changes in the ^{31}P spectra similar to those of vanadate. These compounds inhibit glycolysis by oxidizing the sulfhydryl groups in the enzymes glyceraldehyde 3-phosphate dehydrogenase (19, 27). Inhibition at this step in the glycolysis pathway can explain the decrease in 2,3 DPG and ATP as well as the increase in Pi observed with vanadate. The intracellular concentration of vanadate which is regulated by glutathione (15,28) may be responsible for the control of the activity of the dehydrogenase by a direct effect on the enzyme.

In vitro studies have shown that vanadate may inhibit only phosphoglucomutase and phosphoglycerate mutase at micromolar concentrations (7). This, however, cannot completely explain our findings. Inhibition of phosphoglucomutase would still allow glucose to be metabolized through hexokinase which is only partially inhibited at milimolar concentrations of vanadate (7). Inhibition of monophosphoglycerate mutase cannot explain the decrese in 2,3-DPG nor the increment in Pi. Since 1,3-DPG could still be metabolized to form 2,3-DPG, and Pi could be sequestered in the DPG´s, the cellular levels of 2,3 DPG and Pi would be maintained. In addition, 3-phosphoglycerate could still be metabolized by 2,3-DPG independent phosphoglyceromutase (8). Another alternative to explain the marked fall in 2,3-DPG would be stimulation of the 2,3-diphosphoglycerate phosphatase activity of phosphoglyceromutase (29) or inhibition of 2,3-diphosphoglycerate synthase (7). Nevertheless, other investigators have shown inhibition of the phosphatase at 0.5 mM vanadate and only partial inhibition of the synthase at milimolar concentrations of vanadate (7).

Of the enzymes in the glycolytic pathway, hexokinase, phophofructokinase and pyruvate kinase appear to be particularly important in the regulation of the glycolytic rate (30). However, at alkaline pH, which favors the oxidized vanadium (metavanadate, Vanadium V), the stimulation of hexokinase and phosphofructokinase activity results in the accumulation of triose phosphates because the glyceraldehyde phosphate dehydrogenase reaction becomes rate limiting (31). Under these conditions one would expect that vanadate may play a role as a metabolic regulator of the enzyme.

In summary, we propose that the effects of vanadate in RBC ^{31}P spectrum are best explained by oxidation of glyceraldehyde phosphate dehydrogenase, an effect similar to those of iodoacetamide or ethacrynic acid. Some of the physiologic effects of vanadate may be mediated through regulation of glycolysis.

ACKNOWLEDGMENT

Supported in part by the Veterans Administration Merit Review and by Grants No. 1-R01 GM33940 and RO-1-AM-30201 of the National Institutes of Health.

REFERENCES

1. Beaugé, L.A., and I.M. Glynn. Commercial ATP containing traces of vanadate alters the response of (Na^+, K^+) - ATPase to external potassium. Nature, London. 272:551-552, (1978).

2. Cantley, L.C.Jr., L. Josephson, R. Warner, M. Yanagisawa, C. Lechene and G. Guidotti. Vanadate is a potent (Na^+, K^+) - ATPase inhibitor found in ATP derived from muscle. J. Biol. Chem. 252:7421-7423, (1977).

3. López, V., T. Stevens, and R.N. Lindquist. Vanadium ion inhibition of alkaline phosphatase catalized phosphate ester hydrolysis. Arch. Biochem. Biophys. 175:31-38, (1976).

4. Choata, G.L., and T.E. Mansour. Inhibition of sheep heart phosphofructokinase by orthovanadate. Fed. Proc. 37:1433, (1978).

5. Simons, T.J.B. Vanadate - a new tool for biologist. Nature. 281:337-338, 1979.

6. De Master, E.G., and R.A. Mitchell. A comparison of arsenate and vanadate as inhibitors or uncouplers of mitochondrial and glycolytic energy metabolism. Biochemistry 12:3616-3621, (1973).

7. Climent, F., R. Bartrons, G. Pons, and J. Carreras. Effect of vanadate on phosphoryl transfer enzymes involved in glucose metabolism. Biochem. Biophys. Res. Commun. 101:570-576, (1981).

8. Carreras, J., and R. Bartrons. Vanadate inhibits 2,3-diphosphoglycerate dependent phosphoglycerate mutases but does not affect the 2,3-bisphosphoglycerate indepent phosphoglycerate mutases. Biochem. Biophys. Res. Commun. 96:1267-1273, (1980).

9. Carreras, J., F. Climent, R. Bartrons, and G. Pons. Effect of vanadate on the formation and stability of the phosphoenzyme forms of 2,3-bisphosphoglycerate dependent phosphoglycerate mutase and of phosphoglucomutase. Biochem. Biophys. Acta. 705:238-242, (1982).

10. Vives-Corrons, J. Ll., J.M. Jou, A. Ester, and M. Ibars. Vanadate increases oxygen affinity and affects enzyme activities and membrane properties or erythocytes. Biochem. Biophys. Res. Commun. 103:111-117, (1981).

11. Benabe, J.E., M.A. Cruz-Soto, and M. Martínez-Maldonado. Critical role of extracellular calcium in vanadate-induced renal vasoconstriction. Am. J. Physiol. 246 (Renal Fluid Electrolyte Physiol. 15):F317-F322, (1984).

12. Varecka, L. and E. Carafoli. Vanadate-induced movements of Ca^{2+} and K^+ in human red blood cells. J. Biol. Chem. 257:7414-7421, (1982).

13. Epstein, F.H. Calcium and the kidney. Am. J. Med. 45:700-714, (1968).

14. Dousa, T.P., and H. Valtin. Cellular actions of vasopressin in the mammalian kidney. Kidney Int. 10:46-63, (1976).

15. Macara, I.G., K. Kustin, and L.C. Cantley Jr. Glutathione reduces cytoplasmic vanadate mechanism and physiological implications. Biochem. Biophys. Acta. 629:95-106, (1980).

16. Henderson, T.O., A.J.R. Costello, and A. Omachi. Phosphate metabolism in intact human erythrocytes: determination by ^{31}P Nuclear Magnetic Resonance spectroscopy. Proc. Natl. Acad. Sci. U.S.A. 71:2487-2490, (1974).

17. Moon, R.B., and J.H. Richards. Determination of intracellular pH by ^{31}P magnetic resonance. J. Biol. Chem. 248:7276-7278, (1973).

18. Omachi, A., W.E. Marshall, and T.O. Henderson. Phosphate metabolism in intact human erythrocytes investigated with phosphorus nuclear magnetic resonance (^{31}P NMR) spectroscopy. In. Biomolecular structure and function. P.F. Agris, ed., Academic Press, New York, Pp. 225-231, (1978).

19. Klahr, S. The effect of diuretics on kidney intermediary metabolism. In Methods in Pharmacology. M. Martínez-Maldonado, ed., Plum Press, New York, Pp. 167-197, (1976).

20. Jorgensen, P.L. Sodium and potassium ion pump in kindey tubules. Physiol. Rev. 60:864-917, (1980).

21. Simonoff, M., Y. Llabador, A. MacKenzie-Peers, and G.N. Simonoff. Vanadium in human serum, as determined by neutron activation analysis. Clin. Chem. 30:1700-1703, (1984).

22. Stroop, S.D., G. Helinek, and H.L. Greene. More sensitive flameless atomic absorption analysis of vanadium in tissue and serum. Clin. Chem. 28:79-82, (1982).

23. Chasteen, N.D. The biochemistry of vanadium. In Structure and Bonding. Copper, molybdenum, and vanadium in biological systems. M.J. Clarke, J.B. Goodenough, P. Hemmrich, J.A. Ibers, C.K. Jorgensen, J.B. Neilands, D. Reineir, R. Weiss, and R.J.P. Williams eds. Springer Verlag, Berlin, Pp 103-136, (1983).

24. López-Novoa, J.M., V. Mayol and M. Martínez-Maldonado. Renal actions of orthovanadate in the dog. Proc. Soc. Exp. Biol. Med. 170:418-426, (1982).

25. López-Novoa, J.M., J.C. García, M.A. Cruz-Soto, J.E. Benabe, and M. Martínez-Maldonado. Effect of sodium orthovanadate on renal renin secretion in-vivo. J. Pharmacol. Exp. Ther. 222:447-451, (1982).

26. Cruz-Soto, M.A., J.E. Benabe, J.M. López-Novoa, and M. Martínez-Maldonado. Na^+, K^+ - ATPase inhibitors and renin release: relationship to calcium. Am. J. Physiol. 247 (Renal Fluid Electrolyte Physiol 16:F650-F655, (1984).

27. Gordon, E.E., and de Hartog, M. Localization and characterization of the inhibitory action of ethacrynic acid on glycolysis. Biochem. Pharmacol. 20:2339-234, (1971).

28. Heinz, A., K.A. Rubinson, and J.J. Grantham. The transport and accumulation of oxyvanadium compounds in human erythrocytes in vitro. J. Lab. Clin. Med. 100:593-612, (1982).

29. Ninfali, P., A. Accorsi, A. Fazi, R. Palma, and G. Fornaini. Vanadate affects of glycose metabolism of human erythrocytes. Arch. Biochem. and Biophys. 226:441-447, (1983).

30. Williamson, J.R. General features of metabolic control as applied to the erythrocyte in Advances in Experimental Medicine and Biology Vol. 6. G.J. Brewer, ed., Plenum Press, New York, Pp. 117-136, (1970).

31. Beutler, E. Energy metabolism and maintenance of erythrocytes Chapt. 35 in Hematology 3rd edition. W.J. Wiliams, E. Beutler, A.J. Erslev, and M.A. Lichtman, eds. McGraw Hill, New York, Pp. 331-345, (1983).

REGULATION OF Na^+/H^+ ANTIPORT IN THE INTACT RENAL PROXIMAL TUBULAR CELL

Edward P. Nord, Dalia Goldfarb and Leon G. Fine

Department of Medicine/Division of Nephrology
UCLA School of Medicine
Los Angeles, California 90024

The kinetic properties of the Na^+/H^+ antiporter, present in the brush border membrane of the renal proximal tubule, have been delineated in a series of studies on isolated brush border membrane vesicles (1,2). In order to determine the factors which regulate this transport system using this approach, the system has had to be perturbed either in vivo or in vitro prior to fragmenting the cell for preparation of isolated vesicles. Whether disruption of the cell alters the apparent response to exogenous stimuli; e.g. hormones, neurogenic influences, etc., is not clear.

We have recently characterized the proximal tubular (PT) Na^+/H^+ antiporter in intact PT cells. In this communication we describe the characteristics of the system as well as the effects of hormones which alter its activity. Using a freshly prepared suspension of isolated PT cells and a primary culture of PT cells it is possible to control the access of substrates to the luminal membrane at the same time that the cell is exposed to exogenous agents.

Freshly prepared proximal tubular cells were isolated from a highly purified preparation of proximal tubular fragments by mechanical agitation in the presence of a hypotonic medium devoid of divalent cations (3). The procedure yields a suspension of isolated cells that display the morphological characteristics and functional parameters of polarized proximal tubular cells (3). The same tubular fragments were used to obtain a primary culture of proximal tubular cells (4). Cells reach confluence in 10-12 days and display morphological and functional characteristics consistent with their proximal tubular origin (4). Sodium influx measurements were obtained by suspending the cell preparation in a sodium-free medium and assaying $^{22}Na^+$ trapped within the cells. Ouabain was used to prevent efflux of Na^+ from the cell. Both the fresh PT cells and the primary culture were studied with the cells in suspension. The activity of the antiporter in the primary cultures was also studied by measuring Na^+ uptake into cells in monolayer. In addition antiporter activity was evaluated in suspended primary culture PT cells by measuring the rate of change of intracellular pH (pH_i) in response to external Na^+ in acidified cells preincubated in zero Na^+. For these experiments the cells were loaded with bis(carboxyethyl) carboxyfluorescein (BCECF) for 30 min and acidified with nigericin to an approximate pH_i of 6.5. The fluorescein emission is a linear function of pH_i. Thus the rate of rise of pH_i in response to 100 mM NaCl is a function of Na^+-H^+ antiport activity.

The J_{max} and K_t for Na^+ for the luminal side of the antiporter in freshly prepared cells (Fig 1) and in primary cultures, were derived from the Hanes-Woolf linearization of the Michaelis-Menten equation. K_t for the process in both preparations was 5-10 mM, comparable to the value reported for isolated rabbit renal microvilli vesicles, studied under comparable conditions (2). J_{max} for sodium in the suspend cultured and freshly prepared cells was approximately 1-3 nmoles/10^6 cells/min. In contrast, the J_{max} of PT cells in monolayer was approximately 20 nmoles/10^6 cells/min with a similar K_t.

Fig 2 depicts the effect of guanabenz, an α_2-adrenergic agonist on Na^+-H^+ antiport activity (5). The 3 minute uptake values for ouabain-treated cells are shown. Guanabenz enhanced J_{Na^+} by 30%. Amiloride (10^{-3} M) completely inhibited the guanabenz induced increment in J_{Na^+} indicating antiport activity. Yohimbine, an α_2 adrenergic antagonist completely blocked the guanabenz induced enhanced J_{Na^+} (Fig 3) demonstrating the specificity of the α_2-adrenergic receptor (5). In contrast, prazosin, an α_1-adrenergic antagonist was a poor inhibitor of guanabenz mediated increased J_{Na^+}. Thus α_2 adrenergic agonists appear to activate the proximal tubular Na^+/H^+ activity.

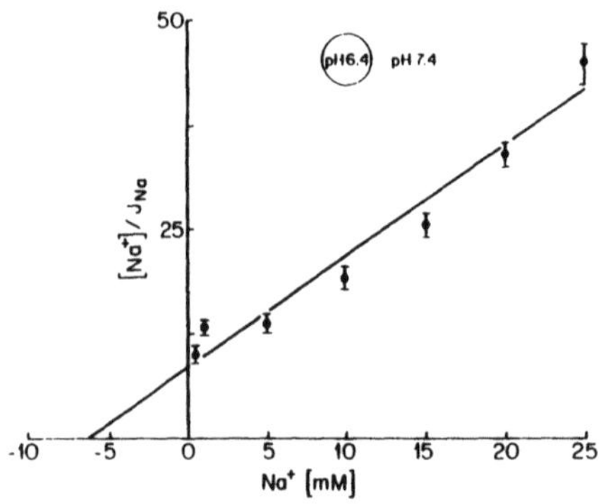

Figure 1: Hanes-Woolf plot for sodium uptake in freshly prepared isolated proximal tubule cells. Cells were suspended in a sodium-free buffer (300 mOsm) at pH 7.4. Na^+ concentration in the transport buffer varied from 0.5 - 25 mM. Units for the ordinate are mM/(nmoles.10^6 cells.min^{-1}).

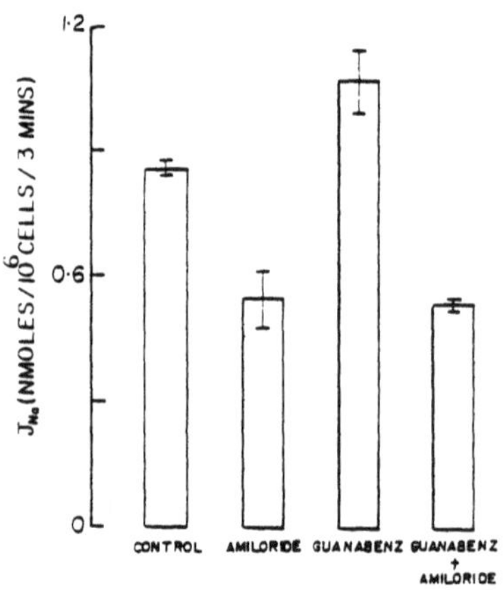

Figure 2: Effect of amiloride on guanabenz-mediated enhanced sodium influx in proximal tubular cells. Three minute uptake values are depicted. The transport buffer contained amiloride (10^{-3} M) guanabenz (10^{-5} M), both or neither of the compounds.

The effects of 10 ug/ml insulin and 1 mM prostaglandin E_1 (PGE_1) on Na^+ uptake (6) are shown in Fig 4. Both agents enhanced Na^+ uptake. This effect was blocked by 10^{-4} M amiloride. Since the concentrations of insulin and PGE_1 required to stimulate Na^+/H^+ antiport activity were extremely high, it is unlikely that these agents play a physiological role in this context.

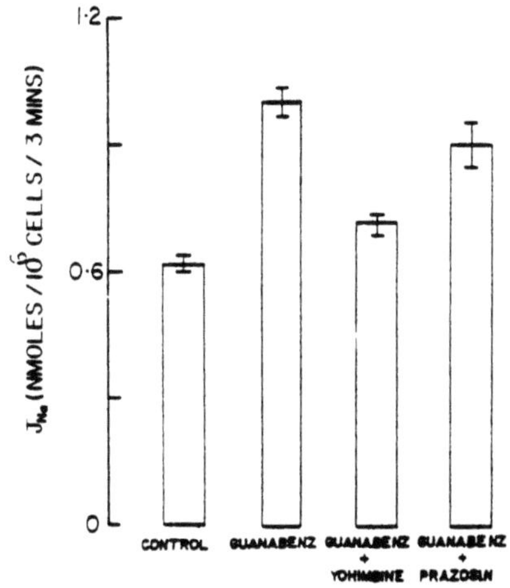

Figure 3: Effect of adrenergic agonists on guanabenz-enhanced sodium influx in proximal tubular cells. Where indicated 10^{-5} M guanabenz, 10^{-5} M yohimbine or 10^{-5} M prazosin were added to the transport buffer.

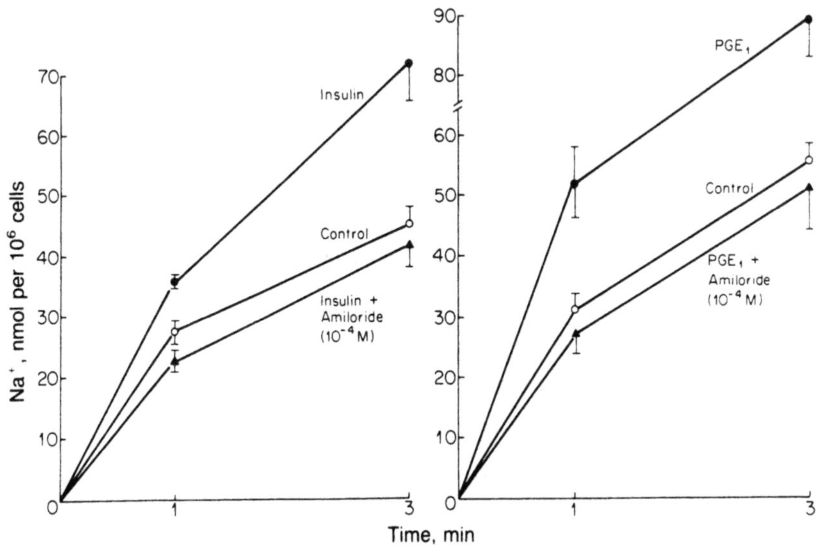

Figure 4: Effect of insulin (10 ug/ml) and PGE (10^{-6} M) on Na uptake by primary cultures of proximal tubular cells in suspension. The stimulatory effect was blocked by amiloride.

Angiotensin II, a hormone which is produced locally within the kidney, has been shown to stimulate proximal tubular volume reabsorption in vitro at physiological concentrations (10^{-11} M) and to inhibit it at higher concentrations (7). When examined on sodium depleted primary cultures in monolayer, AII inhibited the maximum rate of saturable Na^+ uptake by 60% at concentrations from 10^{-11} M to 10^{-9} M. Since angiotensin binding appears to be maximal at high extracellular sodium concentrations (i.e. 200 mM) these experiments were repeated in the absence of a Na^+-depletion step. Under these circumstances no inhibitory effect of angiotensin was evident. Thus there appears to be no stimulatory effect of AII on PT cells in monolayer; an inhibitory effect was evident only in cells exposed to low Na^+ concentrations. These effects were confirmed using BCECF fluorescence to monitor the Na^+-dependent rise in pH_i in control and AII-treated cells.

These studies indicate that modulation of the proximal tubular Na^+/H^+ antiporter by hormones can be studied in the intact cell but that the basal rates of transport and the incubation conditions determine the nature of the elicited response.

REFERENCES

1. J.L. Kinsella and P.S. Aronson, Properties of the Na^+-H^+ exchanger in renal microvillar membrane vesicles. Am J Physiol 238 (Renal Fluid Electrolyte Physiol 7):F461-F469 (1980).
2. E.P. Nord, A. Hafezi, E.M. Wright, and L.G. Fine, Mechanisms of Na^+ uptake into renal brush border membrane vesicles, Am J Physiol (Renal Fluid Electrolyte Physiol 16):F548-F554 (1984).
3. E.P. Nord, D. Goldfarb, N. Mikhail, P. Moradeshaghi, A. Hafezi, S. Vaystub, E.J. Cragoe Jr., and L.G. Fine, Characteristics of the Na^+-H^+ antiporter in the intact renal proximal tubular cell, Am J Physiol (in press) (1985).
4. L.M. Sakhrani, B. Badie-Dezfooly, W. Trizna, N. Mikhail, A.G. Lowe, M. Taub, and L.G. Fine, Transport and metabolism of glucose in renal proximal tubular cells in primary culture. Am J Physiol 246 (Renal Fluid Electrolyte Physiol 15):F757-F764 (1984).
5. P.A. Insel, M.D. Snavely, D.P. Healy, P.A. Munzel, C.L. Potenza, and E.P. Nord, Radioligand binding and functional assays demonstrate postsynaptic $alpha_2$-receptors on proximal tubules of rat and rabbit kidney, J Cardiovasc Pharm (in press) (1985).
6. L.G. Fine, B. Badie-Dezfooly, A.G. Lowe, A. Hamzeh, J. Wells, and S. Salehmoghaddam, Stimulation of Na^+/H^+ antiport is an early event in hypertrophy of renal proximal cells. Proc Natl Acad Sci USA 82:1736-1740 (1985).
7. V.L. Schuster, J.P. Kokko, and H.R. Jacobson, Angiotensin II directly stimulates sodium transport in rabbit proximal convoluted tubules, J Clin Invest 73:507-515 (1984).

INTERACTION OF VITAMIN D-METABOLITES WITH ADENYLATE CYCLASE/CYCLIC AMP SYSTEM: A BIOLOGICAL MODEL OF CONTROLLED REGULATION

Mordecai M. Popovtzer, Hanna Wald, Michael M. Friedlaender and Dvora Rubinger

Nephrology and Hypertension Services
Hadassah University Hospital
Jerusalem, Israel

Parathyroid hormone and other peptide hormones convey their biological messages by combining with membrane-bound receptors and activating them. The message is advanced then by catalitically generated cyclic AMP which carries the signal to the effector components of the cytosol, leading ultimately to the physiological response. The factors which control the operation of this system have not been delineated in detail yet. Such factors if present may exert their effect at any level of the activation cascade. For example inhibitors of phosphodiesterase may enhance the effect of adenylate cyclase agonists by blocking the degradation of cyclic AMP. In contrast, depletion of intracellular ATP which serves as a substrate for the generation of cyclic AMP may abolish the response to peptide hormones. We have recently demonstrated refractoriness of the kidney to parathyroid hormone in rats with maleate-induced Fanconi syndrome.[1] In these animals the absence of phosphaturic response and the absence of augmented excretion of urinary cyclic AMP after the administration of parathyroid hormone was not due to altered function of the receptor complex but due to documented lack of ATP. Addition of ATP to the in vitro system restored to normal the response to parathyroid hormone as reflected by enhanced generation of cyclic AMP. The question which requires further investigations is whether variations in intracellular ATP may play a role in the physiological response to parathyroid hormone and other adenylate cyclase agonists.

Divalent ions exert both an inhibitory and stimulating effect on the activity of the catalytic component of adenylate cyclase, in vitro. Their effect on peptide hormone-stimulated adenylate cyclase is rather small both in vivo and in vitro.[2,3] It has to be emphasized, however, that minimal physiological concentrations of calcium-ion are necessary for the response to parathyroid hormone.

Factors that modulate the physiological action of parathyroid hormone and other peptide hormones at the kidney level and other target sites remain to be defined. Experimental data which have accumulated in our laboratory suggest a possible role for vitamin D-metabolites in regulating renal responses to parathyroid hormone and to other peptide hormones.

We have recently extended our observations on the antiphosphaturic actions of metabolites of vitamin D. We have further characterized the interactions between adenylate cyclase and the derivatives of vitamin D,

Table 1. Interactions of parathyroid hormone (PTH) and other agonists of adenylate cyclase/cyclic AMP system with metabolites of vitamin D in parathyroidectomized rats.

Metabolite of vitamin D	Phosphaturic response to PTH*	Phosphaturic response to dibutyryl cyclic AMP	PTH-induced increase in urinary cyclic AMP	PTH-induced** activation of renal adenylate cyclase in vitro
$25(OH)D_3$	↓	↑	↓	↓
$1,25(OH)_2D_3$	↓	↑	↓	↓
$24,25(OH)_2D_3$	↓	-	↓	↓

*The same applies to calcitonin and glucagon but not to vasopressin
**In vitro system contains a preparation of broken membranes of renal cells
*** ↓ suppression; ↑ enhancement

both in vivo and in vitro.[4-8] Some of the main features are listed in tables 1 and 2. It became apparent that the inhibitory effect of vitamin D derivatives is specific both with regard to 1) the molecular species of the vitamin, and 2) the activating peptide hormone. Thus the presence of group OH in position 25 on the side chain of the sterol is a prerequisite for a biological activity whereas the presence of OH in position 1α (alpha) is not a prerequisite. Similar relationship between structure and function have been suggested by others with regard to vitamin D-induced changes in membrane potential in the proximal tubule of necturus kidney.[9] Vitamin D derivatives were shown to antagonize the effects of 3 peptide hormones; parathyroid hormone, calcitonin and glucagon, but not that of vasopressin even though it activated the adenylate cyclase. The inhibitory action of vitamin D on the response to PTH both in vivo and in vitro is suppressed by cycloheximide (Table 2). This suggests that synthesis of protein, possibly post-transcriptional in nature is required for the antiphosphaturic action and for the suppression of parathyroid hormone-stimulated adenylate cyclase. In contrast, the integrity of microtubular reticulum is not required for the effects of vitamin D derivatives, as indicated by the inability of colchicine, an agent that disrupts the microtubular reticulum, to interfere with their action (table 2). The nature of the protein that mediates the effect of vitamin D-derivatives is unknown. Recently a high affinity between calmodulin and $25(OH)D_3$ has been described by Reddy and Johnson.[10] Our recent findings in vitro assign a possible role to calmodulin in the mechanism of action of vitamin D-derivatives (table 2). This is based on the information that the administration of trifluoperazine, an antagonist of calmodulin, blocks the effect vitamin D-derivatives in vitro. Further experiments will be needed to delineate in greater detail the nature of the relationship between the observed actions of vitamin D-derivatives and calmodulin. The question which arises is whether the synthesis of calmodulin is vitamin D-dependent or/and is the effect of vitamin D mediated by calmodulin.

Table 2

	Antiphosphaturic actions of $25(OH)D_3$ and $1,25(OH)_2D_3$	$25(OH)D_3$ and $1,25(OH)_2D_3$ induced inhibition of PTH-stimulated adenylate cyclase in vitro
Cycloheximide	↓	↓
Colchicine	0	0
Trifluoperazine	-	↓

One of the interesting findings accruing from our experiments is the observation that vitamin D-metabolites exert both an inhibitory and stimulating effect in the activation cascade of the adenylate cyclase/cyclic AMP system. On the one hand vitamin D-metabolites attenuate the activation of adenylate cyclase by its agonists while on the other hand, they enhance the response to cyclic AMP (table 1).[11] Thus 25(OH)D_3 suppresses the parathyroid hormone-stimulated increase in urinary cyclic AMP and the formation of cyclic AMP in vitro, but it enhances the phosphaturic response to exogenous cyclic AMP administered as dibutyryl cyclic AMP (table 1). In most of our experimental settings, the final outcome reflected interference with the activation of adenylate cyclase, yielding an antiphosphaturic response. In recent experiments in our laboratory performed with rats with severe secondary hyperparathyroidism, acute administration of 25(OH)D_3 elicited a phosphaturic response. This observation suggested that in this case augmentation of the response to cyclic AMP was probably the overriding action of the vitamin D-metabolite.[12] Validation of this possibility awaits further experimental work.

It is of interest that our observations on the opposing effects of vitamin D-metabolites on adenylate cyclase/cyclic AMP system in the kidney were recently reproduced in other biological systems. Catherwood and Rubin demonstrated opposing effects of 1,25$(OH)_2D_3$ in the human monoblastic cells' line, U 937.[13] On the one hand 1,25$(OH)_2D_3$ suppressed the ability of the cells

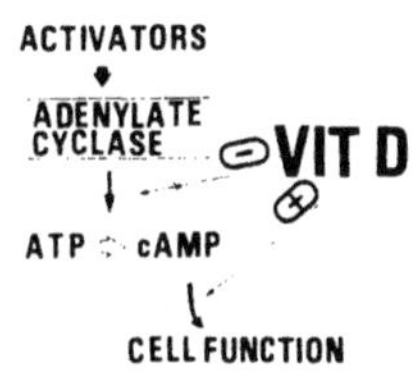

Figure 1. Schematic presentation of the opposing effects of vitamin D-metabolites on adenylate cyclase/cyclic AMP system. Vitamin D metabolites suppress ⊖ the formation of cyclic AMP in response to the adenylate cyclase activators, but they enhance ⊕ the effect of cyclic AMP on cell function.

to generate cyclic AMP in response to adenylate cyclase agonists including isoproterenol and prostaglandin E, but on the other hand it facilitated the cellular differentiation in response to cyclic AMP agonists. These workers also demonstrated that 1,25$(OH)_2D_3$ attenuates the cyclic AMP response to adenylate cyclase activators in three additional cell lines, including rat osteosarcoma, T 84 colon carcinoma and lecithin-activated human lymphocytes. These results indicate that 1,25$(OH)_2D_3$ diminishes the ability of the cells to generate cyclic AMP in response to a hormonal stimulus, but at the same time 1,25$(OH)_2D_3$ is able to facilitate the action of cyclic AMP.

In view of the experimental data from our laboratory and that from others it is likely that the interaction between adenylate cyclase activators and vitamin D-metabolites is not limited to the kidney but exists also in other organ systems. This interaction is depicted schematically in figure 1 and it may represent a biological mechanism for fine regulation of balanced physiological functions. It acts to restrain excessive stimulation of adenylate cyclase/cyclic AMP system, but at the same time it acts to preserve normal function by enhancing the effect of cyclic AMP.

REFERENCES

1. H. Wald, and M. M. Popovtzer, Renal refractoriness to parathyroid hormone in experimental Fanconi syndrome. Evidence for intact adenylate cyclase activation. Pfluegers Arch. 402:116 (1984).
2. M. M. Popovtzer, J. B. Robinette, K. M. McDonald, C. K. Kuruvila, The effect of hypercalcemia on renal handling of phosphorus. Evidence for two reabsorptive mechanisms. Amer. J. Physiol. 229:901 (1975).
3. H. Wald, and M. M. Popovtzer, Effect of divalent ions on basal and hormone-activated renal adenylate cyclase/cyclic AMP system. Min. Electr. Metab. 10:133 (1983).
4. M. M. Popovtzer, J. Robinette, H. DeLuca, and M. F. Holic, The acute effect of 25(OH) cholecalciferon on renal handling of phosphorus. Evidence for parathyroid hormone dependent mechanism. J. Clin. Invest. 53:913 (1974).
5. M. M. Popovtzer, and J. B. Robinette, The effect of 25-OH-vitamin D_3 on renal handling of phosphorus. Evidence for inhibition of cyclic monophosphate formation. Amer. J. Physiol. 229:907 (1975).
6. M. M. Popovtzer, R. S. Flis, and M. Blum, The effect of 25(OH)vit D_3 on renal handling of phosphorus: Evidence for interference with the phosphaturic action of calcitonin. Amer. J. Physiol. 232:515 (1977).
7. M. M. Popovtzer, and H. Wald, Evidence for interference of 25(OH)vitamin D_3 with phosphaturic action of glucagon. Amer. J. Physiol. 240:F269 (1981).
8. M. M. Friedlaender, Z. Kornberg, H. Wald, and M. M. Popovtzer, Renal response to 1α(OH)vitamin D_3 and 24,25$(OH)_2$vitamin D_3: Evidence for the essential role of 25(OH) group. Amer. J. Physiol. 13:F674 (1983).
9. A. Edelman, M. Garabedian, I. Anagnostopoulos, and S. Balsan, Vitamin D and steroids. Rapid effect on membrane potential of hypertrophic chandrocytes and proximal tubular cells, in: "Vitamin D. A Chemical and Biochemical and Clinical Update," W. de Gruyter & Co., Berlin & New York, p. 433 (1985).
10. G. S. Reddy, and J. D. Johnson, in: Proc. VIIth International Conference on Calcium Regulating Hormones. Abstract p. 4, Tokyo, Japan (1983).
11. M. M. Popovtzer, Interactions between PTH, vitamin D metabolites and other factors in tubular reabsorption of phosphate. Adv. Exp. Med. Biol. 103:11 (1978).
12. D. Rubinger, and M. M. Popovtzer, Unpublished observations.
13. B. D. Catherwood, and J. E. Rubin, Interaction of 1,25$(OH)_2$D with cAMP class of biological signals, in: "Vitamin D, A Chemical and Biochemical and Clinical Update," W. de Gruyter & Co., Berlin & New York, p. 229 (1985).

DIBUTRYLADENOSINE 3',5'-CYCLIC MONOPHOSPHATE (dBcAMP) DOES NOT MIMIC THE ACTION OF PARATHYROID HORMONE (PTH) ON CANINE PROXIMAL TUBULAR BASOLATERAL MEMBRANE $Na^{+}:Ca^{2+}$

J.E. Scoble, D. Moskowitz, and K.A. Hruska

Renal Division
Washington University and The Jewish Hospital of St. Louis
St. Louis, Missouri

INTRODUCTION

We have previously demonstrated that the $Na^{+}:Ca^{2+}$ exchanger found by Gmaj et al (1) in rat renal cortical basolateral membrane vesicles (BLMV) is present in canine renal cortical BLMV (2). We showed that its activity is depressed in the thyroparathyroidectomized (TPTX) state and that it was restored by in vivo infusion of parathyroid hormone (PTH). This effect was achieved by a change in V_{max} of the exchanger but no significant change in the K_m. It has been established that the phosphaturic effect of PTH on the proximal tubule can be mimicked by cAMP analogues (3). We sought to examine the effect of dBcAMP on the activation of the $Na^{+}:Ca^{2+}$ exchanger in BLMV prepared from TPTX dogs. The dose of dBcAMP used was chosen to achieve the same phosphaturic effect as the previously used 2 µg/Kg PTH (4).

METHODS

TPTX was performed on female mongrel dogs weighting 15-20 Kg fed on Purina dog chow as previously described (2). 48 hrs post-TPTX the dogs were anesthetized, external jugular catheter inserted, bladder catheterized and the abdominal cavity opened. Saline, 0.9%, was infused at 0.8 mls/min and three successive clearance periods studied: i) control-saline infusion alone; (ii) post nephrectomy-saline infusion after the removal of the control kidney; (iii) dBcAMP-saline infusion containing 50 mg dBcAMP after period (ii). Each clearance period lasted 30 min. The second dBcAMP treated kidney was removed after period (iii). Fractional excretion of Na^+, Ca^{2+} and $PO_4 2-$ was measured as was plasma Ca^{2+} and $PO_4 2-$.

BLMV were prepared by percoll density centrifugation as previously described. The final preparation of the BLMV was in 150 mM KCl, 20 mM Tris, 20 mM Hepes, pH 7.5 and these were aliquoted into 10 λ samples. The BLMV were then preincubated with 40 λ of 150 mM KCl, 20 mM Tris, 20 mM Hepes, 100 µM $^{45}CaCl_2$ pH 7.5 for 30 minutes. Efflux was initiated by the addition of either: i) 150 mM KCl, 20 mM Tris, 20 mM Hepes, 1 mM EGTA pH 7.5; or ii) 150 mM NaCl, 20 mM Tris, 20 mM Hepes, 1 mM EGTA pH 7.5. Efflux was terminated by addition of ice-cold stop solution containing either 150 mM KCl or NaCl, 20 mM Tris, 20 mM Hepes pH 7.5, depending on the efflux solution, followed by rapid filtration as previously described (2). The trapped radioactivity on the filter was measured with a scintillator spectrometer. All results are the mean of at least three experiments.

RESULTS

Figure 1. The infusion of dBcAMP produced a significant phosphaturic effect compared to both the pre-nephrectomy and post-nephrectomy control periods. The phosphaturic effect was not caused by the removal of the control kidney. The phosphaturia was associated with a fall in plasma phosphaturia phosphate concentration. There was no significant effect on plasma Ca^{2+}. The phosphaturic effect was the same as that achieved by 2 µg/Kg PTH reported previously (4).

Figure 2. There was no significant difference in the rate of $^{45}Ca^{2+}$ efflux from BLMV isolated from kidneys of the control or dBcAMP treated kidneys. This was the case of either K^+ equivalent (KCl efflux solution) or Na^+ gradient stimulated (NaCl efflux solution) conditions. This is in marked distinction from the previously reported effect of PTH on both these efflux conditions (2). The results indicate that in a setting where dBcAMP mimiced the effect of PTH on phosphate transport (4) it did not reproduce the effect of PTH on the Na^+:Ca^{2+} exchange activity.

CONCLUSION

These results confirm the phosphaturic effect of dBcAMP on phosphate excretion in the dog. The dose of dBcAMP was sufficient to produce the same phosphaturic effect as the dose of PTH previously shown to modulate $Na^+:Ca^{2+}$ exchange (2). It has been shown that the proximal tubule possesses a PTH sensitive adenyl cyclase (5) but the modulation of basolateral $Na^+:Ca^{2+}$ exchange and Ca^{2+} permeability does not appear

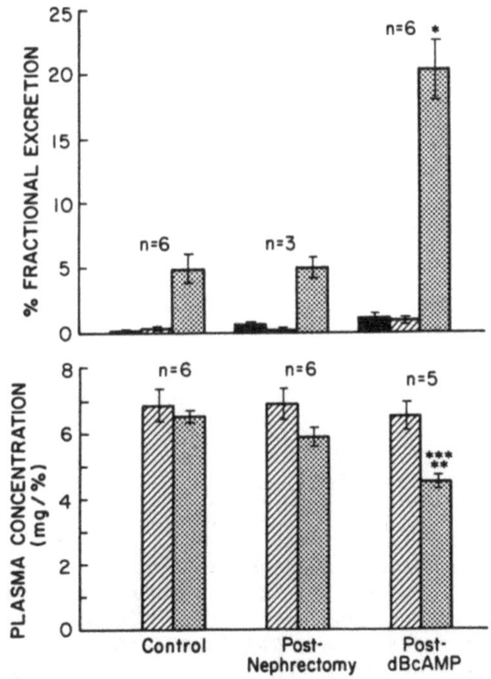

Figure 1 The effect of dBcAMP infusion on the fractional excretion of Na^+ (left), Ca^{2+} (middle), and phosphate (right), and plasma Ca^{2+} and phosphate. The clearance studies were performed as described in Methods. *$p<0.05$ compared with post nephrectomy, ** $p<0.001$ compared with control, *** $p<0.002$ compared with post nephrectomy.

to be through cAMP. This raises the possibility that another second messenger of PTH modulates these actions. It has been shown that PTH increased cytosolic calcium concentration in canine proximal tubular cells (6). It appears possible that cytosolic calcium could modulate both calcium influx and efflux in the basolateral membrane.

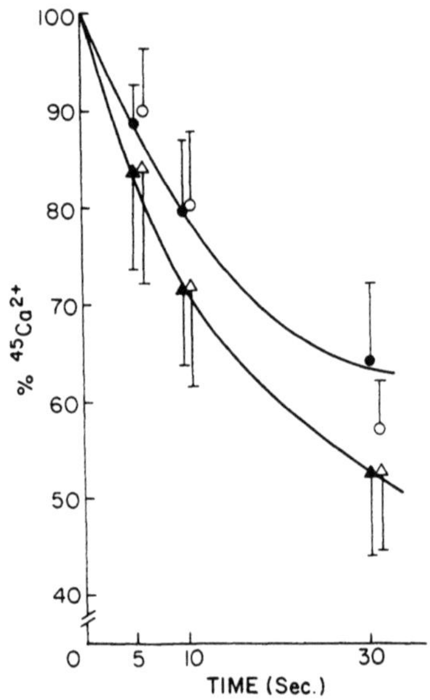

Figure 2 The effect of dBcAMP on $^{45}Ca^{2+}$ efflux from BLMV exposed to symetrical KCl solutions or to Na^+ gradients directed from out to in. BLMV were prepared from kidneys of TPTX (●,▲) and TPTX and dBcAMP (O, Δ) animals using a buffer of 150 mM KCl, 20 mM Tris, 20 mM Hepes, pH 7.5, in the final "vesicular loading" steps as described in Methods. BLMV, 10 µl, were preincubated at 30° with 40 µl of 150 mM KCl, 20 mM Tris, 20 mM Hepes pH 7.5, 100 µM $^{45}CaCl_2$ for 30 minutes, and Ca^{2+} efflux was initiated by dilution with O, ●, KCl 150 mM, 20 mM Tris, 20 mM Hepes pH 7.5, 1 mM EGTA, or Δ, ▲ 150 mM NaCl, 20 mM Tris, 20 mM Hepes pH 7.5, 1 mM EGTA. Efflux was terminated by addition of ice cold stop solution containing either 150 mM KCl or NaCl, 20 mM Tris, 20 mM Hepes pH 7.5, depending on the solutions used to initiate efflux, followed by rapid filtration as described in Methods.

REFERENCES

1. P. Gmaj, H. Murer, and R. Kinne, Calcium ion transport across plasma membranes isolated from rat kidney cortex, Biochem J. 178:549-557, 1979.
2. J. E. Scoble, S. Mills, and K. A. Hruska, Calcium transport in canine renal basolateral membrane vesicles, J. Clin. Invest. 75:1996-1105, 1985.
3. Z. S. Agus, J. B. Puschett, D. Senesky, and M. Goldberg, Mode of action of parathyroid hormone and cyclic adenosine 3',5'-monophosphate on renal tubular phosphate reabsorption in the dog, J. Clin. Invest. 50:617-626, 1971.
4. M. R. Hammerman, I. E. Karl, and K. A. Hruska, Regulation of canine renal vesicle Pi transport by growth hormone and parathyroid hormone, Biochim. Biophys. Acta 603:322-335, 1980.
5. F. Morel, Sites of hormone action in the mammalian nephron, Amer. J. Physiol. 240: F159-F164, 1981.
6. K. A. Hruska, S. Mills, and S. Westbrook, Regulation of proximal tubular (PT) cell cystosolic (Ca^{2+}) by parathyroid hormone (PTH), Kidney Int. 27:117, 1985.

PHOSPHATE TRANSPORT THROUGH PLACENTA BRUSH BORDER MEMBRANE

Michèle G. Brunette, Sylvie Letendre and Serge Allard

Department of Pediatrics, University of Montreal

Maisonneuve Hospital, Montreal, Quebec, Canada

INTRODUCTION

During the whole gestation, and particularly the third trimester, large quantities of inorganic phosphate (PO_4) move from the mother to the foetus against a concentration gradient. Since intracellular PO_4 concentration is believed to be higher than in plasma, it is probable that PO_4 is actively transported through the brush border membrane (BBM) facing the maternal side of placenta. Moreover in total placenta perfusion experiments, Stulc et al. (1982-1984) reported that PO_4 transport is dependent on Na concentration in the perfusate, suggesting that in placenta, as in the intestine and the kidney, PO_4 is co-transported with Na.

The purpose of this study was to investigate PO_4 transport through the placenta BBM, in order to extend our knowledge of the process involved in this active mechanism.

MATERIALS AND METHODS

The placentas were obtained from full term vaginal deliveries. Immediately after delivery, they were perfused with one liter of saline through the superficial ramifications of the umbilical artery. The central part of the placenta was then cut into 4 to 6 mm slices, and gently stirred with a magnetic bar at 4°C for 30 min. (Smith et al 1974). The suspension was filtered through a cotton gauze, and centrifuged at 100,000 X G for 60 minutes. The red blood cells were discarded by gently washing the pellet, which was further homogenized, suspended in 15 ml of 10 mM Tris - 270 mM Mannitol buffer pH 7.4 added with 10 mM $MgCl_2$, and stirred again for 20 min. After a low speed centrifugation at 3,000 X G for 10 minutes, the supernatant was recentrifuged at 27,000 X G for 10 other minutes. The loosely packed white layer was collected and suspended in 10 mM Tris 270 mM Mannitol pH 7.4, and kept on ice until transport experiments. Homogenate of the whole placenta was prepared from the central part of the placenta. The alkaline phosphatase activity was 7.1 ± 0.7 and 189.0 ± 10.1 umol PO_4 mg^{-1}, 15 $min.^{-1}$ for homogenate and BBM preparations, the Na-K ATPase activity was 0.05 ± 0.01 and 0.33 ± 0.04 umol mg^{-1}, 15 $min.^{-1}$, and succinate dehydrogenase, 0.97 ± 0.15 and 0.29 ± 0.06 umol mg^{-1}, 15 $min.^{-1}$.

PO_4 uptake was measured using the rapid filtration technique initially,

RESULTS

The time-course of PO4 uptake by BBM vesicles, incubated in a medium containing no 100 mM sodium, is shown in Fig. 1. During the first 30 sec., uptake is rapid and linear. It reaches a peak at 2 min., to subsequently decrease to a relatively steady state between 15 and 30 min. When Na is absent from the medium, no significant overshoot is observed and the uptake is 5 times lower than when it is present. At 30 min., the uptake is still three fold higher when Na is present than when it is absent.

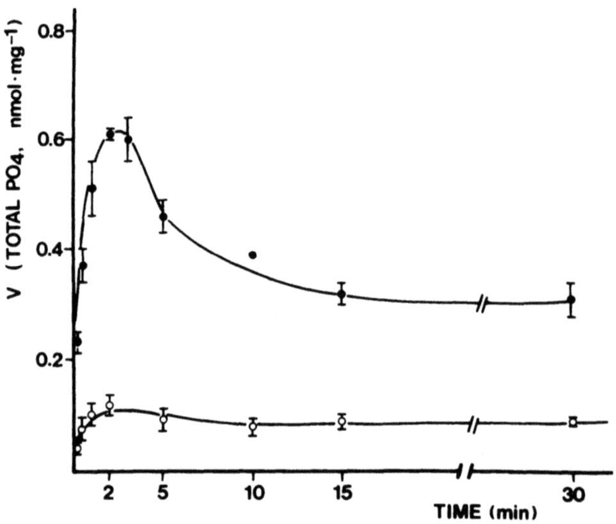

Fig. 1: Time course of PO4 uptake by placenta BBM when 100 mM Na is present (●) or absent (○) in the incubation medium. Substrate concentration was 0.04 mM.

The uptake of PO4 by the BBM vesicles after 30 minutes incubation in the presence of 100 mM Na is strongly dependent on the osmalarity of the incubation medium. As shown in Fig. 2, the addition of sucrose which is not transported by the membranes, decreases the uptake by reducing the intravesicular space. Extrapolation of the curve to its origin, i.e. at 1/osmolarity=zero, osmolarity=∝, indicated that a certain binding to the membranes nes exists, and contributes to the total PO4 uptake. The PO4 binding being independent of the presence of the absence of Na in the incubation medium, it does not contribute to the Na dependent PO4 uptake, which is the only uptake considered in the following figures.

The Na dependent uptake of PO4 by the BBM vesicles increases with the PO4 concentration in the medium, until a saturation level, and this function can be treated according to simple Michaelis-Menten kinetics. Fig. 3 shows the lineweaver-Burk plot obtained in one experiment where 1 mM arsenate was or was not added to the medium. As in BBM from kidney (Hoffman 1976), arsenate strongly inhibits PO_4 uptake, according to a competitive type of inhibition. The mean values of Km and Vmax, when 100 mM Na is present in the medium, are 162 ± 11 µM and 1.56 ± 0.17 nmol mg^{-1} 30 $sec.^{-1}$ respectively.

Fig. 4 illustrates the influence of pH on the PO_4 uptake. The substrate concentration was 0.04 mM and the pH was identical inside and outside the vesicles. pH greatly influences the uptake which is maximal at pH 7.0 or slightly below.

In an attempt to clarify whether Na influences PO_4 uptake by changing or not the affinity of the carrier for the substrate, we studied the influence of Na concentration on the kinetics of PO_4 uptake. As show in Fig. 5, increasing the Na concentration accelerates the maximal initial uptake, but does not influence the Km value (table 1).

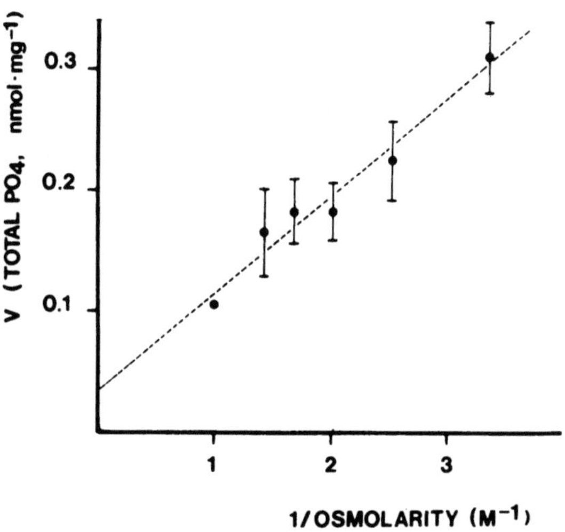

Fig. 2: Influence of osmolarity of the incubation medium on the PO_4 uptake by placenta BBM. Osmolarity was produced by adding sucrose to the medium. Substrate concentration was 0.04 mM, time of incubation 20 min.

DISCUSSION

As already suggested by Stulc et al (1982-1984), PO_4 transport through the placenta syncitium is an active, Na dependent phenomenon, and according to our results, this active transport is situated at the BBM facing the maternal blood compartment. This uptake, as measured at 35°C, is slower than that observed at the same temperature with kidney BBM: the initial velocity is lower in placenta (1.84 ± 0.08 nmol mg^{-1} 30 $sec.^{-1}$) than in the rat kidney (3.56 ± 0.28 nmol mg^{-1} 10 $sec.^{-1}$) (Brunette 1984) and human kidney (2.5 ± 0.2 nmol mg^{-1} 20 $sec.^{-1}$) (unpublished data), and the peak uptake later than for the rat kidney BBM.

As in kidney and intestine, the pH greatly influences PO_4 uptake by placenta BBM. However, the optimal pH is different in each organ: whereas

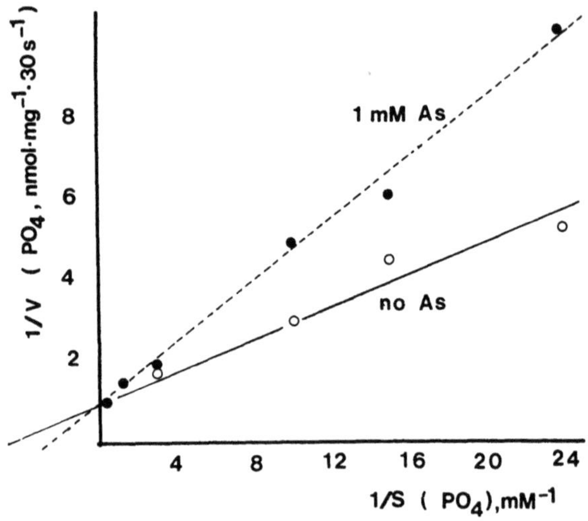

Fig. 3: Lineweaver Burk plot of PO_4 uptake by placenta BBM with (●) and without (○) Na As included in the medium.

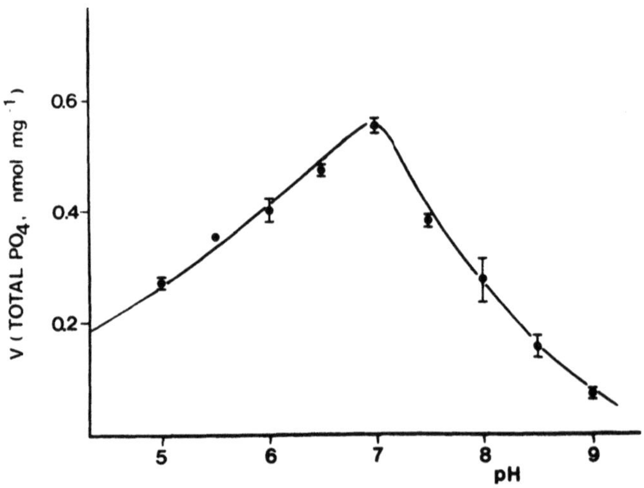

Fig. 4: Influence of pH on PO_4 uptake by placenta BBM. pH was the same inside and outside the vesicles. Substrate concentration was 0.04 mM.

in kidney the optimal pH is within the alkaline range (Hoffman 1976, Sacktor 1981, Brunette 1984) because the divalent form is preferentially transported, and also because pH influences the carrier system (Brunette 1984), in intestine, the situation is opposite: acidosis increases the PO4 transport through the BBM (Kinne 1981, Danisi 1984).

In placenta, the maximal transport was observed at a pH slightly below 7 which exactly corresponds to the pK of the PO4 anions. Then in contrast to what was reported for kidney and intestine BBM, the placenta transports indiscriminately the two forms of PO4.

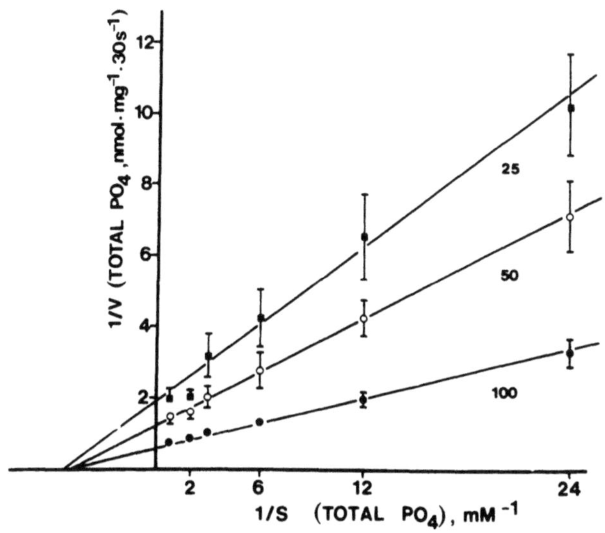

Fig. 5: Influence of Na concentration (25, 50, 100 mM) in the incubation medium on PO4 uptake by the placenta BBM. Substrate concentration was 0.04 mM.

Na dependent transport of PO4 by BBM vesicles was found to be a saturable phenomenon, following the Michaelis-Menten kinetics. The Lineweaver-Burk plot is linear, and in contrast to what was reported for PO4 transport through renal BBM vesicles (Brunette 1984), a single system was observed, whose Km was 162 $\pm$ 11 µM. This value is far below the normal blood concentration, and consequently it is beleived that PO4 transport through placenta in vivo, operates under saturation conditions, independently of any variations in the maternal plasma PO4 concentrations.

Finally PO4 uptake is influenced by Na concentration in the medium in a different manner in placenta and in kidney. In kidney, increasing Na enhances the affinity of the carrier system for PO4, possibly by changing the molecular conformation (Brunette 1985). In contrast, in placenta, Na only increases the maximal velocity of PO4 initial uptake, without changing the Km value. These variations of co-substrate interrelationship reflect the diversity of these transport systems in the various biological membrane.

A summary of the main caracteristics of PO4 uptake by BBM from placenta

Table 1: A comparison of phosphate transport through kidney and placenta brush border membrane.

	PLACENTA (pH 7.5)	KIDNEY (ph 7.5)
Na DEPENDENCY	$\pm$	$\pm$
Km	162 μM $\pm$ 11	167 μM $\pm$ 6
VMax	1.56 $\pm$ 0.17 nmol/mg/30sec.	5.54 $\pm$ 0.4 nmol/mg/10sec.
OPTIMAL pH	6.9	7.5
PREFERENTIAL FORM	NONE	DIVALENT
INFLUENCE of Na T. PO_4	Km	VMax

and kidney is presented in table 1. In both organs PO_4 transport is Na dependent. Whether superficial kidney BBM shows a dual system of transport, a single one was observed in placenta BBM, whose Km is identical to that of the high affinity system in the kidney. PO_4 uptake, however is more rapid in kidney than in placenta, and the optimal pH more alkaline. Whereas the two forms of PO_4, monovalent and divalent are equally transported by placenta BBM, renal preferentially transports the divalent form. Finally in placenta, the affinity of the carrier system varies with concentrations of Na, a situation which is not observed in kidney.

REFERENCES

1. Stulc J, Stulcova B, Svihovec J: Uptake of inorganic phosphate by the maternal border of the guinea pig placenta. Pflügers Arch. 395: 326-330, 1982.

2. Stulc J, Stulcova B: Transport of inorganic phosphate by the foetal border of the guinea pig placenta perfused in situ. Placenta 5: 9-20, 1984.

3. Smith NC, Brush M, Luckett S: Preparation of human placental villous surface membrane. Nature 252: 302-303, 1974.

4. Hoffman N, Thees M, Kinne R: Phosphate transport by isolated renal brush border vesicles. Pflügers Arch. 362: 147-156, 1976.

5. Sacktor B, Cheng L: Proton gradient-dependent phosphate transport in renal brush border membrane vesicles. J. Biol. Chem. 256: 8080-8084, 1981.

6. Brunette MG, Beliveau R, Chan M: Effect of temperature and pH on phosphate transport through brush border membrane vesicles in rats. Can. J. Physiol. & Pharmacol. 62: 229-234, 1984.

7. Danisi G, Murer H, Straub RW: Effect of pH on phosphate transport into intestinal brush border membrane vesicles. Am. J. Physiol. 246: G180-G186, 1984.

8. Brunette MG, El Mernissi G, Doucet A: Renal sodium transport in vitamin D resistant hypophosphatemic rickets. Can. J. Physiol. & Pharmacol. In press, 1985.

INDEX

MIX
Papier aus verantwortungsvollen Quellen
Paper from responsible sources
FSC® C105338

If you have any concerns about our products,
you can contact us on
ProductSafety@springernature.com

In case Publisher is established outside the EU,
the EU authorized representative is:
Springer Nature Customer Service Center GmbH
Europaplatz 3, 69115 Heidelberg, Germany

Printed by Libri Plureos GmbH
in Hamburg, Germany